Mechanical Engineering Series

Series Editor

Vish Prasad, Department of Mechanical and Energy Engineering, University of North Texas, Denton, USA

The Mechanical Engineering Series presents advanced level treatment of topics on the cutting edge of mechanical engineering. Designed for use by students, researchers and practicing engineers, the series presents modern developments in mechanical engineering and its innovative applications in applied mechanics, bio-engineering, dynamic systems and control, energy, energy conversion and energy systems, fluid mechanics and fluid machinery, heat and mass transfer, manufacturing science and technology, mechanical design, mechanics of materials, micro- and nano-science technology, thermal physics, tribology, and vibration and acoustics. The series features graduate-level texts, professional books, and research monographs in key engineering science concentrations.

Maurizio Bottoni • Simone Mantovani •
Gaetano Zanghirati

Molecular Dynamics - Theory and Applications

Springer

Maurizio Bottoni
Meteorological and
Environmental Earth Observation
Ferrara, Italy

Simone Mantovani
Meteorological and
Environmental Earth Observation SRL
(MEEO SRL)
Ferrara, Italy

Gaetano Zanghirati
Department of Mathematics and Computer
Science
University of Ferrara
Ferrara, Italy

ISSN 0941-5122 ISSN 2192-063X (electronic)
Mechanical Engineering Series
ISBN 978-3-031-84566-6 ISBN 978-3-031-84567-3 (eBook)
https://doi.org/10.1007/978-3-031-84567-3

This Springer imprint is published by the registered company Springer Nature Switzerland AG
The registered company address is: Gewerbestrasse 11, 6330 Cham, Switzerland

Text by M. Bottoni

This work is dedicated to the memory of the philosopher Lucius Annaeus Seneca (4 BC—65 AD) who wrote: "soleo enim et in aliena castra transire, non tamquam transfuga sed tamquam explorator" *(I used to go through foreign domains not like a refugee, but like an explorer). (Epistulae: Ad Lucilium—XXII)*

Introduction: A Programmatic Synopsis of the Work

In naturalibus non quaeritur quid semper fiat sed quid in pluribus accidat (St. Thomas of Aquinas, Summa Theologica, I, II, 84, 1,3)

The Atomistic Vision

The advent of electronic computers has allowed applications of the probabilistic methods of statistic mechanics which had been foreseen by philosophers and mathematicians of the past times as an intellectual, though abstract and unrealizable, achievement of the human mind. The methods of molecular dynamics have shown the applicability of numerical computational techniques as an alternative to the numerical solutions of the earlier well established Navier-Stokes equations of fluid-dynamics. Besides, molecular dynamics methods became applicable to flows in the range of intermediate and large Knudsen numbers, where the Navier-Stokes equations, based upon the concept of a continuum, fail. Our work explores the successful application of the methods of the statistical Monte Carlo method, in the frame of molecular dynamics, made mainly in the past three decades, to "particles", which in a generalized sense can coincide either with molecules or cluster of molecules, or with stars or galaxies. The domain of application of the method is thus extended from the micro-cosmos, to the macro-cosmos, and to cosmological problems.

One of the most important heritages of modern science from the world view of the ancients has its roots in the very essence of the Greek philosophical interpretation of nature, as expressed in the atomism of Democritus (about 460–370 B.C.). Twenty centuries later Dante emphasized the intellectual achievement of Democritus and implicitly established a connection to the yet unknown world of chaos quoting "*Democrito, che'l mondo a caso pone*" (Alighieri Dante, n.d.), a statement that assumed the existence of chaotic behaviours.

The concept of the existence of elementary indivisible particles, the "atoms", advanced by the atomistic school around Democritus, held up to the beginning of the past century when the development of subatomic physics was started. In the atomistic vision of the Greeks, interactions and aggregation of atoms allow for the variety of forms sensible to humans. The single particles were conceived rather like mathematical points, too small to be sensed, though a finite tiny mass was assigned to the particles. Atoms were assumed to be in continuous movement, made appreciable by the existence of the "non-being" which we may associate to the modern concept of vacuum, irrespective of the fact that this vacuum could be permeated by an "ether" allowing the propagation of electromagnetic fields, as conceived at the dawn of the theory of electromagnetism, or filled with a yet unknown form of "vacuum energy", hypothesized by modern science.

The atomistic vision of Democritus is essentially deterministic. Atoms move in space with originally random directions and velocities and, in case it were possible to know these a priori, the whole "fieri facias" of the Cosmos, its evolution in unbounded time and space, would be determined. This deterministic philosophical concept of the evolution of the material world is essentially modern and reconnects directly, with an historical jump of some twenty-two centuries, to the modern concepts expressed for instance by Laplace. As noted by Rapaport in the beautiful introduction to his book on the art of molecular dynamic calculations (Rapaport, 2004), the significance of a deterministic solution of the many-body problem was appreciated by Laplace (Laplace, 1840) who wrote (translation from French): *"Given for one instant an intelligence which could comprehend all the forces by which nature is animated and the respective situation of the beings who compose it—an intelligence sufficiently vast to submit these data to analysis—it would embrace in the same formula the movements of the greatest bodies of the universe and those of the lightest atom; for it, nothing would be uncertain and the future, as the past, would be present to its eyes".*

In spite of the deterministic approach favoured by these great precursors of modern science, the advancements of many scientific fields is nowadays based essentially upon probability calculations and statistical methods. One of the most striking examples is provided by the advancement of statistical mechanics, which offers different interpretation and variants of Brownian motion. Therefore, today it is even more true than in the past centuries, what had been stated by Laplace in another passage of his *Essai Philosophique*: *"It is remarkable that a science which started from considering games could rise to involve the most important objects of human knowledge".*

What had been grasped by philosophers and mathematicians of the past times and conceived as an intellectual game, the completion of which was, however, beyond reach of the human mind, has become realizable with the advent, the evolution, and the ever-increasing power of modern electronic computers. Instead of performing by hand a limited number of simple calculations, the computer has allowed scientists to perform full series of complex simulations computing position and speed, and thus trajectories, of an ever-increasing number of simulated particles. In order to enable meaningful and statistically significant ensemble averaging, the number of particles in the ensemble must be large and the ensembles must be numerous. Thanks to subsequent steps in the evolution of computational platforms, we can nowadays perform parallel calculations using clusters of computers to enable the application of the techniques of ensemble averaging.

There are, however, inherent limitations in the numerical simulations of molecular dynamics. Simulations performed in a quasi-deterministic way with the computer are affected by two inherent limitations that denote a lack of attitude to go along with two of the most important advancements of science in the past century, namely the theory of relativity and the concepts of quantum mechanics.

Prescinding for the time being the mesmerizing and yet poorly understood phenomenon of entanglement, according to the theory of relativity information cannot be transferred between two bodies, for instance two particles in a molecular dynamic simulation, with a speed larger than the speed of light. In molecular dynamic simulations however, as applied to then N-body problem, for instance, we assume an infinite speed of propagation of the information between molecules or, more in general, "particles". In this case information is carried by the gravitational force which links the bodies.

The principle of uncertainty in quantum mechanics tells us that it does not make sense to try to determine the position of a particle at the subatomic scale at a given instant of time. It is rather the probability of finding the particle in some location in phase space which can be determined by the Schrödinger equation. Molecular dynamics simulations, however, provide precise information of the position and moment of particles at any given instant of time, thus defying the canons of modern science. In some particular cases, anyway a relativistic concept can be applied even in the frame of molecular dynamics, as it has been justified in the seminal book by Bird (1994).

From Continuum to Discrete Representation

The Navier-Stokes equation were the "paradigm" of fluid-dynamics science describing flows of such a large number of molecules which enables the consideration of the flow field as a continuum. Conversely, the Boltzmann equation was the paradigm of the molecular description of the particle-dynamics, conceived either as molecules or photons, or, after Chadwick's discovery in 1932, neutrons. Analytical or numerical solution of the Boltzmann equation in the general case are impossible because of the complexity of the collision integral. This difficulty motivated the search for approximate solution techniques. Among these one of the most successful is the Bhatnagar, Gross and Krook (BGK) method (Bhatnagar et al., 1954), which set the basis of the BGK equation and subsequent numerical method. The above-quoted original article gives a precise elucidation of the main problems presented by the numerical solution of the kinetic equations describing the space and time dependent behaviour of molecules or particles.

The dynamic theory of gases can be studied from two points of view. The first traditional way starts from consideration of the basic equations of thermal- and fluid-dynamics, the Navier-Stokes equations, and an energy equation, in terms of density, velocities, and enthalpy as dependent variables, and aims at determining the time and space distributions of these variables. In this process an equation of state, relating density to pressure and enthalpy, and physical laws for viscosity, heat transfer coefficients, and momenta exchange (based on a turbulence model) are requested. The second approach relies on basic principles and upon a microscopic formalism, compared to the macroscopic approach of the traditional method. Since the earlier days of the development of the kinetic approach, this formalism was based on particle distribution functions satisfying integro-differential equation of the type of the Boltzmann equation. A distribution function $f\left(\vec{v}, \vec{x}, t\right)$ is implicitly defined by the condition that $f\left(\vec{v}, \vec{x}, t\right) d\vec{v}\, d\vec{x}$ represents the number of particles with velocity in the cell $d\vec{v}$ at point $\vec{v}$ of phase space and with coordinates in the cell $d\vec{x}$ at point $\vec{x}$.

The physical significance of the equations of fluid dynamics gets lost in the Knudsen ($K_n = \lambda/L$) region ($\tau_f < < \tau$) when the characteristic time τ_f of the physical phenomenon to be described becomes smaller than the time interval τ between molecular collisions, or, as an alternative formulation, when the spatial scale L of the physical phenomenon becomes smaller than the mean free path λ of the molecules. The typical example is that of high-frequency sound waves in a rarefied gas. It would be desirable to dispose of a computational method applicable to all ranges of Knudsen numbers, from high density to rarefied gas flows. The developments of the methods of molecular dynamics, combined with the ever-increasing efficiency of computational algorithms and power of computer platforms, has made such tools available.

Regarding the methods of molecular dynamics, based on the Boltzmann equations, or its approximations, we may distinguish between the high-density case and the case of rarefied gases.

In the high-density case ($\tau_f > > \tau$), the Chapman-Enskog theory is used as an alternative to the numerical solution of the Navier-Stokes equations. This theory is essentially based on the assumption of a perturbation of the distribution function by a small amount from the value which would correspond to a Maxwellian distribution. The perturbed distribution function is then expanded in a series of powers of the mean free path λ of the flow. In case of sound waves with wavelength L much larger than λ, the first two terms of the power expansion are sufficient to reproduce all relevant details of the physical process. In case L becomes comparable with λ, it is, however, necessary to consider subsequent terms in power expansion, with great increase in the analytical and numerical difficulties involved.

In the other extreme of rarefied gases ($\tau_f < < \tau$), the BGK equation provides a simple formalism which eliminates the most important difficulties in handling the Boltzmann equation, namely the computation of the collision integrals, which are replaced by simple terms retaining, however, the validity of the conservation laws of mass, momentum, and energy of the molecules. The simplest approximation consists in applying the BGK equation without the collision term, i.e., for collisionless flow. The approximation is applicable in particular cases, as for instance modelling thermophoretic forces in case of particles of extremely small dimensions as for instance aerosol particles in the atmosphere. Numerical simulation of thermophoretic forces will be considered later in Chap. 5 of this book. We used, however, the molecular dynamic approach, rather than the BGK equations.

An example of applications to extremely variable conditions is provided by the simulation of ionized gases as they occur in nature. In the interstellar medium the density of particles may range from 10^{-24} to 10^{-21} (g/cm^3), corresponding to 1–1000 hydrogen atoms per cubic centimetre, with a degree of ionization ranging from almost zero in HI regions to complete ionization in HII regions close to luminous stars. Conversely, in the stellar atmospheres the particle density may range from 10^8 to 10^{11} particles per cubic centimetre with a variable degree of ionization.

Experimental analysis of ionized gases in interstellar domains is possible in principle by applying to the domain similarity transformations. When a linear scale changes, for instance, by a factor γ, thus allowing to reproduce the scale of the physical phenomenon in laboratories, most of the general physical characteristics of the phenomenon to be investigated do not change, provided one also changes time and the other physical quantities according to some general "similarity law". As justified in Alfven and Fälthammar (1963), the laboratory mock-up of interstellar space on galactic scale would imply a transformation of the linear galactic dimension of $\approx 3 \times 10^{22}$ cm to a laboratory scale of 10 cm with a reduction factor $\gamma = 3 \times 10^{-21}$ while the physical density of 1 would change to $\rho = 3 \times 10^{21}$ particles per cubic centimetre. The physical characteristic time, assumed to correspond to the period of galactic rotation, 3×10^8 years or about 3×10^{15} s, would change to 3×10^{-6} s in the laboratory scale. The electric field must be changed by a factor γ^{-1} in order to leave the potential $V \equiv \ell E$ unchanged. Because of the interrelationships among physical quantities implied by the Maxwell's equations, the magnetic field B must change as the electric field E. Hence, in the experimental make-up we would have to simulate interstellar magnetic fields of about 10^{-5} Gauss by experimental fields of the order of magnitude of 3×10^6 Gauss. These high magnetic fields cannot be realized in the laboratory and this sets a limit to the realization of such experimental investigations.

The methods of particle dynamics have fortunately intervened to enable numerical experiments in the conditions sketched above. Both cases of neutral or weakly ionized gases and of highly ionized gases can be tackled with the methods of particle dynamics. In the former case, short-range forces must be simulated, mostly among neutral particles, the range being comparable of smaller than the mean distance between the particles. That is the situation in which the "splitting operator" method can be applied by simulating numerically collisionless paths of the particles, followed by the collisions at the end of the time step. Another implication of the short-range forces is that the probability of a multiple collisions is negligible, so that in practice only binary collisions need to be simulated. Conversely, in the latter case of long-range forces the average number of particles at a distance r, or less than r, from a given particle increases as r^3, while the intermolecular forces decrease only with r^{-2}. As a consequence, every particle is more influenced by the whole of the particles in the ensemble rather than by the few particles in its close proximity. The effective large cross section for interaction between ionized particles must therefore be accounted for even for slightly ionized gases.

Long-range interactions have been studied and numerically simulated essentially with two different methods (Rapaport, 2004). The first is based on the technique of "interaction re-summation", known as Ewald method (Ewald, 1921) developed in the frame of the condensed matter physics. Taking advantage of the periodic structure of the lattice in a crystal, the total energy of the ensemble of atoms can be expressed in a closed mathematical form and easily evaluated, together with the forces acting among them. An alternative method consists in partitioning the full computational domain in a hierarchy of cells, halving each cell in every coordinate direction, till, at the lowermost level, every cell contains at maximum one simulated particle. Interactions among the particles are computed at the lowermost level of the cell hierarchy, while at ascending levels in the hierarchy a coarser representation of the content of the cells is used to compute the interaction among distant particles.

Simulations from Micro-cosmos to Macro-cosmos

In the past decades, a variety of topics have been treated with the methods of molecular dynamics and statistical physics. Among these topics are:

(i) Fundamental studies involving kinetic theory, simulation of Brownian motion, molecular chaos, inner energies of complex molecules (vibration and rotational motions)
(ii) Classical and special hydrodynamics, the latter including hydrodynamics non-Newtonian fluids, properties of amorphous substances like glasses, and properties of polymers
(iii) Phase transition including evaporation or condensation processes and sublimation
(iv) Fundamental properties of solids like elasticity, plasticity, crystal dislocations, micro-fractures, and radiation damage as well as properties of magnetic materials
(v) Properties of macro-molecules in biology
(vi) Simulation of thermophoretic problems (a topic dealt in detail in the following)

In the following we will review just a few of the basic advancements in the application of the methods of molecular dynamics to the topics mentioned above, with particular reference to problems of fluid dynamics which best represent the cultural bias of the authors.

Applications of Molecular Dynamics to Nuclear Reactor Safety

One of the major topics involved in safety analysis of liquid metal fast breeder reactors (LMFBRs) is the concern about material relocation in the course of a hypothetical accident and the associated recriticality. These concerns led to the consideration of the two major accidents initiators for LMFBR safety analysis: (i) the transient over-power (TOP) accident due to positive reactivity insertion; and (ii) the transient undercooling (TUC), following loss of flow (LOF), accident in which the coolant flow is reduced by constant reactor power.

The sequence of events following an initiating perturbation, TOP or TUC, of the normal reactor operation, leading to severe core damage and possibly to core disruption is generally referred to as hypothetical core disruption accident (HCDA), which may eventually lead to the disassembly phase as postulated in the pioneering work of the Bethe-Tait scenario (Smidt, 2012). An intermediate phase between initiating events and the core disassembly is characterized by progressive failure of fuel elements with still intact hexagonal canning structures of the subassemblies. Dominating mechanisms are: (i) voiding dynamics within the subassemblies, with continuous release of positive void reactivity; (ii) cladding melting and motion with possible blockage formation; (iii) fuel melting or sputtering and transport of fuel debris with possible formation of fuel crusts; (iv) fission gas release from molten and slumping fuel; and (v) fuel-coolant interaction (FCI).

In order to investigate the accident scenarios and their consequences on the control of LMFBRs, a series of computer programs has been developed worldwide, mainly in the USA, Germany, France, Japan, and United Kingdom, and verified against the results of in-pile and out of pile test facilities. These codes, which describe the damage propagation from subassembly to subassembly, are generally based on a strict coupling between the thermo-fluid-dynamic part and the neutronic calculations. The latter are generally based on the solution of the point kinetic equations with the calculation of space-dependent reactivity feedbacks due to Doppler broadening, voiding expansion, temperature changes, and relocation of materials (cladding and fuel slumping and possibly blockage formation).

The state of the art of these codes is represented by the series of the SIMMER computer programs developed over twenty years, between about 1970 and 1990, at the Los Alamos National Laboratory (LANL), USA. The prototype code of the SIMMER series, SIMMER-II (Smith & Sheheen, 1980), computes neutronic and fluid-dynamic behaviour in axis-symmetric (two-dimensional) cylindrical geometry. The latest version of the code, SIMMER-III (Kondo et al., 1999), resulted from a tripartite cooperation between the JNC (Japan Nuclear Cycle Development Institute), CEA (Commissariat à l'Energie Atomique, France), and the Forschungszentrum Karlsruhe (FZK), Germany. The main new features available in SIMMER-III are related to two-phase flow models and turbulence in presence of particles produced by fuel fragmentation.

Transport of fuel debris in voided coolant channels is generally modelled by computation of entrainment forces due to turbulent motion of gases (fission gases as well as vapour of coolant). Because in chocked channels the velocity of gases is imposed by the oscillating behaviour of upper and lower liquid slugs and may oscillate around a zero mean value, it has been thought that thermophoretic forces, due to strong temperature gradients between pin surfaces and hexagonal canning, may play an important role in transport and deposition, with possible crust formation, of fuel debris.

As an alternative to the traditional approach, thermophoretic forces have been simulated with the molecular dynamic method in a sector of a seven pin bundle with hexagonal pin arrangement. In the numerical simulations, made with the statistical Direct Simulation Monte Carlo (DSMC) method (Bird, 1994), the principle of uncoupling is applied, by splitting the calculation in the collisionless motion of the molecules, which are assigned positions updated by $x_i^{n+1} = x_i^n + u_i^n \Delta t$, followed by the simulation of collision events among the molecules. A three-dimensional code version, referred to as THEMIS/3D.1 (Bottoni, 2004), which was used for numerical investigations of condensation, rewetting, and vaporization of sodium in a LMFRB bundle under hypothetical accident conditions, was implemented with the numerical simulation of thermophoretic velocities and applied to simulate the transport of a cloud of fuel debris of variable size formed by a postulated fuel fragmentation under hypothetical post-dryout accident conditions. Applications of the resulting code, referred as THEMIS/3D.2, were documented in Bottoni and Bottoni (2005).

The Advent of Monte Carlo Algorithms

The beginnings of the application of statistical methods in computational science are reviewed in a famous article written by Nicholas Metropolis (Metropolis, 1987) and dedicated to the memory of Stanislav Ulam (1909–1984). In 1945 the first large computer, the ENIAC, was built at the School of Electrical Engineering at the University of Pennsylvania, Philadelphia, and applied for simulating thermonuclear fusion processes which eventually led, under the leadership of Edward Teller, to the construction of the H-nuclear weapon. By spring 1946, a technical report about numerical calculations made for the simulation of thermonuclear processes was written and then discussed at the meeting attended, among others, by Edward Teller, Enrico Fermi, John von Neumann, and Stalislaw Ulam. The latter realized in that occasion that statistical sampling techniques, fallen in disuse because tedious and slow when used with simple electro-mechanical computing devises, could be revived for applications with the powerful ENIAC computer. The idea sparkled by Ulam was favourably received by von Neumann who, in March 1947, communicated it to Robert Richtmyer in a letter proposing application of statistical methods for simulation of collision events of neutrons in scattering media. These investigations aimed at determining the neutron distributions and neutron leakage from radiation shielding materials.

The statistical methods used for the simulation of thermonuclear processes and of neutron distributions in scattering media were studied in a project which went to be called "Monte Carlo" because an uncle of Ulam frequented the Monte Carlo Casino in Monaco, where he used to borrow money for gambling.

The method, unpublished and unnamed, had however already been used more than 15 years earlier by Enrico Fermi and his co-operators when studying the slowdown and the distribution of neutrons in moderators at the famous laboratory of Via Panisperna in Rome. The fact is recalled by Emilio Segrè in his biography of Enrico Fermi (Segrè, 1970): "*Fermi had invented, but of course not named, the present Monte Carlo Method when he was studying the moderation of neutrons in Rome. He did not publish anything on the subject, but he used the method to solve many problems with whatever calculating facilities he had, chiefly a small mechanical adding machine*".

Steven Strogatz, professor of mathematics at the Cornell University, wrote in 2004 an article for the *New York Times* in which he stated that Enrico Fermi should be remembered not only for his achievements in the domain of neutron physics but also as the man who transformed the computer in the "*telescope of the mind*", allowing to "*see the invisible*" and to "*imagine the unimaginable*".

Even when using the best available computers in the 1970s and 1980s, applications of Monte Carlo statistical algorithms were plagued by the length of the calculations. Referring to the particular topic of molecular dynamic calculations, with collisions among gas molecules in a given domain, the normal procedure was to calculate the trajectories of impinging molecules up to the very instant of collisions, then applying a collision model to determine the post-collision speed and trajectories of the collisional molecules. A great advantage from the viewpoint of computational time was achieved by replacing the exact calculation of the collision events with a procedure based on the following steps:

(i) Compute for every cell in the simulated domain the total number of collisions per unit time, N_c, which is obtained by multiplying the collision frequency, ν, with the number density, n, and dividing by 2 (to avoid to count twice the same couple of molecules): $N_c = n\nu/2 = n^2\sigma_T c_r/2$. In this formula c_r is the relative velocity of the molecules and σ_T is the geometrical cross section of the molecule.
(ii) A number of $N_{\Delta t}$ collisional events are then simulated by choosing at random couples of molecules within the same computational cell and applying to them the collision model obtaining the post-collision trajectories and velocities.

The exact calculation of the trajectories of the molecules up to the impact point is thus avoided with great computational advantage. The inaccuracy introduced in this way is negligible if the computational cells are small (of the order of magnitude of 10 μm) and imposing the constraint that only molecules within the same cell can collide. Figure 1 illustrates the difference between the traditional and the new approach for computing the molecular collisions.

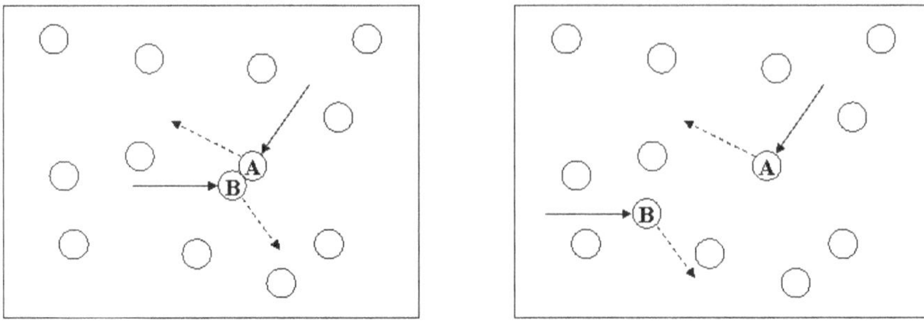

Fig. 1 Calculations of molecular collisions. Left: Traditional approach in which the trajectories are computed till the collision point. Right: Modern approach in which collisional pairs are selected at random within a computational cell

A word of caution is, however, necessary about the use of Monte Carlo methods. A remark attributed to von Neumann is: "*Anyone using Monte Carlo is in a state of sin*". This remark wanted to convey the meaning that Monte Carlo algorithms should be applied as "ultima ratio", when classical approaches based on analytical methods do not exist, or when traditional schemes based on the discretization of differential equations are, for whatever physical reason, not applicable.

Dawn and Development of Thermophoretic Research

In this section we report about numerical simulations of thermophoretic forces acting upon particles in a gas-filled domain with temperature gradients. In some technical domains thermophoretic forces have been used to induce a peculiar process, as for instance in the semiconductor industries where heated surfaces of wafers keep particles of impurities away from the surface and preserve its purity. In other domains, thermophoretic forces are responsible for reduced efficiency, as in boiler surfaces where transport and deposition of particles reduce the efficiency of heat exchange. A good review of both thermophoretic theories and experimental investigations on thermophoretic forces and resulting velocities is given in Zheng (2002).

In view of the subsequent developments, made in one chapter of the book, we summarize briefly the progress made, thanks to the methods of molecular dynamics, in the simulation of thermophoresis, name of evident Greek origin. Thermophoresis refers to the phenomenon in which a particle immersed in a fluid subjected to a temperature gradient drifts in a sense opposite to the vector which visualizes the gradient directed in the sense of increasing temperature. Let us assume that the temperature gradient is due to two horizontal metallic plates placed a few centimetres away and kept at different temperatures, for instance the upper plate at 100 °C and the lower plate at 0 °C. The medium between the plates contains a gas. A particle situated in the medium is struck by gas molecules irradiated from the upper plate at larger temperature and at the same time by molecules irradiated from the lower plate at a lower temperature. Because molecules arising from the upper plate have larger kinetic energy than those coming from the lower plate, the particle is displaced towards the bottom surface, in the sense opposite to the temperature gradient.

To be more precise, one should observe that molecules incident on the particle arising from above, because of intermolecular collisions undergone during their path, will not have the temperature of the upper plate, but a lesser temperature. Likewise, molecules arising from the lower plate will have higher temperature than the plate. This implies that thermophoretic forces are smaller than they would be in the absence of intermolecular collisions.

The phenomenon of thermophoresis was observed at the beginning of the nineteenth century by several researchers who could not give a plausible interpretation of it. In 1870, the Irish physicist John Tyndall (1820–1893) observed that in a room with air stained by minute dust particles regions without particles formed around bodies at a temperature larger than the mean room temperature. Tyndall, however, could not give a physical interpretation of the phenomenon. In 1884, the Scottish physicist John Aitken (1839–1919) could explain the phenomenon observing that the dust molecules were struck by air molecules with larger kinetic energy arising from regions hotter than the regions around them.

The complex of calculations referred to above establishes a method of numerical simulations referred to as "Direct Simulation Monte Carlo" (DSMC), which has been illustrated in several variants and in the basic text by Bird quoted above.

In the atmospheric physics, which we will consider later on, the behaviour of aerosol particles suspended in the atmosphere has an important role in formation and growth of condensation nuclei within clouds and therefore in the formation of hydrometeors. Diffusion and transport of aerosol particles in the atmosphere are strictly connected with the behaviour of the atmosphere in its complex, which is not only due to convective or turbulent motions but also to the impact of thermophoretic forces due to temperature gradients.

Experimental investigations are long and expensive but now they can be integrated by numerical simulations of thermophoretic forces acting on aerosol particles in conditions of microgravity in the higher regions of the atmosphere. Numerical investigations have a double purpose: (i) experimental results must be verified with calculations and, vice versa, good experimental results provide a verification of the consistence of the calculations; and (ii) as soon as a computational program has been verified and has reached a good degree of confidence, it becomes legitimate to apply it to conditions not yet investigated experimentally like, for instance, to simulations of interactions between molecules and cosmic dust particles in rarefied conditions as occurring in the interplanetary space.

Fundamental studies on the application of the DSMC method to problems of fluid dynamics were made by the group headed by Kaburaki at the JAERI (Japan Atomic Energy Research Institute) research centre of Tokaimura in Japan (Kaburaki et al., 1990; Kaburaki & Yokokawa, 1993). Numerical simulations of fluid dynamics concerning the safety of nuclear reactors were conducted with the DSMC method by a research group of the JNC (Japan Nuclear Cycle Development Institute) at the O-Arai in Ibaraki-ken research centre in Japan. These investigations were documented in the computational code THEMIS (Muramatsu, 1995), which was made available to the author during the academic year 1996–1997 spent as guest professor at the Tokyo Institute of Technology (Tokodai). The computer code THEMIS, originally developed for two-dimensional problems, was then extended to the third dimension and adapted to numerical simulations of evaporation and condensation problems for applications to safety analysis of natrium-cooled fast breeder reactors. This code version, documented in Bottoni (2004), was called THEMIS/3D.1. In a subsequent development stage, simulations of thermophoretic forces were added to the code to simulate the transport of fuel particles (uranium oxide) fragmented in case of a hypothetical nuclear incident occurring in a gas mixture with sodium vapour and fixion products. This last code version called THEMIS/3D.2 was documented in the reference Bottoni and Bottoni (2005).

In the last decades, an increasing variety of problems have been treated with the methods of molecular dynamics and statistical physics, as mentioned in Sect. 3.

Multiple Time Scales: N-Body Problems

The cosmologist J. Richard Gott refers to N-Body calculations (Gott, 2016) made in Cambridge with the colleagues Sverre Aarseth and Ed Turner. Aarseth, beyond being one of the best chess players of the world by correspondence, and beside climbing peaks in the Himalaya, had developed a pioneering computer program for the simulation of N-Body problems. In the development of this code, in which tight binary couples of the stars form, it was customary to follow the fast evolution of the binaries by using very tiny time steps, thus strongly reducing the computational speed. Aarseth introduced for the first time the option of using different time steps to follow the evolution of the system: macroscopic time steps were sufficient to follow the evolution of stars moving slowly, while microscopic time steps were applied for quickly evolving systems. Synchronization of the calculations was made at regular intervals. The ideas, advanced by Aarseth in the above-quoted reference, were applied for the first time in cosmological problems to follow the evolution of 4000 particles. In the following years, the number of simulated particles was systematically increased and large parts of the universe were simulated. In this way, Gott and co-operators discovered the large-scale structure of the Universe and the existence of the "Great Barrier" formed by arrays of galaxies.

The author applied Aarseth methods to the simulation of thermophoretic problems, as reported later in this book.

The particular case of the N-Boby problem, in which particles are representative of stars in clusters or representative of entire galaxies moving in cosmic domain, has been treated by the life long work of Sverre Aarseth at the Institute of Astronomy in Cambridge, where he started his career as assistant to Prof. F. Hoyle. The basic principle of separated time scale for the components of an assembly of particles to take in account their different speed and time step needed for efficient numerical integration is explained in Aarseth (1985), and the whole of the experience gained in years of commitment to these investigations has eventually been summarized in the beautiful book published about eighty years thereafter (Aarseth, 2003). In the introduction of this book, Aarseth recalls a quotation by von Hoerner (Wielebinski & Sebastian, 1958) telling that *"the classical N-body problem in the modern spirit can be said to have started with the pioneering effort of von Hoerner (1960)"*.

In the 1960s, with the modest computer facilities available at that time, only groups of a few dozen of bodies could be simulated numerically. Nevertheless, in Wielen (1967) a simulation conducted on star clusters led to the valuable result that the half-life of a star cluster could be estimated in about 2×10^8 yr. By the year 1968, with a dedicated computer, it was possible to increase the number of simulated particles to $N = 250$ (Aarseth, 1968). In 1973, a large cluster model with $N = 10^4$ could be simulated. In this case the Ahmad and Cohen (1973) "neighbour scheme" was applied, in which forces acting upon particles are represented by polynomials, with contributions from neighbouring particles being updated more frequently.

With the advent of the HARP (Hermite AcceleratoR Pipeline) special purpose computer (Makino et al., 1993) tasks like formation of proto-planets from a swarm of low-mass planetesimals were successfully tackled. The state of the art of this evolution was then represented by the parallel computer program NBODY6 (Aarseth, 2003).

Globular clusters provide a unique field of research because they offer the opportunity to study dense stellar systems more accessible than the nuclei of galaxies, where the situation may also be complicated by the presence of a black hole. In computational science, modelling the evolution of globular clusters poses a formidable challenge to modern parallel computers and provide ideal opportunities for testing the performances of teraflop machines. [From the abstract of the reference Hut et al. (1998)].

Simulations of the time evolution of clusters, typically for a time span of 10^8–10^9 years, were made in earlier times by computing the gravitational potential of the system and then updating the star's positions according to the forces originated by the potential field. The time step was limited by the speed of the fastest stars in the ensemble. In case of close encounters of stars, as it occurs more frequently in the numerical simulation of colliding globular clusters, as well as in the simulation of colliding galaxies, couples of stars approach along close trajectories and their gravitational potential is exchanged into increase of the momenta of the stars. Their velocities can be orders of magnitudes larger than the average star velocity. Application of a uniform small time step required very long computational times. The multiple time step technique, widely used in the computer programs developed by Aarseth at the University of Cambridge, allows a much faster computational speed. The idea is to treat fast moving stars with a small time step and slow stars with a large macro-time-step, synchronizing then the situation at the end of the macro-time-step.

Figure 2 shows a snapshot from a numerical simulation of two colliding globular clusters. Although the collisions of stars in the clusters are extremely rare, so that these events can be ruled out, the star trajectories are bent by the mutual gravitational pull. The inflection of the trajectories is shown qualitatively in Fig. 3. The velocities

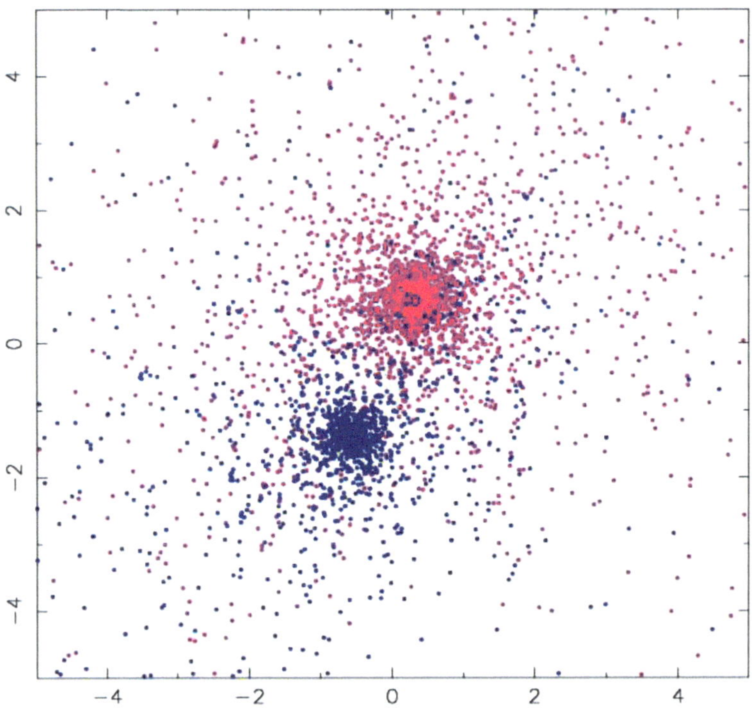

Fig. 2 Numerical simulation of colliding globular clusters. (Courtesy of Dr. S. J. Aarseth, Cambridge University, UK)

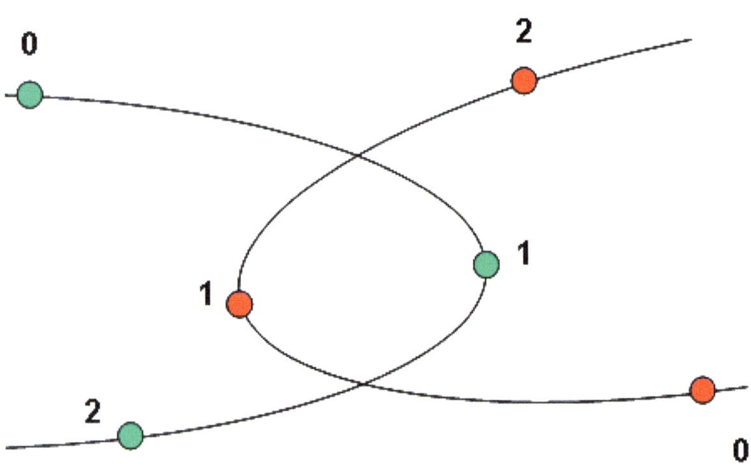

Fig. 3 Sketch representing gravitational interaction of two approaching stars, represented at three time points, labelled 0, 1, 2

of the interacting stars at points 1 may be orders of magnitude larger than the velocities at positions labelled 0 and 2.

Brownian Motion

Since the famous article by Einstein (1905) about the explanation of the Brownian motion of small particles suspended in a fluid, observed and described by the Scottish botanist R. Brown in 1827, few domains of science have led to such a flourishing of studies as the interpretation of Brownian motion with the methods of statistical physics. In the year 2005, the centennial anniversary of Einstein's famous publication, a special issue (No. 15) of the scientific magazine *Chaos* was especially dedicated to articles commemorating that important event in the history of science and summarizing most recent advances in the theoretical interpretation of Brownian motion and diffusion, ranging from stochastic processes to theoretical modelling of chaotic systems. In the introductory article of the series of papers published in *Chaos*, with the significant title "*Introduction: 100 years of Brownian motion*", Haenggi and Marchesoni (2005) trace the summary of the body of theoretical advances made in the century elapsed since Einstein's publication. Subsequent articles give all most modern advances from both viewpoints of theoretical and computational approaches.

Development of the THERCONF Code

In atmospheric physics, migration of aerosol particles suspended in the atmosphere plays an important role in the formation and growth of liquid and ice nuclei within clouds and therefore in the formation and accretion of hydrometeors. Diffusion of aerosol particles is due not only to fluid-dynamic transport and turbulent behaviour of the atmosphere but also to their displacement under impact of thermophoretic forces due to temperature gradients in the atmosphere. These forces may have an important impact upon scavenging of aerosol particles in the clouds. In the past years, a two-dimensional computer program called THEPHOR has been developed to simulate numerically the displacement of aerosol particles in environmental conditions typical of the upper Earth atmosphere. The computational approach followed in the numerical investigations is based upon the application of the Direct Simulation Monte Carlo (DSMC) method (Bird, 1994) in a bi-dimensional (x, z) computational domain with the aerosol particle initially set at the centre of the domain and the thermophoretic forces acting in the negative z-direction, i.e., with the temperature gradient directed upwardly.

Recently the European Space Agency (ESA) in an "Invitation to Tender" (ESA, 2006), illustrated a feasibility study of the Interaction in Cosmic and Atmospheric Particle System (ICAPS) experiment, aiming at confining and trapping cosmic particles in the extremely rarefied atmosphere of interplanetary space. The experiment consists essentially of thermal elements, called "Gas Flow Drivers" contained in a cylindrical box, the centre of which is a trapping area for thermally confined particles floating in a rarefied gas within the chamber. Thermophoretic forces induced from thermal gradients across the bulk of the gas confine particles carried from the gas in the trapping area located in the centre of the chamber. The physical dimensions of the experiment and of its components, the location of the thermal elements within the chamber and their temperature are free parameters that should be experimentally and numerically investigated in order to enhance the efficiency of the experiment in collecting particles in the trapping area.

To enable numerical investigations of the proposed device, the above-mentioned code version THEPHOR has been further developed to simulate a bunch of particles in the geometrical configuration proposed by the ESA. The resulting code version was called THERCONF, an acronym for thermal confinement. Besides simulating a down-scaled geometrical configuration of the experiment proposed by the ESA, the THERCONF computer program also includes options to simulate other physical phenomena like the insurgence of thermal instability in the Bénard problem and thermal creep flow induced by temperature gradients along heated material surfaces. This new code version is therefore rather versatile, being able to simulate a variety of physical situations.

The THERCONF program relies on the strict application of the basic laws of the kinetic theory (Vincenti & Kruger, 1965) and presents the following advantages with respect to traditional approaches based upon numerical solution of the conservation equations of fluid-dynamics: (i) The definition domain is partitioned in cells which are not used to derive difference equations but are used only for determining the inter-molecular collision rates. Therefore, irregular cells do not infirm the accuracy of numerical results. (ii) The DSMC method does not imply the use of heat transfer coefficients or of turbulence models. Heat and momentum transfer are derived directly from the application of the kinetic laws. (iii) The domain of application, in terms of flow regimes, is unbounded, including very rarefied gases.

The microscopic or molecular approach is based upon the Monte Carlo simulation of molecular trajectories, collision rates among the molecules, and collisions of the molecules with ideal (symmetry) or material surfaces. Ideal surfaces are treated as specular reflecting surfaces, while material surfaces are treated as diffusively reflecting surfaces. Diffusive reflections imply that molecules reflected from material surfaces are assigned velocities obtained by sampling a Maxwellian velocity distribution corresponding to the temperature of the surface in the impact point.

Theoretical analysis of thermophoretic forces has been made in case of gases characterized by small or intermediate Knudsen numbers, and led to the analytical expression of thermophoretic velocity provided by the so-called Waldmann formula (Waldmann, 1959). Preliminary results of the application of the THERCONF program have been presented in Bottoni and Mantovani (2008).

Conclusion

In several cases, algorithms devised to improve the computational speed in numerical mathematics are suggested only by numerical considerations, irrespective of the type of physical problem being represented by the systems of equations under consideration. In many other cases, however, as exemplified by the splitting techniques or by the application of standard algorithms to new physical situations, like the alternating direction implicit (ADI) method for two-phase flow (Bottoni 1994), improved algorithms are suggested by deep knowledge of the physical problems to be modelled. Insofar the conclusion can be drawn that interdisciplinary knowledge is of paramount importance in this domain of scientific endeavour.

References

Aarseth, S. J. (1968). Dynamical evolution of simulated N-body systems. *Bulletin Astronomique, 3*, 105–25.

Aarseth, S. J. (1985). Direct methods for N-body simulations. In J. U. Brackbill, & B. J. Cohen (Eds.), *Multiple time scales*. Orlando: Academic Press.

Aarseth, S. J. (2003). *Gravitational N-body simulations: Tools and algorithms*. Cambridge University Press.

Ahmad, A., & Cohen, L. (1973). A numerical integration scheme for the N-body gravitational problem. *Journal of Computational Physics, 12*(3), 389–402.

Alfven, H., & Fälthammar, C. G. (1963). *Cosmical electro-dynamics*, 2nd ed. Oxford: Clarendon Press.

Alighieri Dante (n.d.). *Inferno, 4*, 136.

Bhatnagar, P. L., Gross, E. P., & Krook, M. (1954). A model for collision processes in gases. I. Small amplitude processes in charged and neutral one-component systems. *Physical Review, 94*(3), 511–525.

Bird, G. A. (1994). *Molecular gas dynamics and the direct simulation of gas flows*. Oxford: Clarendon Press.

Bottoni, M. (1994). A variant of the ADI method for two-phase flow calculations. *Computer and Fluids, 23*.

Bottoni, M. (2004). Molecular approach to sodium vapour condensation, rewetting and vaporization in LMFBR bundle under hypothetical accident condition. *Journal of Nuclear Science and Technology, 41*(5).

Bottoni, M., & Bottoni, C. (2005). Direct Monte Carlo simulation of fuel particles relocation in nuclear reactor bundle during hypothetic core disruptive accident. In:*13th International Conference on Nuclear Engineering (ICONE13), Beijing, China, 16–20 May, 2005*.

Bottoni, M., & Mantovani S. (2008) Molecular dynamic simulation of thermal confinement of particles. In *Proceedings of the 2008 ASME International Mechanical Engineering & Exposition, Boston, USA*.

Einstein, A. (1905). Ueber die von der molekularkinetischen Theorie der Waerme geforderte Bewegung von in ruhenden Fluessigkeiten suspendierten Teilchen. *Annals of Physics, 17*, 549–560.

ESA (2006). Invitation to Tender: AO/1-5045/06/NL/VK, Feasibility Study Phase B0 of the Interaction in Cosmic and Atmospheric Particle System (ICAPS) Experiment, August 15, 2006.

Ewald, P. P. (1921). Die Berechnung optischer und elektrostatischer Gitterpotentiale. *Annals of Physics, 369*, 253–287.

Gott, J. R. (2016). *The cosmic web: Mysterious architecture if the universe*. Princeton University Press.

Haenggi, P., & Marchesoni, F. (2005). Introduction: 100 years of Brownian motion. *Chaos, 15*(2).

Hut, P., Makino J., & McMillan S. (1988). Modelling the evolution of globular star clusters. *Nature, 336*.

Kaburaki, H., Yamamoto, H., Yokokawa, M., & Arisawa, T. (1990). A new collision cell system for the direct simulation Monte Carlo method. In *Proceedings of the First International Conference on Supercomputing in Nuclear Applications (SNA' 90), March 12–16, 1990, Mito, Japan* (pp. 51–56).

Kaburaki, H., & Yokokawa, M. (1993). Numerical simulation of near continuum gas flows by the direct simulation Monte Carlo method. In *Proceedings of the Joint International Conference on Mathematical Methods and Supercomputing in Nuclear Applications (M&C + SNA' 93), Karlsruhe, Germany* (Vol. 1).

Kondo, S., et al. (1999). Current status and validation of the SIMMER-III LMFR safety analysis code. In *7th International Conference on Nuclear Engineering, ICONE-7249, Tokyo, Japan*.

Laplace, P. S. (1840). Essai philosophique sur les probalilités, Bachelier.

Makino, J., Kokubo, E., & Taiji, M. (1993). HARP: A special purpose computer for N-body problem. *Publications of the Astronomical Society of Japan, 45*(3), 349–360.

Metropolis, N. (1987). The beginning of the Monte Carlo Method. *Los Alamos Science, Special Issue, 15*, 125–130.

Muramatsu, T. (1995). A direct simulation Monte Carlo code THEMIS for the evaluation of nonstationary heat transfer phenomena in LMFR Thermohydraulics. In *8th IAHR Working Group Meeting on Advanced Nuclear Reactors Thermal Hydraulics, Nuclear Research Institute Rezplc, Czech Republic*.

Rapaport, D. C. (2004). *The art of molecular dynamics simulation* (2nd ed.). Cambridge University Press.

Segré, E. (1970). *Enrico fermi physicist*. The University of Chicago Press.

Smidt Dieter (2012). *Reaktor-Sicherheitstechnik: Sicherheitssysteme und Störfallanalyse für Leichtwasserreaktoren und Schnelle Brüter*. Springer.

Smith, L. L., & Sheheen N. N. (1980). *SIMMER-II: A computer program for LMFBR disassembly core analysis*. NUREG/CR-0453; LA-7515-M.

Vincenti, W. G., & Kruger, C. H. (1965). *Introduction to physical gas dynamics*. Krieger Publishing Company.

Waldmann, L. (1959). Ueber die Kraft eines inhomogenen Gases auf kleine suspendierte Kugeln. *Zeitschrift fuer Naturforschung, 14a*.

Wielebinski Richard, & Sebastian von Hoerner (1958). *Mitteilungen der Astronomischen Gesellschaft*, 89.

Wielen, R. (1967). *Dynamical evolution of star cluster models, I*. Veroeffentligung Astronomisches Rechen-Institut Heidelberg, No. 19.

Zheng, F. (2002). Thermophoresis of spherical and non-spherical particles: A review of theories and experiments. A*dvance in Colloid and Interface Science, 97*.

Contents

Abbreviations

Acronyms

AFM	Atomic Force Microscope
AME	Applied Micro Electronics
ANL	Argonne National Laboratory
BGK	Bhatnagar-Gross-Krook method of solution of the Boltzmann equation
BDS	Brownian Dynamics Simulation
BLAS	Basic Linear Algebra Subroutines
BWR	Boiling Water Reactor
CEA	Commissariat a l' Energie Atomique
CENG	Centre de Etudes Nucleaires de Grenoble
CNR	Consiglio Nazionale delle Ricerche
CLLK	Cercignani-Lampis-Lord model
DLA	Diffusion Limited Aggregation
DLG	Diffusion Limited Growth
DSMC	Direct Simulation Monte Carlo
EAC	European Accident Code
ECC	Error Checking and Correction
EES	Engineering Equation Solver
ESA	European Space Agency
ESSL	Engineering and Scientific Subroutine Library
FBR	Fast Breeder Reactor
FCI	Fuel Coolant Interaction
FZK	Forschungs Zentrum Karlsruhe
HCDA	Hypothetical Core Disruptive Accident
HEM	Homogeneous Equilibrium Model
HACM	High Availability Cluster Multiprocessing
HPC	High-Performance Computing

HPS	High-Performance Switch
ICAPS	Interaction in Cosmic and Atmospheric Particle System
IFR	Integral Fast Reactor
INET	Institute of Nuclear Energy Technology
ISAC	Institute of Atmospheric Science and Climate
ISM	Improved Slip Model
JNC	Japan Nuclear Cycle (Formerly PNC)
JRC	Joint Research Centre (Ispra, Euratom)
KFK	KernForschungszentrum Karlsruhe (now FZK)
LAPACK	Linear Algebra PACKage
LOF	Loss of Flow
LMFBR	Liquid Metal Fast Breeder Reactor
LANL	Los Alamos National Laboratory
LOCA	Loss of Coolant Accident
LWR	Light Water Reactor
MD	Molecular dynamics
MEEO	Meteorological and Environmental Earth Observation
MPI	Message Passing Interface
PBLAS	Parallel BLAS
PE	Parallel Environment
PMDS	Parallel Molecular Dynamics Stencil
PNC	Power and Nuclear reactor Corporation
PWR	Pressurized Water Reactor
RB	Rayleigh-Bénard
SPM	Separated Phase Model
RRT	RRT Reaction-Rate Theory
SCALAPACK	Scalable LAPACK
SMP	Smaller Symmetric Multiprocessing
TOP	Transient Over Power
TP	Type Locality
TST	Transition State Theory
TUC	Transient Under Coolant
ULOF	Unprotected Loss-of-Flow
VHS	Variable Hard Sphere inter-molecular collision model
VPN	Virtual Private Network
VSS	Variable Soft Sphere inter-molecular collision model

Nomenclature

A	Atomic number
A	Avogadro number
b	Distance between trajectories of colliding molecules (m)

$\vec{c}$	Velocity of a molecule (m/s)
$\vec{c}_o$	Mean stream velocity (m/s)
$\vec{c}' = \vec{c} - \vec{c}_o$	Velocity of a molecule relative to the mean stream velocity (m/s)
$\vec{c}_r$	Relative velocity between molecules (m/s)
$\vec{c}_m$	Center of mass velocity (m/s)
c'_m	Most probable thermal velocity (m/s)
$\vec{c}^*$	Post collision velocity (m/s)
C	Collision number between molecules (s^{-1})
$\vec{C}_p$	Velocity of a particle (m/s)
d	Molecule diameter (m)
d_{eff}	Effective molecul diameter (m)
D_1, D_2	Coefficients of the Kramers-Moyal expansion (m/s^2)
E	Energy/Mean Kinetic Energy (J)
E	Young's molulus (Pa)
Etr	Translational Kinetic Energy (J)
F	Intermolecular Force (N)
$F = \frac{u}{\left(g\frac{\Delta\rho}{\rho}L\right)^{1/2}}$	Froude number
F_ℓ	Friction force (N)
F_N	Number of real molecules represented a sample molecule
$F(r) = -[d\Phi(r)/dr]$	Force acting among the molecules (N)
$\lvert g \rvert$	Modulus of relative velocity between two molecules (m/s)
g	Relative velocities of the molecules [m/s]
G	Condensation rate (kg/s)
$Gr = \frac{L^3 g\rho^2 \beta\Delta T}{\mu^2}$	Grashof number (for vertical plates)
$Gr = \frac{D^3 g\rho^2 \beta\Delta T}{\mu^2}$	Grashof number (for pipes)
H	Length scale (m)
k	Boltzmann Constant (J/K)
k_x, k_y	Wave number
K	Thermal conductivity (W/m K)
K_{n_L}	Knudsen system number
L	Length scale (m)
Lj	Normalized Lennard-Jones potential
M	Molecular weigth (kg/mol)
Ma	Mass in computational domain (kg)
m_r	Reduced mass (kg)
m	Mass of a molecule (kg)
M_p	Mass of a particle (kg)
n	Number density (m^3)
$\vec{n}$	Unit vector $\vec{n}$
N	Number of sample molecules
N_0	Avogadro number (kg mol^{-1})

$Nu = \frac{\alpha D_h}{\lambda}$	Nusselt number
p	Pressure (N/m^2)
P	Probability distribution
P	Pressure Tensor (N/m^2)
$Pe = \mathrm{Re} \cdot \mathrm{Pr} = \frac{wD_h}{\alpha^2}$	Peclet number
$\mathrm{Pr} = \frac{\nu}{\alpha^2} = \frac{\mu c_p}{\lambda}$	Prandtl number
Q	Source term in diffusion equation for aerosol particle (W/m^3)
Q = G/ρ	Volumetric condensation rate (m^3/s)
R	Gas constant (J/K kg)
R	Domain in phase space
$\mathbb{R}$	Universal gas constant (J/K-mole)
$Ra = \frac{G\beta g H^4}{\nu\alpha^2} = \frac{\partial T}{\partial z}\frac{\beta g H^4}{\nu\alpha^2}$	Rayleigh number (H = distance between plates)
$\mathrm{Re} = \frac{wD_h}{\nu} = \frac{\rho\nu D_h}{\mu}$	Reynolds number
$Ri = -\frac{\lvert g\rvert}{\rho}\frac{\partial\rho/\partial z}{(\partial u/\partial z)^2}$	Gradient Richardson number
$Ri = -\frac{\Delta\rho}{\rho}\frac{gh}{u^2}$	Richardson number
R_p	Radius of a particle (m)
R_1, R_2, R_3	Uniform deviates in the open interval [0 − 1]
S	Surface (m^2)
$Sc = \frac{\nu}{D}$	Schmidt number (D = mass diffusion rate)
$St = \frac{Lf}{u}$	Strouhal number (f = frequency of vortex shedding)
$S(\omega)$	Spectral density (W/kg)
t	Time (s)
tc	Mean Collision Time (s)
T	Temperature (K)
Tr	Translational Kinetic Temperature (K)
u	Velocity component of air molecules in the x coordinate direction (m/s)
u′	Thermal component of velocity u (m/s)
u0	Thermal component of velocity u (m/s)
u_	Post collision velocity component in the y coordinate direction (m/s)
u*	Post collision velocity component in the y coordinate direction (m/s)
u_f^+	Universal velocity profile (over film thickness)
U_p	Velocity component of a particle in the x coordinate direction (m/s)
v	Velocity component of air mlecules in the y coordinate direction (m/s)
v′	Thermal component of velocity v (m/s)
v0	Thermal component of velocity v (m/s)

v*	Post collision velocity component in the x coordinate direction (m/s)
V	Volume of computational domain (m^3)
w	Velocity component in the z coordinate direction (m/s)
w′	Thermal component of velocity w (m/s)
w0	Thermal component of velocity w (m/s)
w*	Post collision velocity component in the z coordinate direction (m/s)
W_f	Mass flow (kg/s)
W_p	Velocity component of a particle in the x coordinate direction (m/s)
x	Coordinate direction (m)
y	Coordinate direction (m)
z	Coordinate direction (m)
X	Point in phase space

Greek

α	Thermal diffusivity (m^2/s)
$\beta = 1/c'_m = \sqrt{2kT/m}$	Inverse of most probable thermal velocity (s/m)
$\beta = -\frac{1}{\rho}\left(\frac{\partial \rho}{\partial T}\right)_p$	Volumetric expansion coefficient at constant presssure (K^{-1})
$\gamma = \alpha/m = \frac{1}{\tau}$	Reciprocal of relaxation time (s^{-1})
Γ	Langevin force (m/s^2)
δ	Film thickness (m)
δ_m	Mean spacing between molecules (m)
ε	Internal energy of polyatomic molecules (J)
$\varepsilon = K \cdot 0.77T_c$	Characteristic energy of interaction among molecules (J/molecule)
ε	Strain (m)
ζ	Number of inner degrees of freedom of polyatomic molecules
θ	Co-latitude in a spherical coordinate system (Azimuthal angle)
λ	Mean free path (m)
λ_x, λ_y	Wave number
Λ	Collision frequency (1/s)
μ	Dynamic viscosity (kg/ms)
Δ	Difference operator
∇	Nabla operator
ν	Kinematic viscosity (m^2/s)
ρ	Density (kg/m^3)

π_{ij}	Pressure strain tensor (m^2/s^3)
σ	Differential cross section (m^2)
σ	$2.44\sqrt{T_c/p_c}$, Characteristc molecular diameter (Angstrom)
σ	Sress (N/m^2)
σ_g, σ_T	Geometrical cross section (m^2)
σ_{eff}	Effective cross section (m^2)
τ	Relaxation time (s)
φ	Relative humidity
Φ	Azimuth
Φ	Velocity potential (m)
$\Phi(r)$	Lennard-Jones potential (J)
Φ_{tt}	Lcockart-Martinelli parameter
χ	Deflection angle in a binary collision
ψ_i	Two-phase multiplier
ω	Temperature exponent of the viscosity coefficient
ω	Mass fraction (in mixture of species)

Indices

a	Air
c	Collision/cell
d	Effective
h	Hydrogen
i	Inlet/dummy index
ℓ	Local
max	Maximum
min	Minimum
p	particle
r	Relative/rotational
s	Saturation
tr	Translational
v	Steam
* (superscript)	Post-collision

List of Figures

List of Tables

Chapter 1
Basic Concepts of Molecular Dynamics

Contents

Text by M. Bottoni

M. Bottoni et al., *Molecular Dynamics - Theory and Applications*, Mechanical Engineering Series, https://doi.org/10.1007/978-3-031-84567-3_1

This chapter is dedicated to preliminary and basic concepts of molecular dynamics with details on interactions of molecules with ideal or material surfaces and with particles carried by a fluid. The fundamentals of the Boltzmann equation and of the Maxwellian distribution function and the basic ideas about the Larsen-Borgnakke phenomenological method are also revised. This chapter, however, is not intended to be a textbook, but rather to be a compendium of the major topics dealt with in the subsequent chapters, thus sparing to the un-experienced young reader a tedious search of topics which would involve consulting several textbooks with different backgrounds. In thus far, it is appropriate to state that there is nothing original in this chapter, apart from the presentation which is intended to be essential and suited for the beginner. For this purpose one section has been dedicated to a review of several techniques available for sampling probability distributions. The chapter closes with a topic concerning future applications of aerosol particles in an environment simulating the upper Earth environment conditions. For this purpose is also presented the treatment of the equation of motion for droplets in a cloud and the impinging of droplets upon a hydrometeor of spherical shape in atmospheric conditions.

1.1 The Molecular Dynamic Approach

Fluid-dynamic calculations are traditionally based on the numerical solution of the Navier-Stokes equations in terms of primitive variables (velocity components of the fluid) and pressure. In most cases these solutions are obtained by integrating the equations of mass and momentum conservation upon cell volumes and then deriving from the volume averaged equations an algebraic system of finite difference equations defined on the grid mesh, which partitions the definition domain of the problem. In case the definition domain is bounded by curved surfaces, which cannot be represented rigorously in a convenient (f. i. spherical or cylindrical) coordinate system, these bounding surfaces must be approximated by stepwise surfaces, which introduce numerical error and increase the so-called numerical diffusion.

In the case of heated fluids, an energy conservation equation is added to the mass and momentum balance equations. In turbulent flow regimes, a turbulence model, usually based on at least two transport equations, for the turbulent kinetic energy and its dissipation, is added to the previous system. This traditional approach, which might be called "macroscopic approach", is based on the assumption of a continuum medium and fails in the case of rarefied gases in which the scale of the phenomena to be described, like eddies in the flow, becomes comparable in size or smaller than the mean free path of the molecules. The rarefaction of a gas is expressed by the Knudsen number defined by $Kn = \lambda/L$, where λ is the mean free path of the molecules and L is a characteristic dimension. The limit of applicability of the macroscopic approach is usually assumed to be $Kn = 0.1$. Flow regimes characterised by a system Knudsen number just above this threshold, in the range $0.1 < Kn_L < 10$, are referred to as "intermediate regime flows", while larger Knudsen numbers characterise very rarefied flows, almost collisionless flows or "free-molecule regimes".

In the case of heated flows, the traditional macroscopic approach relies upon the application of fluid-structure heat transfer coefficients which are generally known with sufficient approximations under usual temperature and pressure conditions, but only with crude approximation in the intermediate and rarefied flow regimes.

One might therefore conclude that the traditional approach is affected by three inherent limitations, regarding: (1) the rigorous representation of the geometry; (2) the use of semi-empirical heat transfer and momentum transfer (turbulence model) correlations; and (3) the domain of applicability restricted to continuum medium, which rules out applications to rarefied gases.

As an alternative to the macroscopic approach there is a "microscopic approach" based on the application of the Direct Simulation Monte Carlo (DSMC) methodology in the frame of the molecular dynamic (MD) method (Bird, 1994) which relies upon the strict application of the basic laws of the kinetic theory (Euken, 1922). The main advantages of the microscopic method, in comparison with the traditional one, can be summarised as follows. The geometry can be represented rigorously, in whatever coordinate system, as long as the equations defining the surfaces bounding the definition domain are known. The definition domain is partitioned in cells, like in the traditional approach, but the cells are not used to derive difference equations. Therefore irregular cells do not infirm the accuracy of numerical results. Cells are defined for the only purpose of determining the collision rates among the molecules within the cells. Cumbersome representation of geometry is therefore avoided in the microscopic approach. The DSMC method does not imply the use of heat transfer coefficients or of turbulence models. Heat and momentum transfer are derived directly from the application of the kinetic laws. The domain of application, in terms of flow regimes, is unbounded, including very rarefied gases.

The microscopic or molecular approach is based on the Monte Carlo simulation of molecular trajectories, collision rates among the molecules, and collisions of the molecules with ideal (symmetry) or material surfaces. Ideal surfaces are treated as specularly reflecting surfaces, while material surfaces are treated as diffusively reflecting surfaces. Diffusive reflections imply that molecules reflected from material surfaces are assigned velocities obtained by sampling a Maxwellian velocity distribution corresponding to the temperature of the surface in the impact point.

1.2 Theoretical Background

1.2.1 Preliminary Concepts

In this section we introduce the main physical quantities, like mean free path of molecules and collision frequency, which need to be defined in the frame of a microscopic description of a fluid flow. As a second step we derive from the microscopic quantities the thermodynamic quantities, like pressure and temperature, which characterise the fluid from the macroscopic viewpoint and which are of interest for practical applications of computational results. In this frame we also derive from the microscopic quantities the transport coefficients of momentum, energy, and mass, which show in detail the physical significance of the usual concepts of dynamic viscosity, thermal conductivity, and mass diffusion coefficient.

1.2.1.1 Basic Molecular Dynamics Approach

We consider a fluid with number density (number of molecules per unit volume) n. If δ is the mean average spacing between the molecules of the fluid, then it holds

$$\delta^3 = \frac{1}{n} \tag{1.1}$$

hence

$$\delta = n^{-1/3}. \tag{1.2}$$

The mass of a molecule is

$$m = \frac{M}{N_\circ} \tag{1.3}$$

where M is the molecular weight and $N_\circ \approx 6.02525 \times 10^{23}$ is the Avogadro number. It holds

$$n = \rho \frac{N_\circ}{A}, \tag{1.4}$$

where A is the atomic number. The density of the fluid is

$$\rho = nm. \tag{1.5}$$

Let d be the diameter of the molecules. The geometrical collision cross section is

$$\sigma_T = \pi d^2. \tag{1.6}$$

Let us define:
$\vec{c}$ = Velocity of a molecule,
$\vec{c}_\circ$ = Mean stream velocity,
$\vec{c}' = \vec{c} - \vec{c}_\circ$ = Velocity of a molecule relative to the mean stream velocity,
$\vec{c}_r$ = Relative velocity between one chosen molecule ("test" or "sample" molecule) and the field molecules.

The mean collision rate per molecule, or collision frequency, ν, is obtained multiplying the mean volume swept by a molecule in the unit time ($\sigma_T c_r$) with the number density, n:

$$\nu = n\sigma_T c_r = \frac{1}{t_c}, \tag{1.7}$$

where t_c is the mean collision time, or mean time interval between collisions.

The total number of collisions per unit time, N_c, is obtained by multiplying the collision frequency, ν, with the number density, n, and dividing by 2 (to avoid to count twice the same couple of molecules):

$$N_c = \frac{1}{2} n\nu = \frac{1}{2} n^2 \sigma_T c_r. \tag{1.8}$$

The mean free path of a molecule is obtained by dividing its velocity relative to the mean stream with the collision frequency:

$$\lambda = \frac{c'}{\nu} = \frac{c'}{n\sigma_T c_r}. \tag{1.9}$$

If Eq. (1.6) holds, one obtains

$$\lambda = \frac{c'}{n\pi d^2 \overline{c}_r}. \tag{1.10}$$

In the case of a gas in thermodynamic equilibrium, the mean relative velocity is given (Bird, 1994) by

$$\overline{c}_r = \sqrt{2}\overline{c}', \tag{1.11}$$

hence, from Eq. (1.9)

$$\lambda = \frac{1}{\sqrt{2}n\sigma_T} = \frac{m}{\sqrt{2}\rho\sigma_T}. \tag{1.12}$$

The pressure tensor is defined by

$$\ldots P = nm\vec{c}'\vec{c} = \rho\vec{c}'\vec{c}', \tag{1.13a}$$

or, referring to its components, by

$$P_{ij} = \rho c_i' c_j' = nmc_i' c_j'. \tag{1.13b}$$

The components of the vector $\vec{c}'$ will be denoted u', v', w'. The scalar pressure is given by

$$p = \frac{1}{3}\rho\left(u'^2 + v'^2 + w'^2\right) = \frac{1}{3}\rho c'^2 = \frac{1}{3}nmc'^2. \tag{1.14}$$

The viscous stress tensor $\ldots\tau$ is defined by the equation

$$\ldots P = -\ldots\tau + p\ldots\delta, \tag{1.15}$$

where $\ldots\delta$ is the unit tensor. Hence

$$\tau_{ij} = -P_{ij} + p\delta_{ij}. \tag{1.16}$$

The average kinetic energy of a molecule of mass m is

$$E = \frac{1}{2}mc'^2. \tag{1.17}$$

The translational kinetic energy, or specific energy, is defined by

$$e_{tr} = \frac{1}{2}c'^2. \tag{1.18}$$

Combining Eqs. (1.14) and (1.18) one derives

$$p = \frac{2}{3}\rho e_{tr}. \tag{1.19}$$

The translational kinetic temperature T_{tr} is defined for a gas by

$$\frac{3}{2}RT_{tr} = e_{tr} = \frac{1}{2}c'^2, \tag{1.20}$$

where R is the gas constant. Using the Boltzmann constant

$$k = mR, \tag{1.21}$$

Equation (1.20) becomes

$$\begin{aligned}
\frac{3}{2}kT_{tr} &= me_{tr} = \frac{1}{2}mc'^2 = \frac{1}{2}m\left(u'^2 + v'^2 + w'^2\right) = \\
&= \frac{1}{2}m\Big((u-u_0)^2 + (v-v_0)^2 + (w-w_0)^2\Big) = \\
&= \frac{1}{2}m\left[u^2 + v^2 + w^2 - \left(u_0^2 + v_0^2 + w_0^2\right)\right] = \\
&= \frac{1}{2}m\left(c^2 - c_0^2\right).
\end{aligned} \tag{1.22}$$

The translational kinetic temperature can be computed using Eq. (1.22) by

$$\begin{aligned}
T_{tr} &= \frac{m}{3k}\left\langle c'^2\right\rangle = \frac{m}{3k}\left(\left\langle c^2\right\rangle - c_0^2\right) = \frac{m}{3k}\left(\frac{1}{N}\sum_{i=1}^{N} c_i^2 - \frac{1}{N}\sum_{i=1}^{N} c_0^2\right) \\
&= \frac{m}{3k}\left(\frac{1}{N}\sum_{i=1}^{N} c_i^2 - c_0^2\right),
\end{aligned} \tag{1.23}$$

where the symbol $\langle\rangle$ denotes mean value and

$$c_0 = \frac{1}{N}\sum_{i=1}^{N} c_i. \tag{1.24}$$

In the applications we consider a number N of sample molecules per unit volume, each one representative of a number F_N of real molecules. Thus it holds $n = NF_N$.

With reference to the Earth atmosphere, the number of molecules per cubic mean free path can be approximated (Bird, 1994) by

$$n\lambda^3 = \frac{3856}{\left(\frac{n}{n_0}\right)^2}, \tag{1.25}$$

where n/n_0 is the density ratio. The standard number density is given by $n_0 = p/kT = 2.6866 \times 10^{19}\text{cm}^3$.

Table 1.1 shows some useful gas properties at $t = 298.15$ K.

Some useful additional data are given in Table 1.2:

Table 1.1 Gas properties at $T = 298.15$ K (M = molecular weight)

Gas properties at T = 298.15 K

Specie	M	R (J/gm-K)	Cv (J/gm-K)
C	12.01100	0.00000	0.017708
CO	28.01040	0.29683	0.743551
CO2	44.00980	0.18892	0.654727
H2	2.01588	4.12416	10.179479
H2O	18.01528	0.46150	1.402215
N2	28.01340	0.29680	0.742878
N2O	44.01280	0.18889	0.688477
O2	31.99880	0.25983	0.658068
air	28.85033	0.28819	0.723125
AR	39.94800	0.20813	0.312192
HE	4.00260	2.07705	3.115522
NH3	17.03052	0.48820	1.603728
CH4	16.04276	0.51825	1.703236
OH	17.00734	0.48886	1.274294
H	1.00794	8.24856	12.373200
HO2	33.00674	0.25189	0.805014
NO	30.00610	0.27707	0.717481
O	15.99940	0.51964	0.849832
NH	15.01464	0.55374	1.387406
HNO	31.01404	0.26807	0.848757
H2O2	34.01468	0.24442	1.023102
NH2	16.02258	0.51891	1.576402
LG	3.70459	2.24411	3.690719
Xenon	131.29000	0.06333	0.094993
H2OL	18.01528		4.179300

Table 1.2 Main physical data, geometrical data, and constants

Avogadro number $N_0 = 6.02252 \times 10^{23}$[g - mol^{-1}]
Loschmidt number $n_0 = 2.69699 \times 10^{19}$[cm^{-3}]
Boltzmann constant $k = 1.38054 \times 10^{-16}$[erg/K]
Universal gas constant $\widehat{R} = 8.3143 \times 10^{-7}$[erg/K mol]
Molecular weight of air [averaged over the constituents O_2 (21%) and N_2 (79%)]: $0.21 \times 32 + 0.79 \times 28 = 28.96$ [g/mol]
Mass or air molecule [averaged over the constituents O_2 (21%) and N_2 (79%)]: $m = (0.21 \times 53.12 + 0.79 \times 46.5) \times 10^{-24} = 47.89 \times 10^{-24}$ [g]
Diameter of air molecule [averaged over the constituents O_2 (21%) and N_2 (79%)]: $d = 4.15 \times 10^{-8}$ [cm]
Air density [all following data at $T = 273$ K, $p = 10^5$[N/m^2]: $\rho_0 = mn_0 = 4.789 \times 10^{-23}$ $x2.68699 \times 10^{19} = 1.2868 \times 10^{-3}$[gr/cm^3]
Geometrical cross section of air molecules $\sigma_T = \pi d^2 = 54.11 \times 10^{-16}$ [cm^2]
Mean thermal velocity of air molecules $\overline{c}' = 4.469 \times 10^4$ [cm/s]
Mean relative velocity of air molecules $\overline{c}_r = \sqrt{2}\overline{c}' = 6.32 \times 10^4$ [cm/s]
Root mean thermal velocity of air molecules $c'_s = \sqrt{\overline{c}'^2} = 4.852 \times 10^4$ [cm/s]
Most probable thermal velocity of air molecules $c'_m = 3.96 \times 10^4$ [cm/s]

Table 1.3 Macroscopic transport parameters derived from molecular kinetics

Macroscopic unbalance	Molecular transport	Macroscopic parameters and physical laws
Nonuniform flow (u)	Momentum	Viscosity $\mu\left[\frac{\text{kg}}{\text{ms}}\right]$; $\tau = \mu\frac{du}{dy}\left[\frac{\text{N}}{\text{m}^2}\right]$
Nonuniform temperature (T)	Energy	Thermal conductivity $K\left[\frac{\text{W}}{\text{mK}}\right]$; $q = -K\frac{dT}{dy} \times \left[\frac{\text{W}}{\text{m}^2}\right]$
Nonuniform composition (n_A)	Mass	Diffusion coefficient $D\left[\frac{\text{m}^2}{\text{s}}\right]$; $\Gamma_A = -D\frac{dn_A}{dy}\left[\frac{1}{\text{m}^2\text{s}}\right]$

1.2.1.2 Transport Coefficients Derived from Molecular Dynamics

When a fluid is not in thermodynamic equilibrium conditions, molecular motion tends to smooth out the gradients of the inhomogeneous quantities and therefore tends to relax the fluid towards equilibrium conditions. From a macroscopic viewpoint, unbalance in a fluid can be due to nonuniform flow or nonuniform temperature or inhomogeneous concentration of a species of molecules. In these cases of unbalance, molecular motion tends to smooth nonuniformities through transport of momentum or energy or mass of the different species of molecules. From the macroscopic thermodynamic viewpoint the net transport is described by defining transport coefficients named viscosity, thermal conductivity, and mass diffusion. These concepts are summarized in Table 1.3. Beside the transport coefficients, the transport equations are recalled with reference to the simplest case of a two-dimensional domain with a fluid flowing in the x-direction and having

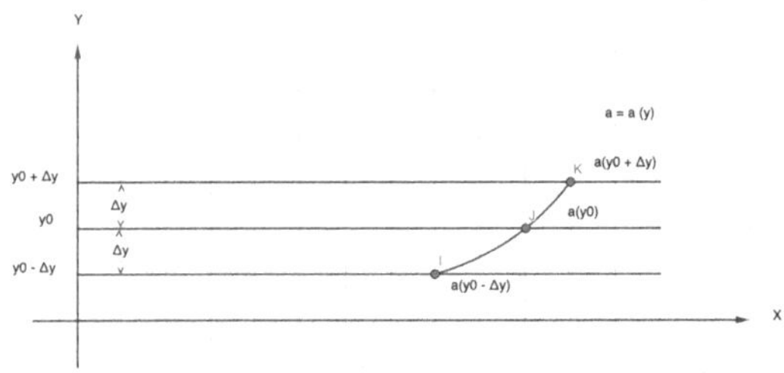

Fig. 1.1 Distribution $a = a(y)$ of a physical quantity in a two-dimensional flow

distribution of physical quantities in the y-direction only. n_A is the concentration of a species "A" of molecules.

Let us refer to a two-dimensional domain, as sketched in Fig. 1.1, with a fluid flowing in the x direction and characterized by a distribution of physical quantities in the y direction only. Let $a = a(y)$ be any physical macroscopic quantity with a distribution in the y direction as shown in the sketch. Let $a(y_0)$ be a reference value of this physical quantity at $y = y_0$. Let us consider two more values of a at a distance Δy from y_0, one below and one above the line $y = y_0$, namely $a(y_0 - \Delta y)$ and $a(y_0 + \Delta y)$. The physical quantity a has a distribution in the y direction because molecular collisions imply an exchange of microscopic quantities, essentially molecular velocity, among the molecules. The value $a(y_0 + \Delta y)$ differs from the value $a(y_0 - \Delta y)$ because molecules collided at level $y_0 - \Delta y$, for any x, before reaching level $y_0 + \Delta y$. We may choose the value Δy to be representative of the mean free path of the molecules in the fluid, just setting

$$\Delta y = k_a \lambda, \tag{1.26}$$

where k_a is a constant depending on the physical macroscopic quantity a.

The flow F_a of molecules across the line $y = y_0$, in any direction, is proportional to the molecular concentration n and to the mean velocity $\bar{c}$ of the molecules, thus

$$F_a = n\bar{c} \tag{1.27}$$

It will be proved later that the mean value of molecular velocities at equilibrium, i.e. when the molecular distribution of velocities is a Maxwellian distribution at the mean temperature of the fluid, is

$$\bar{c} = \sqrt{\frac{8RT}{\pi}}. \tag{1.28}$$

The flux of property a across the line $y = y_0$ is proportional to the difference $[a(y_0 - \Delta y) - a(y_0 + \Delta y)]$ and to the flow F_a of molecules undergoing the transition, hence

$$\Lambda_a \equiv n\bar{c}[a(y_0 - \Delta y) - a(y_0 + \Delta y)]. \tag{1.29}$$

Expanding in Taylor series the terms within square brackets one has

$$a(y_0 - \Delta y) = a(y_0) - \Delta y{\cdot}a'(y_0) + \frac{(\Delta y)^2}{2}a''(y_0) - \ldots \tag{1.30a}$$

$$a(y_0 + \Delta y) = a(y_0) + \Delta y{\cdot}a'(y_0) + \frac{(\Delta y)^2}{2}a''(y_0) + \ldots, \tag{1.30b}$$

hence

$$a(y_0 - \Delta y) - a(y_0 + \Delta y) = -2\Delta y a'(y_0) = -2\Delta y\left(\frac{da}{dy}\right)_0 = -2k_a\lambda\left(\frac{da}{dy}\right)_0, \tag{1.31}$$

Introducing Eq. (1.31) into (1.29) one has

$$\Lambda_a \equiv -n\bar{c}2k_a\lambda\left(\frac{da}{dy}\right)_0 = -\beta_a n\bar{c}\lambda\left(\frac{da}{dy}\right)_0, \tag{1.32}$$

with the definition

$$\beta_a = 2k_a. \tag{1.33}$$

According to the physical property being transported, we have three different cases:

(i) **Non-uniform flow** with velocity distribution $u = u(y)$ and momentum transport: $a = mu$. Letting $-\Lambda_a = \tau$, and $\beta_a = \beta_\mu$, from Eq. (1.32) one obtains

$$\tau = \beta_\mu n\bar{c}\lambda m\left(\frac{du}{dy}\right)_0 = \beta_\mu \rho\bar{c}\lambda\left(\frac{du}{dy}\right)_0 = \mu\frac{du}{dy}, \tag{1.34}$$

with the definition of the dynamic viscosity coefficient

$$\mu = \beta_\mu \rho\bar{c}\lambda. \tag{1.35}$$

(ii) **Non-uniform temperature** with distribution $T = T(y)$ and energy transport $a = \frac{\xi}{2}kT = \frac{\xi}{2}mRT$, where ξ is the degree of freedom of the molecules. Letting $\Lambda_a = q$, and $\beta_a = \beta_K$, from Eq. (1.32) one obtains

$$q = \beta_K n\bar{c}\lambda\frac{\xi}{2}mR\left(\frac{dT}{dy}\right)_0 = \beta_K \rho\bar{c}\lambda c_v\left(\frac{dT}{dy}\right)_0 = K\frac{dT}{dy}, \tag{1.36}$$

having used $c_v = \frac{\xi}{2}R$ and introduced the definition of the thermal conductivity

$$K = \beta_K \rho c_v \bar{c}\lambda. \tag{1.37}$$

(iii) **Non-uniform composition** or concentration of a molecular species with $a = \frac{n_A}{n}$. Letting $\Lambda_a = \Gamma_A$, and $\beta_a = \beta_D$, from Eq. (1.32) one obtains

$$\Gamma_A = -\beta_D n\bar{c}\lambda\frac{1}{n}\left(\frac{dn_A}{dy}\right)_0 = -\beta_D\bar{c}\lambda\frac{dn_A}{dy} = -D\frac{dn_A}{dy}, \tag{1.38}$$

with the definition of the diffusion coefficient

$$D = \beta_D\bar{c}\lambda. \tag{1.39}$$

The transport coefficients thus introduced are summarized in Table 1.4.

Combining Eqs. (1.35) and (1.37) one obtains

$$\bar{c}\lambda = \frac{\mu}{\beta_\mu \rho} = \frac{K}{\rho c_v \beta_K}, \tag{1.40}$$

Table 1.4 Macroscopic transport coefficients derived from molecular kinetics

$\mu = \beta_\mu \rho\bar{c}\lambda \equiv \rho\cdot\bar{c}\lambda$ (1.35)
$K = \beta_K \rho c_v \bar{c}\lambda \equiv \rho c_v\cdot\bar{c}\lambda$ (1.37)
$D = \beta_D\bar{c}\lambda \equiv \bar{c}\lambda$ (1.39)

$$K = \frac{\beta_K}{\beta_\mu} \mu c_v. \tag{1.41}$$

For mono-atomic gases one has $\beta_\mu \approx 1/2$ and $\beta_K \approx 5/4$, hence

$$K = \frac{\beta_K}{\beta_\mu} \mu c_v = \frac{5}{2} \mu c_v \text{ (mono-atomic gases)}. \tag{1.42}$$

For polyatomic gases theoretically derived values are not known. The different values of β_μ and β_K in Eq. (1.41) show that even for mono-atomic gases, the mechanisms of momentum transfer and of energy transfer are different, because momentum depends on velocity and energy depends on the square of velocity. If momentum and energy were transferred in the same way one would have $\beta_\mu = \beta_K$ and

$$K = \mu c_v. \tag{1.43}$$

Equation (1.43) is actually used for the energy transfer associated with the inner energy modes, which are rotational modes, if excitation energy of molecules can be neglected at low temperatures.

Using the diffusion coefficient $\alpha^2 = \frac{K}{\rho c_p}$ and $\gamma = c_p/c_v$, the Prandtl number is given by

$$\mathrm{P}_r = \frac{\nu}{\alpha^2} = \frac{\nu \rho c_p}{K} = \frac{\mu c_p}{K} = \frac{\mu \gamma c_v}{K} = \text{(for mono-atomic gases)} \frac{2}{5} \gamma. \tag{1.44}$$

Thus for mono-atomic gases, with $\gamma = 5/3$, one obtains $P_r = 2/3$.

In the case of diatomic gases ($\gamma = 7/5$), one knows that $P_r \approx 3/4 = 0.75$. This value does not fit with formula (1.44) because in that formula the values of β_μ and β_K refer to mono-atomic gases.

A fair estimation of the Prandtl number for poly-atomic molecules is made with the approach by Eucken (1922). The total energy of a molecule is considered as the sum of translational energy and of inner (rotational) energy (a topic which will be further discussed in following sections):

$$e = e_{tr} + e_{\text{int}}. \tag{1.45}$$

Correspondingly, we may assume that the specific heat at constant volume and the thermal conductivity of the gas can also be split into two contributions and therefore given by

$$c_v = c_{v,tr} + c_{v,\text{int}}, \tag{1.46}$$

$$K = K_{tr} + K_{\text{int}}. \tag{1.47}$$

The assumption is now made that for the translational contribution of the thermal conductivity Eq. (1.42) can be applied, while the contribution from the inner energy is given by Eq. (1.43). Thus one obtains

$$K = K_{tr} + K_{\text{int}} = \frac{5}{2}\mu c_{v,tr} + \mu c_{v,\text{int}} = \mu\left(\frac{5}{2}c_{v,tr} + c_{v,\text{int}}\right)$$
$$= \mu\left(\frac{5}{2}c_{v,tr} + c_v - c_{v,tr}\right) = \mu\left(\frac{3}{2}c_{v,tr} + c_v\right). \tag{1.48}$$

Using $c_{v,tr} = \frac{3R}{2}$ and $R = c_v(\gamma - 1)$ one obtains

$$K = \mu\left(\frac{3}{2}c_{v,tr} + c_v\right) = \mu\left(\frac{9}{4}R + c_v\right) = \mu\left(\frac{9}{4}c_v(\gamma - 1) + c_v\right) = \mu c_v \frac{9\gamma - 5}{4}. \tag{1.49}$$

The Prandtl number can now be expressed by

$$P_r = \frac{\mu c_p}{K} = \frac{\mu c_p}{\mu c_v}\frac{4}{9\gamma - 5} = \frac{4\gamma}{9\gamma - 5}. \tag{1.50}$$

Equation (1.50) is known as Euken's relation. It gives:

- For mono-atomic gases ($\gamma = 5/3$), $P_r = 2/3$ (confirming the previous value);
- For diatomic gases ($\gamma = 7/5$), $P_r = 28/38 \approx 0.737$ (close to the above given value of 3/4).

The above-obtained relationships allow to express the dynamic viscosity coefficient in terms of the fluid temperature and of the diameter of the molecules. Using Eq. (1.28) for the mean velocity of molecules and Eq. (1.12) for the mean free path, Eq. (1.35) becomes

$$\mu = \beta_\mu \rho \bar{c} \lambda \equiv \rho\sqrt{\frac{8RT}{\pi}}\frac{1}{\sqrt{2}n\pi d^2} \equiv \frac{\sqrt{T}}{d^2}. \tag{1.51}$$

According to this simplified theoretical approach, the dynamic viscosity coefficient is proportional to the square of the absolute temperature of the gas and inversely proportional to the square of the diameter of the molecules. Experiments show that in reality the increase of the viscosity with the temperature is more rapid than predicted by this simplified approach. The reason is that the simple approach does not take into account the molecular attraction which may be described by the Lennard-Jones potential, as explained in the following of this treatise.

1.2.2 Calculation of Rotational Temperature of Diatomic Molecules

1.2.2.1 Introduction to the Topic

From the strict theoretical viewpoint calculation of the rotational temperature of diatomic or polyatomic molecules would require simulation of all internal rotational energy levels of the molecules. The number of these rotational energy levels is proportional to the ratio between the temperature of the gas and the characteristic temperature of rotation. The characteristic temperatures of rotation of the diatomic molecules of oxygen and nitrogen are 2.07 K and 2.88 K, respectively. At a reference gas temperature of 300 K to apply the theoretical approach of describing transitions between all theoretical energy levels would imply consideration of more than one hundred levels with all possible transitions among them, a task impossible at the actual state of computational capabilities.

The alternative to the rigorous theoretical approach is to apply a phenomenological description of all internal states of rotational energy, considered as a unique band of energy levels, which interacts with translational energy and shares with it all available energy, leading eventually, for an equilibrium gas, to equipartition of energy between translational and rotational modes. In the phenomenological approach collisions between diatomic or polyatomic molecules are treated essentially as spherically symmetric collisions between mono-atomic molecules. Most collisions are however inelastic. In elastic collisions the share of translational and rotational energies of the molecules remains unchanged. In inelastic collisions the total available energy is shared among the translational and rotational energies of the collisional partners.

Phenomenological models can be divided into two classes according to the way of treating the elastic and inelastic collisions. In the first class of methods a fixed percentage of elastic collisions is allowed in the total number of mostly inelastic collisions. In the second class of methods all collisions are considered as inelastic, but the actual collision frequency modelled, Λ, is some fraction of the maximum theoretical collision frequency. The rotational relaxation time of real gases, i.e. the time necessary to reach an equipartition of energy between translational and rotational modes, depends on temperature. This can be modelled in principle letting the parameter Λ vary with the overall temperature of the gas.

We introduced in our computational program the phenomenological model of Larsen and Borgnakke (1974), following the theoretical treatment presented in Bird (1994).

1.2.2.2 The Larsen-Borgnakke Phenomenological Method

Let ω be the temperature exponent of the coefficient of viscosity of a gas. The translational energy of the gas is

$$E_{tr} = \frac{3}{2} RT_{tr} = \frac{1}{2} c'^2. \tag{1.52}$$

Equation (1.52) states that the translational energy associated with every translational degree of freedom is $(1/2)RT_{tr}$. This equation can be considered as the definition of the translational temperature T_{tr}.

The distribution function of the translational energy E_{tr} is given by

$$f(E_{tr}) = E_{tr}^{3/2-\omega} \exp[-E_{tr}/(kT)]. \tag{1.53}$$

Let E_i be the internal energy of the gas and T_i the temperature associated with this internal energy. The vibrational energy of polyatomic molecules is negligible in the low temperature ranges, but it should be considered at higher temperatures. For instance the characteristic vibrational temperatures of the diatomic molecules of oxygen and nitrogen are 2256 K and 3371 K, respectively. Therefore, for the temperature ranges we consider in the applications, the inner energy of the molecules is simply rotational energy. In analogy with the definition of the translational temperature and energy, we may state that the inner energy of molecules is (1/2) RT_i for every degree of freedom ζ. Therefore

$$E_i = \frac{\zeta}{2} RT_i. \tag{1.54}$$

In the case of diatomic molecules, like oxygen O_2 and nitrogen N_2, a molecule behaves like a narrow cylinder joining the nuclei of the two atoms. The rotational energy associated with the rotation around the axis of this cylinder, passing through the nuclei, is negligible, because the momentum of inertia of the molecule respect to this axis is very small compared to the momenta of inertia with respect to the other two axes normal to the axis of the cylinder. Therefore, for diatomic molecules $\zeta = 2$.

Let us denote with indices 1 and 2 two molecules undergoing a collision. In a more general sense the two indices represent two classes of molecules such that for every collision one partner is taken from the first class and the other partner from the second class. The total internal energy of one couple is the sum of the internal energies of the collisional partners. Hence, using capital letters for the pair and small Greek letters for the single molecules, we may write

$$E_i = \varepsilon_{i1} + \varepsilon_{i2}. \tag{1.55}$$

The distribution function of the internal energy ε_i of molecules with ζ internal degrees of freedom is given in Hinshelwood (1940), by

$$f_{\varepsilon_i} = \varepsilon_i^{\zeta/2-1} \exp[-\varepsilon_i/(kT)]. \tag{1.56}$$

Applying Eq. (1.56) to the collisional partners and using Eq. (1.55) to eliminate ε_{i2}, one has

$$f_{\varepsilon_{i1}} = \varepsilon_{i1}^{\zeta/2-1} \exp[-\varepsilon_{i1}/(kT)], \tag{1.57a}$$

$$f_{\varepsilon_{i2}} = \varepsilon_{i2}^{\zeta/2-1} \exp[-\varepsilon_{i2}/(kT)] = (E_i - \varepsilon_{i1})^{\zeta/2-1} \exp[-(E_i - \varepsilon_{i1})/(kT)]. \tag{1.57b}$$

Let us now consider ε_{i1} as a fixed value of internal energy for a molecule of class 1. A molecule of class 2 colliding with that of class 1 has energy $\varepsilon_{i2} = E_i - \varepsilon_{i1}$.

The fraction, say $f_{1,2}(\varepsilon_{i1}, \varepsilon_{i2})d\varepsilon_{i1}d\varepsilon_{i2}$, of collision pairs with the first molecule of class 1 having internal energy in the range $(\varepsilon_{i1}, \varepsilon_{i1} + d\varepsilon_{i1})$ and the second molecule of class 2 having internal energy in the range $(\varepsilon_{i2}, \varepsilon_{i2} + d\varepsilon_{i2} = \varepsilon_{i2} + dE_i)$ (having used $d\varepsilon_{i2} = dE_i$) is proportional to the product of the respective distribution functions, multiplied by the differential increments. Hence

$$\begin{aligned} f_{1,2}(\varepsilon_{i1},\varepsilon_{i2})d\varepsilon_{i1}d\varepsilon_{i2} &= f_{\varepsilon_{i1}} f_{\varepsilon_{i2}} d\varepsilon_{i1}d\varepsilon_{i2} \\ &= \varepsilon_{i1}^{\zeta/2-1}\exp[-\varepsilon_{i1}/(kT)]\varepsilon_{i2}^{\zeta/2-1}\exp[-\varepsilon_{i2}/(kT)]d\varepsilon_{i1}d\varepsilon_{i2} \\ &= \varepsilon_{i1}^{\zeta/2-1}\exp[-\varepsilon_{i1}/(kT)](E_i-\varepsilon_{i1})^{\zeta/2-1}\exp[-(E_i-\varepsilon_{i1})/(kT)]d\varepsilon_{i1}dE_i \\ &= \varepsilon_{i1}^{\zeta/2-1}(E_i-\varepsilon_{i1})^{\zeta/2-1}\exp[-E_i/(kT)]d\varepsilon_{i1}dE_i. \end{aligned} \tag{1.58}$$

The total fraction, say $f_{E_i}dE_i$, of collision pairs with internal energy in the range $(E_i, E_i + dE_i)$ is obtained by integrating the contribution represented by Eq. (1.58) over all ranges of energies ε_{i1} from 0 to the maximum possible value E_i:

$$\begin{aligned} f_{E_i} &= \left[\int_0^{E_i} f_{1,2}(\varepsilon_{i1},\varepsilon_{i2})d\varepsilon_{i1}\right]dE_i \\ &= \left[\int_0^{E_i} \varepsilon_{\varepsilon_{i1}}^{\zeta/2-1}(E_i-\varepsilon_{i1})\exp[-E_i/(kT)]d\varepsilon_{i1}\right]dE_i \\ &= \left[\int_0^{E_i} E_i^{\zeta-1}\exp[-E_i/(kT)]\right]dE_i. \end{aligned} \tag{1.59}$$

The total energy of a couple of molecules in a collision,E_c, is the sum of the translational energy E_{tr} and of the internal energy E_i:

$$E_c = E_{tr} + E_i. \tag{1.60}$$

The probability density f_{E_{tr},E_i} of the combined values of E_{tr} and E_i in a collision is the product of the respective distribution functions, given by Eqs. (1.53) and (1.59):

$$\begin{aligned} f(E_{tr})f(E_i) &= E_{tr}^{3/2-\omega}\exp[-E_{tr}/(kT)]E_i^{\zeta-1}\exp[-E_i/(kT)] \\ &= E_{tr}^{3/2-\omega}E_i^{\zeta-1}\exp[-(E_{tr}+E_i)/(kT)] \\ &= E_{tr}^{3/2-\omega}E_i^{\zeta-1}\exp[-E_c/(kT)]. \end{aligned} \tag{1.61}$$

For a given couple of molecules undergoing a collision E_c is a constant and the temperature T in the above equations, which is the overall temperature corresponding to the total energy E_c, is also a constant. The exponential in Eq. (1.61) is therefore a constant and the equation can be written as

$$f_{E_{tr},E_i} = f(E_{tr})f(E_i) = CE_{tr}^{3/2-\omega}E_i^{\zeta-1} = CE_{tr}^{3/2-\omega}(E_c - E_{tr})^{\zeta-1}. \tag{1.62}$$

Being E_c fixed for a given collision, Eq. (1.62) can be interpreted as the distribution function of a given value of the translational energy in the collision event:

$$P(E_{tr}) = CE_{tr}^{3/2-\omega}(E_c - E_{tr})^{\zeta-1}. \tag{1.63}$$

The derivative of $P(E_{tr})$ with respect to the translational energy E_{tr} is given by

$$\frac{dP(E_{tr})}{dE_{tr}} = C[(3/2-\omega)]E_{tr}^{1/2-\omega}(E_c - E_{tr})^{\zeta-1} + E_{tr}^{3/2-\omega}(\zeta-1)(E_c - E_{tr})^{\zeta-2}. \tag{1.64}$$

Equating the derivative to zero and solving with respect to E_{tr} one finds a maximum for

$$E_{tr} = E_c\frac{3/2-\omega}{\zeta+1/2-\omega}, \tag{1.65}$$

where the fraction is a constant.

The value of the maximum is obtained inserting Eq. (1.65) in Eq. (1.63):

$$\begin{aligned} P(E_{tr})_{\max} &= C\left(E_c\frac{3/2-\omega}{\zeta+1/2-\omega}\right)^{3/2-\omega}\left(E_c - E_c\frac{3/2-\omega}{\zeta+1/2-\omega}\right)^{\zeta-1} \\ &= CE_c^{1/2+\zeta-\omega}\left(\frac{3/2-\omega}{\zeta+1/2-\omega}\right)^{3/2-\omega}\left(1-\frac{3/2-\omega}{\zeta+1/2-\omega}\right)^{\zeta-1} \\ &= CE_c^{1/2+\zeta-\omega}\left(\frac{3/2-\omega}{\zeta+1/2-\omega}\right)^{3/2-\omega}\left(\frac{\zeta+1/2-\omega-3/2+\omega}{\zeta+1/2-\omega}\right)^{\zeta-1} \\ &= CE_c^{1/2+\zeta-\omega}\frac{(3/2-\omega)^{3/2-\omega}(\zeta-1)^{\zeta-1}}{(\zeta+1/2-\omega)^{1/2+\zeta-\omega}}. \end{aligned} \tag{1.66}$$

The ratio of the probability $P(E_{tr})$ to its maximum value is therefore

$$\begin{aligned}\frac{P(E_{tr})}{P(E_{tr})_{\max}} &= \frac{E_{tr}^{3/2-\omega}(E_c-E_{tr})^{\zeta-1}(\zeta+1/2-\omega)^{1/2+\zeta-\omega}}{E_c^{\zeta+1/2-\omega}(\zeta-1)^{\zeta-1}}\\ &= \frac{E_{tr}^{3/2-\omega}(E_c-E_{tr})^{\zeta-1}(\zeta+1/2-\omega)^{3/2-\omega+\zeta-1}}{E_c^{3/2-\omega+\zeta-1}(3/2-\omega)^{3/2-\omega}(\zeta-1)^{\zeta-1}}\\ &= \left(\frac{\zeta+1/2-\omega}{3/2-\omega}\right)^{3/2-\omega}\left(\frac{E_{tr}}{E_c}\right)^{3/2-\omega}\left(\frac{E_c-E_{tr}}{E_c}\right)^{\zeta-1}\left(\frac{\zeta+1/2-\omega}{\zeta-1}\right)^{\zeta-1}\\ &= \left(\frac{\zeta+1/2-\omega}{3/2-\omega}\frac{E_{tr}}{E_c}\right)^{3/2-\omega}\left[\frac{\zeta+1/2-\omega}{\zeta-1}\left(1-\frac{E_{tr}}{E_c}\right)\right]^{\zeta-1}.\end{aligned} \tag{1.67}$$

Equation (1.67) is used to apply the acceptance-rejection method to the collision under consideration. Let us denote with a star $*$ as a superscript the post-collision energy values. A post-collision translational energy E_{tr}^* is chosen at random in the range 0 to E_c. Let R_1 be a random deviate in the open interval $(0-1)$. If

$$R_1 < \frac{P(E_{tr})}{P(E_{tr})_{\max}} \tag{1.68}$$

the collision leading to the translational energy E_{tr}^* is accepted, otherwise it is rejected. Let us assume that the collision of the considered couple of molecules has been accepted, and that the above randomly determined post-collision value E_{tr}^* has therefore been accepted. Because the total energy is conserved, beyond Eq. (1.60), in the collision one has:

$$E_c = E_c^* \tag{1.69}$$

$$E_c^* = E_{tr}^* + E_i^*, \tag{1.70}$$

hence

$$E_i^* = E_c^* - E_{tr}^* = E_c - E_{tr}^*. \tag{1.71}$$

Having determined the post-collision internal energy E_i^* of the couple, next task is to divide this internal energy between the two collision partners. From Eq. (1.58) one deduces that the probability of molecule 1 to have post-collision internal energy ε_{i1}^* is

$$\begin{aligned}P(\varepsilon_{i1}^*) &= (\varepsilon_{i1}^*)^{\zeta/2-1}(E_i^*-\varepsilon_{i1}^*)^{\zeta/2-1}\exp\left[-E_i^*/(kT)\right]\\ &= D(\varepsilon_{i1}^*)^{\zeta/2-1}(E_{i1}^*-\varepsilon_{i1}^*)^{\zeta/2-1},\end{aligned} \tag{1.72}$$

because the exponential term is a constant, say D.

Deriving Eq. (1.72) with respect to ε_{i1}^* and equating to zero the derivative, one finds that $P(\varepsilon_{i1}^*)$ has a maximum for $\varepsilon_{i1}^* = \varepsilon_{i2}^* = E_i^*/2$ and this maximum is

$$P(\varepsilon_{i1}^*)_{\max} = D\left[\left(\frac{\varepsilon_i^*}{2}\right)^2\right]^{\zeta/2-1} = D\left(\frac{\varepsilon_i^*}{2}\right)^{\zeta-2}. \tag{1.73}$$

The ratio of the probability $P(\varepsilon_{i1}^*)$ to its maximum is therefore

$$\frac{P(\varepsilon_{i1}^*)}{P(\varepsilon_{i1}^*)_{\max}} = \frac{(\varepsilon_{i1}^*)^{\zeta/2-1}(E_i^* - \varepsilon_{i1}^*)^{\zeta/2-1}}{(E_i^*/2)^{\zeta-2}} = \left(\frac{\varepsilon_{i1}^*}{E_i^*}\right)^{\zeta/2-1}\left(1-\frac{\varepsilon_{i1}^*}{E_i^*}\right)^{\zeta/2-1} 2^{\zeta-2}. \tag{1.74}$$

Equation (1.74) is applied with the acceptance-rejection method to determine a value for ε_{i1}^*, hence one obtains

$$\varepsilon_{i2}^* = E_i^* - \varepsilon_{i1}^*. \tag{1.75}$$

There remains to compute the post-collision translational energies of the collisional partners, from the knowledge of the resulting post-collision translational energy of the couple

$$E_{tr}^* = \frac{1}{2} m_r (c_r^*)^2, \tag{1.76}$$

where $m_r = m_1 m_2/(m_1 + m_2)$ is the reduced mass and c_r^* is the post-collision centre of mass velocity. Unlike in the case of elastic collisions the post-collision centre of mass velocity is different from the pre-collision centre of mass velocity.

From Eq. (1.76) the new post-collision centre of mass velocity c_r^* is derived. As in the case of elastic collisions, the components of the post-collision relative velocity c_r^* can be computed by multiplying a unit vector generated with the Marsaglia technique (Marsaglia, 1972), with the module of the relative velocity:

$$\begin{aligned} u_r^* &= S c_r \cos\beta, \\ v_r^* &= S c_r \sin\beta, \\ w_r^* &= C_s c_r, \end{aligned} \tag{1.77}$$

using two uniform deviates R_1 and R_2 and computing

$$\begin{aligned} \beta &= 2\pi R_1, \\ C_s &= 1 - 2R_2, \\ S &= \sqrt{1 - C_s^2}. \end{aligned} \tag{1.78}$$

The centre of mass velocity, which does not change in the collision, is given by

$$\overline{c}_m = \frac{m_1 c_1 + c_2 m_2}{m_1 + m_2}. \tag{1.79}$$

The translational velocity of the collisional partners is hence obtained by:

$$\overline{c}_1 = \overline{c}_m + \overline{c}_r^* \frac{m_r}{m_1}, \tag{1.80a}$$

$$\overline{c}_2 = \overline{c}_m - \overline{c}_r^* \frac{m_r}{m_2}. \tag{1.80b}$$

In the simple case of a diatomic gas $\zeta = 2$. Assuming a gas with viscosity temperature coefficient $\omega = 1/2$, Eq. (1.68) simplifies to

$$P(\varepsilon_{i1}^*) = CE_{tr}(E_c - E_{tr}). \tag{1.81}$$

Taking the derivative with respect to E_{tr}, and equating to zero, one finds that a maximum occurs for $E_{tr} = E_c/2$ and this maximum is

$$P(E_{tr})_{\max} = C\frac{E_c^2}{4}. \tag{1.82}$$

The ratio of the probability $P(E_{tr})$ to its maximum value therefore becomes

$$\frac{P(E_{tr})}{P(E_{tr})_{\max}} = 4\frac{E_{tr}}{E_c}\left(1 - \frac{E_{tr}}{E_c}\right). \tag{1.83}$$

1.2.2.3 Computation of Rotational and Overall Temperature

In the rotational mode of one molecule the rotational energy T_i provides a definition of the rotational temperature T_r. From Eq. (1.54), using $k = mR$, one has

$$T_r = \frac{2E_i}{\zeta R} = \frac{2mE_i}{\zeta k}. \tag{1.84}$$

Let $\varepsilon_r = \varepsilon_i = mE_i$ be the rotational energy of the molecule of mass m and let us denote with an overbar its mean value. The rotational temperature becomes

$$T_r = \frac{2\bar{\varepsilon}_r}{\zeta k}. \tag{1.85}$$

The overall temperature is then defined by

$$T_{ov} = \frac{3T_{tr} + \zeta T_r}{3 + \zeta}. \tag{1.86}$$

1.2.2.4 The Cercignani-Lampis-Lord (CLL) Model

When simulating the diatomic molecules, beyond the calculation of the inner rotational energy, it is also necessary to apply new models for (i) the change of molecular velocities due to the molecule-surface interactions and (ii) the jump (increase of decrease) of rotational energy suffered by molecules in the collisions with the material surfaces. Both the change of velocity components and of rotational energy may be computed on the basis of the Cercignani-Lampis-Lord model (Cercignani & Lampis, 1974; Lord, 1991). The essence of this model is outlined in the following.

Let u_i be the normal velocity component and v_i, w_i be the tangential velocity components of one molecule colliding with a material wall. Let u_r, v_r and w_r be the reflected normal and tangential velocity components. Let in the following all velocity components be normalized by dividing them with the most probable thermal velocity c'_m corresponding to the wall temperature T_w.

The velocity of a molecule upon reflection does not necessarily adjust perfectly to the temperature T_w of the wall. To account for this we define an accommodation coefficient

$$\alpha_n = \frac{q_i - q_r}{q_i - q_w}, \tag{1.87}$$

where q_i, q_r are the incident and reflected energy fluxes and q_w is the energy flux which would correspond to a perfect diffuse reflection in which the temperature of the reflected molecules equals in average the wall temperature, so that $T_r = T_w$. The subscript n in the above equation refers to the definition of the accommodation coefficient for the normal velocity component. Similar definitions can be used for the accommodation coefficients α_t referring to the tangential components, but we assume for simplicity that $\alpha_t = \alpha_n$. The definition (1.87) implies that for $q_r = q_i$, equal incident and reflected fluxes, results $\alpha_n = 0$, hence there is no accommodation at all. Conversely, if $q_r = q_w$, $\alpha_n = 1$, hence there is a perfect accommodation to the temperature of the wall.

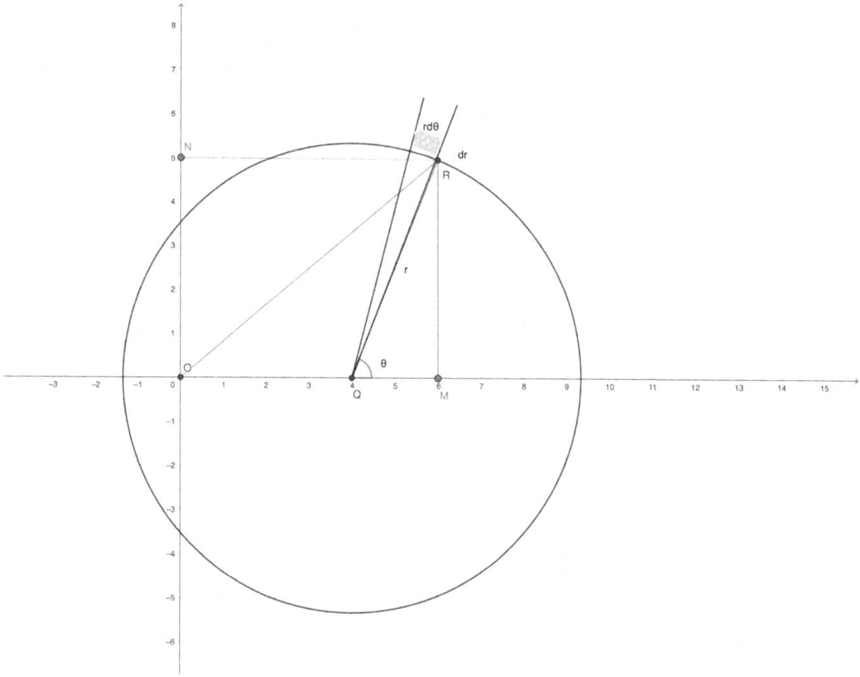

Fig. 1.2 Geometrical representation of the CLL model

Let us consider a point P on the x-axis of Fig. 1.2. It represents the initial normalized normal velocity component $\left(u_i/c'_m\right)$ of a colliding molecule. In other words, point P represents the initial state (with reference to the normal component) before collision. The average state of the reflected molecules is represented by point Q which is chosen so that

$$\overline{OQ} = \left(u_i/c'_m\right)\sqrt{1-\alpha_n}. \tag{1.88}$$

The actual state of the reflected molecule is represented by the point R. The probability that a point R falls in the shaded differential area $dA = dr \cdot rd\theta$ is given by a two-dimensional Gaussian distribution centred on point Q

$$P(r,\theta)dr{\cdot}rd\theta = \frac{1}{\pi\alpha_n}\exp\left(-r^2/\alpha_n\right)dr{\cdot}rd\theta. \tag{1.89}$$

This probability distribution satisfies the normalization condition

$$\int_0^{2\pi}\int_0^{\infty} P(r,\theta)dr\cdot rd\theta = \int_0^{2\pi}\left[\int_0^{\infty}\frac{1}{\pi\alpha_n}\exp\left(-r^2/\alpha_n\right)rdr\right]d\theta$$
$$= \frac{2}{\alpha_n}\int_0^{\infty}\exp\left(-r^2/\alpha_n\right)rdr =$$
$$= 2\int_0^{\infty}\exp\left(-r^2/\alpha_n\right)\left(r/\sqrt{\alpha_n}\right)d\left(r/\sqrt{\alpha_n}\right) \tag{1.90}$$
$$= = \int_0^{\infty}\exp\left(-x^2\right)xdx = 1,$$

having used $x = r/\sqrt{\alpha_n}$.

Appendix C of Bird (1994) proves that the position of point R on a circle centred in point Q can be sampled by choosing at random both the radius r of the circle and the azimuthal angle θ with:

$$\theta = 2\pi R_1, \tag{1.91a}$$

$$r = \left[-\alpha_n \ln(R_2)\right], \tag{1.91b}$$

where R_1 and R_2 are uniform deviates in the open interval (0, 1).

The length of the segment OR represents the normalised normal velocity component after reflection $\left(u_r/c'_m\right)$, while OM and ON represent the normalized tangential velocity components after reflection. Applying the Carnot theorem one has therefore

$$u_r/c'_m = \left[r^2 + (1-\alpha_n)\left(u_i/c'_m\right)^2 + 2r\sqrt{(1-\alpha_n)}\left(u_i/c'_m\right)\cos\theta\right]^{1/2}, \tag{1.92}$$

and

$$v_r/c'_m = \left(u_i/c'_m\right)\sqrt{(1-\alpha_n)} + r\cos\theta, \tag{1.93}$$

$$w_r/c'_m = r\sin\theta. \tag{1.94}$$

The last two equations can be exchanged for the calculation of the tangential velocity components.

The calculation of the reflected rotational energy $(E_{rot})_r$ from the value $(E_{rot})_i$ before impact with the surface is computed with the same model explained above, applied to the rotational energy as if it were a normal velocity component, and normalizing to the energy kT_w. In this case it is suggested to use the accommodation coefficient

$$\alpha_{rot} = \alpha_n(2 - \alpha_n). \tag{1.95}$$

One has therefore

$$(E_{rot})_r/(kT_w) = \left[r^2 + (1-\alpha_n)\left((E_{rot})_i/(kT_w)\right)^2 + 2r\sqrt{(1-\alpha_n)}\left((E_{rot})_i/(kT_w)\right)\cos\theta\right]^{1/2}. \tag{1.96}$$

1.2.3 Gas Mixtures

In the case of mixture of gases, one may define an equivalent mono-component gas with physical properties averaged over the constituents, as done earlier for the air assumed to consist only of molecules of oxygen and nitrogen. When the mixture has more than two chemical components, is preferable to define the properties of the gas mixture as explained in the following.

Let us consider a gas mixture formed by "s" chemical components and use to "p" and "q" as indices referring to any one component (p, q = 1, 2, ... s). If n_p is the number density of the chemical components, the number density n of the mixture is given by

$$n = \sum_{p=1}^{s} n_p. \tag{1.97}$$

Let m_p be the mass of the molecules of the components. The macroscopic density of the mixture is

$$\rho = \sum_{p=1}^{s} m_p n_p = \sum_{p=1}^{s} \rho_p = n m_{eq}, , \tag{1.98}$$

which implicitly defines the equivalent molecular mass

$$m_{eq} = \frac{\rho}{n} = \frac{1}{n}\sum_{p=1}^{s} m_p n_p. \tag{1.99}$$

The collision cross section between two molecules of diameters d_p and d_q (in the following referred to as "species p and q") is defined by

$$\sigma_{T,pq} = \pi d_{pq}^2 = \pi \left(\frac{d_p + d_q}{2} \right)^2. \tag{1.100}$$

Let $c_{r,\,pq}$ be the relative velocity between two molecules of species p and q. The collision frequency between molecules of these species is

$$\nu_{pq} = n_q \langle \sigma_{T,pq} c_{r,pq} \rangle, \tag{1.101}$$

where the symbol $\langle\rangle$ denotes mean value.

The collision frequency for the species p is obtained summing up Eq. (1.101) for every q species:

$$\nu_p = \sum_{q=1}^{s} \nu_{pq} = \sum_{q=1}^{s} n_q \langle \sigma_{T,pq} c_{r,pq} \rangle. \tag{1.102}$$

The mean collision frequency of the mixture is defined by

$$\nu = \frac{1}{n} \sum_{p=1}^{s} \nu_p n_p. \tag{1.103}$$

The total number of collisions in the gas mixture per unit time and volume is

$$N_c = \frac{1}{2} n\nu = \frac{1}{2} \sum_{p=1}^{s} \nu_p n_p. \tag{1.104}$$

Let the mass average velocity of the mixture be defined by

$$c_0 = \frac{1}{\rho} \sum_{p=1}^{s} \rho_p \bar{c}_p. \tag{1.105}$$

The thermal velocity of a molecule of species p is then

$$c'_p = c_p - c_0. \tag{1.106}$$

The mean free path of species p is defined by

$$\lambda_p = \frac{c'_p}{\nu_p} = \frac{c'_p}{\sum_{p=1}^{s} n_q \langle \sigma_{T,pq} c_{r,pq} \rangle}. \tag{1.107}$$

The pressure of the gas mixture is defined by

$$p = \frac{1}{3}\sum_{p=1}^{s}\rho_p\left\langle c_p'^2\right\rangle = \frac{1}{3}\sum_{p=1}^{s}m_p n_p\left\langle c_p'^2\right\rangle = \frac{n}{3}\sum_{p=1}^{s}\frac{m_p n_p}{n}\left\langle c_p'^2\right\rangle = \frac{n}{3}m_{eq}\left\langle c'^2\right\rangle, \tag{1.108}$$

having used definition (1.99) for the equivalent mass of the gas mixture. Equation (1.108) implicitly defines the mean square velocity of the mixture by

$$m_{eq}\left\langle c'^2\right\rangle = \sum_{p=1}^{s}\frac{m_p n_p}{n}\left\langle c_p'^2\right\rangle.. \tag{1.109}$$

The translational or kinetic temperature of species p is defined by

$$\frac{3}{2}kT_{tr,p} = \frac{1}{2}m_p\left\langle c_p'^2\right\rangle. \tag{1.110}$$

Averaging Eq. (1.110) over the molecular densities one obtains

$$\frac{3}{2}k\sum_{p=1}^{s}\frac{n_p}{n}T_{tr,p} = \frac{1}{2}\sum_{p=1}^{s}\frac{n_p}{n}m_p\left\langle c_p'^2\right\rangle. \tag{1.111}$$

Equation (1.111) can be written as

$$\frac{3}{2}kT_{tr} = \frac{1}{2}m_{eq}\left\langle c'^2\right\rangle, \tag{1.112}$$

with the definition of overall translational temperature

$$T_{tr} = \sum_{p=1}^{s}\frac{n_p}{n}T_{tr,p}. \tag{1.113}$$

1.3 Binary Elastic Collisions

We consider at first the case of binary elastic collisions, in which there is no exchange between kinetic energy and inner (rotational or vibrational) energy of the molecules.

1.3.1 General Case of Unequal Molecular Masses

Let us denote with $\vec{c}_1$, $\vec{c}_2$ the pre-collision velocities of two molecules undergoing a collision. We refer to (u, v, w) as the components of the velocity vectors, with the indices or superscripts as appropriate. The post-collision velocities are denoted by $\vec{c}_1^{\,*}$, $\vec{c}_2^{\,*}$. The pre- and post-collision relative velocities are given by

$$\vec{c}_r = \vec{c}_1 - \vec{c}_2, \tag{1.114}$$

$$\vec{c}_r^{\,*} = \vec{c}_1^{\,*} - \vec{c}_2^{\,*}. \tag{1.115}$$

Conservation of momenta and energies of the molecules are expressed by the equations

$$\vec{M} = m_1 \vec{c}_1 + m_2 \vec{c}_2 = m_1 \vec{c}_1^{\,*} + m_2 \vec{c}_2^{\,*} = (m_1 + m_2) \vec{c}_m, \tag{1.116}$$

$$E_k = m_1 \vec{c}_1^{\,2} + m_2 \vec{c}_2^{\,2} = m_1 \vec{c}_1^{\,*2} + m_2 \vec{c}_1^{\,*2}, \tag{1.117}$$

where the centre of mass velocity $\vec{c}_m$ is defined by

$$\vec{c}_m = \frac{m_1 \vec{c}_1 + m_2 \vec{c}_2}{m_1 + m_2}. \tag{1.118}$$

Equations (1.115) and (1.116) are four scalar equations in which the six components of the post-collision velocity vectors $\vec{c}_1^{\,*}$ and $\vec{c}_2^{\,*}$ are the unknowns. In the deterministic way of solving the problem of determining these unknowns, it is necessary to supply two more parameters, namely the orientation of one pre-collision velocity vector ($\vec{c}_1$ for instance) by means of azimuthal and latitude angles in a given reference system. We proceed however in a different way. Considering that in molecular dynamic simulations one sample molecule is representative of a large number of real molecules and therefore one collision is representative of a large number of collisions, the problem of determining the post-collision velocities is approached in a statistical way, as it will be explained below.

Eliminating $\vec{c}_2$ between Eqs. (1.114) and (1.116), and similarly eliminating $\vec{c}_1$ between the same equations, one obtains

$$\vec{c}_1 = \vec{c}_m + \frac{m_2}{m_1 + m_2} \vec{c}_r = \vec{c}_m + \frac{m_r}{m_1} \vec{c}_r = \vec{c}_m + \vec{\tilde{c}}_1, \tag{1.119}$$

$$\vec{c}_2 = \vec{c}_m - \frac{m_1}{m_1 + m_2}\vec{c}_r = \vec{c}_m - \frac{m_r}{m_2}\vec{c}_r = \vec{c}_m - \vec{\tilde{c}}_2, \tag{1.120}$$

where we used the definition of the reduced mass

$$m_r = \frac{m_1 m_2}{m_1 + m_2}, \tag{1.121}$$

and defined

$$\vec{\tilde{c}}_1 = \frac{m_r}{m_1}\vec{c}_r, \tag{1.122}$$

$$\vec{\tilde{c}}_2 = \frac{m_r}{m_2}\vec{c}_r. \tag{1.123}$$

Similar equations hold for the post-collision velocities, namely

$$\vec{c}^{\,*}_1 = \vec{c}_m + \frac{m_2}{m_1 + m_2}\vec{c}^{\,*}_r = \vec{c}_m + \frac{m_r}{m_1}\vec{c}^{\,*}_r = \vec{c}_m + \vec{\tilde{c}}^{\,*}_1, \tag{1.124}$$

$$\vec{c}^{\,*}_2 = \vec{c}_m - \frac{m_1}{m_1 + m_2}\vec{c}^{\,*}_r = \vec{c}_m - \frac{m_r}{m_2}\vec{c}^{\,*}_r = \vec{c}_m - \vec{\tilde{c}}^{\,*}_2. \tag{1.125}$$

$$\vec{\tilde{c}}^{\,*}_1 = \frac{m_r}{m_1}\vec{c}^{\,*}_r, \tag{1.126}$$

$$\vec{\tilde{c}}^{\,*}_2 = \frac{m_r}{m_2}\vec{c}^{\,*}_r. \tag{1.127}$$

Equations (1.119) and (1.120) show that in the centre of mass system the velocities of the two molecules are anti-parallel.

Next we use the energy conservation equation, Eq. (1.117), to prove that the modulus of the relative velocity remains unchanged in the collision. With reference to Fig. 1.3, by applying the Carnot theorem, we have

$$c_1^2 = c_m^2 + \left(\frac{m_r}{m_1}c_r\right)^2 - 2c_m\frac{m_r}{m_1}c_r\cos(\pi - \alpha) = c_m^* + \left(\frac{m_r}{m_1}c_r\right)^2 + 2c_m\frac{m_r}{m_1}c_r\cos\alpha \tag{1.128}$$

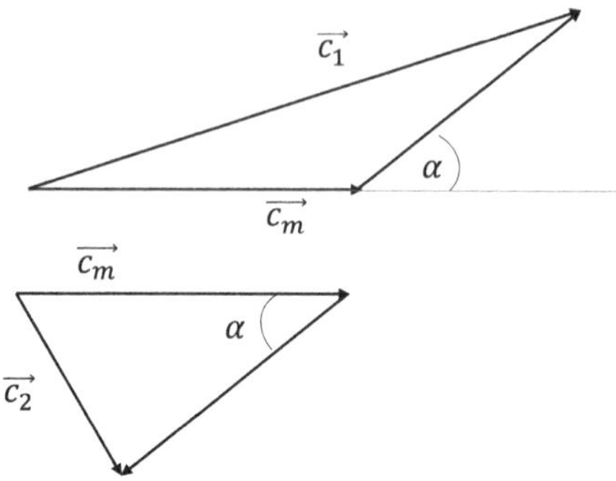

Fig. 1.3 Composition of pre-collision velocities of molecules 1 and 2

$$c_2^2 = c_m^2 + \left(\frac{m_r}{m_2} c_r\right)^2 - 2c_m \frac{m_r}{m_1} c_r \cos\alpha. \tag{1.129}$$

Multiplying Eq. (1.128) by m_1, Eq. (1.129) by m_2 and adding one obtains

$$m_1 c_1^2 + m_2 c_2^2 = (m_1 + m_2) c_m^2 + \left(\frac{m_r^2}{m_1} + \frac{m_r^2}{m_2}\right) c_r^2 = (m_1 + m_2) c_m^2 + m_r c_r^2. \tag{1.130}$$

Similarly, with reference to the post-collision velocities, one obtains

$$m_1 c_1^{*2} + m_2 c_2^{*2} = (m_1 + m_2) c_m^2 + m_r c_r^{*2}. \tag{1.131}$$

The first members of Eqs. (1.130) and (1.131) are equal because of the energy conservation equation. The first terms at the second member of the equations are identical, and hence follows

$$c_r = c_r^*. \tag{1.132}$$

In an isotropic "hard sphere" scattering model the direction of the post-collision relative velocity vector is uniformly distributed on as sphere of unit radius, centred on the collision point. In this model the components of the post-collision relative velocity can be computed by multiplying a unit vector generated with the Marsaglia technique (Marsaglia, 1972), with the modulus of the relative velocity:

$$\begin{aligned} u_r^* &= S c_r \cos\beta, \\ v_r^* &= S c_r \sin\beta, \\ w_r^* &= C_s c_r, \end{aligned} \tag{1.133}$$

using two uniform deviates R_1 and R_2 and computing

$$\begin{aligned}\beta &= 2\pi R_1,\\ C_s &= 1 - 2R_2,\\ S &= \sqrt{1 - C_s^2}.\end{aligned} \tag{1.134}$$

Because the components u_m, v_m and w_m of the centre of mass velocity are known from the pre-collision velocity components, the components of the post-collision velocities of the two molecules are given by

$$\begin{aligned}u_1^* &= u_m + \frac{m_r}{m_1} u_r,\\ u_2^* &= u_m - \frac{m_r}{m_2} u_r,\end{aligned} \tag{1.135a}$$

$$\begin{aligned}v_1^* &= v_m + \frac{m_r}{m_1} v_r,\\ v_2^* &= v_m - \frac{m_r}{m_2} v_r,\end{aligned} \tag{1.135b}$$

$$\begin{aligned}w_1^* &= w_m + \frac{m_r}{m_1} w_r,\\ w_2^* &= w_m - \frac{m_r}{m_2} w_r.\end{aligned} \tag{1.135c}$$

1.3.2 Differential and Total Cross Sections

Writing Eqs. (1.124) and (1.125) in the form

$$m_1\left(\vec{c}_1 - \vec{c}_m\right) = m_r \vec{c}_r = m_1 \vec{\vec{c}}_1, \tag{1.136}$$

$$m_2\left(\vec{c}_2 - \vec{c}_m\right) = -m_r \vec{c}_r = -m_2 \vec{\vec{c}}_2, \tag{1.137}$$

makes clear that in the centre of mass reference system the momenta of the colliding molecules are equal in modulus and of opposite sign

$$m_1\left|\vec{\vec{c}}_1\right| = m_2\left|\vec{\vec{c}}_2\right|. \tag{1.138}$$

Because the forces acting between the molecules are equal in magnitude and of opposite sign, also the post-collision momenta are equal in modulus and of opposite sign

$$m_1\left|\overrightarrow{\widetilde{c}}_1^{\,*}\right| = m_2\left|\overrightarrow{\widetilde{c}}_2^{\,*}\right|. \tag{1.139}$$

Besides, because of the conservation of angular momentum, the distance, say "b", between the trajectories of the molecules does not change in the collision.

Figure 1.4 shows the collision plane and the deflection angle χ between the relative velocity vectors $\vec{c}_r$ and $\vec{c}_r^{\,*}$ before and after collision. To determine the position of the collision plane in the space an angle ε is defined between the collision plane and an arbitrary reference plane. Let us assume the origin of the post-collision relative velocity vector $\vec{c}_r^{\,*}$ in point A and call ℓ the distance AL. An increment $d\chi$ of the deflection angle χ moves the point A to point B by an arc length $AB = \ell \; d\chi$. An

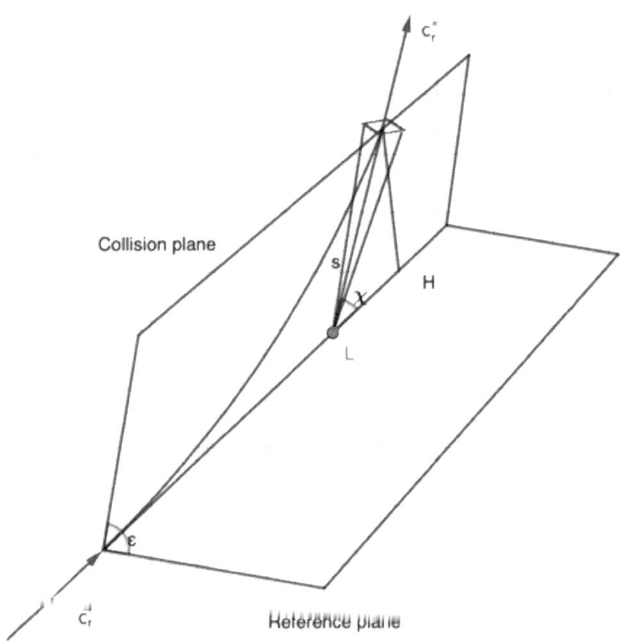

Fig. 1.4 Definition of collisional parameters ε and χ. Path s shows the deflected trajectory

increment $d\varepsilon$ of the angle ε brings point A to a point D, with an arc length AD = AH $d\varepsilon = \ell \sin \chi d\varepsilon$. We define a solid angle $d\Omega$ by

$$d\Omega = \frac{\mathrm{AB}\cdot\mathrm{AD}}{\ell^2} = \sin\chi d\chi d\varepsilon. \tag{1.140}$$

A differential cross section can be defined in the following way. Calling d the molecular diameter [$d = d_{1,\,2} = (d_1 + d_2)/2$ in the case of different molecules] and b the distance of the closest approach of the trajectories in the centre of mass reference system, one has $b = d \sin \psi$,

$$bdbd\varepsilon = d^2 \sin\psi \cos\psi d\psi d\varepsilon, \tag{1.141}$$

with $\sin\psi = b/d$, $db = d \cos \psi d\psi$ (see Bird, 1994, p. 32 for details).

The second member of Eq. (1.141) appears in the collision integral of the Boltzmann equation and has the physical meaning of a differential cross section $\sigma d\Omega$ (Ref. Vincenti & Kruger, 1965). Using Eq. (1.140) we can therefore write

$$\sigma d\Omega = \sigma \sin\chi d\chi d\varepsilon = bdbd\varepsilon, \tag{1.142}$$

hence, taking the absolute value of the ratio of the differentials because the cross section must be positive,

$$\sigma = \frac{b}{\sin\chi}\left|\frac{db}{d\chi}\right|. \tag{1.143}$$

Because the range of the deflection angle χ is $(0 - \pi)$ and that of the angle ε is $(0 - 2\pi)$, the total cross section for the collision is given by

$$\sigma_T = \int\limits_{\Omega} d\Omega = \int_0^{2\pi}\int_0^{\pi} \sigma \sin\chi d\chi d\varepsilon = 2\pi \int_0^{\pi} \sigma \sin\chi d\chi. \tag{1.144}$$

As shown in Fig. 1.5, in the case of unequal molecules, the geometrical diameter of the sphere of influence is $d = d_{1,\,2} = (d_1 + d_2)/2$. From $\pi = 2\psi + \chi$ one has $\psi = \pi/2 - \chi/2$, hence

$$b = d_{1,2} \sin\psi = d_{1,2} \sin(\pi/2 - \chi/2) = d_{1,2} \cos(\chi/2), \tag{1.145}$$

$$\left|\frac{db}{d\chi}\right| = \frac{d_{1,2}}{2} \sin(\chi/2). \tag{1.146}$$

Inverting Eq. (1.145) yields

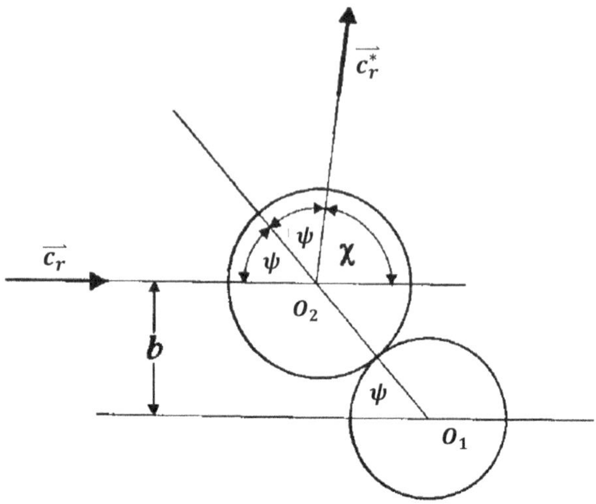

Fig. 1.5 Binary collision of unequal molecules

$$\chi = 2\arcsin(b/d_{1,2}). \tag{1.147}$$

The microscopic cross section becomes

$$\sigma = \frac{b}{\sin\chi}\left|\frac{db}{d\chi}\right| = \frac{d_{1,2}\cos(\chi/2)}{\sin\chi} d_{1,2}\sin(\chi/2) = \frac{d_{1,2}^2}{4}, \tag{1.148}$$

having used the identity sinχ = 2 sin (χ/2) cos (χ/2). The macroscopic cross section becomes the geometrical cross section

$$\sigma_T = \int_0^{4\pi} \sigma d\Omega = \pi d_{1,2}^2. \tag{1.149}$$

In the "Hard Sphere" (HS) model referred to above the total cross section is constant, while experiments show that for real gases the cross section depends on the relative translational energy $E_{r,tr} = mc_r^2/2$, hence on the relative velocity of the molecules, displaying a "1/c_r" behaviour. To cope with this deficiency of the hard sphere model, a more realistic "Variable Hard Sphere" (VHS) model is often used in computational analysis. The VHS model is based on the assumption of an effective molecule diameter, related to a constant reference value d_{ref} and to a constant reference relative velocity $c_{r,\,ref}$ by a law of the type

$$d_{eff} = d_{ref} \frac{c_{r,ref}}{c_r} \tag{1.150}$$

Table 1.5 Parameter α for the "Variable Soft Sphere" model (VSS). [Crf. Eq. (1.151)]

	α
Ar	1.40
H_2	1.35
N_2	1.36
O_2	1.40
CO	1.49
CO_2	1.61

A further improvement is obtained in the "Variable Soft Sphere" model (VSS) in which one assumes the same "$1/c_r$" dependence of the cross section on the relative velocity of the molecules, but the deflection angle χ is given by

$$\chi = [2\arcsin(b/d_{1,2})]^{1/\alpha}, \tag{1.151}$$

where the parameter α is determined experimentally. Some values of α are given in Table 1.5 (from Bird, 1994):

1.3.3 Equivalent Cross Section Based on the Lennard-Jones Potential

We assume that the air molecules, having characteristics weighted among those of the constituents nitrogen and oxygen, can be represented as elastic spheres. Their interaction can be described by a potential energy function, the gradient of which determines the inter-molecular forces (see Fig. 1.6). The potential energy function is given by the Lennard-Jones model and is represented, for air, by Fig. 1.7, as a function of the distance r between the geometrical centres of the molecules. The plot shows a minimum at $r = r_m$. For distances smaller than this value there is a strong repulsive force among molecules, for larger distances a weak attractive force. The attractive forces assure that the system does not spread indefinitely; the repulsive forces imply that the system does not collapse to a single macro-molecule.

The weak attractive force, however, inflects the trajectories of approaching molecules and contributes to increase the effective cross section σ_{eff}, with respect to the geometrical cross section given by $\sigma_g = \pi d^2$. The effective cross section is computed in the following way.

The inter-molecular potential, according to the Lennard-Jones model, is given by

$$\Phi(r) = 4\varepsilon\left[\left(\frac{\sigma}{r}\right)^{12} - \left(\frac{\sigma}{r}\right)^{6}\right], \tag{1.152}$$

where:

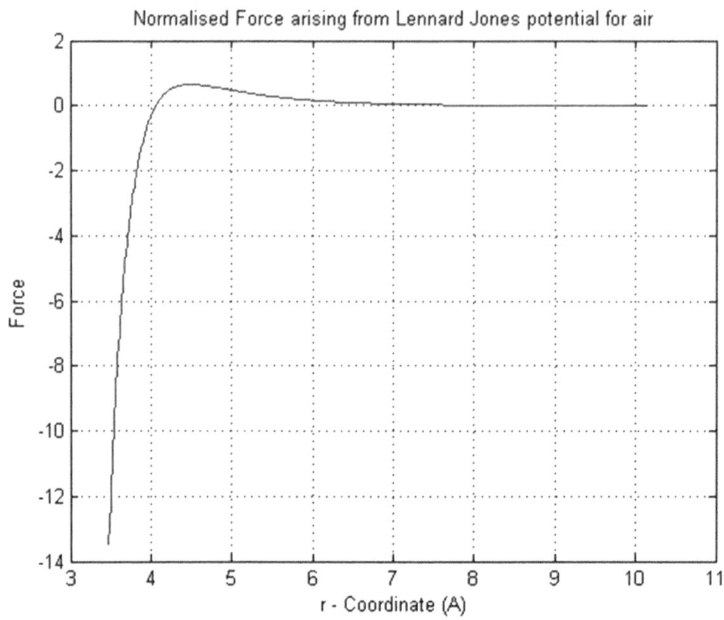

Fig. 1.6 Attractive (positive) and repulsive (negative) forces among molecules

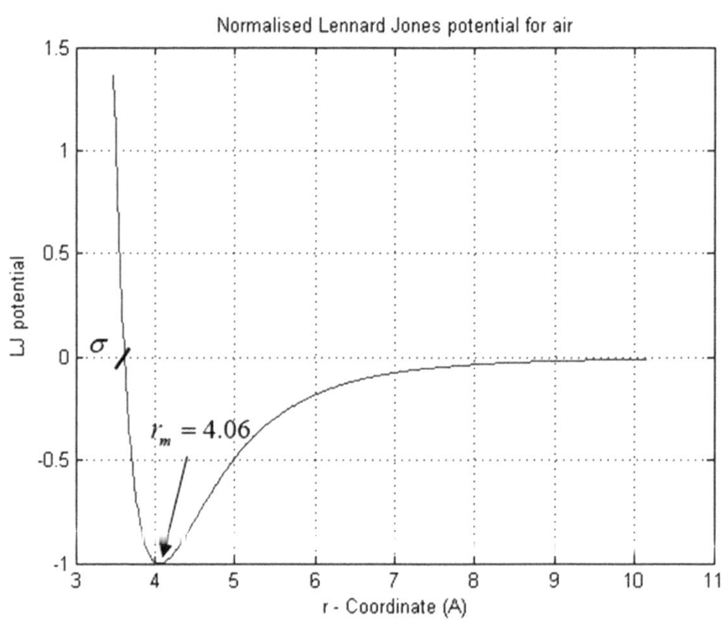

Fig. 1.7 Normalized Lennard-Jones potential

r = distance among molecules;

$\sigma = 2.44\sqrt{T_c/p_c}$ = improperly called "characteristic diameter" of the molecules, expressed in Angstrom. It is computed from the absolute temperature and pressure (in atmosphere units) of the fluid at the critical point. In fact σ represents the abscissa for which $\Phi(r) = 0$;

$\varepsilon = K \cdot 0.77T_c$ = characteristic energy of interaction between molecules [J/molecule]. ε is the minimum of $\Phi(r)$ for $r \approx r_m$;

K = Boltzmann constant [J/°K] = 1.38×10^{-23} [J/K].

The effective diameter for molecular collisions can be computed by (Vincenti & Kruger, 1965),

$$d_{eff}^2 = d^2 \left[1 + \frac{4}{m} \overline{\frac{1}{g^2}} \int_{r_m}^{\infty} (-F(r)dr) \right], \tag{1.153}$$

where:

m = mass of the molecules [kg];

$\overline{\frac{1}{g^2}}$ = mean value of the reciprocal of the square of relative velocities (g) of the molecules [s^2/m^2];

$F(r) = -[d\Phi(r)/dr]$ = force acting among the molecules [N].

Equation (1.153) shows that the effective molecular diameters differ from the geometrical one more in case of slow molecules than in case of fast molecules, for which the value $\overline{\frac{1}{g^2}}$ is smaller. This is, in plane words, because slow molecules have more time to attract each other.

Differentiating Eq. (1.152) with respect to the variable r one obtains

$$F(r) = -24\varepsilon \left(-\frac{2\sigma^{12}}{r^{13}} + \frac{\sigma^6}{r^7} \right). \tag{1.154}$$

The lower limit of the integral in Eq. (1.153) corresponds to the value r_m obtained by looking for the minimum of the function (1.152). The result is

$$r_m = \sigma 2^{1/6} \approx \sigma \cdot 1.12246. \tag{1.155}$$

The integral in Eq. (1.153) can be computed numerically by using Legendre polynomial interpolation of order 40, dividing the interval ($r_m - \infty$) in five or more partial intervals. The limit at infinity is actually replaced by a cut-off limit usually taken at $r_c = 2.5r_m$.

After computing d_{eff}^2 one obtains the effective cross section is given by

$$\sigma_{eff} = \pi d_{eff}^2. \tag{1.156}$$

For the simulated air molecule Bird (1994) gives the following values:
$d = 4.150$ [°A], hence $\sigma_g = \pi d^2 = 54.11 \times 10^{-20}$ [m^2],
$\sigma = 3.617$ [°A],
$\varepsilon = 1.3386 \times 10^{-21}$ [°K],
$r_m = 4.06$ [°A],
ε/K = 97.0 [°K],
T_c= 132.0 [K],
p_c= 36.4 [atm] = 36.88 [N/m^2].
With these values one obtains $d_{eff} = 11.64$ [°A], hence $\sigma_{eff} = 425.65 \times 10^{-20}$ [m^2]. The ratio between effective diameter and geometrical diameter results $d_{eff}/d = 11.64/4.15 = 2.80$. The ratio between effective cross section and geometrical cross section results to be $\sigma_{eff}/\sigma_g = 7.87$.

Figure 1.8 shows the difference between the geometrical and the effective molecular diameters and the trajectories of the centres of two colliding molecules. The continuous lines show the real trajectories of attracting molecules which collide in point Q. The dotted lines refer to equivalent non-attracting molecules for which the sphere of influence has diameter d_{eff} and the collision point is F. This figure represents the collision of two molecules A and B, from the viewpoint of an observer at rest with molecule B, while molecule A is moving.

Equation (1.153) can be used to obtain an estimation of the dependence of dynamic molecular viscosity on temperature, better than that obtained in Sect. 1.2.1.2. Because it can be proved that $\left\langle \frac{1}{g^2} \right\rangle \equiv \frac{1}{T}$, Eq. (1.153) can be written as

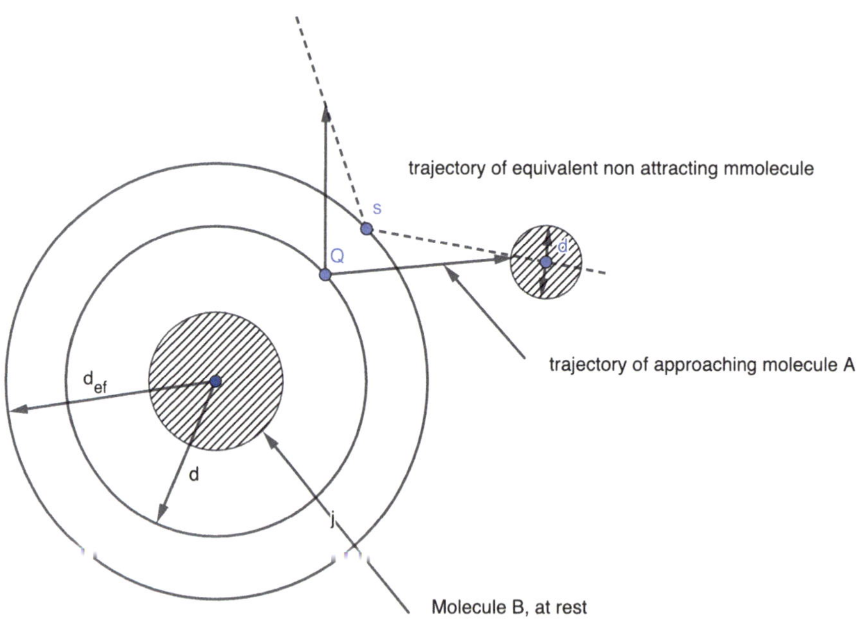

Fig. 1.8 Sphere of influence of attracting molecules with comparison between real (continuous line) and simulated (dotted lines) trajectories

$$d_{eff}^2 = d^2\left[1 + \frac{\chi}{T}\right], \tag{1.157}$$

with the definition

$$\chi = \frac{4}{m}\int_{r_m}^{\infty}(-F(r)dr). \tag{1.158}$$

Recalling Eq. (1.51) and using the effective molecular diameter, instead of the geometrical one, one has

$$\mu \equiv \frac{\sqrt{T}}{d_{eff}^2} \equiv \frac{\sqrt{T}}{d^2\left(1 + \frac{\chi}{T}\right)}. \tag{1.159}$$

Writing Eq. (1.159) with a reference temperature T_0 and dividing one obtains

$$\frac{\mu}{\mu_0} = \sqrt{\frac{T}{T_0}}\frac{1 + \frac{\chi}{T_0}}{1 + \frac{\chi}{T}}. \tag{1.160}$$

Equation (1.160) is referred to a Sutherland's formula and predicts a more rapid increase of the dynamic viscosity than the simple $\sqrt{T}$ dependence. It thus simulates better experimental results.

1.3.4 Equation of Motion and Thermodynamic Quantities Based on the Lennard-Jones Potential

Let us consider now a couple of molecules with indices i and j in a gas domain containing N molecules. Let $\vec{r}_{ij}$ be the distance between the molecules. Using Eq. (1.154), the force among the molecules can be written as

$$\vec{F}_{ij} = -\frac{24\varepsilon}{\sigma^2}\left[2\left(\frac{\sigma}{r_{ij}}\right)^{14} - \left(\frac{\sigma}{r_{ij}}\right)^{8}\right]\vec{r}_{ij}. \quad (i \neq j) \tag{1.161}$$

The equation of motion for molecule i is obtained summing up the forces contributed by all molecules $j = 1, 2, \ldots N$ in the domain:

$$m\frac{d^2\vec{r}_i}{dt^2} = \sum_{j=1}^{N}\vec{F}_{ij} = -\sum_{j=1}^{N}\frac{24\varepsilon}{\sigma^2}\left[2\left(\frac{\sigma}{r_{ij}}\right)^{14} - \left(\frac{\sigma}{r_{ij}}\right)^{8}\right]\vec{r}_{ij}. \quad (i \neq j) \tag{1.162}$$

From the computational viewpoint one may observe:

(i) In Eq. (1.162), because $\vec{F}_{ij} = -\vec{F}_{ji}$, only one half of all possible couples of molecules needs to be considered, thus saving computational time. Nevertheless, the total CPU time for a calculation is proportional to N^2.
(ii) A cut-off value $r_c (r_c \approx 3r_m)$ can be set for the distance between molecules, such that the force among molecules at a distance larger than the cut-off value can be neglected. The sum in Eq. (1.162) can be made only for molecules with $r_{ij} \leq r_c$.
(iii) All considerations made above can be applied also for the calculation of the repulsive force among molecules. To do this the cut-off value can be set $r_c = r_{ij}$.

Referring to Eq. (1.18), the average translational kinetic energy per unit of mass of the system can be written as:

$$\langle E_{tr} \rangle = \frac{1}{N} \sum_{i=1}^{N} e_{tr,i} = \frac{1}{2N} \sum_{i=1}^{N} c_i'^2. \tag{1.163}$$

From Eq. (1.22) the translational temperature of molecule i is

$$\frac{3}{2} k T_{tr,i} = \frac{1}{2} m c'^2_i. \tag{1.164}$$

The average translational temperature of the system becomes

$$\langle T_{tr} \rangle = \frac{1}{N} \sum_{i=1}^{N} T_{tr,i} = \frac{1}{3Nk} \sum_{i=1}^{N} \left(m c'^2_i \right). \tag{1.165}$$

The internal energy of the system, U, takes into account the contribution due to the kinetic energy of the molecules and that due to the intermolecular forces:

$$U = \frac{3}{2} Nk \langle T_{tr} \rangle + \sum_{i=1}^{N} \sum_{j=1}^{N} \Phi(r_{ij}) = \frac{3}{2} Nk \langle T_{tr} \rangle + \Phi, (j > i) \tag{1.166}$$

with the definition

$$\Phi = \sum_{i=1}^{N} \sum_{j}^{N} \Phi(r_{ij}) = \sum_{i=1}^{N} \sum_{j=1}^{N} \left\{ 4\varepsilon \left[\left(\frac{\sigma}{r_{ij}} \right)^{12} - \left(\frac{\sigma}{r_{ij}} \right)^{6} \right] \right\} . (j > i) \tag{1.167}$$

The mean value of the potential Φ is

$$\langle \Phi \rangle = \frac{1}{N} \Phi \frac{1}{N} \sum_{i=1}^{N} \sum_{j=1}^{N} \left\{ 4\varepsilon \left[\left(\frac{\sigma}{r_{ij}} \right)^{12} - \left(\frac{\sigma}{r_{ij}} \right)^{6} \right] \right\} . (j > i) \tag{1.168}$$

The mean value of the internal energy is

$$\langle U\rangle = \frac{U}{N} = \frac{3}{2}k\langle T_{tr}\rangle + \langle\Phi\rangle. \tag{1.169}$$

In case one prefers making computations using dimensionless units, it is customary to normalize the above equations using m, ε, σ as reference units of mass, energy and length, respectively. Thus, denoting with a tilde dimensionless quantities, one uses:

$$\widetilde{m} = m/m = 1, \tag{1.170}$$

$$\widetilde{\varepsilon} = \varepsilon/\varepsilon = 1, \tag{1.171}$$

$$\widetilde{\sigma} = \sigma/\sigma = 1, \tag{1.172}$$

$$\widetilde{t} = t/t_0, \tag{1.173}$$

$$\widetilde{c} = c/c_0, \tag{1.174}$$

$$\widetilde{T} = T/T_0, \tag{1.175}$$

$$\widetilde{\Phi} = \Phi/\varepsilon, \tag{1.176}$$

with the definitions

$$t_0 = \sigma\sqrt{\frac{m}{\varepsilon}}, \tag{1.177}$$

$$c_0 = \frac{\sigma}{t_0}, \tag{1.178}$$

$$T_0 = \frac{\varepsilon}{k}. \tag{1.179}$$

Using dimensionless quantities one obtains the following equations:
The dimensionless equation of motion for molecule is

$$\frac{d^2\widetilde{r}_i}{dt^2} = \sum_{j=1}^{N}\left[48\left(\widetilde{r}_{ij}^{-14} - \frac{1}{2}\widetilde{r}_{ij}^{-8}\right)\vec{r}_{ij}\right], \quad (j \neq i) \tag{1.180}$$

with $\widetilde{r} = r/\sigma$.

The average dimensionless translational kinetic energy per unit of mass of the system is

$$\langle \widetilde{E}_{tr} \rangle = \frac{1}{2N}\sum_{i=1}^{N}\widetilde{c}_i'^2. \tag{1.181}$$

The average dimensionless translational temperature of the system:

$$\langle \widetilde{T}_{tr} \rangle = \frac{1}{3N}\sum_{i=1}^{N}\left(m\widetilde{c}_i'^2\right). \tag{1.182}$$

The average dimensionless potential is

$$\left\langle \widetilde{\Phi} \right\rangle = \frac{1}{N}\widetilde{\Phi} = \frac{4}{N}\left(\widetilde{r}_{ij}^{-12} - \widetilde{r}_{ij}^{-6}\right). \tag{1.183}$$

The average dimensionless internal energy of the system is

$$\langle \widetilde{U} \rangle = \langle \widetilde{E}_{tr} \rangle + \left\langle \widetilde{\Phi} \right\rangle = \frac{1}{2N}\sum_{i=1}^{N}\widetilde{c}_i^2 + \frac{4}{N}\left(\widetilde{r}_{ij}^{-12} - \widetilde{r}_{ij}^{-6}\right). \tag{1.184}$$

1.3.5 The "Soft Spheres" Model

When applying the Lennard-Jones potential with the cut-off distance $r_c = 2.5r_m$ there might be oscillations in the positions of molecules distant about r_c, because of the discontinuity of the potential function at r = r_c. The discontinuity is eliminated in a variant of the model referred to as "Soft Spheres" model. In this model only the strong repulsive forces for $r \leq r_m$ are considered while the weak attractive forces are neglected. The equations describing the potential in the "Soft Spheres" model insure that the minimum of the potential is 0 at $r = r_m$. These equations are:

$$\Phi(r) = 4\varepsilon\left[\left(\frac{\sigma}{r}\right)^{12} - \left(\frac{\sigma}{r}\right)^{6}\right] + \varepsilon, \quad r \leq r_m \tag{1.185}$$

$$\Phi(r) = 0, r > r_m$$

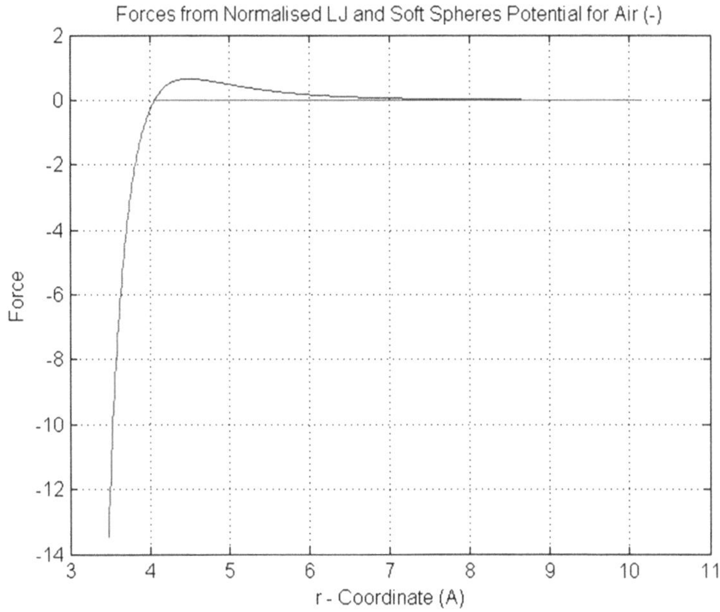

Fig. 1.9 Sketch showing attractive (positive) and repulsive (negative) forces among molecules in the Lennard-Jones model and in the Soft Spheres model

The dimensionless form of this equation becomes:

$$\Phi(\widetilde{r}) = 4\left[\widetilde{r}^{-12} - \widetilde{r}^{-6}\right] + 1, \widetilde{r} \leq r_m/\sigma \tag{1.186}$$

$$\Phi(r) = 0. \ \widetilde{r} > r_m/\sigma.$$

Figure 1.9 shows forces arising from the Soft Spheres model in comparison with the forces in the Lennard-Jones model. For $r \leq r_m$ the models coincide, while for $r > r_m$ the forces in the Soft Spheres model are zero.

Figure 1.10 shows the potential in the Soft Spheres model (upper plot) in comparison with the Lennard-Jones potential.

1.3.6 *Molecules of Equal Mass*

In the case of molecules of equal mass, the conservation of momenta and energies of the molecules upon collision are given by the equations

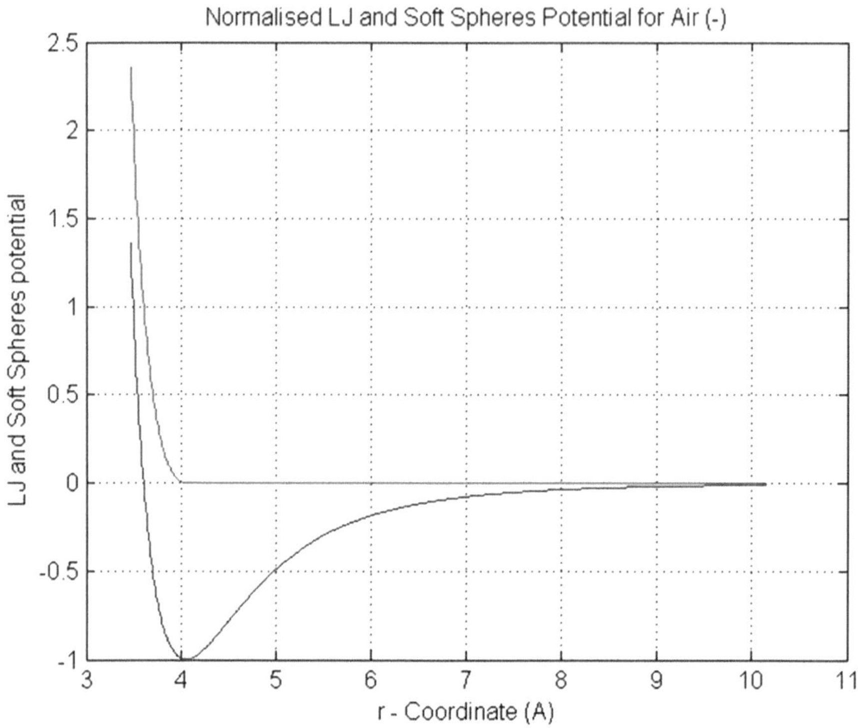

Fig. 1.10 Potential in the Soft Spheres model (upper plot) in comparison with the Lennard-Jones potential

$$\vec{M} = \vec{c}_1 + \vec{c}_2 = \vec{c}_1^{\,*} + \vec{c}_2^{\,*}, \tag{1.187}$$

$$E_k = \vec{c}_1^{\,2} + \vec{c}_2^{\,2} = \vec{c}_1^{\,*2} + \vec{c}_1^{\,*2}. \tag{1.188}$$

Let us define

$$\vec{n}\,k = \vec{c}_1 - \vec{c}_1^{\,*}, \tag{1.189}$$

where $\vec{n}$ is a unit vector and the scalar k has to be determined on the basis of the conservation of energy, Eq. (1.188).

From Eqs. (1.189) and (1.187) one has

$$\vec{c}_1^{\,*} = \vec{c}_1 - \vec{n}\,k, \tag{1.190}$$

$$\vec{c}_2^{\,*} = \vec{c}_2 + \vec{n}\,k. \tag{1.191}$$

Scalar product of $\vec{c}_1^{\,*}$ and $\vec{c}_2^{\,*}$ with themselves yields

$$c_1^{*2} = c_1^2 - 2\,\vec{c}_1 \cdot \vec{n}\,k + k^2, \tag{1.192}$$

$$c_2^{*2} = c_2^2 + 2\,\vec{c}_2 \cdot \vec{n}\,k + k^2. \tag{1.193}$$

Inserting Eqs. (1.192) and (1.193) into Eq. (1.188) one obtains

$$k = \vec{n} \cdot \left(\vec{c}_1 - \vec{c}_2\right) = \vec{n} \cdot \vec{c}_r. \tag{1.194}$$

Equations (1.190) and (1.191) hence yield

$$\vec{c}_1^{\,*} = \vec{c}_1 - \vec{n}\left[\vec{n} \cdot \left(\vec{c}_1 - \vec{c}_2\right)\right] = \vec{c}_1 - \vec{n}\left(\vec{n} \cdot \vec{c}_r\right), \tag{1.195}$$

$$\vec{c}_2^{\,*} = \vec{c}_2 + \vec{n}\left[\vec{n} \cdot \left(\vec{c}_1 - \vec{c}_2\right)\right] = \vec{c}_2 + \vec{n}\left(\vec{n} \cdot \vec{c}_r\right). \tag{1.196}$$

The post-collision velocities $\vec{c}_1^{\,*}$ and $\vec{c}_2^{\,*}$ can be determined from the last two equations by choosing, in the frame of an isotropic scattering model, a unit vector $\vec{n}$ sampling a uniform distribution on a sphere of unit radius. For this purpose the Marsaglia technique explained above can again be applied.

Subtracting Eq. (1.196) from Eq. (1.195) one obtains

$$\vec{c}_r^{\,*} = \vec{c}_r - 2\vec{n}\left(\vec{n} \cdot \vec{c}_r\right). \tag{1.197}$$

Taking the scalar products of the first and second member of the above equation with themselves and using Eq. (1.194), one obtains again the result that the modulus of the relative velocity does not change in the collision ($c_r = c_r^*$).

The unit vector $\vec{n}$ bisects the angle formed by the vectors $-\vec{c}_r$ and $\vec{c}_r^{\,*}$.

Example 1.1

In a two-dimensional case, let us assume the pre-collision velocity vectors

$\vec{c}_1 = -3\,\vec{i} + \vec{j}$, $\vec{c}_2 = 2\,\vec{i} + 5\,\vec{j}$, hence $\vec{c}_r = -5i - 4\,\vec{j}$. The amplitudes of these vectors are: $c_1 = \sqrt{10}$, $c_2 = \sqrt{29}$, $c_r = \sqrt{41}$.

Let us assume to have determined by random choice $\vec{n} = \vec{i}$.

The post-collision vectors are: $\vec{c}_1^{\,*} = 2\,\vec{i} + \vec{j}$, $\vec{c}_2^{\,*} = -3\,\vec{i} + 5\,\vec{j}$ hence $\vec{c}_r^{\,*} = 5\,\vec{i} - 4\,\vec{j}$. The amplitudes of these vectors are: $c_1^* = \sqrt{5}$, $c_2^* = \sqrt{34}$, $c_r^* = \sqrt{41}$.

One verifies easily that the momentum $\vec{M} = -\vec{i} + 6\vec{j}$ and the energy $E_k = 10 + 29 = 5 + 34 = 39$ are conserved.

Denoting $\left\langle \vec{a}, \vec{b} \right\rangle$ the angle between two vectors $\vec{a}$ and $\vec{b}$, one can verify that:

$$\alpha = \left\langle \vec{c}_1^{\,*}, -\vec{c}_1 \right\rangle = 0.785,$$

$$\beta = \left\langle \vec{c}_1^{\,*}, \vec{n} \right\rangle = 0.463,$$

$$\gamma = \left\langle -\vec{c}_1, \vec{n} \right\rangle = 0.322,$$

hence $\alpha = \beta + \gamma$,

$$\psi = \left\langle -\vec{c}_r, \vec{n} \right\rangle = \left\langle \vec{c}_r^{\,*}, \vec{n} \right\rangle = 0.675.$$

Example 1.2

In a two-dimensional case, let us assume that the pre-collision velocity vectors are, as in Example 3.1, $\vec{c}_1 = -3\vec{i} + \vec{j}$, $\vec{c}_2 = 2\vec{i} + 5\vec{j}$, hence $\vec{c}_r = -5i - 4\vec{j}$. The amplitudes of these vectors are: $c_1 = \sqrt{10}$, $c_2 = \sqrt{29}$, $c_r = \sqrt{41}$.

Let us assume to have determined by random choice $\vec{n} = \vec{j}$.

The post-collision vectors are: $\vec{c}_1^{\,*} = -3\vec{i} + 5\vec{j}$, $\vec{c}_2^{\,*} = 2\vec{i} + \vec{j}$ hence $\vec{c}_r^{\,*} = -5\vec{i} + 4\vec{j}$. The amplitudes of these vectors are: $c_1^* = \sqrt{34}$, $c_2^* = \sqrt{5}$, $c_r^* = \sqrt{41}$.

One verifies easily that the momentum $\vec{M} = -\vec{i} + 6\vec{j}$ and the energy $E_k = 10 + 29 = 5 + 34 = 39$ are conserved (and are the same as in Example 1.1).

Denoting $\left\langle \vec{a}, \vec{b} \right\rangle$ the angle between two vectors $\vec{a}$ and $\vec{b}$, one can verify that:

$$\alpha = \left\langle \vec{c}_1^{\,*}, -\vec{c}_1 \right\rangle = 2.432,$$

$$\beta = \left\langle \vec{c}_1^{\,*}, \vec{n} \right\rangle = 0.540,$$

$$\gamma = \left\langle -\vec{c}_1, \vec{n} \right\rangle = 1.892,$$

hence $\alpha = \beta + \gamma$,

$$\psi = \left\langle -\vec{c}_r, \vec{n} \right\rangle = \left\langle \vec{c}_r^{\,*}, \vec{n} \right\rangle = 0.896.$$

1.4 Interaction Between Air Molecules and Particles

1.4.1 *Kinetic Energy Incident upon and Reflected from a Particle*

We compute the kinetic energy transferred from an air molecule to a particle upon impact, the kinetic energy reflected, associated with a molecule emerging from the particle, and their difference which is the net energy transferred from the air molecule to the particle.

The kinetic energy of a sample molecule incident with velocity c_i upon a particle is

$$e_{kin,i} = \frac{1}{2} m c_i^2. \tag{1.198}$$

Let F_N be the number of real molecules represented by a sample air molecule. The real kinetic energy associated with a sample molecule is

$$E_{kin,i} = F_N e_{kin,i} = \frac{1}{2} m F_N c_i^2. \tag{1.199}$$

Similarly, the kinetic energy associated with a molecule reflected from the particle (subscript r) with velocity c_r is

$$E_{kin,r} = F_N e_{kin,r} = \frac{1}{2} m F_N c_r^2. \tag{1.200}$$

The components u_r, w_r of the velocity c_r are computed sampling a Maxwellian velocity distribution corresponding to the temperature of the impact point.

1.4.2 *Momentum Exchange*

Let $\vec{C}_p$ and M_p denote velocity and mass of a particle and indices 1 and 2 refer to the velocities of an air molecule and of the particle before and after a collision. The momentum conservation implies that

$$m \vec{c}_1 + M_p \vec{C}_{p1} = m \vec{c}_2 + M_p \vec{C}_{p2}. \tag{1.201}$$

From Eq. (1.201) the post-collision velocity components of the particle are derived as

$$W_{p2} = W_{p1} + \frac{m}{M_p}(w_1 - w_2), \tag{1.202}$$

$$U_{p2} = U_{p1} + \frac{m}{M_p}(u_1 - u_2). \tag{1.203}$$

1.4.3 *Computation of Inlet Flow*

1.4.3.1 Inlet Flow Rate

We assume that the mean stream velocity vector, $\vec{c}_o$, lies on the (y, z) plane and is inclined an angle ϑ with respect to the z axis. From the definition of the instantaneous molecular velocity

$$\vec{c} = \vec{c}_o + \vec{c}\,', \tag{1.204}$$

we derive

$$\begin{aligned} u &= u', \\ v &= v' + c_o \sin\vartheta, \\ w &= w' + c_o \cos\vartheta. \end{aligned} \tag{1.205}$$

At the inlet plane of the computational domain the inward velocity flux is

$$N_i = n\overline{w} = n \int_{-\infty}^{+\infty} \int_{-\infty}^{+\infty} \int_{0}^{+\infty} w f_o du dv dw. \tag{1.206}$$

In Eq. (1.206) f_o is the equilibrium distribution function given by

$$f_o = \frac{\beta^3}{\pi^{3/2}} \exp\left(-\beta^2 c'^2\right). \tag{1.207}$$

Inserting Eq. (1.207) into (1.206) one obtains

$$N_i = n\overline{w} = \frac{n\beta^3}{\pi^{3/2}} \int_{-\infty}^{+\infty} \int_{-\infty}^{+\infty} \int_{0}^{+\infty} w \exp\left(-\beta^2 c'^2\right) du dv dw. \tag{1.208}$$

Using Eq. (1.205) and changing the integration variables from u, v, w to u', v', w', one obtains

$$N_i = \frac{n\beta^3}{\pi^{3/2}} \int_{-\infty}^{+\infty} \int_{-\infty}^{+\infty} \int_{-c_o \cos\vartheta}^{+\infty} (w' + c_o \cos\vartheta) \exp\left[-\beta^2\left(u'^2 + v'^2 + w'^2\right)\right] du' dv' dw'. \tag{1.209}$$

The variables in Eq. (1.209) can be separated to give

$$N_i = \frac{n\beta^3}{\pi^{3/2}} \int_{-\infty}^{+\infty} \exp\left(-\beta^2 u'^2\right) du' \int_{-\infty}^{+\infty} \exp\left(-\beta^2 v'^2\right) dv' \int_{-c_o \cos\vartheta}^{+\infty} (w' + c_o \cos\vartheta) \exp\left(-\beta^2 w'^2\right) dw'. \tag{1.210}$$

Using the definition of the error function and the integrals

$$\int_{-\infty}^{+\infty} \exp\left(-\beta^2 u'^2\right) du' = \int_{-\infty}^{+\infty} \exp\left(-\beta^2 v'^2\right) dv' = \frac{\sqrt{\pi}}{\beta}, \tag{1.211}$$

$$\int_{0}^{+\infty} w' \exp\left(-\beta^2 w'^2\right) dw' = \frac{1}{2\beta^2}, \tag{1.212}$$

Equation (1.210) becomes

$$N_i = \frac{n}{2\beta\sqrt{\pi}} \left\{ \exp(-c_o^2\beta^2 \cos^2\vartheta) + \sqrt{\pi} c_o \beta \cos\vartheta [1 + erf(c_o \beta \cos\vartheta)] \right\}. \tag{1.213}$$

Let

$$s = c_o \beta = \frac{c_o}{c'_m} = \frac{c_o}{\sqrt{2RT}} \tag{1.214}$$

be the molecular speed ratio, or ratio of the mean stream velocity c_o to the most probable molecular velocity c'_m. Equation (1.213) can therefore be written as

$$N_i = \frac{n}{2\beta\sqrt{\pi}}\left\{\exp\left(-s^2\cos^2\vartheta\right) + \sqrt{\pi}s\cos\vartheta[1 + erf(s\cos\vartheta)]\right\}. \qquad (1.215)$$

Letting

$$\tilde{z} = s\cos\vartheta, \qquad (1.216)$$

one derives

$$N_i = \frac{n}{2\beta\sqrt{\pi}}\left\{\exp\left(-\tilde{z}^2\right) + \sqrt{\pi}\tilde{z}[1 + erf(\tilde{z})]\right\} = \frac{n\sqrt{2RT}}{2\sqrt{\pi}}K(\tilde{z}), \qquad (1.217)$$

with the definition

$$K(\tilde{z}) = \exp\left(-\tilde{z}^2\right) + \sqrt{\pi}\tilde{z}[1 + erf(\tilde{z})]. \qquad (1.218)$$

If the mean stream velocity is zero $[c_o = s = 0; K(\tilde{z}) = 1]$, Eq. (1.217) gives

$$N_i = \frac{n}{2\beta\sqrt{\pi}}. \qquad (1.219)$$

1.4.3.2 Inlet Flow Velocity

We now assume that the mean stream velocity $\vec{c}_o = (0, 0, w_o)$ is parallel to the z axis so that

$$w = w' + w_o. \qquad (1.220)$$

In this case one has

$$\beta w = \beta(w' + w_o) = \beta w' + s_n, \qquad (1.221)$$

with

$$s_n = \beta w_o = \frac{w_o}{c'_m}. \qquad (1.222)$$

From Eq. (1.210) we see that the distribution function of the thermal velocity component w' of the inlet flow is

$$f_{w'} = w\exp\left(-\beta^2 w'^2\right). \qquad (1.223)$$

Hence

$$f_{\beta w'} = f_{(w'/c'_m)} = \beta w \exp\left(-\beta^2 w'^2\right) = (\beta w' + s_n) \exp\left(-\beta^2 w'^2\right), \tag{1.224}$$

having used Eq. (1.221).

Equating to zero the derivative $\partial f_{\beta w'}/\partial w'$ one finds that the distribution has a maximum for

$$\beta w' = \frac{1}{2}\left(-s_n + \sqrt{s_n^2 + 2}\right), \tag{1.225}$$

and the maximum is

$$[f_{\beta w'}]_{\max} = \left[\frac{1}{2}\left(\sqrt{s_n^2 + 2} - s_n\right) + s_n\right] \exp\left[-\frac{1}{4}\left(\sqrt{s_n^2 + 2} - s_n\right)\right]$$

$$= \frac{1}{2}\left(\sqrt{s_n^2 + 2} + s_n\right) \exp\left[-\left(\frac{1}{2} + \frac{s_n^2}{2} - \frac{s_n}{2}\sqrt{s_n^2 + 2}\right)\right]. \tag{1.226}$$

Hence we compute the ratio of the probability P of a molecule entering through the inlet surface with thermal velocity w' to the maximum of its distribution function $P_{\max} = [f_{\beta w'}]_{\max}$ as

$$\frac{P}{P_{\max}} = \frac{f_{\beta w'}}{[f_{\beta w'}]_{\max}} = \frac{2(\beta w' + s_n) \exp\left[-\left(\beta^2 w'^2\right)\right]}{\left(\sqrt{s_n^2 + 2} + s_n\right) \exp\left[-\left(\frac{1}{2} + \frac{s_n^2}{2} - \frac{s_n}{2}\sqrt{s_n^2 + 2}\right)\right]}$$

$$= \frac{2(\beta w' + s_n)}{\sqrt{s_n^2 + 2} + s_n} \exp\left[\frac{1}{2} + \frac{s_n}{2}\left(s_n - \sqrt{s_n^2 + 2}\right) - \beta^2 w'^2\right]. \tag{1.227}$$

Letting

$$WN = \beta w = \beta w' + s_n, \tag{1.228}$$

$$W = \beta w', \tag{1.229}$$

$$FS1 = s_n + \sqrt{s_n^2 + 2}, \tag{1.230}$$

$$FS2 = \frac{1}{2}[1 + s_n(2s_n - FS1)] = \frac{1}{2}\left(1 + s_n^2 - s_n\sqrt{s_n^2 + 2}\right), \tag{1.231}$$

Equation (1.227) becomes

$$\frac{P}{P_{\max}} = \frac{2 \cdot WN}{FS1} \exp\left(FS2 - W^2\right). \tag{1.232}$$

Equation (1.232) is used in the following way: Let R_1 and R_2 be two uniform deviates. Let us assume that the limits of the distributions of $\beta w'$ are $\pm(\beta w')_{\max}$, replacing infinity. A random value

$$\beta w' = -(\beta w')_{\max} + 2(\beta w')_{\max} R_1, \tag{1.233}$$

is generated, with uniform distribution in the interval $\pm(\beta w')_{\max}$. Then the ratio $P/P_{\max}$ is computed from Eq. (1.232). The value $\beta w'$ from Eq. (1.233) is accepted if

$$R_2 \leq \frac{P}{P_{\max}}, \tag{1.234}$$

otherwise $\beta w'$ is rejected. Finally

$$w = \frac{WN}{\beta} = WN \cdot c'_m \tag{1.235}$$

is computed. This procedure generates inlet molecule velocities under the condition that $\beta w = \beta w' + s_n \geq 0$, hence that the velocity component w is directed into the computational domain.

1.5 Wall Reflection

Let us consider a computational domain bounded by two material surfaces (top and bottom) treated as diffusive reflecting surfaces and by two symmetry surfaces (lateral surfaces) treated as specularly reflecting surfaces. Besides, there is the surface of a particle which also behaves as a diffusive reflecting surface. In the following part of this section we explain the kind of interaction of an air molecule with these surfaces.

1.5.1 *Diffuse Reflection from Material Walls*

1.5.1.1 Diffuse Reflection from Material Surfaces

We must first of all determine the coordinates of the collision point of an air molecule with the upper material surface located at $z = z_{\max}$. Let $P_1(x_1, z_1)$ be the point of a molecule before the collision and let $\widetilde{P}_2(\widetilde{x}_2, \widetilde{z}_2)$ be the point which would be attained by the molecule if there were not collision, which means beyond the surface. The equation of the trajectory of the particle determined by these two points is

$$\frac{x - x_1}{\widetilde{x}_2 - x_1} = \frac{z - z_1}{\widetilde{z}_2 - z_1}. \tag{1.236}$$

Letting

$$m_z = \frac{z_{\max} - z_1}{\widetilde{z}_2 - z_1}, \tag{1.237}$$

one derives from Eq. (1.236) that the x and z coordinates of the collision point $P_c(x_c, z_c)$ are given by

$$\begin{aligned} x_c &= x_1 + m_z(\widetilde{x}_2 - x_1), \\ z_c &= z_1 + m_z(\widetilde{z}_2 - z_1) = z_{\max}. \end{aligned} \tag{1.238}$$

We now consider a local coordinate system (x_ℓ, z_ℓ) with origin in P_c and defined as follows: Axis x_ℓ is horizontal and oriented in the opposite direction of the x axis; axis z_ℓ is normal to the wall and oriented inwardly. We also consider a local polar system (ρ, ϑ) with the same origin in P_c and defines as follows: Given a vector $\rho =$ OP, ϑ is the angle between the vector OP and the z_ℓ axis.

The velocity components in the local coordinate system after reflection are computed by

$$\begin{aligned} w_{2\ell} &= c'_m \sqrt{-\ln R_1}, \\ u_{2\ell} &= c'_m r \cos\zeta, \end{aligned} \tag{1.239}$$

where

$$\begin{aligned} \zeta &= R_2 2\pi, \\ r &= \sqrt{-\ln R_3}, \end{aligned} \tag{1.240}$$

and R_1, R_2, R_3 are uniform deviates. c'_m is the most probable thermal velocity corresponding to the Maxwellian distribution of molecular velocities at the temperature of the surface. Notice that $w_{2\ell}$ is always positive while the component $u_{2\ell}$ may be positive or negative. The transformation from the local system to the main (x, z) system yields $u_2 = -u_{2\ell}$; $w_2 = -w_{2\ell}$. The component w_2 is always negative.

The velocity of the particle after reflection is then

$$V_2^{n+1} = \sqrt{u_2^2 + w_2^2}. \tag{1.241}$$

The particle travels, after collision with the surface, a distance

$$R_{P2} = \Delta t_{ac} V_2^{n+1}, \tag{1.242}$$

where Δt_{ac} is the fraction of the time interval after the collision. The orientation of the vector R_{P2} is determined, in the local polar system, randomly by

$$\vartheta = R_1 \pi. \tag{1.243}$$

The coordinates of the particle after diffuse reflection in the local system are

$$\begin{aligned} x_{2\ell} &= R_{P2} \cos \vartheta, \\ z_{2\ell} &= R_{P2} \sin \vartheta. \end{aligned} \tag{1.244}$$

The coordinates of the particle after diffuse reflection in the main coordinate system are

$$\begin{aligned} x_2 &= xx(m, 1) = x_c - x_{2\ell} \\ z_2 &= xx(m, 2) = z_c - z_{2\ell}. \end{aligned} \tag{1.245}$$

Similar considerations hold for the collision of an air molecule with the lower material surface.

1.5.1.2 Diffuse Reflection from the Surface of a Particle

Let $P_1(x_1, z_1)$ be the position of a molecule at time point t_n, and $P_2(x_2, z_2)$ its position at time t_{n+1}. If, during the time interval $\Delta t = t_{n+1} - t_n$, the molecule hits the surface of the particle at a point $P_c(x_c, z_c)$, before being deflected, the coordinates of the point P_c must be computed first and then the coordinates of P_2 can be computed.

Let $\widetilde{P}_2(\widetilde{x}_2, \widetilde{z}_2)$ be the point where the molecule would arrive at time point t_{n+1}, if it were not deflected by the surface of the particle. The equation of the line $L = P_1\widetilde{P}_2$ is

$$\frac{x - x_1}{\widetilde{x}_2 - x_1} = \frac{z - z_1}{\widetilde{z}_2 - z_1}, \tag{1.246}$$

or

$$z = z_1 + (x - x_1)\frac{\widetilde{z}_2 - z_1}{\widetilde{x}_2 - x_1} = z_1 + (x - x_1)tg\alpha, \tag{1.247}$$

with

$$tg\alpha = \frac{\widetilde{z}_2 - z_1}{\widetilde{x}_2 - x_1}. \tag{1.248}$$

Equation (1.247) can be written in the normal form

$$ax + bz + c = 0, \tag{1.249}$$

with

$$\begin{aligned} a &= tg\alpha, \\ b &= -1, \\ c &= z_1 - x_1 tg\alpha. \end{aligned} \tag{1.250}$$

The distance $d(L)$ of the line $L = P_1\widetilde{P}_2$ from the centre (x_p, z_p) of the particle is

$$d(L) = \left|\frac{ax_p + bz_p + c}{\sqrt{a^2 + b^2}}\right| = \left|\frac{tg\alpha(x_p - x_1) + z_1 - z_p}{\sqrt{1 + tg^2\alpha}}\right|. \tag{1.251}$$

The equation of the circle C cross section of the particle of radius R with the (x, z) plane is

$$(x - x_p)^2 + (z - z_p)^2 = R^2. \tag{1.252}$$

Combining Eq. (1.247) with Eq. (1.252) one obtains

$$(x - x_p)^2 + (z_1 + (x - x_1)tg\alpha - z_p)^2 = R^2. \tag{1.253}$$

Equation (1.253) can be written as

$$\overline{\alpha}x^2 + \beta z^2 + \gamma = 0, \tag{1.254}$$

with

$$\overline{\alpha} = 1 + tg^2\alpha, \tag{1.255a}$$

$$\beta = -x_p - x_1 tg^2\alpha + z_1 tg\alpha - z_p tg\alpha, \tag{1.255b}$$

$$\gamma = x_p^2 + z_p^2 + z_1^2 + x_1^2 tg^2\alpha - 2x_1 z_1 tg\alpha - 2z_1 z_p + 2x_1 z_p tg\alpha - R^2. \tag{1.255c}$$

Equation (1.254) has the two solutions

$$X_1 = \frac{-\beta + \sqrt{\beta^2 - \overline{\alpha}\gamma}}{\overline{\alpha}}, \tag{1.256a}$$

$$X_2 = \frac{-\beta - \sqrt{\beta^2 - \overline{\alpha}\gamma}}{\overline{\alpha}}, \tag{1.256b}$$

which correspond to the z-coordinates, from Eq. (1.247),

$$Z_1 = z_1 + (X_1 - x_1)tg\alpha, \tag{1.257a}$$

$$Z_2 = z_1 + (X_2 - x_1)tg\alpha. \tag{1.257b}$$

Thus the two intersections between line L and the circle C are the points A and B with coordinates $A = (X_1, Z_1)$ and $B = (X_2, Z_2)$. To decide which one of these two points is the collision point P_c we compute the distances

$$d_1 = AP_1 + A\widetilde{P}_2 = \sqrt{(x_1 - X_1)^2 + (z_1 - Z_1)^2} + \sqrt{(\widetilde{x}_2 - X_1)^2 + (\widetilde{z}_2 - Z_1)^2}, \tag{1.258a}$$

$$d_2 = BP_1 + B\widetilde{P}_2 = \sqrt{(x_1 - X_2)^2 + (z_1 - Z_2)^2} + \sqrt{(\widetilde{x}_2 - X_2)^2 + (\widetilde{z}_2 - Z_2)^2}. \tag{1.258b}$$

If $d_1 < d_2$, A is the collision point ($P_c = A$), otherwise B is collision point ($P_c = B$).

In the case of small particles, a molecule can cross the particle during the time step, so that both points $P_1(x_1, z_1)$ and $P_2(x_2, z_2)$ can be outside of the particle. This occurrence must be distinguished from the case in which both points are outside of the particle but on the same side so that collision did not occur. To do this we consider the line $L^{'}$ passing through the centre (x_p, z_p) of the particle and normal to the line L. The equation of line $L^{'}$ is

$$z' = z_p + (x' - x_p)m', \tag{1.259}$$

with

$$m' = -1/m = -1/tg\alpha. \tag{1.260}$$

The distances of the points $P_1(x_1, z_1)$ and $P_2(x_2, z_2)$ from the line L' are

$$h_1 = \frac{m'\left(x_1 - x_p\right) + z_p - z_1}{\sqrt{1 + m'^2}}, \tag{1.261a}$$

$$h_2 = \frac{m'\left(x_2 - x_p\right) + z_p - z_2}{\sqrt{1 + m'^2}}. \tag{1.261b}$$

If h_1 and h_2 have the same sign the points $P_1(x_1, z_1)$ and $P_2(x_2, z_2)$, both outside of the particle, are on the same side of the line L' and therefore there was no collision. If h_1 and h_2 have different sign there was collision. In this case the coordinates of the collision points are computed as explained above.

Next one has to compute position and velocity of the molecule after diffuse reflection from the surface of the particle. We consider first a coordinate system $(\widetilde{x}, \widetilde{z})$ with origin in the collision point P_c and defined as follows: the $\widetilde{x}$ axis is tangent to the circle C and directed in clockwise direction; the $\widetilde{z}$ axis is directed along the line from the centre of the circle C to the collision point. A second coordinate system, referred to as local (index ℓ) system (x_ℓ, z_ℓ), is defined as follows: Axis x_ℓ is coincident with the $\widetilde{z}$ axis; the z_ℓ axis is coincident with the $\widetilde{x}$ axis but oriented in opposite direction. Finally, the coordinate system (x, y, z), with $y = y_p = c$, is the main Cartesian coordinate system in which the main code calculations are made.

Let

$$\alpha = ar\cos\left[\left(x_c - x_p\right)/R\right] \tag{1.262}$$

be the angle between the x_ℓ and the x axes.

The components of the molecule velocity and the components of its position vector after diffuse reflection are computed in the following steps:

Step 1: Calculation of velocity components in the local system (x_ℓ, z_ℓ)

The velocity components in the local coordinate system are computed by

$$\begin{aligned} u_{2\ell} &= c'_m \sqrt{-\ln R_1}, \\ w_{2\ell} &= c'_m r \cos\zeta, \end{aligned} \tag{1.263a}$$

$$\begin{aligned} \zeta &= R_2 2\pi, \\ r &= \sqrt{-\ln R_3}, \end{aligned} \tag{1.263b}$$

and R_1, R_2, R_3 are uniform deviates. c'_m is the most probable velocity correspondent to the temperature of the surface of the particle in the point of the impact. Notice that $u_{2\ell}$ is always positive while $w_{2\ell}$ may be positive or negative.

Step 2: Calculation of velocity components in the main system (x, z)

A coordinate transformation between the local system (x_ℓ, z_ℓ) and the main Cartesian system (x, z) yields:

$$\begin{aligned} u_2 &= u_{2\ell} \cos \alpha - w_{2\ell} \sin \alpha, \\ w_2 &= u_{2\ell} \sin \alpha + w_{2\ell} \cos \alpha. \end{aligned} \tag{1.264}$$

The velocity of the particle after reflection is then

$$V_2^{n+1} = \sqrt{u_2^2 + w_2^2}. \tag{1.265}$$

Step 3: Calculation of the components of position of a particle in the $(\widetilde{x}, \widetilde{z})$ system

Let $P^{n+1} = P_2(x_2, z_2)$ be the position of a particle after diffuse reflection. The length of the vector $R_{P2} = P_c P^{n+1}$ is computed by

$$R_{P2} = P_c P^{n+1} = V_2^{n+1} \Delta t\left(P_c\widetilde{P}_2 / P_1\widetilde{P}_2\right). \tag{1.266}$$

$\Delta t\left(P_c\widetilde{P}_2 / P_1\widetilde{P}_2\right)$ is the time interval, or fraction of the time step, after the collision. Hence

$$\begin{aligned} \widetilde{x} &= R_{P2} \cos \Phi, \\ \widetilde{z} &= R_{P2} \sin \Phi, \end{aligned} \tag{1.267}$$

with $\Phi = R_1 \pi$, where R_1 is a random deviate.

Step 4: Calculation of the components of the position of the particle in the local coordinate system (x_ℓ, z_ℓ).

From the definition of the coordinate systems it is evident that

$$\begin{aligned} x_{2\ell} &= \widetilde{y}, \\ z_{2\ell} &= -\widetilde{x}. \end{aligned} \tag{1.268}$$

Step 5: Calculation of the components of the position of the particle in the main coordinate system (x, z).

The coordinate transformation between the local system (x_ℓ, z_ℓ) and the main system (x, z) yields:

$$\begin{aligned} x_2 &= x_c + x_{2\ell} \cos \alpha - y_2 \sin \alpha, \\ z_2 &= z_c + x_{2\ell} \sin \alpha + z_2 \cos \alpha. \end{aligned} \tag{1.269}$$

1.5.2 *Specular Reflection from the Symmetry Planes*

When the computational domain is bounded by a symmetry plane inclined by an angle α with respect to the x-axis, the symmetry plane is treated as perfectly reflecting surface. If the point $P(x, y)$ defines the position of a molecule before reflection from a symmetry plane, the position of the molecule after reflection, $P^*(x^*, z^*)$ and the velocity components are given by

$$\begin{aligned} x^* &= x\cos 2\alpha + y\sin 2\alpha, \\ y^* &= x\sin 2\alpha - y\cos 2\alpha, \end{aligned} \tag{1.270a}$$

$$\begin{aligned} u^* &= u\cos 2\alpha + v\sin 2\alpha, \\ v^* &= u\sin 2\alpha - v\cos 2\alpha. \end{aligned} \tag{1.270b}$$

For $\alpha = (\pi/2)$ the above formulae simplifies to

$$\begin{aligned} x^* &= x\cos 2\alpha = -x, \\ y^* &= -y\cos 2\alpha. = y, \end{aligned} \tag{1.271a}$$

$$\begin{aligned} u^* &= u\cos 2\alpha = -u, \\ v^* &= -v\cos 2\alpha = v. \end{aligned} \tag{1.271b}$$

In the following we give a general three-dimensional treatment for the case of molecules with equal mass. Let us refer to Fig. 1.11 which shows the collision of a molecule with a material surface in the point P_c. We consider the plane tangent to the surface in P_c and refer to a local coordinate system centred in P. The x-axis is in the tangent plane and y axis normal to the tangent plane and directed outwardly. The unit

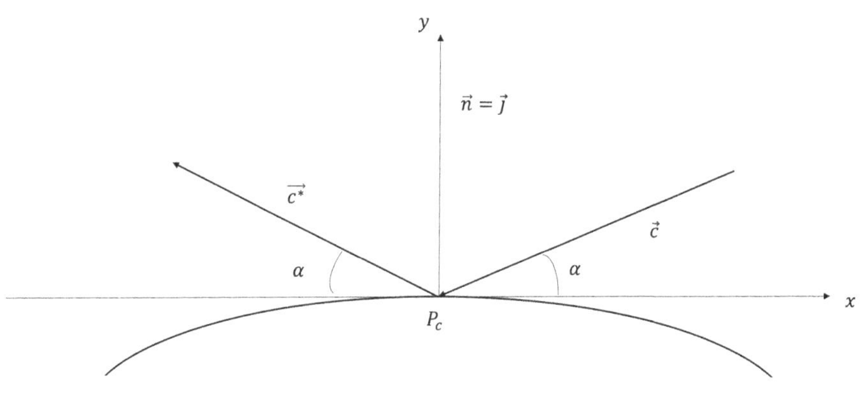

Fig. 1.11 Schematic representation of elastic collision of a molecule with a material surface

vector $\vec{n}$ coincides with the unit vector $\vec{j}$ directed along the y-axis. The z-axis (not drawn) is normal to the plane of the figure.

The above Eq. (1.270) is given in the general form by

$$\vec{c}^{\,*} = \vec{c} - 2\vec{n}\left(\vec{n}\cdot\vec{c}\right). \tag{1.272}$$

The projections on the coordinate axes yield

$$\vec{c}^{\,*}\cdot\vec{i} = \vec{c}\cdot\vec{i} - 2\vec{n}\left(\vec{n}\cdot\vec{c}\right)\cdot\vec{i} = \vec{c}\cdot\vec{i}, \text{ or } u^* = u, \tag{1.273a}$$

$$\vec{c}^{\,*}\cdot\vec{j} = \vec{c}\cdot\vec{j} - 2\vec{n}\left(\vec{n}\cdot\vec{c}\right)\vec{j} = v - 2v, \text{ or } v^* = v - 2v = -v, \tag{1.273b}$$

$$\vec{c}^{\,*}\cdot\vec{k} = \vec{c}\cdot\vec{k} - 2\vec{n}\left(\vec{n}\cdot\vec{c}\right)\cdot\vec{k} = \vec{c}\cdot\vec{k}, \text{ or } w^* = w. \tag{1.273c}$$

Equation (1.272) is easily inverted by multiplying it scalarly with the unit vector $\vec{n}$, yielding the reflection law for the inverse collision

$$\vec{c} = \vec{c}^{\,*} - 2\vec{n}\left(\vec{n}\cdot\vec{c}^{\,*}\right). \tag{1.274}$$

In case the material wall moves with a velocity $\vec{c}_w$ the reflection law becomes

$$\vec{c}^{\,*} = \vec{c} - 2\vec{n}\left[\vec{n}\cdot\left(\vec{c} - \vec{c}_w\right)\right]. \tag{1.275}$$

The change of coordinates (x, y, z) upon specular reflection is given by

$$\begin{aligned} x^* &= -x, \\ y^* &= y, \\ z^* &= z. \end{aligned} \tag{1.276}$$

Example 1.3

Avoiding intermolecular collisions, one can devise a test that allows for a precise control of the models of specular reflection from the material walls. The trajectories of a sample molecule between the collision points with the walls must be straight lines with constant inclinations. The equations of a sample molecule between collisions with the walls are

$$x = x_0 + ct\cos\alpha_0 = x_0 + ut \tag{1.277a}$$

$$z = z_0 + ct\sin\alpha_0 = z_0 + wt, \tag{1.277b}$$

where x_0 e z_0 are the coordinates of the initial position of the molecule, $c(u, w)$ is the velocity vector and α_0 is the angle between the velocity vector and the x axis.

Let us introduce two non-dimensional parameters

$$U = \sin\left(\frac{\pi x}{X_{\max}}\right) = \sin\left(\frac{\pi}{X_{\max}}(x_0 + ut)\right), \tag{1.278a}$$

$$W = \sin\left(\frac{\pi z}{Z_{\max}}\right) = \sin\left(\frac{\pi}{Z_{\max}}(z_0 + wt)\right), \tag{1.278b}$$

where $X_{\max}$ e $Z_{\max}$ are the dimensions of the domain. In phase space the non-dimensional variables U and W describe trajectories which can be obtained by combination of two orthogonal harmonics of frequencies

$$f_u = \frac{u}{2X_{\max}}, \tag{1.278a}$$

$$f_w = \frac{w}{2Z_{\max}}. \tag{1.279b}$$

The resulting trajectories in phase space are Lissajous figures. See Figs. 1.12 and 1.13.

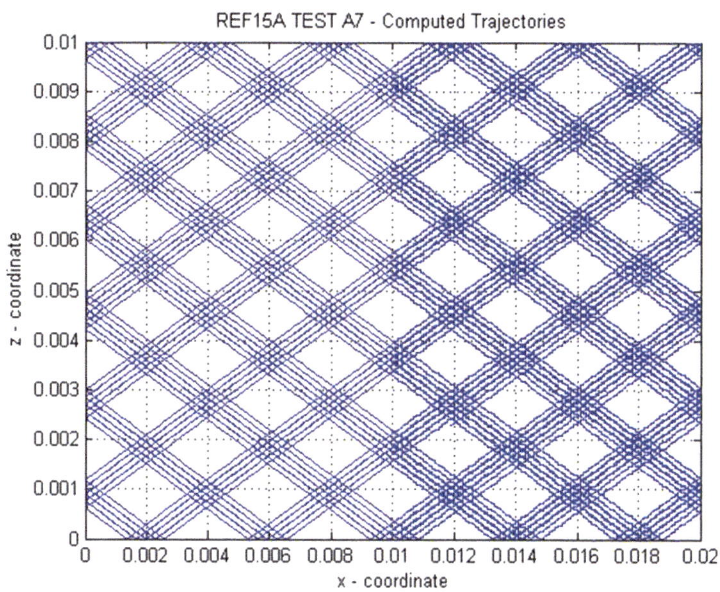

Fig. 1.12 Computed trajectories of a sample molecule in a domain with specularly reflecting surfaces

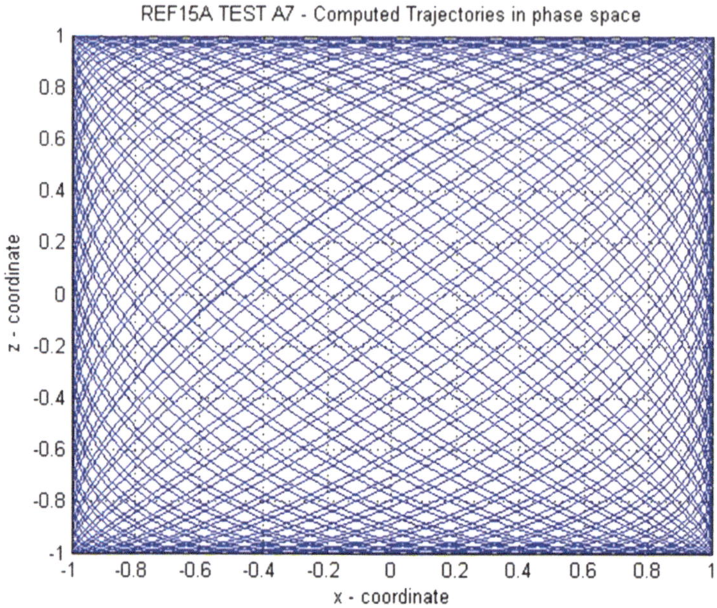

Fig. 1.13 Computed trajectories of a sample molecule in phase space

1.6 Elastic Collisions in Phase Space

1.6.1 Phase Space and the Liouville Theorem

Let us consider a set of N points in the 6N dimensional phase space

$$X = X(x, y, z, u, v, w) = X(\overline{x}, \overline{c}) = [\overline{x}_i, \overline{c}_i], (i = 1, 2, 3, \ldots N) \tag{1.280}$$

$$\overline{c} = \overline{x}' = d\overline{x}/dt. \tag{1.281}$$

The derivative of vector X with respect to time is

$$Z = \frac{dX}{dt} = \left(\frac{d\overline{x}}{dt}, \frac{d\overline{c}}{dt}\right) = (\overline{x}', x'') = (\overline{c}, \overline{F}), \tag{1.282}$$

where $\overline{F}$ are forces acting on the particles.

Given an initial point X_0 in phase space, Eq. (1.282) can be integrated, in principle, determining exact trajectories in phase space. Because of the large number of particles needed to model physical problems and especially because initial conditions are affected by errors, Eq. (1.282) cannot however be integrated rigorously. To cope with this problem, we define a probability $P(X)dX$ of finding the point X, representative of the system in phase space, within X and $X + dX$. The discrete system of particles is thus represented, like an idealised fluid of density ρ, by the distribution probability $P(X)$. The vector Z, defined by Eq. (1.282), behaves as the velocity of the idealised fluid. In analogy with the continuity equation of the idealised fluid, we can write the Liouville equation

$$\frac{\partial P}{\partial t} + \nabla \cdot (PZ) = 0, \tag{1.283}$$

where, for any vector $\overline{a}$ in phase space, the divergence is

$$\nabla \cdot \overline{a} = \frac{\partial \overline{a}}{\partial X} = \sum_{i=1}^{6N} \frac{\partial a_i}{\partial X_i}. \tag{1.284}$$

Because

$$\nabla \cdot (PZ) = P(\nabla \cdot Z) + Z \cdot \nabla P, \tag{1.285}$$

Equation (1.283) becomes

$$\frac{\partial P}{\partial t} + P(\nabla \cdot Z) + Z \cdot \nabla P = 0. \tag{1.286}$$

The divergence of the vector $Z = \left(\overline{c}, \overline{F}\right)$ is

$$\nabla \cdot Z = \frac{\partial Z}{\partial X} = \sum_{i=1}^{6N} \frac{\partial Z_i}{\partial X_i} = \sum_{i=1}^{6N} \left(\frac{\partial c_i}{\partial x_i} + \frac{\partial F_i}{\partial c_i} \right). \tag{1.287}$$

The derivatives $\partial c_i / \partial x_i$ in the above equation are zero, because the x_i and c_i variables are independent. The derivatives $\partial F_i / \partial c_i$ are the derivatives of the forces acting on the particle with respect to the velocity components. In the case of the Lorentz forces, $\vec{F} = q \vec{c} \times \vec{B}$ acting upon an electric charge q moving with speed $\vec{c}$ in the field of the magnetic induction vector $\vec{B}$, these derivatives are different from zero, but we consider at present only velocity-independent forces so that the terms $\partial F_i / \partial c_i$ also vanish, hence $\nabla \cdot Z = 0$. The Liouville equation therefore simplifies to

$$\frac{\partial P}{\partial t} + Z \cdot \nabla P = 0, \tag{1.288}$$

which can also be written as

$$\frac{\partial P}{\partial t} + \sum_{i=1}^{6N} \left(c_i \frac{\partial P}{\partial x_i} + F_i \frac{\partial P}{\partial c_i} \right) = 0. \tag{1.289}$$

If we assume that the problem represented by the evolution Eq. (1.282) is well posed and admits a unique solution, then a point $X_0 = X(t = 0)$ from a domain R_0 of phase space describes a trajectory and maps in a point $X(t)$ of a domain R. Trajectories do not intersect because of the assumption of a unique solution. The whole motion of the system of the N particles is a mapping of the domain R_0 in the domain R.

The Liouville theorem states that if $\nabla \cdot Z = 0$, the volume of a region in phase space is invariant with respect to the mapping represented by the motion of the system of particles. Mathematically this implies that

$$\int_R dX = \int_{R_0} J[X \Leftarrow X_0] dX_0 = \int_{R_0} dX_0, \tag{1.290}$$

where $J[X \Leftarrow X_0]$ is the Jacobian of the transformation, for which it must hold

$$J[X \Leftarrow X_0] = 1. \tag{1.291}$$

The Liouville theorem implies that the probability P takes at time t, at the point $X(t)$ of the domain R, the same value that it had at time $t = 0$ in the point $X_0 = X(t = 0)$ of the domain R_0. Therefore if the probability P is a constant in the domain R_0 it will be the same constant in the domain $R_t = R(t)$ in which R_0 is mapped by the flow induced in phase space by the evolution Eq. (1.282).

Vice versa, if, in a transformation of coordinates in phase space, the Jacobian has value 1, the Liouville theorem holds. This is, for instance, the case when coordinates and velocity components change as a result of elastic collisions between molecules or of elastic collision of a molecule with a surface. This will be shown in the following sections.

1.6.2 Specular Reflection from a Wall

The Jacobian of the velocity transformation is given by

$$J_c = \frac{\partial(u^*, v^*, w^*)}{\partial(u, v, w)} = \begin{vmatrix} \frac{\partial u^*}{\partial u} & \frac{\partial u^*}{\partial v} & \frac{\partial u^*}{\partial w} \\ \frac{\partial v^*}{\partial u} & \frac{\partial v^*}{\partial v} & \frac{\partial v^*}{\partial w} \\ \frac{\partial w^*}{\partial u} & \frac{\partial w^*}{\partial v} & \frac{\partial w^*}{\partial w} \end{vmatrix} = \begin{vmatrix} 1 & 0 & 0 \\ 0 & -1 & 0 \\ 0 & 0 & 1 \end{vmatrix} = -1. \tag{1.292}$$

The Jacobian of the coordinates transformation is given by

$$J_x = \frac{\partial(x^*, y^*, z^*)}{\partial(x, , y, z)} = \begin{vmatrix} \frac{\partial x^*}{\partial x} & \frac{\partial x^*}{\partial y} & \frac{\partial x^*}{\partial z} \\ \frac{\partial y^*}{\partial x} & \frac{\partial y^*}{\partial y} & \frac{\partial y^*}{\partial z} \\ \frac{\partial z^*}{\partial x} & \frac{\partial z^*}{\partial y} & \frac{\partial z^*}{\partial z} \end{vmatrix} = \begin{vmatrix} -1 & 0 & 0 \\ 0 & 1 & 0 \\ 0 & 0 & 1 \end{vmatrix} = -1. \tag{1.293}$$

Denoting with $X = X(x, y, z, u, v, w)$ the point determined by the six coordinates in phase space, the Jacobian of the whole (velocities and coordinates) transformation is given by

$$J = \frac{\partial X^*}{\partial X} = \begin{vmatrix} J_x & 0 \\ 0 & J_c \end{vmatrix} = (-1)(-1) = 1. \tag{1.294}$$

According to the Liouville theorem, being the Jacobian of the transformation equal to 1, volume elements are preserved in phase space.

1.6.3 Binary Elastic Collision

The transformation from the pre-collision velocities $\vec{c}_1$, $\vec{c}_2$ to the post-collision velocities $\vec{c}^{\,*}_1$, $\vec{c}^{\,*}_2$ can be thought of as consisting of the following steps:

(i) From $\vec{c}_1$, $\vec{c}_2$ to $\vec{c}_m$, $\vec{c}_r$,
(ii) from $\vec{c}_m$, $\vec{c}_r$ to $\vec{c}^{\,*}_m \left(= \vec{c}_m\right)$, $\vec{c}^{\,*}_r$,
(iii) from $\vec{c}^{\,*}_m \left(= \vec{c}_m\right)$, $\vec{c}^{\,*}_r$ to $\vec{c}^{\,*}_1$, $\vec{c}^{\,*}_2$.

Correspondingly, the Jacobian of the transformation from $\vec{c}_1$, $\vec{c}_2$ to $\vec{c}^{\,*}_1$, $\vec{c}^{\,*}_2$ can be written applying the three transformations represented by the J operator in the following sequence:

$$\begin{aligned} J_c &= J_c\left[c_1^*, c_2^* \Leftarrow c_1, c_2\right] \\ &= J_{c3}\left[c_1^*, c_2^* \Leftarrow c_m, c_r^*\right] J_{c2}\left[c_m, c_r^* \Leftarrow c_m, c_r\right] J_{c1}\left[c_m, c_r \Leftarrow c_1, c_2\right]. \end{aligned} \tag{1.295}$$

The transformation represented by the Jacobian J_{c2} has the Jacobian given by Eq. (1.292), equal to -1. The Jacobians J_{c1} and J_{c3} represent transformations which are inverse of each other and the corresponding values are the reciprocals of each other. Therefore, if we set $J_{c1} = J_0$, $J_{c3} = 1/J_0$ one has

$$J_c = J_{c1} \cdot J_{c2} \cdot J_{c3} = J_0 \cdot J_{c2} \cdot 1/J_0 = -1. \tag{1.296}$$

Similarly, for the coordinate transformations one has $J_x = -1, J_c = -1$, hence

$$J = \frac{\partial X^*}{\partial X} = \begin{vmatrix} J_x & 0 \\ 0 & J_c \end{vmatrix} = (-1)(-1) = 1. \tag{1.297}$$

1.6.4 *Uncertainties of Initial Conditions*

Uncertainties of initial conditions, position and velocities of the simulated particles imply deformations of volume elements in phase space. To exemplify the problem, let us consider a single particle moving on an x-axis between coordinates $x = \ell$ and $x = -\ell$ and undergoing specular reflections at these boundaries. Assume that at instant $t = 0$ position x and velocity c of the particle are known with uncertainties Δx and Δc, respectively. In the phase space, the plane spanned by the axes x and c, the point $P(x, c)$ representative of the particle lies in the rectangle with base Δx and height Δc. After a time t the point $P(x, c)$ has moved to $P^*(x + ct, c)$ having undergone the transformation

$$\begin{aligned} x^* &= x + ct, \\ c^* &= c. \end{aligned} \tag{1.298}$$

The Jacobian of the transformation is

$$J = \frac{\partial(x^*, c^*)}{\partial(x, c)} = \begin{vmatrix} \dfrac{\partial x^*}{\partial x} & \dfrac{\partial x^*}{\partial c} \\ \dfrac{\partial c^*}{\partial x} & \dfrac{\partial c^*}{\partial c} \end{vmatrix} = \begin{vmatrix} 1 & t \\ 0 & 1 \end{vmatrix} = 1. \tag{1.299}$$

Because the Jacobian of the transformation is 1 the volume of a domain in phase space is preserved. The initial rectangle, representing the area of uncertainties in the location of the moving particle in phase space, retains its initial area, but is deformed because the points of the rectangle move with different speed. The initial rectangle is deformed into a parallelogram with base Δx, which becomes slender and slender as time progresses. When the particle reaches the point $x = \ell$, specular reflection changes the coordinates (ℓ, c) into (ℓ, $-c$). When it reaches point $x = -\ell$ specular reflection changes the coordinates ($-\ell$, $-c$) into ($-\ell$, c).

1.7 Boltzmann Equation and Maxwellian Distribution Function

1.7.1 Phase Space and Distribution Function

Let $d\overline{x} = dxdydz$ be a volume element in physical space and $d\overline{c} = dudvdw$ be a volume element in the velocity space spanned by the velocity vector $\vec{c} = (u, v, w)$. A volume element in phase space is then given by $d\overline{x}d\overline{c} = dxdydz\ dudvdw$. Time is the seventh coordinate in phase space. Let N be the number of molecules of a fluid in the volume $d\overline{x}$. The molecular density will be

$$n = \frac{N}{|d\overline{x}|}, \tag{1.300}$$

where $|d\overline{x}|$ is the measure of the volume element in physical space. Let $f\left(\vec{c}\right)$ be the probability that a given molecule has a velocity in the range ($\vec{c}$, $\vec{c}$ + d$\vec{c}$). The number of molecules which, at a given time point, are in the element $d\overline{c}$ is

$$dN = Nf\left(\vec{c}\right)d\vec{c} = nf\left(\vec{c}\right)d\overline{x}d\vec{c}. \tag{1.301}$$

Alternatively, the number of molecules which, at time point t, are in the element $d\overline{c}$ can be written as

$$dN = F(\overline{x}, \overline{c}, t)d\overline{x}d\vec{c}, \tag{1.302}$$

which introduces the definition of the non-normalized distribution function $F(\overline{x}, \overline{c}, t)$. Comparing Eq. (1.301) and Eq. (1.302) one derives

$$nf\left(\vec{c}\right) = F(\overline{x}, \overline{c}, t). \tag{1.303}$$

The integrals

$$\int_{-\infty}^{+\infty} f\left(\vec{c}\right)d\vec{c} = \frac{1}{N}\int dN = \frac{N}{N} = 1, \tag{1.304}$$

$$\int_{-\infty}^{+\infty}\int_{-\infty}^{+\infty} F(\overline{x}, \overline{c}, t)d\overline{x}d\overline{c} = \int dN = N, \tag{1.305}$$

show that the distribution function $f\left(\vec{c}\right)$ is normalized, whilst the distribution function $F(\overline{x}, \overline{c}, t)$ is not.

For every physical quantity Q we define the mean value $\overline{Q}$ over the velocity space by

$$\overline{Q} = \int_{-\infty}^{+\infty} Qf\left(\vec{c}\right)d\overline{c}. \tag{1.306}$$

For instance, the mean value of the velocity $\vec{c}$ is given by

$$\overline{c} = \int_{-\infty}^{+\infty} \vec{c}f\left(\vec{c}\right)d\overline{c}. \tag{1.307}$$

1.7.2 The Boltzmann Equation

The Boltzmann equation gives a balance of the number of molecules $nf\left(\vec{c}\right)d\overline{x}d\vec{c} = F(\overline{x}, \overline{c}, t)d\overline{x}d\vec{c}$ contained in an element $d\overline{x}d\overline{c}$ of phase space, making a balance between its time variation, the leakage (positive or negative) across the surfaces of the element, and the contribution of scattering of molecules into the element or away from the element. The terms of the Boltzmann equation which form the balance are as follows:

Term I $= \frac{\partial(fn)}{\partial t}d\overline{x}d\overline{c}$, time-dependent term;

Term II $= -\vec{c}\cdot\frac{\partial(fn)}{\partial\overline{x}}d\overline{x}d\overline{c}$, net influx of molecules of class $\vec{c}$ through the surface of the space element $d\overline{x}$. If the divergence $\nabla\cdot\left(nf\vec{c}\right)$ is negative the inflow of molecules into the space element is larger than the outflow and the element acts as a sink of molecules. Vice versa, if the divergence is positive the element is a source of molecules, or, generally speaking, of particles released by some physical process (for instance of neutrons released by nuclear fission).

Term III $= -\vec{F}\cdot\frac{\partial(fn)}{\partial\overline{c}}d\overline{x}d\overline{c}$, net influx of molecules of class $\vec{c}$ through the surface of the velocity space element $d\overline{c}$. In this term $\vec{F}$ is a force acting on the molecules within the velocity space element.

Term IV. This term takes into account the contribution of scattering of molecules into the element or away from the element and deserves a more detailed derivation. We refer to the molecules in the velocity range ($\vec{c}$, $\vec{c} + d\vec{c}$) as molecules "of class $\vec{c}$", and to molecules in the range ($\vec{c}_1$, $\vec{c}_1 + d\vec{c}_1$) as molecules "of class $\vec{c}_1$". Then we have

$nfdc$ = number of molecules of class $\vec{c}$ per unit volume in velocity space;
nf_1dc_1 = number of molecules of class $\vec{c}_1$ per unit volume in velocity space.

Let $\vec{c}_r$ be the relative velocity between a sample molecule of class $\vec{c}$ and a molecule of class $\vec{c}_1$. Let $\sigma d\Omega$ be the differential cross section for a collision, with

$d\Omega$ an element of solid angle. The elementary volume spaced by the sample molecule in the unit time is $\sigma d\Omega \vec{c}_r$. Then $\left(\sigma d\Omega \vec{c}_r\right) nf_1 dc_1 =$ number of collisions per unit time of the sample molecule with molecules of class $\vec{c}_1$ in the volume $\sigma d\Omega \vec{c}_r$;

Multiplying the above term with $nfd\bar{x}d\bar{c}$ (= number of molecules of class $\vec{c}$ in $d\bar{x}d\bar{c}$) we have $\left(\sigma d\Omega \vec{c}_r\right) nf_1 dc_1\ nfd\bar{x}d\bar{c} = \left(\sigma d\Omega \vec{c}_r\right) n^2 f f_1 dc_1\ d\bar{x}d\bar{c} =$ number of collisions of molecules of class $\vec{c}$ with molecules of class $\vec{c}_1$ in $d\bar{x}d\bar{c}$;

Integrating the above term over all classes $\vec{c}_1$ and over the full space directions one obtains $\int_0^{4\pi}\int_{-\infty}^{+\infty}\left\{\left(\sigma d\Omega \vec{c}_r\right) n^2 f f_1 dc_1\ d\bar{x}d\bar{c}\right\} =$ number of molecules scattered from class $\vec{c}$ to every possible class $\vec{c}_1$;

The above number of molecules scattered from class $\vec{c}$ to every possible class $\vec{c}_1$ represents a depletion of the molecules of class $\vec{c}$ and therefore is as a negative contribution to the balance of molecules in the phase space element $d\bar{x}d\bar{c}$. In the same way the number of molecules scattered from every class $\vec{c}_1$ to class $\vec{c}$ can be calculated on the basis of probability distributions which, to be distinguished from the previous ones, will be denoted by a star (*) as superscript. This second term must be considered as positive because it replenishes the element $d\bar{x}d\bar{c}$ in phase space. The two terms together give

$$\text{Term IV} = \int_0^{4\pi}\int_{-\infty}^{+\infty}\left\{ n^2\left(f^* f_1^* - f f_1\right)\left(\sigma d\Omega c_r\right) dc_1 \right\} d\bar{x}d\bar{c}.$$

Collecting all terms and dividing by the phase space element $d\bar{x}d\bar{c}$ one obtains the **Boltzmann equation**

$$\frac{\partial (fn)}{\partial t} - \vec{c}\cdot\frac{\partial (fn)}{\partial \bar{x}} - \vec{F}\cdot\frac{\partial (fn)}{\partial \bar{c}} = \int_0^{4\pi}\int_{-\infty}^{+\infty}\left\{ n^2\left(f^* f_1^* - f f_1\right)\left(\sigma d\Omega c_r\right) dc_1 \right\}. \tag{1.308}$$

The right-hand side of Eq. (1.308) is referred to as "collision integral".

Useful developments of the molecular theory are obtained by multiplying the Boltzmann equation by a physical quantity Q and integrating over all molecules of class $\vec{c}$. We assume that Q is either a constant or a function of the molecular speed $\vec{c}$ only. Thus one obtains

$$\underbrace{\int_{-\infty}^{+\infty} Q\frac{\partial (fn)}{\partial t} d\bar{c}}_{\text{Term (i)}} - \underbrace{\int_{-\infty}^{+\infty} Q\,\vec{c}\cdot\frac{\partial (fn)}{\partial \bar{x}} d\bar{c}}_{\text{Term (ii)}} - \underbrace{\int_{-\infty}^{+\infty} Q\vec{F}\cdot\frac{\partial (fn)}{\partial \bar{c}} d\bar{c}}_{\text{Term (iii)}} =$$

$$= \int_{-\infty}^{+\infty}\int_0^{4\pi}\int_{-\infty}^{+\infty} Q\left\{ n^2\left(f^* f_1^* - f f_1\right)\left(\sigma d\Omega c_r\right) dc_1 \right\} d\bar{c} \tag{1.309}$$

Term (iv)
The first three terms in Eq. (1.309) are computed as follows:

$$\text{Term (i)} = \int_{-\infty}^{+\infty} Q \frac{\partial (fn)}{\partial t} d\overline{c} = \frac{\partial}{\partial t} \int_{-\infty}^{+\infty} Qfn \, d\overline{c} = \frac{\partial}{\partial t} \left(n\overline{Q} \right), \tag{1.310a}$$

having used the definition (1.306) of the mean value $\overline{Q}$.

Because Q does not depend on $\overline{x}$, one has $\nabla \cdot \left(Qnf \vec{c} \right) = Qnf \left(\nabla \cdot \vec{c} \right) + \vec{c} \cdot \nabla (Qnf) = Q \vec{c} \cdot \nabla (nf)$, hence

$$\begin{aligned} \text{Term(ii)} &= \int_{-\infty}^{+\infty} Q \vec{c} \cdot \frac{\partial (fn)}{\partial \overline{x}} d\overline{c} = \int_{-\infty}^{+\infty} \nabla \cdot \left(nQf \vec{c} \right) d\overline{c} \\ &= \nabla \cdot \int_{-\infty}^{+\infty} \left(nQf \vec{c} \right) d\overline{c} = \nabla \cdot n\overline{Q \vec{c}}. \end{aligned} \tag{1.310b}$$

Because again Q does not depend on $\overline{x}$, one has

$$\frac{\partial (Qnf)}{\partial \overline{c}} = nf \frac{\partial (Q)}{\partial \overline{c}} + Q \frac{\partial (nf)}{\partial \overline{c}},$$

$$Q\vec{F} \cdot \frac{\partial (nf)}{\partial \overline{c}} = \vec{F} \cdot Q \frac{\partial (nf)}{\partial \overline{c}} = \vec{F} \cdot \frac{\partial (Qnf)}{\partial \overline{c}} - \vec{F} \cdot nf \frac{\partial (Q)}{\partial \overline{c}},$$

$$\int_{-\infty}^{+\infty} Q\vec{F} \cdot \frac{\partial (fn)}{\partial \overline{c}} d\overline{c} = \int_{-\infty}^{+\infty} \vec{F} \cdot \frac{\partial (Qnf)}{\partial \overline{c}} d\overline{c} - \int_{-\infty}^{+\infty} \vec{F} \cdot nf \frac{\partial (Q)}{\partial \overline{c}} d\overline{c}.$$

The first integral at the right-hand side of the above equation is zero because the probability distribution f goes to zero at infinity. Hence

$$\begin{aligned} \text{Term(iii)} &= \int_{-\infty}^{+\infty} Q\vec{F} \cdot \frac{\partial (fn)}{\partial \overline{c}} d\overline{c} = \int_{-\infty}^{+\infty} \vec{F} \cdot \frac{\partial (Qnf)}{\partial \overline{c}} d\overline{c} - \int_{-\infty}^{+\infty} \vec{F} \cdot nf \frac{\partial (Q)}{\partial \overline{c}} d\overline{c} \\ &= -\vec{F} \cdot n \left\langle \frac{\partial (Q)}{\partial \overline{c}} \right\rangle, \end{aligned} \tag{1.310c}$$

where the symbol $\langle \rangle$ denotes mean value.

Term (iv) in Eq. (1.309) will be denoted in the following for short

$$\int_{-\infty}^{+\infty}\int_{0}^{4\pi}\int_{-\infty}^{+\infty} Q\{\, n^2\left(f^* f_1^* - f f_1\right)(\sigma d\Omega c_r)\, dc_1 \,\}d\overline{c} = \Delta(Q). \tag{1.310d}$$

Collecting terms Eq. (1.309) becomes

$$\frac{\partial}{\partial t}\left(n\overline{Q}\right) + \nabla \cdot n\overline{Q\vec{c}} - \vec{F}\cdot n\left\langle \frac{\partial(Q)}{\partial \overline{c}} \right\rangle = \Delta(Q). \tag{1.311}$$

Equation (1.311) is called "transfer equation" or "equation of change". It also represents a "momentum equation for Q". It can be shown (Bird, 1994) that in case the physical quantity Q is either the mass "m" of a molecule or the momentum "m $\vec{c}$" or the kinetic energy $mc^2/2$, the collision integral is zero, hence $\Delta(Q) = 0$. For this reason, mass, momentum and kinetic energy of a molecule are called "collisional invariants". The transfer equations for the collisional invariants are derived in the next section.

1.7.3 Collisional Invariants

(i) **Mass Conservation Equation**
In case the physical quantity Q is the mass m of a molecule, using $\rho = nm$, the transfer Eq. (1.311) becomes the mass conservation equation

$$\frac{\partial \rho}{\partial t} + \nabla \cdot \rho\overline{\vec{c}} = 0. \tag{1.312}$$

(ii) **Momentum Conservation Equation**
In case the physical quantity Q is the momentum "m $\vec{c}$ ", the transfer Eq. (1.311) becomes

$$\frac{\partial}{\partial t}\left(nm\overline{\vec{c}}\right) + \nabla \cdot nm\overline{\vec{c}\,\vec{c}} - \vec{F}\cdot n\left\langle \frac{\partial\left(m\vec{c}\right)}{\partial \overline{c}} \right\rangle = 0. \tag{1.313}$$

Written for the three components in the coordinate directions Eq. (1.313) gives

$$\frac{\partial}{\partial t}\left(nm\overline{c_j}\right) + \frac{\partial}{\partial x_i} nm\overline{c_i c_j} - nF_i\left\langle \frac{\partial\left(mc_j\right)}{\partial c_i} \right\rangle = 0, \; (j = 1, 2, 3) \tag{1.314}$$

Because $\partial c_j / \partial c_i = \delta_{ij}$, Eq. (1.314) becomes

$$\frac{\partial}{\partial t}(nmc_j) + \frac{\partial}{\partial x_i} nmc_i c_j - nmF_j = 0, (j = 1, 2, 3) \tag{1.315}$$

hence

$$\frac{\partial}{\partial t}(\rho c_j) + \frac{\partial}{\partial x_i} \rho c_i c_j - \rho F_i = 0.(j = 1, 2, 3) \tag{1.316}$$

Again, as a vector equation, one has

$$\frac{\partial}{\partial t}\rho \vec{c} + \nabla \cdot \left(\rho \vec{c}\,\vec{c}\right) - \rho \vec{F} = 0. \tag{1.317}$$

Recalling $\vec{c} = \vec{c}_0 + \vec{c}\,'$ and taking into account that the mean value of fluctuating quantities are zero one has

$$\left\langle \vec{c}\,\vec{c} \right\rangle = \left\langle \left(\vec{c}_0 + \vec{c}\,'\right)\left(\vec{c}_0 + \vec{c}\,'\right) \right\rangle = \left\langle \vec{c}_0 \vec{c}_0 \right\rangle + \left\langle \vec{c}\,'\vec{c}\,' \right\rangle. \tag{1.318}$$

The second term in Eq. (1.317) becomes

$$\begin{aligned} \nabla \cdot \left(\rho \vec{c}\,\vec{c}\right) &= \nabla \cdot \left[\rho \vec{c}_0 \vec{c}_0 + \rho \vec{c}\,'\vec{c}\,'\right] \\ &= \vec{c}_0 \nabla \cdot \left[\rho \vec{c}_0\right] + \rho \vec{c}_0 \nabla \cdot \left[\vec{c}_0\right] + \nabla \cdot \left[\rho \vec{c}\,'\vec{c}\,'\right], \end{aligned} \tag{1.319}$$

hence

$$\rho \frac{\partial \vec{c}_0}{\partial t} + \vec{c}_0 \frac{\partial \rho}{\partial t} + \vec{c}_0 \nabla \cdot \left[\rho \vec{c}_0\right] + \rho \vec{c}_0 \nabla \cdot \left[\vec{c}_0\right] + \nabla \cdot \left[\rho \vec{c}\,'\vec{c}\,'\right] - \rho \vec{F} = 0. \tag{1.320}$$

The sum of the second and third terms in Eq. (1.320) is zero because of the mass conservation equation. Using $d/dt = \partial/\partial t + \vec{c}_0 \cdot \nabla$, Eq. (1.13a) $[\ldots P = \rho \vec{c}\,'\vec{c}\,']$ and Eq. (1.15) $[\ldots P = -\ldots \tau + p \ldots \delta]$ one obtains the time averaged momentum equation

$$\rho \frac{d\vec{c}_0}{dt} = -\nabla \ddot{P} + \rho \vec{F} = -\nabla p + \nabla \ddot{\tau} + \rho \vec{F}. \tag{1.321}$$

(III) **Energy (Kinetic) Conservation Equation**

In case the physical quantity Q is the kinetic energy $\rho c^2/2$, the transfer Eq. (1.311) becomes

$$\frac{\partial}{\partial t}\left(n\frac{1}{2}mc^2\right) + \nabla\cdot n\frac{1}{2}mc^2\vec{c} - \vec{F}\cdot n\left\langle\frac{\partial(mc^2/2)}{\partial\overline{c}}\right\rangle = 0. \quad (1.322)$$

The last term of Eq. (1.322) can be written as

$$\vec{F}\cdot n\left\langle\frac{\partial(mc^2/2)}{\partial\overline{c}}\right\rangle = F_i\frac{\partial}{\partial c_i}\frac{1}{2}\left(\rho c_i^2\right) = \rho F_i c_i = \rho F_i c_{0i} = \rho\vec{F}\cdot\vec{c}_0, \quad (1.323)$$

having dropped the symbol $\langle\rangle$ and set to zero the mean value of the fluctuating components c'. Then Eq. (1.322) becomes

$$\frac{\partial}{\partial t}\left(n\frac{1}{2}\rho c^2\right) + \nabla\cdot n\frac{1}{2}\rho c^2\vec{c} - \rho\vec{c}_0\cdot\vec{F} = 0. \quad (1.324)$$

1.7.4 H-Theorem and Maxwellian Distribution Function

In case of spatially homogeneous gas without an external force, the second and third terms at the left-hand side of the Boltzmann equation, Eq. (1.308), are zero. Furthermore, the number density is constant and the equation simplifies to

$$\frac{\partial f}{\partial t} = n\int_0^{4\pi}\int_{-\infty}^{+\infty}\{\left(f^* f_1^* - f f_1\right)(\sigma d\Omega c_r)dc_1\}. \quad (1.325)$$

During a small time interval Δt the distribution function changes from f to $f + \Delta f$ undergoing a fractional change $\Delta f/f = \Delta(nf)/(nf) \approx \Delta[\ln(nf)]$. Boltzmann's H-function is defined as the mean value of $\ln(nf)$:

$$H = \langle\ln(nf)\rangle = \int_{-\infty}^{+\infty} f\langle\ln(nf)\rangle d\overline{c}. \quad (1.326)$$

Boltzmann's H-theorem states that the H-function can never increase:

$$\frac{\partial H}{\partial t} \leq 0. \quad (1.327)$$

For every initially given distribution of molecular velocities, the distribution evolves in time and relaxes along a path of velocity space which implies a monotonical decrease of the H-function, till a lower bound of H is reached. At subsequent times H does not change any longer and therefore $\partial H/\partial t = 0$. This situation implies that an equilibrium state has been reached in which the average number of molecules remains constant in every element of velocity space. The final result is that an equilibrium velocity distribution has been reached, which is referred to as Maxwellian distribution and is given by Bird (1994)

$$f_o = \frac{\beta^3}{\pi^{3/2}} e^{-\beta^2 c'^2}, \tag{1.328}$$

with

$$\beta = \frac{1}{\sqrt{2RT}} = \sqrt{\frac{m}{2kT}}. \tag{1.329}$$

Because in the theoretical derivation of the Maxwellian distribution no assumption was made about the type of gas involved, the Maxwellian distribution holds both for monoatomic and polyatomic gases.

Using Eq. (1.308) and the definition of thermal velocity $\vec{c}' = \vec{c} - \vec{c}_0$, the fraction dN/N of molecules in the element $d\bar{c}$ of velocity space can be written as

$$\begin{aligned} \frac{dN}{N} &= \frac{dn}{n} = f_0 d\bar{c} = \frac{\beta^3}{\pi^{3/2}} e^{-\beta^2 c'^2} d\bar{c} \\ &= \frac{\beta^3}{\pi^{3/2}} \exp\left\{ -\beta^2 \left[(u-u_o)^2 + (v-v_o)^2 + (w-w_o)^2 \right] \right\} du\, dv\, dw. \end{aligned} \tag{1.330}$$

In case the main stream velocity $\vec{c}_0$ is zero, one obtains

$$\begin{aligned} \frac{dN}{N} &= \frac{dn}{n} = f_0 d\bar{c} = \frac{\beta^3}{\pi^{3/2}} e^{-\beta^2 c'^2} d\bar{c} \\ &= \frac{\beta^3}{\pi^{3/2}} \exp\left\{ -\beta^2 \left[u^2 + v^2 + w^2 \right] \right\} du\, dv\, dw. \end{aligned} \tag{1.331}$$

1.7.5 Equilibrium Gas Properties

Let us consider the thermal velocity vector $\vec{c}'$ represented in velocity space (see Fig. 1.14) with reference to a spherical coordinate system with azimuthal angle Φ and co-latitude ϑ. A surface element in this coordinate system is given by $dS = c'^2 \sin\vartheta\, d\vartheta\, d\Phi$ and a volume element is given by $d\bar{c}' = dS\, dc' = c'^2 \sin\vartheta\, d\vartheta\, d\Phi\, dc'$. The fraction of molecules with speed in the range $(c', c' + dc')$, crossing the elementary surface dS = AB · AC is $dn = nf_0 dc'$, hence

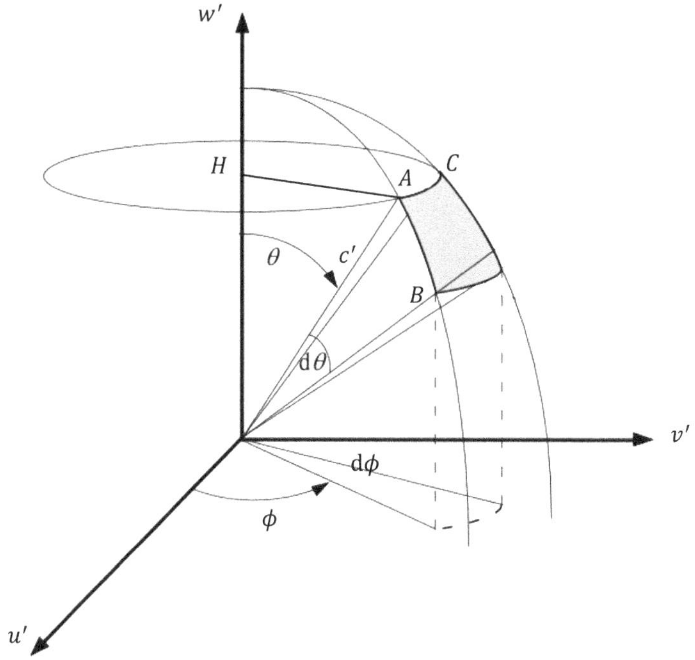

Fig. 1.14 Representation of thermal velocity vector $\vec{c}'$ in spherical coordinate system

$$\frac{dn}{n} = f_0 d\vec{c}' = \frac{\beta^3}{\pi^{3/2}} c'^2 e^{-\beta^2 c'^2} \sin\vartheta \, d\vartheta \, d\Phi \, dc'. \tag{1.332}$$

The fraction of molecules with speed in the range $(c', c' + dc')$, irrespective of their direction, hence crossing any point of the sphere of radius $\vec{c}'$, is obtained integrating Eq. (1.332) over the full solid angle. Because $\int_0^{\pi} \sin\vartheta \, d\vartheta = 2$ and $\int_0^{2\pi} d\Phi = 2\pi$, the result is

$$\frac{dn}{n} = \frac{\beta^3}{\pi^{3/2}} \int_0^{\pi} \int_0^{2\pi} c'^2 e^{-\beta^2 c'^2} \sin\vartheta \, d\vartheta \, d\Phi \, dc' = \frac{4\beta^3}{\sqrt{\pi}} c'^2 e^{-\beta^2 c'^2} \, dc' = f_{c'} \, dc', \tag{1.333}$$

with the definition of the distribution function

$$f_{c'} = \frac{4\beta^3}{\sqrt{\pi}} c'^2 e^{-\beta^2 c'^2}. \tag{1.334}$$

The maximum of the distribution corresponds to the most probable velocity

$$c'_m = \frac{1}{\beta} = \sqrt{2RT} = \sqrt{\frac{2kT}{m}}. \tag{1.335}$$

The average molecular speed is given by

$$\overline{c}' = \int_0^\infty c' f_{c'}\, dc' = \frac{4\beta^3}{\sqrt{\pi}} \int_0^\infty c'^3 e^{-\beta^2 c'^2}\, dc' = \frac{2}{\sqrt{\pi}\beta} = \frac{2}{\sqrt{\pi}} c'_m = \sqrt{\frac{8RT}{\pi}}. \tag{1.336}$$

The definite integral in Eq. (1.336) is equal to $1/(2\beta^4)$. It is one of the fundamental integrals given in Table 1.6.

The root mean thermal velocity is defined by

$$c'_s = \sqrt{\overline{c'^2}} = \sqrt{\int_0^\infty c'^2 f_{c'}\, dc'} = \sqrt{\frac{4\beta^3}{\sqrt{\pi}} \int_0^\infty c'^4 e^{-\beta^2 c'^2}\, dc'} = \sqrt{3RT} = \sqrt{\frac{3}{2}\frac{1}{\beta}}$$
$$= \sqrt{\frac{3}{2}} c'_m = \sqrt{\frac{3\pi}{8}} \overline{c}'. \tag{1.337}$$

The mean relative velocity between molecules is given by Bird (1994)

$$\overline{c}_r = \sqrt{2}\, \overline{c}'. \tag{1.338}$$

One has therefore $c'_m < \overline{c}' < c'_s < \overline{c}_r$.

Let us refer now to one coordinate direction, for instance the x-coordinate with thermal velocity component u' in that direction. The fraction $F_{u'}$ of molecules with velocity in the range $(u', u' + du')$, irrespective of any other velocity v', w' in the other coordinate directions, is obtained by integrating the distribution $f_{c'}$ over the velocity components v', w':

$$F_{u'} = \int_{-\infty}^{+\infty} \int_{-\infty}^{+\infty} \{f_0\, du'\}\, dv'\, dw' = \frac{\beta^3}{\pi^{3/2}} \int_{-\infty}^{+\infty} \int_{-\infty}^{+\infty} e^{-\beta^2 \left(u'^2 + v'^2 + w'^2\right)}\, du'\, dv'\, dw'$$
$$= \frac{\beta^3}{\pi^{3/2}} \left\{ e^{-\beta^2 u'^2} du' \int_{-\infty}^{+\infty} e^{-\beta^2 v'^2}\, dv' \int_{-\infty}^{+\infty} e^{-\beta^2 w'^2}\, dw' \right\} = \frac{\beta}{\sqrt{\pi}} e^{-\beta^2 u'^2} du' = f_{u'} du', \tag{1.339}$$

with the definition

$$f_{u'} = \frac{\beta}{\sqrt{\pi}} e^{-\beta^2 u'^2}. \tag{1.340}$$

The mean value of $f_{u'}$ taken over all molecules with positive and negative velocities is zero. The mean value of $f_{u'}$ taken over molecules with positive velocity only is

$$\bar{f}_{u'} = \int_0^\infty f_{u'} du' = \int_0^\infty \frac{\beta}{\sqrt{\pi}} e^{-(\beta u')^2} du' = \frac{1}{2}. \tag{1.341}$$

The normalization condition is satisfied, because

$$\int_{-\infty}^\infty f_{u'} du' = \int_{-\infty}^\infty \frac{\beta}{\sqrt{\pi}} e^{-(\beta u')^2} du' = 1. \tag{1.342}$$

The mean value of u' evaluated only for velocities in the positive x-direction is

$$\overline{u'} = \int_0^\infty u' f_{u'} du' = \frac{1}{2\beta\sqrt{\pi}}. \tag{1.343}$$

The average of the thermal velocity components in the x-direction, taken over only those molecules with positive u' velocity is

$$\frac{\overline{u'}}{\bar{f}_{u'}} = \frac{\int_0^\infty u' f_{u'} du'}{\int_0^\infty f_{u'} du'} = \frac{1}{2\beta\sqrt{\pi}} 2 = \frac{1}{\beta\sqrt{\pi}} = \frac{\overline{c'}}{2}. \tag{1.344}$$

By sampling the Maxwellian distribution (1.334) one usually applies a cut-off of the extreme tail at some value, say $\beta c_x'$. Typically one may choose $\beta c_x' = 3$ or $\beta c_x' = 3.5$. It is important to know which fraction of molecules, represented by the tail of the distribution, is cut-off. This fraction is given by

$$\begin{aligned} F_x &= \int_{c_x'}^\infty f_{c'} dc' = \frac{4\beta^3}{\sqrt{\pi}} \int_{c_x'}^\infty c'^2 e^{-\beta^2 c'^2} dc' = \frac{4\beta^3}{\sqrt{\pi}} \left[\frac{\sqrt{\pi}}{4\beta^3} \left(1 - erf\left(\beta c_x'\right) + \frac{\beta c_x'}{2\beta^3} e^{-\beta^2 c_x'^2} \right] \\ &= \left(1 - erf\left(\beta c_x'\right)\right) + \frac{2}{\sqrt{\pi}} \left(\beta c_x'\right) e^{-\beta^2 c_x'^2}, \end{aligned} \tag{1.345}$$

with

$$erf(a) = \frac{2}{\sqrt{\pi}} \int_0^a e^{-x^2} dx. \tag{1.346}$$

Example 1.4
With $T = 300$ [K], $R = 0.28819 \times 10^3$[J/kg K], $c_x' = 1500$ [m/s], one obtains

$$\beta = \frac{1}{\sqrt{2RT}} = 2.40 \times 10^{-3}[\mathrm{s/m}],$$

$$\beta c'_x = 2.40 \times 10^{-3} \times 1500 = 3.6,$$

$$erf(\beta c'_x) = erf(3.6) = 0.999\ 999\ 644,$$

$$F_x = (1 - erf(\beta c'_x)) + \frac{2}{\sqrt{\pi}}(\beta c'_x)e^{-\beta^2 c'^2_x} = 9.55^* 10^{-6} \approx 10^{-5}.$$

Therefore the fraction of molecules neglected, corresponding to the cut-off tail of the Maxwellian distribution, is of about one molecule in 100,000.

Table 1.6 Important definite integrals

Let $I(n) = \int_0^\infty x^n e^{-\beta^2 x^2} dx$, with β given by Eq. (1.329). One has:
$\mathrm{I}(0) = \frac{\sqrt{\pi}}{2\beta}$, (1.347)
$\mathrm{I}(1) = \frac{1}{2\beta^2}$, (1.348)
$\mathrm{I}(2) = \frac{\sqrt{\pi}}{4\beta^3}$, (1.349)
$I(3) = \frac{1}{2\beta^4}$, (1.350)
$\mathrm{I}(4) = \frac{3\sqrt{\pi}}{8\beta^5}$, (1.351)
$\mathrm{I}(5) = \frac{1}{\beta^6}$, (1.352)
$\mathrm{I}(6) = \frac{15\sqrt{\pi}}{16\beta^7}$, (1.353)
$\mathrm{I}(7) = \frac{3}{\beta^8}$, (1.354)
$\int_a^\infty x^2 e^{-\beta^2 x^2} dx = \frac{\sqrt{\pi}}{4\beta^3}[1 - erf(\beta a)] + \frac{(\beta a)}{2\beta^3} e^{-(\beta a)^2}$, (1.355)
$\int_{-a}^\infty x^2 e^{-\beta^2 x^2} dx = \frac{\sqrt{\pi}}{4\beta^3}[1 + erf(\beta a)] - \frac{(\beta a)}{2\beta^3} e^{-(\beta a)^2}$. (1.356)

1.8 Sampling Probability Distributions

1.8.1 Generalities

Let x be a random deviate in the open interval (0, 1) defined by

$$\begin{aligned} p(x)dx &= dx, \qquad 0 < x < 1, \\ &= 0, \qquad x < 0, \quad x > 1. \end{aligned} \tag{1.357}$$

The probability of having a value of the random deviate in the interval $(x, x + dx)$ is proportional to the width dx of the interval. Hence $p(x) = 1$ and it holds

$$\int_{-\infty}^{+\infty} p(x)dx = \int_0^1 p(x)dx = \int_0^1 dx = 1, \tag{1.358}$$

Let $y(x)$ be a function of the random deviate x. The transformation law of probabilities states that the probability distribution of the function $y(x)$ is related to the probability $p(x)$ of the random deviate by

$$|p(y)dy| = |p(x)dx|, \tag{1.359}$$

hence

$$p(y) = p(x)\left|\frac{dx}{dy}\right|. \tag{1.360}$$

Example 1.5
Let $y = -\ln x$, hence $x = e^{-y}$, $dx = -e^{-y}dy$. From Eq. (1.360) one has

$$p(y) = p(x)\left|\frac{dx}{dy}\right| = p(x)e^{-y} = e^{-y}. \tag{1.361}$$

Integration of the probability distribution $y(x)$ and application of the definition of mean value of $y(x)$ yield

$$\int_0^{\infty} p(y)dy = \int_0^{\infty} e^{-y}dy = 1, \tag{1.362}$$

$$\bar{y} = \int_0^{\infty} yp(y)dy = \int_0^{\infty} ye^{-y}dy = 1. \tag{1.363}$$

The distribution (1.361) is therefore an exponential positive random deviate of unit mean.

Example 1.6
Let $y = -\ln x/\lambda$, hence $x = e^{-\lambda y}$, $dx = -\lambda e^{-\lambda y}dy$. From Eq. (1.360) one has

$$p(y) = p(x)\left|\frac{dx}{dy}\right| = \lambda e^{-\lambda y}. \tag{1.364}$$

Equation (1.364), with appropriate parameter λ, describes the law of decay of radioactive nuclei.

1.8.2 *Constructing Normal (Gaussian) Deviates with the Box-Muller Method*

Let now x_1, x_2, x_3. …x_N be a set of random deviates defined in the open interval (0, 1). Because $p(x_1) = p(x_2) = p(x_3) = \ldots p(x_N) = 1$, for the cumulative probability distribution it also holds $p(x_1, x_2, x_3 \ldots x_N) = 1$. Let $y(x_1)$, $y(x_2)$, $y(x_3) \ldots y(x_N)$ be a set of functions of the given random deviates. For the cumulative probability distribution function $p(y_1, y_2, y_3 \ldots y_N)$ holds a generalization of the transformation law of probabilities given by Eq. (1.359), namely

$$p(y_1,y_2,y_3\ldots y_N)dy_1dy_2dy_3\ldots dy_N = p(x_1,x_2,x_3\ldots x_N)\left|\frac{\partial(x_1,x_2,x_3\ldots x_N)}{\partial(y_1,y_2,y_3\ldots y_N)}\right|dy_1dy_2dy_3\ldots dy_N. \tag{1.365}$$

Let $N = 2$ and define two functions of two random deviates x_1, x_2:

$$y_1 = \sqrt{-2\ln x_1}\cos(2\pi x_2), \tag{1.366}$$

$$y_2 = \sqrt{-2\ln x_1}\sin(2\pi x_2). \tag{1.367}$$

From Eqs. (1.366) and (1.367) one obtains

$$x_1 = e^{-\frac{y_1^2+y_2^2}{2}}, \tag{1.368}$$

$$x_2 = \frac{1}{2\pi}\,arctg\left(\frac{y_2}{y_1}\right). \tag{1.369}$$

The Jacobian determinant of the transformation law of probabilities given by Eqs. (1.368) and (1.369) becomes

$$\frac{\partial(x_1,x_2)}{\partial(y_1,y_2)} = \begin{vmatrix} \frac{\partial x_1}{\partial y_1} & \frac{\partial x_1}{\partial y_2} \\ \frac{\partial x_2}{\partial y_1} & \frac{\partial x_2}{\partial y_2} \end{vmatrix} = \begin{vmatrix} -y_1 e^{-\frac{y_1^2+y_2^2}{2}} & -y_2\, e^{-\frac{y_1^2+y_2^2}{2}} \\ \frac{1}{2\pi}\frac{1}{1+\left(\frac{y_2}{y_1}\right)^2}\left(-\frac{y_2}{y_1^2}\right) & \frac{1}{2\pi}\frac{1}{1+\left(\frac{y_2}{y_1}\right)^2}\left(\frac{1}{y_1}\right) \end{vmatrix} =$$

$$= e^{-\frac{y_1^2+y_2^2}{2}}\frac{1}{2\pi}\frac{1}{1+\left(\frac{y_2}{y_1}\right)^2}\begin{vmatrix} -y_1 & -y_2 \\ -\frac{y_2}{y_1^2} & \frac{1}{y_1} \end{vmatrix}$$

$$= e^{-\frac{y_1^2+y_2^2}{2}}\frac{1}{2\pi}\frac{1}{1+\left(\frac{y_2}{y_1}\right)^2}\left[-1-\left(\frac{y_2}{y_1}\right)^2\right] =$$

$$= -\left(\frac{1}{\sqrt{2\pi}}e^{-\frac{y_1^2}{2}}\right)\left(\frac{1}{\sqrt{2\pi}}e^{-\frac{y_2^2}{2}}\right). \tag{1.370}$$

Because Eq. (1.370) is the product of a function of y_1 alone and of y_2 alone, these two functions y_1 and y_2 are independent. Their probability distributions are, from Eq. (1.365)

$$p(y_1,y_2)dy_1dy_2 = p(y_1)p(y_2)dy_1dy_2 = \left(\frac{1}{\sqrt{2\pi}}e^{-\frac{y_1^2}{2}}\right)\left(\frac{1}{\sqrt{2\pi}}e^{-\frac{y_2^2}{2}}\right)dy_1dy_2. \tag{1.371}$$

Equation (1.371) shows that the functions y_1 and y_2 have a Gaussian distribution with mean value $\mu = 0$ and variance $\sigma^2 = 1$.

In case one wishes to generate a Gaussian distribution

$$p(y) = \frac{1}{\sqrt{2\pi}\sigma}e^{-\frac{(y-\mu)^2}{2\sigma^2}}, \tag{1.372}$$

with mean value μ different from 0 and variance σ^2, one sets $z = (y-\mu)/\sigma$ and one applies the above method, referred to as Box-Muller method.

1.8.3 Inverse Cumulative Method

Let $p(y)$ be an arbitrary probability distribution of the function $y(x)$ with

$$\int_0^\infty p(y)dy = 1. \tag{1.373}$$

Equation (1.360), with $p(x) = 1$ for an uniform deviate in the open interval (0, 1), yields

$$\frac{dx}{dy} = p(y). \tag{1.374}$$

Indefinite integration of Eq. (1.374) gives

$$x = \int_0^y p(y')dy' = F(y), \tag{1.375}$$

where $F(y)$ is a primitive function of $p(y)$. The integral in Eq. (1.375) is called the cumulative distribution function. To the maximum value of y, say $y_{\max}$, corresponds

$$x_{\max} = \int_0^{y_{\max}} p(y')dy' = 1. \tag{1.376}$$

If the function $F(y)$ can be inverted we may write

$$y = F^{-1}(x). \tag{1.377}$$

Choosing a uniform deviate x_i and applying Eq. (1.377) yield a value $y_i = F^{-1}(x_i)$ with probability distribution $p(y)$. Figure 1.15 exemplifies the procedure.

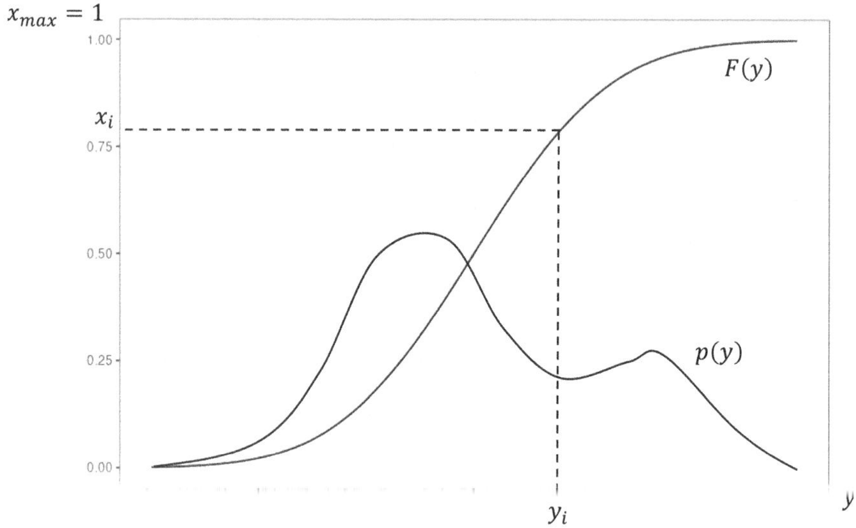

Fig. 1.15 Schematic representation of the "Inverse Cumulative Method" for sampling a probability distribution

The geometrical interpretation of the procedure is straightforward. The integral

$$F(y_i) = \int_0^{y_i} p(y')dy' \tag{1.378}$$

represents the area below the plot of $F(y)$ to the left of point y_i.

The inverse cumulative method is very simple, but can be applied only if one can compute analytically the integral (1.375) and also the inverse function (1.377). In case this is not possible, one must apply, in the general case, the acceptance-rejection method explained below.

As an example of function for which the inverse cumulative method cannot be applied we consider the distribution function $f_{u'}du' = \frac{\beta}{\sqrt{\pi}} e^{-(\beta u')^2} du'$ of the thermal velocity component u'. In this case indefinite integration yields

$$\int_0^{u'} f_{u'}du' = \int_0^{u'} \frac{\beta}{\sqrt{\pi}} e^{-(\beta u')^2} du' = \frac{1}{2} \frac{2}{\sqrt{\pi}} \int_0^{u'} e^{-(\beta u')^2} d(\beta u') = \frac{1}{2} erf(u'). \tag{1.379}$$

The error function cannot be inverted analytically, hence the inverse cumulative method cannot be applied.

1.8.4 Acceptance-Rejection Technique

Let $p(y)$ be an arbitrary probability distribution of the function $y(x)$, but assume that analytical integration of $p(y)$ is either not possible or very cumbersome, so that the inverse cumulative method for sampling $p(y)$ cannot be applied. In this case one can apply the acceptance-rejection method. We must choose a function $f(y)$, called "comparison function", such that in the full definition domain it holds $f(y) > p(y)$, and such that it is integrable and its primitive function $F(y)$ can be inverted. Notice that, unlike in the case of the inverse cumulative method, now $F(y)$ is the primitive of the comparison function and not of the function $p(y)$. Thus we admit to know

$$x = \int_0^{y} p(y')dy' = F(y), \tag{1.380}$$

with

$$x_{\max} = \int_0^{y_{\max}} p(y')dy'. \tag{1.381}$$

The value $x_{\max}$ could be normalized to 1, but this is not compulsory.

With reference to Fig. 1.16, let us now choose a random deviate in the interval $(0, x_{\max})$ and correspondingly determine the point $y_i = F^{-1}(x_i)$. A second uniform

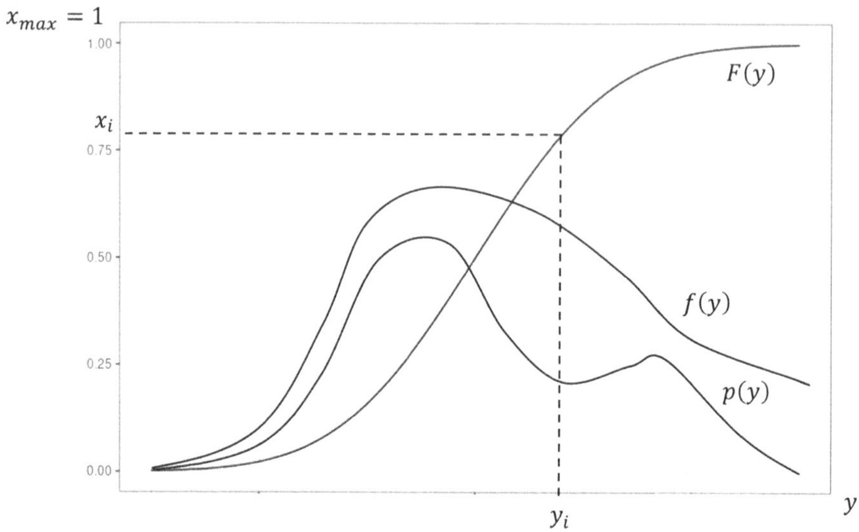

Fig. 1.16 Schematic representation of the "Acceptance-Rejection Method" for sampling a probability distribution. $p(y)$ = Given probability distribution; $F(y)$ = Cumulative Distribution function; $f(y)$ = Comparison function

deviate, say x_2, is now chosen in the interval $[0 - f(y_i)]$. If x_2 is smaller than $p(y_i)$, the value y_i is accepted. Otherwise, if x_2 is larger than $p(y_i)$, therefore is in the interval $[p(y_i), f(y_i)]$ the value y_i is rejected.

The efficiency of this acceptance-rejection method requires that the comparison function $f(y)$ be only slightly larger than $p(y)$ for every y. In fact the efficiency of the method can be measured by the ratio $f_{acc} = A[p(y)]/A[f(y)]$, where $A[p(y)]$ and $A[f(y)]$ are the areas below the plots of $p(y)$ and $f(y)$, respectively.

1.8.5 Application of the Acceptance-Rejection Method: Relaxation to a Maxwellian Distribution

Let us consider the function

$$F(c', \tau) = \frac{4}{\sqrt{\pi}\alpha^{3/2}} \left[\beta\alpha_1 (\beta c')^2 + \alpha_2 (\beta c')^4 \right] e^{-\frac{(\beta c')^2}{\alpha}}, \tag{1.382}$$

with $\beta = \frac{1}{\sqrt{2RT}} = \sqrt{\frac{m}{2KT}}$ (Eq. 1.329) and

$$\alpha = \alpha(\tau) = 1 - \frac{2}{5} e^{-\tau}, \tag{1.383}$$

$$\alpha_1 = \frac{5\alpha - 3}{2\alpha}, \tag{1.384}$$

$$\alpha_2 = \frac{1-\alpha}{\alpha^2}. \tag{1.385}$$

The parameter τ in the above equations is a dimensionless time parameter. One could set, for instance,$\tau = t/t_c$, where t_c is the mean collision time between molecules. For $\tau = 0$, the above Eqs. (1.383) through (1.385) give $\alpha = 3/5$, $\alpha_1 = 0$, $\alpha_2 = 10/9$, hence

$$F(c', 0) = \frac{4}{\sqrt{\pi}\left(\frac{3}{5}\right)^{3/2}} \frac{10}{9} (\beta c')^4 e^{-\frac{5(\beta c')^2}{3}}. \tag{1.386}$$

As shown in Fig. 1.17, plot (a), Eq. (1.386) represents a probability distribution of molecules with thermal velocity c', which differs substantially from a Maxwellian distribution shown in plot (b).

At the limit for $\tau = \infty$ the above Eqs. (1.383) through (1.385) give $\alpha = 1$, $\alpha_1 = 1$, $\alpha_2 = 0$, hence

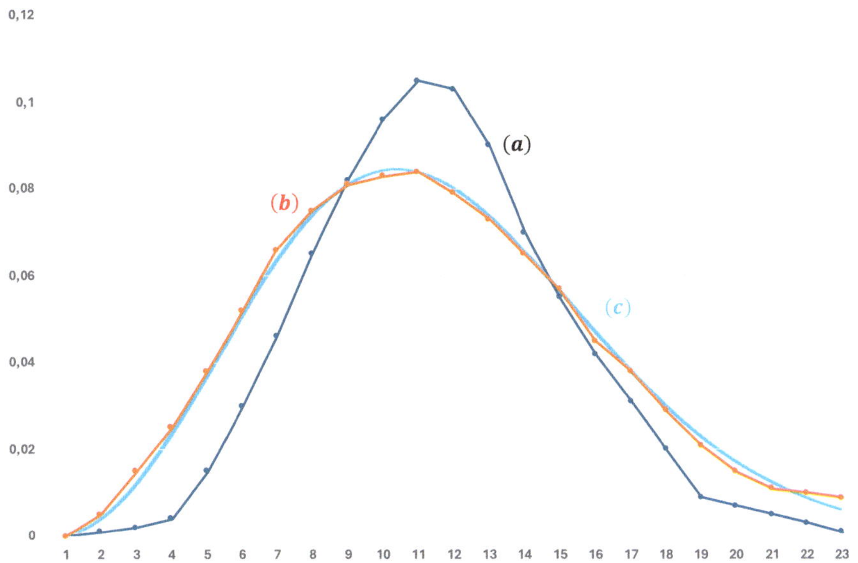

Fig. 1.17 Comparison between initial distribution of molecular velocities and Maxwellian distribution. Plot a (black): initial distribution; Plot b (red): computed Maxwellian; Plot c (Blue): exact Maxwellian

$$F(c',0) = \frac{4}{\sqrt{\pi}}\beta^3 c'^2 e^{-(\beta c')^2} = f_{c'}. \tag{1.387}$$

Equation (1.387) is the well-known Maxwellian distribution of the thermal velocity c'.

A way of testing a computational code describing molecular collisions is to sample the distribution (1.382) for an arbitrary dimensionless time parameter τ and then check that molecular collisions relax the original distribution to a Maxwellian one. Plot (c) in Fig. 1.17 shows the result of the numerical relaxation computed with a test code. In the following we therefore show, as an example, the application of the acceptance-rejection method to sampling distribution (1.382).

Example 1.7 Sampling Distribution (1.382)
Let us consider the domain ($0 < \beta c' < x_c = 3$) (subscript c denotes a cut-off value) and choose a comparison function

$$Y(x) = \frac{2}{\sqrt{3}} \sin\left(\frac{\pi x}{3}\right), \tag{1.388}$$

which satisfies the condition $F(c', \tau) < Y(x)$. The indefinite integral of the comparison function,

$$F(x) = \int_0^x Y(x')dx' = \frac{2\sqrt{3}}{\pi}\left[1 - \cos\left(\frac{\pi x}{3}\right)\right] \tag{1.389}$$

is defined in the interval $[0 \le F(x) \le F(x_c) = 2.2053]$. The inverse of the function $F(x)$ is

$$x = \frac{3}{\pi}\arccos\left[1 - \frac{\pi}{2\sqrt{3}}F(x)\right]. \tag{1.390}$$

Let R_1 be a uniform deviate in the interval $[0, F(x_c) = 2.2053]$. Equation (1.390) allows to find

$$x_1 = \frac{3}{\pi}\arccos\left[1 - \frac{\pi}{2\sqrt{3}}R_1\right]. \tag{1.391}$$

We take now a second uniform deviate R_2 in the open interval (0, 1). The value x_1 is accepted if

$$R_2 \le \frac{F(x_1, \tau)}{Y(x_1)}, \tag{1.392}$$

otherwise it is rejected. The acceptance rate will be

$$f_{acc} = \frac{1}{Z(x_c)} \int_0^{x_c} F(x,\tau)dx = 1/2.2053 = 0.453. \tag{1.393}$$

Example 1.8 Sampling a Normal Distribution for a Thermal Velocity Component

We show the application of the acceptance-rejection method for sampling the normal distribution for a thermal velocity component

$$f_{u'} = \frac{\beta}{\sqrt{\pi}} e^{-(\beta u')^2}. \tag{1.394}$$

Let us consider the domain $(0 < \beta u^{'} < x_c = 2.5)$ and choose a comparison function

$$Y(x) = \cos\left(\frac{\pi x}{5}\right), \tag{1.395}$$

which satisfies the condition $f_{u'} < Y(x)$. The indefinite integral of the comparison function,

$$F(x) = \int_0^x Y(x')dx' = \frac{5}{\pi}\left[\sin\left(\frac{\pi x}{5}\right)\right] \tag{1.396}$$

is defined in the interval $[0 \leq F(x) \leq F(x_c) = 1.59155]$. The inverse of the function $F(x)$ is

$$x = \frac{5}{\pi} \text{ar}\sin\left[\frac{\pi}{5} F(x)\right]. \tag{1.397}$$

Let R_1 be a uniform deviate in the interval $[0, F(x_c)]$. Equation (1.397) allows to find

$$x_1 = \frac{5}{\pi} \text{ar}\sin\left[\frac{\pi}{4} R_1\right]. \tag{1.398}$$

We take now a second uniform deviate R_2 in the open interval (0, 1). The value x_1 is accepted if

$$R_2 \leq \frac{f_{x_1}}{Y(x_1)}, \tag{1.399}$$

otherwise it is rejected. The acceptance rate will be

$$f_{acc} = \frac{1}{F(x_c)} \int_0^{x_c} f_{x_1} dx = \frac{1}{2F(x_c)} = 1/3.1831 = 0.3142. \tag{1.400}$$

1.8.6 Sampling a Gamma Distribution

Let us consider the Gamma distribution

$$p_a(x)dx = \frac{1}{\Gamma(a)} x^{a-1} e^{-x}, \tag{1.401}$$

where

$$\Gamma(a) = \int_0^\infty x^{a-1} e^{-x} dx \tag{1.402}$$

is the Gamma function, introduced as to satisfy the normalization condition

$$\int_0^\infty p_a(x)dx = \frac{1}{\Gamma(a)} \int_0^\infty x^{a-1} e^{-x} dx = 1. \tag{1.403}$$

For large values of the parameter a, the distribution (1.401) has a typical "bell-shaped" form with a peak at $x = a$. We consider the comparison function

$$Y(x) = \frac{K}{\pi} \frac{1}{1 + x^2}, \tag{1.404}$$

where K is a constant to be chosen as to satisfy, for every x, the condition $Y(x) > p_a(x)$. Apart from the factor K, Eq. (1.404) represents a so-called Lorentzian distribution which is useful as a comparison function in case of bell-shaped distributions. Indefinite integration of $Y(x)$ yields

$$F(x) = \int_0^x Y(x')dx' = \frac{K}{\pi} \int_0^x \frac{1}{1 + x'^2} dx' = \frac{K}{\pi} artg(x). \tag{1.405}$$

The inverse of the function $F(x)$ is

$$x = tg\left[\frac{\pi F(x)}{K}\right]. \tag{1.406}$$

Let x_c be a cut-off value and let R_1 be a uniform deviate in the interval $[0, F(x_c)]$. Equation (1.406) allows to find

$$x_1 = tg\left[\frac{\pi R_1}{K}\right]. \tag{1.407}$$

We take now a second uniform deviate R_2 in the open interval (0, 1). The value x_1 is accepted if

$$R_2 \leq \frac{p_a(x_1)}{Y(x_1)}, \tag{1.408}$$

otherwise it is rejected. The acceptance rate is

$$f_{acc} = \frac{1}{F(x_c)} \int_0^{x_c} p_a(x_1)dx \approx \frac{1}{F(x_c)}. \tag{1.409}$$

1.9 Simulation of Droplets in a Cloud

1.9.1 Generalities

The water content in a cloud, w_L, usually increases with the height in the cloud and ranges from about 0.2 to $1 - 2$ [g/cm^3]. Higher values up to about $w_L = 5$ [g/cm^3] can also be reached. The number concentration of droplets is about $n_d = 1 - 100$ [cm^{-3}]. Whilst the size of aerosol particles is around 0.1–100 μm (the largest figures holding for aggregates of aerosol particles), the diameter of water droplets may range from 80 to 1000 μm. According to their size, droplets are usually given different names, as shown in the following Table 1.7, which also gives the settling velocities and the corresponding Reynolds numbers.

Example 1.9
Choosing a drizzle droplet with diameter $d_p = 200$ μm $= 2 \times 10^{-4}$ m, $W = 2$ m/s, $\nu = 1.42 \times 10^{-5}$ [m^2/s] one obtains $\mathrm{Re} = \frac{Wd_p}{\nu} = 28$.

Aerosol particles may act as condensation nuclei for very small droplets, which are referred to as "haze particles". Continental clouds are characterised by a narrow spectrum of droplets of small size, hence high droplet concentration n_d. Conversely, maritime clouds are characterised by a broad spectrum of droplets of larger size, hence lower droplet concentration. The spectrum of droplets (number of droplets per unit volume, versus droplets size) may present a duplet of maxima, referred to as a "bimodal distribution". In the following, however, we will refer to a single-maximum distribution, which may be approximated well by a Gamma-distribution, with given mean value of droplets size.

The mean distance of droplets D_m in a cloud is about 100 times the droplets diameter: $D_m/d_p \approx 100$. Clustering of droplets, mainly in pairs, is also observed.

Table 1.7 Characteristics of aerosol and water droplets

	Aerosol	Mist	Drizzle	Raindrops
Diameter d_p [μm]	$0.1 - 100$	$80 - 100$	$100 - 500$	$500 - 1000$
Settling velocity W [m/s]	$10^{-4} - 0.3$	$0.3 - 1$	$1 - 3$	$3 - 20$
$\mathrm{Re} = \frac{Wd}{\nu}$	$10^{-5} - 1$	$1 - 10$	$10 - 100$	$100 - 10{,}000$

Small droplets with $d_p \leq 280$ μm are generally of spherical shape and may be treated as rigid spheres in numerical simulations. Droplets of medium size, with 280 μm $\leq d_p \leq 1000$ μm, are generally oblate spheroids with inner flows. Droplets of large size, with 1000 μm $\leq d_p$, are oblate spheroids with flat base. Very large droplets generally break up.

1.9.2 The Khrgian-Mazin Distribution

Let us assume we want to simulate a water content w_L [kg/m^3] in a cloud by means of a suitable number N_p of droplets of mean radius $\overline{r}_p$, at a given temperature with corresponding water density ρ_p. The number N_p of droplets to be simulated will be

$$N_p = \frac{w_L}{\frac{4}{3}\pi r_p^3 \rho_p}. \tag{1.410}$$

A suitable droplets distribution is obtained by sampling the Khrgian-Mazin distribution

$$n(r) = Ar^2 e^{-Br}. \tag{1.411}$$

The parameters A and B in Eq. (1.411) must be determined imposing the conditions that the distribution gives the requested number N_p of droplets with mean radius $\overline{r}_p$. The first condition implies

$$\begin{aligned} N_p &= \int_0^\infty n(r)dr = \int_0^\infty Ar^2 e^{-Br} dr = \frac{A}{B^3} \int_0^\infty (Br)^2 e^{-Br} d(Br) \\ &= \frac{A}{B^3} \int_0^\infty z^2 e^{-z} dz = \frac{A\Gamma(3)}{B^3} = \frac{2A}{B^3}, \end{aligned} \tag{1.412}$$

having used $z = \mathrm{Br}$ and $\Gamma(3) = \int_0^\infty z^2 e^{-z} dz = 2$.

The second condition implies

$$\begin{aligned} \overline{r}_p &= \frac{1}{N_p} \int_0^\infty r\, n(r) dr = \frac{1}{N_p} \int_0^\infty Ar^3 e^{-Br} dr \\ &= \frac{A}{N_p B^4} \int_0^\infty (Br)^3 e^{-Br} d(Br) = \frac{A}{N_p B^4} \int_0^\infty z^3 e^{-z} dz = \frac{A\Gamma(4)}{N_p B^4} \\ &= \frac{6A}{N_p B^4} = \frac{3}{B}, \end{aligned} \tag{1.413}$$

having used $\Gamma(4) = \int_0^\infty z^3 e^{-z} dz = 3! = 6$.

From the above conditions one derives

$$B = \frac{3}{\bar{r}_p}, \tag{1.414}$$

$$A = \frac{N_p B^3}{2}. \tag{1.415}$$

The normalized probability distribution to be sampled therefore becomes

$$p(r) = \frac{1}{N_p} n(r) = \frac{1}{N_p} A r^2 e^{-Br}, \tag{1.416}$$

which satisfies the normalization condition

$$\begin{aligned} \int_0^\infty p(r) dr &= \frac{1}{N_p} \int_0^\infty A r^2 e^{-Br} dr = \frac{A}{N_p B^3} \int_0^\infty (Br)^2 e^{-Br} d(Br) \\ &= \frac{A}{N_p B^3} \int_0^\infty z^2 e^{-z} dz = \frac{A\Gamma(3)}{N_p B^3} = \frac{2A}{N_p B^3} = \frac{N}{N} = 1. \end{aligned} \tag{1.417}$$

Example 1.10
Let $w_L = 1.25 \times 10^{-3}[\text{kg/m}^3]$, $\bar{r}_p = 100\ \mu\text{m} = 10^{-4}$ m, $\rho_p = 10^3[\text{kg/m}^3]$. One obtains from Eq. (1.410) $N_p = \frac{w_L}{\frac{4}{3}\pi r_p^3 \rho_p} = 2.98 \times 10^5 \approx 3 \times 10^5[\text{m}^{-3}]$. This number of droplets can be sampled using Eq. (1.416).

1.9.3 Equation of Motion for Droplets

Between binary collisions, sample molecules have ballistic trajectories. Their positions and speeds are updated according to the laws

$$x_i^{n+1} = x_i^n + v_i^n \Delta t + \frac{1}{2} g_i \Delta t^2, \tag{1.418}$$

$$v_i^{n+1} = v_i^n + g_i \Delta t. \tag{1.419}$$

In the above equations n is a time discretization index and g is the acceleration due to gravity. Averaging the updated velocities for all molecules in a cell over time, one obtains the mean value of the fluid velocity in the cell, which we will refer to as

V_f. Water droplets are subjected to entrainment, which can be simulated at different levels of accuracy.

I. First approximation.

The entrainment force acting upon a particle may be written as

$$\left|\vec{F}_p\right| = A_p c_D \frac{1}{2}\rho_f V_{Sl}^2. \tag{1.420}$$

In Eq. (1.420) A_p is the cross section of a particle normal to the direction of motion, c_D is a drag coefficient and $V_{Sl} = u_f - u_p$ is the slip velocity (difference between fluid and particle velocity).

An equation of motion for droplets of spherical shape, with $A_p = \pi r_p^2$, and mass $m_p = (4/3)\pi r_p^3 \rho_p$, may therefore be written as

$$m_p \frac{d\vec{u}_p}{dt} = \pi r_p^2 c_D \frac{1}{2}\rho_f \left|\vec{u}_f - \vec{u}_p\right| \left(\vec{u}_f - \vec{u}_p\right), \tag{1.421}$$

or

$$\frac{d\vec{u}_p}{dt} = \frac{1}{\tau_p}\left(\vec{u}_f - \vec{u}_p\right), \tag{1.422}$$

with the definition of the "particle relaxation time"

$$\tau_p = \frac{m_p}{\pi r_p^2 c_D \frac{1}{2}\rho_f \left|\vec{u}_f - \vec{u}_p\right|} = \frac{8}{3}\frac{r_p \rho_p}{c_D \rho_f \left|\vec{u}_f - \vec{u}_{pp}\right|}.. \tag{1.423}$$

The particle relaxation time is representative of the time necessary for a particle, when it enters a domain with different fluid velocity, for instance a new cell, to adjust its speed to the new flow conditions.

II. Second Approximation.

A more general equation of motion for the droplets entrained by the fluid may be written for a particle of volume V_p

$$\underbrace{V_p \rho_p \frac{d\vec{u}_p}{dt}}_{\text{Term I}} = \underbrace{6\pi\tilde{\mu}_f r_p \left(\vec{u}_f - \vec{v}_p\right)}_{\text{Term II}} + \underbrace{V_p \rho_f \frac{d\vec{u}_f}{dt}}_{\text{Term III}} + \underbrace{\chi V_p \rho_f \left(\frac{d\vec{u}_f}{dt} - \frac{d\vec{u}_p}{dt}\right)}_{\text{Term IV}} +$$

$$+ 6 r_p^2 \sqrt{\pi \rho_f \tilde{\mu}_f} \int_0^t \frac{\frac{d\vec{u}_f}{dt'} - \frac{d\vec{u}_p}{dt'}}{\sqrt{t - t'}} dt' + \left(m_p - m_f\right)\vec{g}. \tag{1.424}$$

Term V Term VI

Equation (1.424), originally derived for a slow motion of particles in a stationary fluid, at low Reynolds number, can be applied to turbulent flow regimes replacing the molecular viscosity with a turbulent viscosity

$$\widetilde{\mu}_f = c_\mu \rho_f \frac{k^2}{\varepsilon}, \tag{1.425}$$

where $c_\mu = 0.09$, k is the turbulent kinetic energy and ε the dissipation rate of turbulent kinetic energy. The physical meaning of the terms at the right-hand side of Eq. (1.424) is as follows:

Term II is a drag viscous force, as a generalization of Stoke's law;

Term III accounts for the pressure gradient around the body, due to fluid acceleration;

Term IV is a virtual mass force. We assume a virtual mass coefficient $\chi = ½$ for droplets of spherical shape;

Term V is the Basset or transient drag force;

Term VI is the force due to gravity acceleration.

In the following we neglect terms V and VI, which are much smaller than the other ones in turbulent flows. Terms I to IV of Eq. (1.424) can be rewritten as

$$\left(V_p\rho_p + \frac{1}{2}V_p\rho_f\right)\frac{d\vec{u}_p}{dt} = 6\pi\widetilde{\mu}_f r_p\left(\vec{u}_f - \vec{v}_p\right) + (1+\chi)V_p\rho_f\frac{d\vec{u}_f}{dt}. \tag{1.426}$$

Using the virtual mass coefficient $\chi = ½$ and $d_p = 2r_p$, Eq. (1.426) becomes

$$\frac{d\vec{u}_p}{dt} + a\vec{v}_p = a\vec{u}_f + b\frac{d\vec{u}_f}{dt}, \tag{1.427}$$

with

$$a = \frac{6\pi\widetilde{\mu}_f r_p}{V_p\rho_p + \frac{1}{2}V_p\rho_f} = \frac{36\widetilde{\mu}_f}{d_p^2(2\rho_p + \rho_f)}, \tag{1.427}$$

$$b = \frac{3\rho_f}{2\rho_p + \rho_f}. \tag{1.429}$$

In case the density of the droplets is much larger than the density of the fluid, the coefficient "a", which is the inverse of the "particle relaxation time", denoted now t_p, simplifies to

$$a = \frac{18\widetilde{\mu}_f}{d_p^2 \rho_p} = \frac{1}{t_p}. \tag{1.430}$$

When the droplets move in a turbulent flow field with eddies, a characteristic time of the eddies is defined as the ratio of the typical eddy size ℓ_e (the macroscopic length of turbulence), to the characteristic velocity U_e of the eddy

$$t_e = \frac{\ell_e}{U_e}. \tag{1.431}$$

The ratio of the characteristic times of the particles to that of the eddies gives, by definition, the Stokes number of the particles

$$St = \frac{t_p}{t_e} = \frac{\rho_p d_p^2}{18\widetilde{\mu}_f} \frac{U_e}{\ell_e}. \tag{1.432}$$

The Stokes number of the particles thus represents the time required by a particle to adjust to a variation of the flow characteristic, normalized by the characteristic eddy time.

We define the Reynolds number of the droplets, with reference to the slip velocity, by

$$\mathrm{Re}_p = \frac{V_{Sl} d_p}{\nu_f}. \tag{1.433}$$

The interaction of a carrier fluid with the carried droplets is twofold. From one side one has, in general, a dispersion of droplets due to the fluid turbulence. However, under certain circumstances there may be a concentration of droplets in some regions within a fluid. From the other side one has an impact of the presence of droplets upon the fluid turbulence, which depends on the Stokes number of the droplets.

From the point of view of the impact of droplets or particles upon the fluid turbulence, it is practical to distinguish the particles in "small" and "large".

Small particles are characterized by: $t_p \ll t_e$; $St \ll 1$; $\mathrm{Re}_p < 110$;

Large particles are characterized by: $t_p \gg t_e$; $St \gg 1$; $\mathrm{Re}_p \geq 400$.

One can say that, in general, small particles behave as large macromolecules and oscillate following, with some lag, the fluid fluctuations. They do not shed eddies but tend to act as a damping factor for the turbulent fluctuations of the fluid velocity components. As a consequence the kinetic energy of turbulence is decreased. Large particles, because of their inertia, tend to follow a trajectory independent of the fluid fluctuations, are deflected to a minor degree by the large eddies, but they also shed small eddies thus increasing the effective kinetic viscosity of the fluid. Obviously, particles with St ≈ 1 have intermediate characteristics and behaviour.

1.9.4 Impinging of Droplets upon an Hydrometeor

Droplets are moved in a simulated atmosphere domain according to Eq. (1.424), where usually the fifth and sixth terms are neglected. When droplets impinge upon the surface of an hydrometeor, mass and enthalpy of the droplets are transferred to the hydrometeor. Enthalpy is taken into account as source of the diffusion equation solved to compute the temperature distribution in the hydrometeor. Upon impingement, the mass of the droplet spreads upon the surface of the hydrometeor. To describe the flattening of the droplet, a so-called "spreading factor "S" is defined as the ratio of the radius of the flattened droplet after impingement, to the radius before impingement (Sadhal et al., 1997). The spreading factor S depends strongly on the impact speed of the droplet and on the surface temperature of the target. S increases with impact velocity and with temperature of the target. The dependence of S on the temperature of the droplet and on its radius is small and in most cases negligible.

The following Table 1.8 gives the dependence of the splashing factor on the impact speed and temperature of the target.

When a droplet freezes on the icy surface of a hydrometeor, the density of the ice accretion is far below the theoretical density of ice. The density of the ice accretion forming on an hydrometeor can be approximated by introducing an empirical Macklin parameter defined by

$$R_M = -\frac{r_p V}{T_S}, \tag{1.434}$$

where $r_p[\mu m]$ is the radius of the droplet in micrometres, V [m/s] is the impact velocity and $T_S[C]$ is the surface temperature of the hydrometeor in the impact point, given in Celsius. Because T_S is negative, the Macklin parameter is positive. An empirical formula for the density of the ice accretion, applicable only in the range $1.5 \leq R_M \leq 7$, is

$$\rho_{ice} = 280 \cdot R_M^{0.6}. \tag{1.435}$$

Table 1.8 Dependence of the splashing factor S on the impact speed and temperature of the target

		Impact speed [m/s]			
		4.3	9.2	18	26
Surface temperature of ice [K]	268.15	1.7	2.2	3.4	5.2
	258.15	1.6	2.0	2.9	3.9
	258.15	1.4	1.7	2.3	3.2
	258.15	1.3	1.6	2.1	2.7

References

Bird, G. A. (1994). *Molecular gas dynamics and the direct simulation of gas flow*. Clarendon Press.
Cercignani, C., & Lampis, M. (1974). *Rarefied gas dynamics (Ed. Karamcheti)*. Academic Press.
Euken, A. (1922). Grundriss der physikalischen chemie. Leipzig.
Hinshelwood, C. N. (1940). *The kinetic of chemical change*. Clarendon Press.
Larsen, P. S., & Borgnakke, C. (1974). In M. Becker & M. Fiebig (Eds.), *Rarefied gas dynamic*. DFVLR Press.
Lord, R. G. (1991). *Rarefied gas dynamics (Ed. Beylich A. E.)*. VCH Verlagsgesellschaft mBH.
Marsaglia, G. (1972). Choosing a point from the surface of a sphere. *Annals of Mathematical Statistics, 43*, 645–646.
Sadhal, S. S., Ayyaswamy, P. S., & Chung, J. N. (1997). *Transport phenomena with drops and bubbles*. Springer-Verlag.
Vincenti, W. G., & Kruger, C. H. (1965). *Introduction to physical gas dynamics*. Krieger Publishing Company.

Chapter 2
Examples of Applications

Contents

This chapter is dedicated to classical problems of thermal instability like the Bénard-Rayleigh problem occurring in the Earth atmosphere, which can be simulated in the laboratory in a square cavity filled with a fluid and bounded by parallel plates. The problem of the thermal instability is treated at first from the classical viewpoint based on the linear convection theory, deriving from the conservation equations for mass, momentum, and energy linear dimensionless equations which can then be solved in case of appropriate boundary conditions (mainly free bounding surfaces or rigid bounding surfaces). The analytical background presented relies upon the treatment given in Koschmieder (1993) and Chandrasekhar (1961). Applications of the molecular dynamics to the Bénard-Rayleigh are preceded by information about the computer program SQUARE CAVITY built for the purpose. Applications of this code are then made for the classical Bénard-Rayleigh problem with rigid surfaces and for a problem of concentric surfaces in a 2D representation. A section is then

Text by M. Bottoni

M. Bottoni et al., *Molecular Dynamics - Theory and Applications*, Mechanical Engineering Series, https://doi.org/10.1007/978-3-031-84567-3_2

dedicated to a two-dimensional flow around a circular obstacle at high Reynolds number. The chapter closes with a simulation of thermal creep with a fluid initially at rest, but bounded by surfaces displaying a temperature gradient. The chapter does not present any novelty from the theoretical viewpoint but it intends to offer young scientist a help by discussing typical classical examples.

2.1 From Atmospheric Physics to Tiny Experiments of Molecular Dynamics

2.1.1 Thermal Instability in the Bénard-Rayleigh Problem

2.1.1.1 The Bénard-Rayleigh Experiments

In this section we review the linear theory which enables a simulation of thermal instabilities in the Bénard-Rayleigh problem, i.e. in a thin fluid layer heated from below. The first analysis of the problem was made by Bénard, who gave a first description of the phenomena involved in his two famous articles Bénard (1900) and Bénard (1901). While Bénard gave basically a phenomenological interpretation of the insurgence of thermal instabilities, observing and describing for the first time the formation of hexagonal convection cells, Lord Rayleigh gave in the article Rayleigh (1916), the first analytical description of the basic fluid-dynamics involved.

We first recall some of the observations made by Bénard and remark the analogy between Bénard cells and some cloud patterns forming in the atmosphere. Then we summarize the linear theory which allows describing the thermal instabilities in a fluid heated from below.

Bénard experiments were first made in a device consisting of a thin plate of iron or brass, covered with a very thin layer, 1 mm or less in thickness, of spermaceti, an oil recovered from whale's body, which melts at the temperature of 46 °C. The metal plate had typically a linear dimension of 20 cm so that the aspect ratio was of the order of 200. The thermal conductivity of the metal plate is about 1000 times larger than that of spermaceti, which assures a pretty uniform thermal boundary condition when the plate is heated. In the original experiments by Bénard the plate was heated to the temperature of 100 °C by a flow of water vapour, while the upper surface of the spermaceti layer was at ambient temperature of about 20 °C.

With this temperature difference of 80 °C between heated plate and upper surface of the spermaceti layer, assumed to have the temperature of the air above it, Bénard observed for the first time the insurgence of convective cells, hexagonal in shape (Fig. 2.1) and characterized by ascending fluid in the centre of the cells and descending fluid at the border of the cells. It was also observed that the free surface of the oil layer is slightly depressed compared to the border, the depression d being however very small, about 1 μm compared with the layer H of the fluid, of about 1 mm. Calling λ the distance between two adjacent cells, it was observed that the ratio λ/H was roughly constant, implying that the cell pattern remains geometrically similar when changing the thickness H. It is now known that actually the size of the

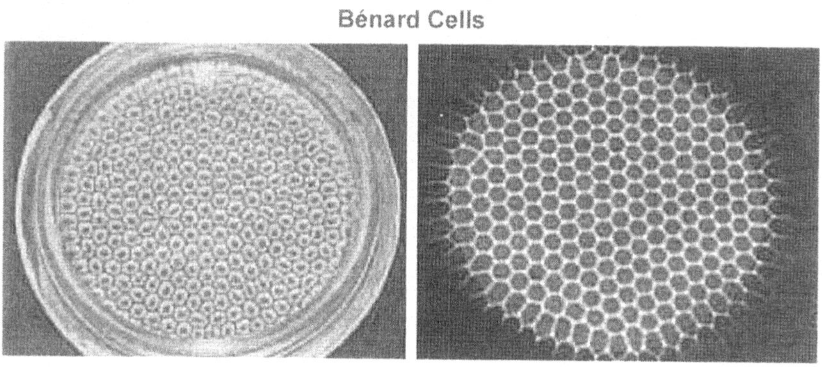

Fig. 2.1 Bénard cells in pan

cells increases slightly with H and also the ratio λ/H increases slightly. For instance one has $\lambda/H = 3.378$ for $H = 0.440$ mm and $\lambda/H = 4.049$ for $H = 0.853$ mm.

Bénard performed also experiments with smaller temperature difference ΔT between lower and upper surfaces. He observed that, decreasing ΔT, the cell size, hence λ, slightly decreases. Most important is, however, the observation that when the temperature difference decreases below a given threshold, which is referred to as critical temperature difference ΔT_c, the cell pattern completely vanishes. When this happens, the ratio λ/H, now referred to as critical ratio $(\lambda/H)_c$, is independent of H. Starting from a liquid at rest, and increasing gradually the temperature difference, instability sets on, with formation of the cellular pattern, when the buoyancy forces overcome the resistance due to the viscosity of the fluid.

In the Bénard experiments the lower surface was "rigid", i.e. a rigid plate, while the upper surface of the liquid was "free", in contact with the air and thus affected by surface tension and also by variation of surface tension with the temperature. In the experiments analysed by Rayleigh, on the contrary, both bounding surfaces were rigid plates and thus both boundary conditions were "rigid". In the following sections we will examine both conditions and derive the analytical solution for the free-free case.

2.1.1.2 Analogy with Bénard Cells in the Atmosphere

Since satellite images have provided direct visual evidence of cloud patterns, Bénard cells have been observed frequently. One can easily distinguish between "*open*" and "*closed*" Bénard cells, as exemplified in Fig. 2.2. Open cells are formed when warm moist air rises on the sides of the cells and water vapour condenses in the upper regions forming a tiny cloud rim at the cell boundaries. Cold and clear air sinks in the middle of the cells. Conversely, closed cells are formed when warm moist air rises in the middle of the cells, forming, by condensation of the water vapour in the upper

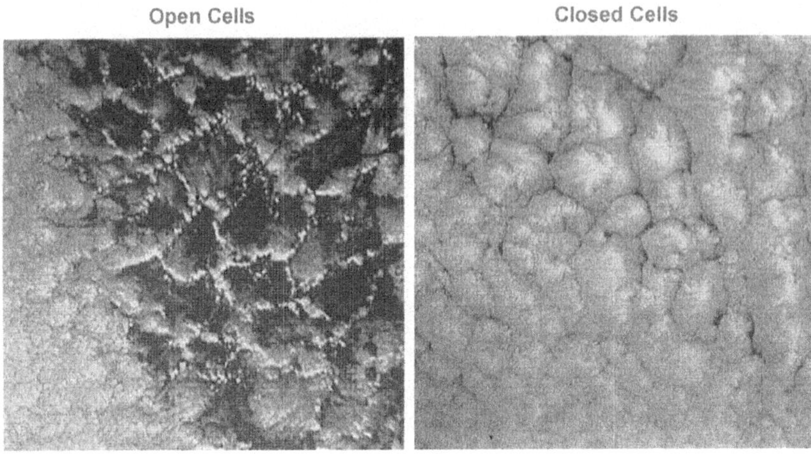

Fig. 2.2 Open and closed cells

regions, a cloud pattern in the middle of the cells. Cold clear air sinks at the cell boundaries leaving more or less clear boundaries.

We will give later an explanation of the instabilities in the atmosphere, leading to the formation of Bénard cells. For the time being, we just recall that the formation of Bénard cells in the atmosphere is due to thermal instabilities (also called "Convective instabilities" or "Rayleigh-Taylor instabilities") arising in case of thermal stratifications with more or less constant Brunt-Vaisala frequency N. In case of stable stratification, characterized by positive derivative $(\partial\theta/\partial z)$, where θ is the potential temperature, the phase speed is real and instability does not occur. In case of unstable stratification, with $(\partial\theta/\partial z) < 0$, the phase speed is imaginary and instabilities occur. Ascending streams develop which lead to the formation of Bénard cells pattern.

Closed Bénard cells are visible in Fig. 2.3 taken by the Meteosat-7 satellite on 23 August 2001.

Because of the amazing analogy between cellular patterns forming in the Bénard experiments and in the atmosphere, we start our analysis of fluid instabilities by recalling the linear theory of the Bénard-Rayleigh convection.

2.1.1.3 Linear Theory of Bénard-Rayleigh Convection

(i) **Conservation Equations**

We consider the following system of conservation equations for a fluid between two horizontal infinite plates, initially at rest:

Continuity equation:

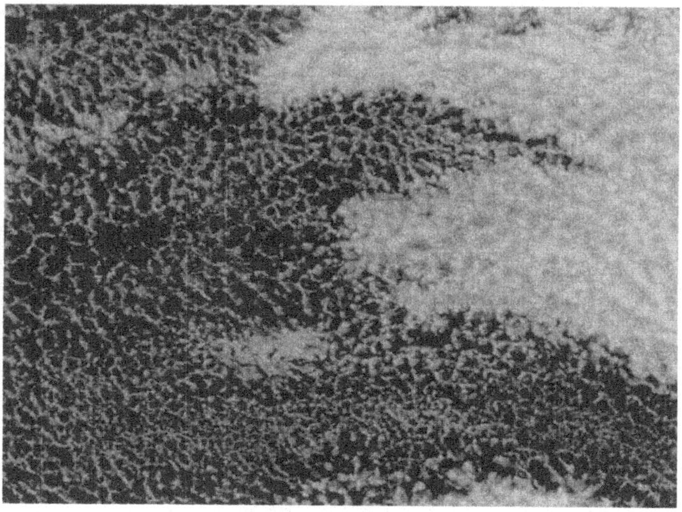

Fig. 2.3 Bénard cells as seen by Meteosat-7 off the West Coast of Africa

$$\frac{\partial \rho}{\partial t} + \nabla \cdot \left(\rho \vec{v}\right) = 0; \tag{2.1}$$

Momentum equation

$$\rho \frac{d\vec{v}}{dt} = -\nabla p + \rho \vec{g} + \mu \nabla^2 \vec{v}; \tag{2.2}$$

Energy equation (or equation of thermal conductivity)

$$\frac{dT}{dt} = \alpha^2 \nabla^2 T; \tag{2.3}$$

Equation of state

$$\rho = \rho_0 (1 - \beta \Delta T) \ (\Delta T = T(z) - T_0 \leq 0). \tag{2.4}$$

In the above equations, we assume $\vec{v} = \left(u\vec{i} + v\vec{j} + w\vec{k}\right)$ is the fluid velocity; δ and μ are density and dynamic viscosity, respectively; p is the pressure; t denotes time; and T is the fluid temperature. The subscript o refers to the lower plate of temperature T_0 where the density has the reference value ρ_0.

$\vec{g} = g\vec{k}$ is the acceleration of gravity. $\alpha^2 = \lambda/(\rho c_p)$ is the thermal diffusivity (λ = thermal conductivity and c_p specific heat at constant pressure).

The thermal expansion coefficient β in Eq. (2.4) is defined by

$$\beta = -\frac{1}{\rho}\left(\frac{\partial \rho}{\partial T}\right). \tag{2.5}$$

For most fluids the derivative $\partial\rho/\partial T$ in Eq. (2.5) is negative, hence the definition chosen yields $\beta > 0$.

In the following we will make the Boussinesq approximation which consists in assuming that all physical properties are constant (independent of temperature), with exception of the density in the buoyancy term.

We will consider at first a fluid at rest and then, imposing a perturbation to this state at rest, we will examine the conditions in which a fluid instability, hence fluid motion, sets in. The initial conditions of the fluid at rest are given by:

$$\text{(a)} \quad \vec{v} = 0 \tag{2.6a}$$

$$\text{(b)} \quad \frac{dT}{dt} = 0, \text{hence, } \nabla^2 T = 0 \tag{2.6b}$$

$$\text{(c)}\,\nabla p = \frac{dp}{dz} = \rho\vec{g} = \vec{g}\rho_0(1 - \beta\Delta T).\text{(hydrostatic condition)} \tag{2.6c}$$

(ii) **Linearization**

The governing Eqs. (2.1)–(2.4) can be linearized assuming that instantaneous values of the dependent variables are given by the sum of a mean value and of a perturbation:

$$\vec{v} = \vec{\overline{v}} + \vec{v'} \; (w = \overline{w} + w'), \tag{2.7a}$$

$$p = \overline{p} + p', \tag{2.7b}$$

$$T = \overline{T} + \vartheta = T_0 - \left(\frac{\partial T}{\partial z}\right) z + \vartheta = T_0 - Gz + \vartheta, \tag{2.7c}$$

$$\rho = \overline{\rho} + \rho' = \overline{\rho} + \rho_0\beta\vartheta. \tag{2.7d}$$

Regarding the fluid velocity we assume a perturbation only for its vertical component w. Regarding the temperature we assume a mean value $\overline{\mathrm{T}} = \mathrm{T}_0 - (\partial \mathrm{T}/\partial z)z = \mathrm{T}_0 - Gz$ which is dependent on the vertical temperature gradient $G = \partial T/\partial z$. The fluctuation of density $\rho' = \rho_0\beta\theta$ is justified by the definition (2.5). According to the Boussinesq approximation, Eq. (2.7d) is used only in the buoyancy term of the momentum equation, otherwise we assume ρ = constant.

Linearization of continuity Eq. (2.1). Using Eq. (2.7a) in Eq. (2.1) yields

$$\frac{\partial \rho}{\partial t} + \nabla \cdot \rho \overrightarrow{\overline{v}} + \nabla \cdot \rho \overrightarrow{v'} = 0. \tag{2.8}$$

The first two terms of Eq. (2.8) are satisfied identically, because of Eq. (2.1). Because ρ = constant, one obtains

$$\nabla \cdot \overrightarrow{v} = 0. \tag{2.9}$$

Linearization of momentum Eq. (2.2). Using Eqs. (2.7a), (2.7b), and (2.7d) in Eq. (2.2) yields

$$\left[\rho \frac{d\overrightarrow{\overline{v}}}{dt}\right] + \rho \frac{d\overrightarrow{v'}}{dt} = [-\nabla \overline{p}] - \nabla p' + \left[\rho \overrightarrow{g}\right] + \rho' \overrightarrow{g} + \left[\mu \nabla^2 \overrightarrow{\overline{v}}\right] + \mu \nabla^2 \overrightarrow{v'}. \tag{2.10}$$

The terms in square brackets satisfy Eq. (2.2) and can be dropped. Let us insert $\rho' = \rho_0 \beta \theta$ in the buoyancy term and divide by $\rho = \rho_0$ (= constant). Expanding the total differential $d/dt = \partial/\partial t + v' \cdot \nabla$ yields

$$\frac{\partial \overrightarrow{v'}}{\partial t} + \overrightarrow{v'} \cdot \nabla \overrightarrow{v'} = -\frac{1}{\rho_0} \nabla p' + \beta \vartheta \overrightarrow{g} + \nu \nabla^2 \overrightarrow{v'}. \tag{2.11}$$

Conforming to the idea of linearizing the equations, we drop the non-linear term $\overrightarrow{v}' \cdot \nabla \overrightarrow{v}'$ obtaining

$$\frac{\partial \overrightarrow{v'}}{\partial t} = -\frac{1}{\rho_0} \nabla p' + \beta \vartheta \overrightarrow{g} + \nu \nabla^2 \overrightarrow{v'}. \tag{2.12}$$

Linearization of the energy Eq. (2.3). Using Eqs. (2.7a) and (2.7c) and expanding the differential $dT/dz = \partial T/\partial z + \overrightarrow{v} \cdot \nabla T$ in Eq. (2.3) yield

$$\frac{\partial}{\partial t}(\overline{T} + \vartheta) + \left(\overrightarrow{\overline{v}} + \overrightarrow{v'}\right) \cdot \nabla (\overline{T} + \vartheta) = \alpha^2 \nabla^2 (\overline{T} + \vartheta), \tag{2.13}$$

hence

$$\left[\frac{\partial \overline{T}}{\partial t}\right] + \frac{\partial \vartheta}{\partial t} + \left[\overrightarrow{\overline{v}} \cdot \nabla \overline{T}\right] + \overrightarrow{\overline{v}} \cdot \nabla \vartheta + \overrightarrow{v'} \cdot \nabla \overline{T} + \overrightarrow{v'} \cdot \nabla \vartheta = \left[\alpha^2 \nabla^2 \overline{T}\right] + \alpha^2 \nabla^2 \vartheta. \tag{2.14}$$

The terms in square brackets can be dropped, because they satisfy Eq. (2.3). The fourth term at the left-hand side, which implies a transfer of a gradient of temperature perturbation, is dropped because it is negligible with respect to $\partial \theta/\partial t$. The sixth term at the left hand side is dropped because is the product of two small perturbations.

Because $\nabla\overline{T}$ has only a vertical component, hence $\vec{v}'\cdot\nabla\overline{T} = w'\left(\partial\overline{T}/\partial z\right)$ and using $\overline{T} = T_0 - Gz$ one obtains

$$\frac{\partial\vartheta}{\partial t} = w'G + \alpha^2\nabla^2\vartheta. \tag{2.15}$$

In the following, for simplicity of notation, we drop the apex denoting fluctuating quantities. The linearized equations are summarized as follows:

$$\nabla\cdot\vec{v} = 0, \tag{2.16}$$

$$\frac{\partial\vec{v}}{\partial t} = -\frac{1}{\rho_0}\nabla p + \beta\vartheta\vec{g} + \nu\nabla^2\vec{v}, \tag{2.17}$$

$$\frac{\partial\vartheta}{\partial t} = wG + \alpha^2\nabla^2\vartheta. \tag{2.18}$$

2.1.2 Derivation of Dimensionless Equations

It is convenient to transform the system of Eqs. (2.16)–(2.18) in dimensionless equations, to obtain relationships independent of the scale of the experimental device. Denoting with a subscript D dimensionless quantities, we define:

$$x_{iD} = \frac{x_i}{H}, (i = 1, 2, 3)\ \left[\widetilde{L} = H\right] \tag{2.19}$$

$$t_D = \frac{\alpha^2 t}{H^2}, \left[\tau = \frac{H^2}{\alpha^2}\right] \tag{2.20}$$

$$\vec{v}_D = \frac{H}{\alpha^2}\vec{v}, \left[\widetilde{\vec{v}} = \frac{\alpha^2}{H}\right] \tag{2.21}$$

$$\vartheta_D = \frac{\beta g H^3}{\nu\alpha^2}\vartheta\ \left[\widetilde{\Theta} = \frac{\nu\alpha^2}{\beta g H^3}\right], \tag{2.22}$$

$$p_D = \frac{H^2}{\rho_0(\alpha^2)^2}p\ \left[\widetilde{p} = \frac{\rho_0(\alpha^2)^2}{H^2}\right]. \tag{2.23}$$

The choice of the characteristics lengths $\widetilde{L} = H$ and time $\tau = H^2/\alpha^2$ is suggested by the geometry of the system and by the internal diffusion coefficient α^2. The characteristic velocity is implicitly defined by $\widetilde{v} = \widetilde{L}/\tau = \alpha^2/H$. The characteristic temperature is suggested by $\widetilde{\theta} = \Delta T/R_a$ where R_a is the Rayleigh number defined later. This particular choice of the dimensionless parameters offers the advantage that the onset of convection turns out to depend only on the R_a number, as will be seen in the subsequent sections.

When deriving dimensionless equations, differentials are transformed according to the rule:

$$\frac{\partial}{\partial \ell} \Rightarrow \frac{1}{H} \frac{\partial}{\partial \ell_D}, \tag{2.24}$$

$$\frac{\partial^2}{\partial \ell^2} \Rightarrow \frac{1}{H^2} \frac{\partial^2}{\partial \ell_D^2}, \tag{2.25}$$

where ℓ denotes any coordinate direction.

Dimensionless continuity equation. Equation (2.16) yields

$$\frac{1}{H} \nabla_D \cdot \vec{v}_D \frac{\alpha^2}{H} = 0 \tag{2.26}$$

hence, being α^2 constant,

$$\nabla_D \cdot \vec{v}_D = 0. \tag{2.27}$$

Dimensionless momentum equation. Using $\vec{g} = g\vec{k}$, Eq. (2.17) yields

$$\partial\left(\frac{\alpha^2}{H}\vec{v}_D\right)\frac{1}{\partial\left(\frac{H^2}{\alpha^2}t_D\right)} = -\frac{1}{\cancel{\rho_0}}\frac{1}{H}\nabla_D\left[\frac{\cancel{\rho_0}(\alpha^2)^2}{H^2}p_D\right] + \beta\frac{\nu\alpha^2}{\beta g H^3}\vartheta_D g\vec{k}$$
$$+\frac{\nu}{H^2}\nabla_D^2\left(\frac{\alpha^2}{H}\vec{v}_D\right). \tag{2.28}$$

Dividing every term by $(\alpha^2)^2/H^3$ yields

$$\frac{\partial \vec{v}_D}{\partial t_D} = -\nabla_D p_D + \frac{\nu}{\alpha^2}\vartheta_D \vec{k} + \frac{\nu}{\alpha^2}\nabla_D^2 \vec{v}_D. \tag{2.29}$$

Using definition of the Prandtl number

$$\Pr = \frac{\nu}{\alpha^2} \tag{2.30}$$

one obtains

$$\frac{\partial \vec{v}_D}{\partial t_D} = -\nabla_D p_D + \Pr \vartheta_D \vec{k} + \Pr \nabla_D^2 \vec{v}_D. \tag{2.31}$$

Dimensionless energy equation. Equation (2.18) yields

$$\partial\left(\frac{\nu\alpha^2}{\beta g H^3}\vartheta_D\right)\frac{1}{\partial\left(\frac{H^2}{\alpha^2}t_D\right)} = \frac{\alpha^2}{H}w_D G + \frac{\alpha^2}{H^2}\nabla_D^2\left(\frac{\nu\alpha^2}{\beta g H^3}\vartheta_D\right). \tag{2.32}$$

Dividing every term by $\nu(\alpha^2)^2/(\beta g H^5)$ yields

$$\frac{\partial \vartheta_D}{\partial t_D} = \frac{G\beta g H^4}{\nu\alpha^2}w_D + \nabla_D^2\vartheta_D. \tag{2.33}$$

With the definition of the Rayleigh number

$$Ra = \frac{G\beta g H^4}{\nu\alpha^2} = \frac{\partial T}{\partial z}\frac{\beta g H^4}{\nu\alpha^2} \tag{2.34}$$

one obtains

$$\frac{\partial \vartheta_D}{\partial t_D} = Ra w_D + \nabla_D^2\vartheta_D. \tag{2.35}$$

In the following we drop for simplicity the subscript D. The dimensionless perturbation equations can therefore be summarized by:

$$\nabla\cdot\vec{v} = 0, \tag{2.36}$$

$$\frac{\partial \vec{v}}{\partial t} = -\nabla p + P_r\vartheta\vec{k} + P_r\nabla^2\vec{v}, \tag{2.37}$$

$$\frac{\partial \vartheta}{\partial t} = Raw + \nabla^2\vartheta. \tag{2.38}$$

One should, however, keep in mind that the dependent variables ($\vec{v}$, p, θ) in Eqs. (2.36)–(2.38) are dimensionless fluctuations. Besides, the linearized form of the momentum equation has been obtained by neglecting the non-linear term $\vec{v}'\cdot\nabla\vec{v}'$;

the linearized energy equation has been obtained by neglecting the terms $\vec{v}\cdot\nabla\theta$ and $\vec{v}'\cdot\nabla\theta$.

2.1.3 Alternative Forms of Perturbation Equations

In this section we obtain alternative form of the perturbation Eqs. (2.37) and (2.38) by assuming some "Ansatz" or analytical expression for the dependent variables. First, by taking the rotational of Eq. (2.37) we obtain

$$\frac{\partial}{\partial t}\left(\nabla\times\vec{v}\right) = -\nabla\times\nabla p + P_r\left(\nabla\times\vartheta\vec{k}\right) + \mathrm{Pr}\nabla^2\left(\nabla\times\vec{v}\right). \tag{2.39}$$

Because $\nabla\times(\nabla p) = 0$, taking again the rotational of Eq. (2.39) yields

$$\frac{\partial}{\partial t}\left(\nabla\times\nabla\times\vec{v}\right) = \mathrm{Pr}\left(\nabla\times\nabla\times\vartheta\vec{k}\right) + \mathrm{Pr}\nabla^2\left(\nabla\times\nabla\times\vec{v}\right). \tag{2.40}$$

Using the identity (for any vector $\vec{A}$) $\nabla\times\nabla\times\vec{A} = \nabla\left(\nabla\cdot\vec{A}\right) - \nabla^2\vec{A}$ applied to the vectors $\vec{v}$ and $\vartheta\vec{k}$ yields

$$\nabla\times\nabla\times\vec{v} = \nabla\left(\nabla\cdot\vec{v}\right) - \nabla^2\vec{v} = -\nabla^2\vec{v} \tag{2.41}$$

$$\nabla\times\nabla\times\left(\vartheta\vec{k}\right) = \nabla\left(\nabla\cdot\vartheta\vec{k}\right) - \nabla^2\left(\vartheta\vec{k}\right). \tag{2.42}$$

The second member of Eq. (2.42) is computed as follows:

$$\nabla\left(\nabla\cdot\vartheta\vec{k}\right) - \nabla^2\left(\vartheta\vec{k}\right) = \nabla\left(\frac{\partial\vartheta}{\partial z}\right) - (\nabla^2\vartheta)\vec{k}$$

$$= \frac{\partial}{\partial x}\left(\frac{\partial\vartheta}{\partial z}\right)\vec{i} + \frac{\partial}{\partial y}\left(\frac{\partial\vartheta}{\partial z}\right)\vec{j} + \frac{\partial}{\partial z}\left(\frac{\partial\vartheta}{\partial z}\right)\vec{k} - \left(\frac{\partial^2\vartheta}{\partial x^2} + \frac{\partial^2\vartheta}{\partial y^2} + \frac{\partial^2\vartheta}{\partial z^2}\right)\vec{k}. \tag{2.43}$$

We assume that the perturbation $(\partial\vartheta/\partial z)$ is uniform in the x and y directions so that the mixed derivatives can be neglected. The second member of Eq. (2.42) reduces therefore to

$$\nabla\left(\nabla\cdot\vartheta\vec{k}\right)-\nabla^2\left(\vartheta\vec{k}\right)=-\left(\frac{\partial^2\vartheta}{\partial x^2}+\frac{\partial^2\vartheta}{\partial y^2}\right)\vec{k}=-\left(\nabla_2^2\vartheta\right)\vec{k}, \tag{2.44}$$

where the subscript 2 denotes that the second-order differential ∇^2 is to be taken with respect to the x and y-directions only.

Introducing Eqs. (2.41) and (2.44) in Eq. (2.40) yields

$$\frac{\partial}{\partial t}\left(\nabla^2\vec{v}\right)=\mathrm{Pr}\left(\nabla_2^2\vartheta\right)\vec{k}+\mathrm{Pr}\left(\nabla^4\vec{v}\right). \tag{2.45}$$

We consider a perturbation of the velocity in the vertical direction only and therefore set $\vec{V}=o\,\vec{i}+o\,\vec{j}+w\,\vec{k}$. The vertical component of Eq. (2.45) becomes

$$\frac{\partial}{\partial t}\left(\nabla^2 w\right)=\mathrm{Pr}\nabla_2^2\vartheta+\mathrm{Pr}\nabla^4 w. \tag{2.46}$$

At time $t=0$ we assumed $u=v=0$, hence $\vec{V}=w\,\vec{k}$. When the fluid motion sets on through insurgence of instabilities, the components u and v become different from zero and therefore we may consider the vorticity $\zeta=\partial v/\partial x-\partial u/\partial y$. For the dependent variables w, θ and for the vorticity ζ we make the following Ansatz:

$$w=W(z)e^{i\left(k_x x+k_y y\right)}e^{pt}, \tag{2.47}$$

$$\partial=\Theta(z)e^{i\left(k_x x+k_y y\right)}e^{pt}, \tag{2.48}$$

$$\zeta=Z(z)e^{i\left(k_x x+k_y y\right)}e^{pt}. \tag{2.49}$$

Every dependent variable is therefore of the form $f=F(z)e^{i\left(k_x x+k_y y\right)}e^{pt}=F(z)\,Ee^{pt}$, where we denote for simplicity the exponential by E. i is the imaginary unit and k_x, k_y are wave numbers defined by

$$k_x=\frac{2\pi}{\lambda_x}, \tag{2.50a}$$

$$k_y=\frac{2\pi}{\lambda_y}, \tag{2.50b}$$

where λ_x, λ_y are wave lengths in the x and y-directions, respectively.

We also define the wave number k by

$$k^2 = k_x^2 + k_y^2. \tag{2.51}$$

Derivatives of the function $f = F(z)Ee^{pt}$ are given by:

$$\frac{\partial f}{\partial t} = pF(z)Ee^{pt} = pf, \tag{2.52a}$$

$$\frac{\partial f}{\partial x} = F(z)(ik_x)Ee^{pt} = (ik_x)f, \tag{2.52b}$$

$$\frac{\partial^2 f}{\partial x^2} = -k_x^2 F(z)Ee^{pt} = -k_x^2 f, \tag{2.52c}$$

$$\frac{\partial f}{\partial y} = F(z)\left(ik_y\right)Ee^{pt} = \left(ik_y\right)f, \tag{2.52d}$$

$$\frac{\partial^2 f}{\partial y^2} = -k_y^2 F(z)Ee^{pt} = -k_y^2 f, \tag{2.52e}$$

$$\frac{\partial^2 f}{\partial z^2} = \frac{\partial^2 F(z)}{\partial z^2}Ee^{pt}, \tag{2.52f}$$

$$\nabla_{2}^2 f = \frac{\partial^2 f}{\partial x^2} + \frac{\partial^2 f}{\partial y^2} = -\left(k_x^2 + k_y^2\right)F(z)Ee^{pt} = -k^2 F(z)Ee^{pt} = -k^2 f, \tag{2.52g}$$

$$\nabla^2 f = \frac{\partial^2 f}{\partial x^2} + \frac{\partial^2 f}{\partial y^2} + \frac{\partial^2 f}{\partial z^2} = \nabla_{2}^2 f + \frac{\partial^2 f}{\partial z^2} = \left[\frac{\partial^2 F(z)}{\partial z^2} - k^2 F(z)\right]Ee^{pt}, \tag{2.52h}$$

$$\begin{aligned}
\nabla^4 f = \nabla^2\left(\nabla^2 f\right) &= \frac{\partial^2\left(\nabla^2 f\right)}{\partial x^2} + \frac{\partial^2\left(\nabla^2 f\right)}{\partial y^2} + \frac{\partial^2\left(\nabla^2 f\right)}{\partial z^2} \\
&= \frac{\partial^2}{\partial x^2}\left[\frac{\partial^2 F(z)}{\partial z^2} - k^2 F(z)\right]Ee^{pt} + \frac{\partial^2}{\partial y^2}\left[\frac{\partial^2 F(z)}{\partial z^2} - k^2 F(z)\right]Ee^{pt} \\
&\quad + \frac{\partial^2}{\partial z^2}\left[\frac{\partial^2 F(z)}{\partial z^2} - k^2 F(z)\right]Ee^{pt} \\
&= \left[\frac{\partial^2 F(z)}{\partial z^2} - k^2 F(z)\right]\left(-k_x^2\right)Ee^{pt} + \left[\frac{\partial^2 F(z)}{\partial z^2} - k^2 F(z)\right]\left(-k_y^2\right)Ee^{pt} \\
&\quad + \left[\frac{\partial^4 F(z)}{\partial z^4} - k^2\frac{\partial^2 F(z)}{\partial z^2}\right]Ee^{pt}
\end{aligned}$$

$$= \left[\frac{\partial^2 F(z)}{\partial z^2} - k^2 F(z)\right](-k^2)Ee^{pt} + \left[\frac{\partial^4 F(z)}{\partial z^4} - k^2 \frac{\partial^2 F(z)}{\partial z^2}\right] Ee^{pt}$$

$$= \left[\frac{\partial^4 F(z)}{\partial z^4} - 2k^2 \frac{\partial^2 F(z)}{\partial z^2} + k^4 F(z)\right] Ee^{pt}$$

$$= \left[\frac{\partial^4}{\partial z^4} - 2k^2 \frac{\partial^2}{\partial z^2} + k^4\right] F(z)Ee^{pt} = \left\{\left[\frac{d^2}{dz^2} - k^2\right]^2 F(z)\right\} Ee^{pt}, \tag{2.52i}$$

$$\begin{aligned} \frac{\partial}{\partial t}(\nabla^2 f) &= \frac{\partial}{\partial t}\left[\frac{\partial^2 F(z)}{\partial z^2} - k^2 F(z)\right] Ee^{pt} = \left[\frac{\partial^2 F(z)}{\partial z^2} - k^2 F(z)\right] pEe^{pt} \\ &= \left[\frac{\partial^2}{\partial z^2} - k^2\right] pf. \end{aligned} \tag{2.52j}$$

The operator $\left(\frac{d^2}{dz^2} - k^2\right)^2$ applied to $F(z)$ in Eq. (2.52i) is the square of the operator applied to $F(z)$ in Eq. (2.52h).

Using Eqs. (2.52i) and (2.52j) for $w = W(z)Ee^{pt}$, and Eq. (2.46) for $\theta = \theta(z)Ee^{pt}$, Eq. (2.46) after dividing by Ee^{pt}, yields

$$p\left[\frac{d^2 W(z)}{dz^2} - k^2 W(z)\right] = \Pr(-k^2)\theta(z) + \Pr\left[\frac{d^2 W(z)}{dz^2} - k^2 W(z)\right]^2. \tag{2.53}$$

Dividing by Pr and rearranging one obtains from Eq. (2.53)

$$\left[\frac{d^2 W(z)}{dz^2} - k^2 W(z)\right]^2 - \frac{p}{\Pr}\left[\frac{d^2 W(z)}{dz^2} - k^2 W(z)\right] = k^2 \theta(z). \tag{2.54}$$

Using the operator $D^2 = d^2/dz^2$, Eq. (2.54) can be written in the compact form

$$\left[(D^2 - k^2)^2 - \frac{p}{\Pr}(D^2 - k^2)\right] W(z) = k^2 \theta(z), \tag{2.55}$$

or

$$\left[(D^2 - k^2)\left(D^2 - k^2 - \frac{p}{\Pr}\right)\right] W(z) = k^2 \theta(z). \tag{2.56}$$

Likewise, using Eqs. (2.52a) and (2.52h) for $\theta = \theta(z)Ee^{pt}$ and Eq. (2.52a) for $w = W(z)Ee^{pt}$, after division by Ee^{pt}, and rearranging, yield

$$\left[\frac{d^2\theta(z)}{dz^2} - k^2\theta(z)\right] - p\theta(z) = -RaW(z) \tag{2.57}$$

or

$$\left[D^2 - k^2 - p\right]\theta(z) = -RaW(z). \tag{2.58}$$

Solving Eq. (2.58) for $W(z)$ and inserting into Eq. (4.18) yield

$$\left(D^2 - k^2\right)\left(D^2 - k^2 - \frac{p}{\mathrm{Pr}}\right)\left(D^2 - k^2 - p\right)\theta(z) = -Rak^2\theta(z). \tag{2.59}$$

Vice versa, eliminating $\theta(z)$ between Eq. (2.56) and (2.58) yields

$$\left(D^2 - k^2\right)\left(D^2 - k^2 - \frac{p}{\mathrm{Pr}}\right)\left(D^2 - k^2 - p\right)W(z) = -Rak^2W(z). \tag{2.60a}$$

Equation (2.60a) can be written in an alternative form with explicit time dependence. Using $\nabla_2^2 w = -k^2 w$ (4.14g) and $\nabla^2 w = (D^2 - k^2)w$ we obtain from Eq. (2.60a):

$$\nabla^2\left(\nabla^2 - \frac{1}{\mathrm{Pr}}\frac{\partial}{\partial t}\right)\left(\nabla^2 w - \frac{\partial w}{\partial t}\right) = Ra\nabla_2^2 w,$$

$$\nabla^2\left[\nabla^4 w - \frac{\partial}{\partial t}\nabla^2 w - \frac{1}{\mathrm{Pr}}\frac{\partial}{\partial t}\left(\nabla^2 w\right) + \frac{1}{\mathrm{Pr}}\frac{\partial^2 w}{\partial t^2}\right] = Ra\nabla_2^2 w,$$

$$\nabla^6 w - \frac{\partial}{\partial t}\nabla^4 w - \frac{1}{\mathrm{Pr}}\frac{\partial}{\partial t}\left(\nabla^4 w\right) + \frac{1}{\mathrm{Pr}}\frac{\partial^2}{\partial t^2}\left(\nabla^2 w\right) = Ra\nabla_2^2 w,$$

$$\nabla^6 w - \left(1 + \frac{1}{\mathrm{Pr}}\right)\frac{\partial}{\partial t}\left(\nabla^4 w\right) + \frac{1}{\mathrm{Pr}}\frac{\partial^2}{\partial t^2}\left(\nabla^2 w\right) = Ra\nabla_2^2 w. \tag{2.60b}$$

Equation (2.60b) can also be written as

$$\left[\frac{\partial}{\partial t} - \nabla^2\right]\left[\frac{1}{\mathrm{Pr}}\frac{\partial}{\partial t} - \nabla^2\right]\nabla^2 w = Ra\nabla_2^2 w. \tag{2.60c}$$

The state of the fluid at rest is time-independent and is obtained by letting $p = 0$. In this case Eq. (2.60a) becomes

$$\left(D^2 - k^2\right)^3 W(z) = -Rak^2 W(z), \tag{2.61a}$$

or

$$\left(D^2 - k^2\right)^3 w = -Rak^2 w. \tag{2.61b}$$

Equation (2.61) is therefore the equation which describes the "neutral stability", or the transition between stable (stationary) and unstable (transient) states. Using Eq. (2.52g) and Eq. (2.52h) the two terms of Eq. (2.61) can be written:

$$\nabla^6 w = Ra \nabla_2^2 w, \tag{2.62}$$

which corresponds to Eq. (2.60b) for the time-independent case.

Once Eqs. (2.56) and (2.58), or their alternative forms, have been solved for θ and w, the components u and v of the velocity perturbation (assuming fluid instability sets on) can be computed as follows.

Any function of the type $f = F(z)e^{i\left(k_x x + k_y y\right)}\, e^{pt} = F(z)Ee^{pt}$ [or simply $f = e^{i\left(k_x x + k_y y\right)} e^{pt}$] satisfies the so-called "membrane equation"

$$\nabla_2^2 f + k^2 f = 0. \tag{2.63}$$

Let us consider two functions $\Phi(x, y)$ and $\Psi(x, y)$, satisfying the membrane equation and such that

$$u = \frac{\partial \Phi}{\partial x} - \frac{\partial \Psi}{\partial y}, \tag{2.64}$$

$$v = \frac{\partial \Phi}{\partial y} + \frac{\partial \Psi}{\partial x}. \tag{2.65}$$

Because $\nabla \cdot \vec{V} = 0$ one derives

$$-\frac{\partial w}{\partial z} = \frac{\partial u}{\partial x} + \frac{\partial v}{\partial y} = \frac{\partial^2 \Phi}{\partial x^2} - \frac{\partial^2 \Psi}{\partial x \partial y} + \frac{\partial^2 \Phi}{\partial y^2} + \frac{\partial^2 \Psi}{\partial y \partial x} = \nabla_2^2 \Phi = -k^2 \Phi \tag{2.66}$$

hence

$$\Phi = \frac{1}{k^2} \frac{\partial w}{\partial z}. \tag{2.67}$$

From the definition of vorticity (z-component) we obtain

$$\zeta = \frac{\partial v}{\partial x} - \frac{\partial u}{\partial y} = \frac{\partial^2 \Phi}{\partial x \partial y} + \frac{\partial^2 \Psi}{\partial x^2} - \frac{\partial^2 \Phi}{\partial y \partial x} + \frac{\partial^2 \Psi}{\partial y^2} = \nabla_2^2 \Psi = -k^2 \Psi \tag{2.68}$$

hence

$$\Psi = -\frac{\zeta}{k^2}. \tag{2.69}$$

Using Eqs. (2.67) and (2.69), Eq. (2.64) and Eq. (2.65) yield

$$u = \frac{1}{k^2}\frac{\partial^2 w}{\partial x \partial z} + \frac{1}{k^2}\frac{\partial \zeta}{\partial y} = \frac{i}{k^2}\left[\frac{\partial W(z)}{\partial z}k_x + Z(z)k_y\right]Ee^{pt}, \tag{2.70}$$

$$v = \frac{1}{k^2}\frac{\partial^2 w}{\partial y \partial z} - \frac{1}{k^2}\frac{\partial \xi}{\partial x} = \frac{i}{k^2}\left[\frac{\partial W(z)}{\partial z}k_y - Z(z)k_x\right]Ee^{pt}. \tag{2.71}$$

2.1.4 Solution of Perturbation Equations

Equation (2.60) is an ordinary differential equation of order six and requires in general six boundary conditions, three at $z = 0$ and three at $z = 1$. A general solution cannot be found, but particular solutions can be obtained for some types of boundary conditions, for examples:

(a) Both surfaces are free. This is the simplest case to treat analytically, although in practice it can be realized with some difficulty, for instance letting the fluid float upon a thin layer of heavier fluid.
(b) Both surfaces are rigid. This case, which is the most difficult case to treat analytically, was considered by Rayleigh.
(c) One surface, say $z = 0$, is rigid, the other ($z = 1$) free. This is the case treated by Bénard, which allows for direct observation of the free surface from above.

In the following we will consider the analytical treatment of case (a) and give the result of case (b).

The main solution procedure consists in finding, for a given value of k^2, the minimum value of the Rayleigh number Ra which represents the critical conditions for the onset of instability.

2.1.4.1 Free Surfaces

In case of free-free surfaces the following boundary conditions are appropriate at $z = 0$ and $z = 1$:

$$\text{(a) } w = 0 \text{ [or } W(z) = 0] \tag{2.72a}$$

$$\text{(b) } \theta = 0 \tag{2.72b}$$

$$\text{(c)}\ \frac{d^2w}{dz^2}=0\left[\text{or } \frac{d^2W(z)}{dz^2}=0\right] \tag{2.72c}$$

$$\text{(d)}\ \frac{d^4w}{dz^2}=0\left[\text{or } \frac{d^4W(z)}{dz^2}=0\right] \tag{2.72d}$$

Boundary condition (2.72c) arises from the requirement that the shear stresses p_{xz} and p_{yz} (normal to z-axis and directed as the x, y-axes, respectively) vanish on the free surfaces, hence

$$p_{xz}=\mu\left(\frac{\partial u}{\partial z}+\frac{\partial w}{\partial x}\right)=0, \tag{2.73a}$$

$$p_{yz}=\mu\left(\frac{\partial v}{\partial z}+\frac{\partial w}{\partial y}\right)=0. \tag{2.73b}$$

Because $w=0$, also $\partial w/\partial x=\partial w/\partial y=0$ on the free surface and Eq. (2.73) yields

$$\frac{\partial u}{\partial z}=\frac{\partial v}{\partial z}=0. \tag{2.74}$$

Dividing the continuity equation with respect to z we obtain

$$\frac{\partial}{\partial z}\left(\frac{\partial u}{\partial x}+\frac{\partial v}{\partial y}+\frac{\partial w}{\partial z}\right)=\frac{\partial}{\partial x}\left(\frac{\partial u}{\partial z}\right)+\frac{\partial}{\partial y}\left(\frac{\partial v}{\partial z}\right)+\frac{\partial^2 w}{\partial z^2}=\frac{\partial^2 w}{\partial z^2}, \tag{2.75}$$

which justifies the boundary condition (2.72c).

Boundary condition (2.61d) is a consequence of $\theta=0$ (2.72b). In fact it is obtained by letting $\theta=0$, $W(z)=0$ and $d^2W(z)/dz^2=0$ (all conditions implied for $z=0$ and $z=1$) in Eq. (2.53).

We try to satisfy Eq. (2.61b) with the following "Ansatz", which clearly conforms to boundary condition (2.72a)

$$W(z)=\sin(n\pi z),\ (n=1,2,\ldots) \tag{2.76}$$

hence

$$w=W(z)e^{i(k_x x+k_y y)}e^{pt}=\sin(n\pi z)e^{i(k_x x+k_y y)}e^{pt}. \tag{2.77}$$

Letting for simplicity of notation $S=\sin(n\pi z)$, $C=\cos(n\pi z)$, $E=e^{i(k_x x+k_y y)}$ one finds

$$\frac{\partial w}{\partial z} = (n\pi)CEe^{pt}, \tag{2.78}$$

$$\frac{\partial^2 w}{\partial z^2} = -(n\pi)^2 SEe^{pt} = -(n\pi)^2 w. \tag{2.79}$$

Inserting (2.79) into (2.61b) and solving for the Rayleigh number, one obtains

$$Ra = \frac{\left[(n\pi)^2 + k^2\right]^3}{k^2} \tag{2.80}$$

The derivative of Ra with respect to the wavenumber k is

$$\frac{\partial Ra}{\partial k} = \frac{3\left[(n\pi)^2 + k^2\right]^2 2kk^2 - \left[(n\pi)^2 + k^2\right]^3 2k}{k^4}. \tag{2.81}$$

Letting $\partial Ra/\partial k = 0$ yields $2k^2 = (n\pi)^2$, from which one obtains the critical wavenumber (for $n = 1$)

$$k_c = \left[\frac{n\pi}{\sqrt{2}}\right]_{n=1} = \frac{\pi}{\sqrt{2}} \cong 2.221. \tag{2.82}$$

The corresponding critical Rayleigh number becomes

$$Ra_c = \frac{\left[\pi^2 + k_c^2\right]^3}{k_c^2} = \left[\frac{27}{4}(n\pi)^4\right]_{n=1} = \frac{27}{4}\pi^4 \cong 657.5. \tag{2.83}$$

The dimensionless critical wavelength is

$$\lambda_c = \frac{2\pi}{k_c} = 2\sqrt{2} \cong 2.828. \tag{2.84}$$

The real wavelength results therefore equal to $H\lambda_c$. The critical Rayleigh number determines the critical ΔT_c necessary to set on the instability of the fluid:

$$\Delta T_c = \frac{Ra_c \nu d^2}{\beta g H^3}. \tag{2.85}$$

As shown by Eq. (2.85) ΔT_c can be very high for thin layers and very small for thick layers. Thus far we used Eq. (2.61b), which is time independent, to determine the critical parameters k_c, λ_c, Ra_c, and ΔT_c. Next we use the time-dependent

Eq. (2.60b) to determine the time behaviour of a perturbation, characterized by the parameter p in Eq. (2.77). To this purpose we first collect the derivatives obtained from Eq. (2.77) for the normal mode $n = 1$:

$$\nabla_2^2 w = \frac{\partial^2 w}{\partial x^2} + \frac{\partial^2 w}{\partial y^2} = -\left(k_x^2 + k_y^2\right) SEe^{pt} = -k^2 SEe^{pt} = -k^2 w, \tag{2.86a}$$

$$\nabla^2 w = \nabla_2^2 w + \frac{\partial^2 w}{\partial z^2} = -\left(k^2 + \pi^2\right) SEe^{pt} = -\left(k^2 + \pi^2\right) w = -Fw, \tag{2.86b}$$

$$\nabla^4 w = \left(k^2 + \pi^2\right)^2 w = F^2 w, \tag{2.86c}$$

$$\nabla^6 w = -\left(k^2 + \pi^2\right)^3 w = -F^3 w, \tag{2.86d}$$

$$\frac{\partial^2 \left(\nabla^2 w\right)}{\partial t^2} = -p^2 \left(k^2 + \pi^2\right) w = -p^2 F w, \tag{2.86e}$$

$$\frac{\partial \left(\nabla^4 w\right)}{\partial t} = p\left(k^2 + \pi^2\right)^2 w = p^2 F w, \tag{2.86f}$$

where, for simplicity of notation, we set $F = (k^2 + \pi^2)$.

Using the derivatives (2.86) in Eq. (2.60b) yields

$$\frac{F}{\mathrm{Pr}} p^2 + F^2 \left(1 + \frac{1}{\mathrm{Pr}}\right) p + F^3 - k^2 Ra = 0. \tag{2.87}$$

Notice that this Eq. (2.87) can be obtained directly from (2.60c) letting $\partial/\partial t \equiv p$; $\nabla^2 \equiv -F$; and $\nabla_2^2 \equiv -k^2$.

Eq. (2.87) has two solutions for p, given by:

$$2p_{1,2} = -F(1 + \mathrm{Pr}) \pm \sqrt{F^2 (1 - \mathrm{Pr})^2 + \frac{4}{F} \mathrm{Pr} Ra k^2}. \tag{2.88}$$

If $Ra > 0$ (as it is for the case of the fluid heated from below) the two roots are real. The case of p complex is called "over-stability" and occurs in convection in rotating systems. We will not consider here this case. It is easy to prove that for $Ra = Ra_c = (\pi^2 + k_c^2)^3 / k_c^2 = F^3 / k_c^2$ one of the roots (2.88) becomes 0. In fact for $Ra_c k_c^2 = F^3$ one has from Eq. (2.88)

$$\begin{aligned} 2p_{1,2} &= -F(1 + \mathrm{Pr}) \pm \sqrt{F^2 (1 - \mathrm{Pr})^2 + 4\mathrm{Pr} F^2} = \\ &= -F(1 + \mathrm{Pr}) \pm F\sqrt{(1 + \mathrm{Pr})^2} = -F(1 + \mathrm{Pr}) + F(1 + \mathrm{Pr}) \end{aligned} \tag{2.89}$$

hence

$$p_1 = 0, \tag{2.90a}$$

$$p_2 = -2\,\mathrm{F}(1 + \mathrm{Pr}). \tag{2.90b}$$

The root $p_1 = 0$ represents the limit of neutral stability. The root $p_2 = -2F(1 + Pr)$ represents a stable solution in which every perturbation of the fluid is damped out and the fluid, initially at rest, remains stable. Instability occurs then for $Ra > Ra_c$ in which case at least one of the roots (2.88) is positive. For the analysis of the onset of stability one is interested in the minimum positive root of the couple of roots (2.88).

2.1.4.2 Rigid Surfaces

The case of rigid surfaces is more difficult to treat analytically. An approximation of the analytical results for the Rayleigh number in the normal mode ($n = 1$), is given, according to Chandrasekhar (1961), by

$$Ra = \frac{\left(\pi^2 + k^2\right)^3}{k^2\left[1 - \frac{16k\pi^2 \cos h(k/2)}{\left(\pi^2+k^2\right)^2 (\sin h(k)+k)}\right]}. \tag{2.91}$$

According to this formula the minimum (critical) Rayleigh number results $Ra_c = 1715.1$, which compares fairly well with the experimental value of $Ra_c = 1707.8$.

Plots showing the Rayleigh numbers versus wavenumber k are given in Fig. 2.4 for the case of free-free boundary conditions (Eq. 2.80) and for the rigid-rigid boundary conditions (Eq. 2.91). For each of the two boundary conditions the domain above the plot represents an unstable region, while the domain below the plot represents a stable region.

In the language of chaos theory one would say that in the point (k_c, Ra_c) the plot shown in the above figure displays a bifurcation and, for $Ra > Ra_c$, two values of k (like k_A, k_B) define the instability range. When bifurcation occurs the systems turn to chaotic following the so-called Feigenbaum route.

2.1.4.3 Physical Interpretation of Results

The most important results obtained with the linear analysis explained in the above sections can be summarized as follows:

(i) A minimum temperature difference between the plates must be exceeded in order to set on fluid instabilities and start convection.
(ii) Convection takes place in the form of a pattern characterized by a specific critical wavelength λ_c and critical wavenumber $k_c = 2\pi/\lambda_c$.

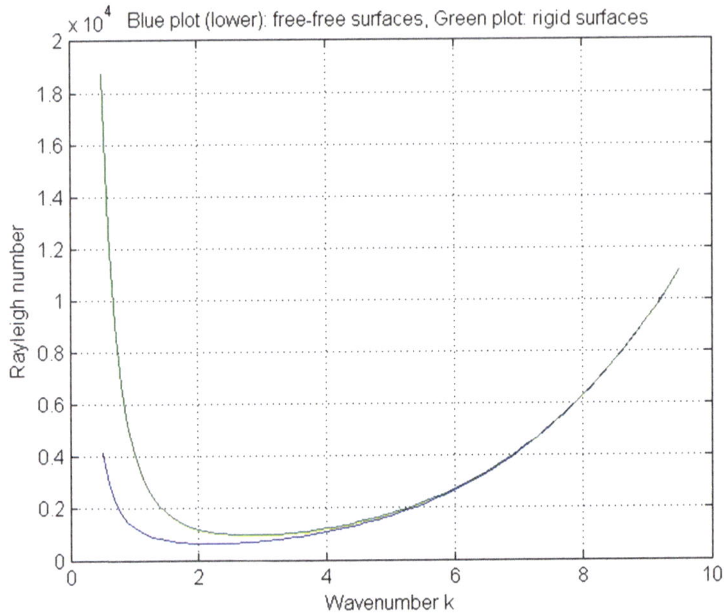

Fig. 2.4 Rayleigh numbers versus wavenumber k

Table 2.1 Critical parameters for the types of boundary conditions

Boundaries	Ra_c	k_c	λ_c
Rigid-rigid	1707.8	3.117	2.016
Rigid-free	1100.7	2.682	2.342
Free-free	657.6	2.221	2.828

Table 2.1 gives the critical parameters Ra_c, k_c, λ_c for the three types of boundary conditions considered.

The minimum critical wavelength is in case of rigid-rigid boundary conditions. In this case a single convection roll (of half wavelength width) is about 1.08 times the depth of the fluid. In case of free-free boundary conditions one has on the contrary the maximum wavelength and a single roll is flatter, about 1.4 H.

These results can be interpreted in terms of the so-called "*balance theorem*", due to Chandrasekhar (1961). The balance theorem states that: "*Instability occurs at a minimum temperature gradient at which a balance can be steadily maintained between the kinetic energy dissipated by viscosity and the internal energy released by the buoyancy force*".

Generally speaking narrow convection cells dissipate more kinetic energy than flatter cells. In case of free-free boundary conditions instability occurs at the lowest possible Ra number, hence at the smaller temperature difference. In this case the minimum kinetic energy is generated by buoyancy and must be dissipated by viscosity. Thus in this case a flatter cell is sufficient to satisfy the balance condition.

Conversely, in case of rigid-rigid boundary conditions instability occurs at the highest Ra_c and therefore at the highest temperature difference. This requires a narrow convection cell to dissipate the higher kinetic energy produced. The critical wavelength in the rigid-rigid case ($\lambda_c = 2.016$) must therefore be shorter than in the free-free case ($\lambda_c = 2.828$).

2.1.5 *Cellular Patterns*

The linear theory explained in the previous section gives information about the critical parameters, mainly k_c and Ra_c, characterizing the onset of instability. In practice a variety of cellular patterns can form and have been observed experimentally. The formation of a particular wave pattern, instead of another one, is not always well understood, but may depend on irregularities in the experimental conditions or on other physical parameters not accounted for in the simple linear theory. Thus, for instance, the hexagonal cellular pattern observed by Bénard depends on surface tension and surface tension gradients due to temperature gradients. A variety of cellular patterns is possible in principle: there are parallel rolls, square cells, rectangular cells, hexagonal cells formed by a combination of elementary triangular cells, axis-symmetric ring cells. Cell boundaries are usually defined by the condition that $\partial w/\partial n = 0$, where n is a direction normal to the cell boundary. This condition implies that a cell boundary is defined as a surface through which there is continuity of the vertical velocity component.

Some examples of cellular patterns are given by Koschmieder (1993).

The simplest cellular pattern is that of parallel rolls given by the equation

$$w = W(z)\cos kx, \tag{2.92}$$

where x is the direction normal to the axes of the rolls which stretch in the y-direction. It is easy to verify that Eq. (2.92) satisfies the membrane equation, Eq. (2.63). Applying Eqs. (2.70) and (2.71) to Eq. (2.92) (with $\zeta = 0$) one derives the velocity components in the x and y-directions:

$$u = \frac{1}{k^2}\frac{\partial^2 w}{\partial x \partial z} = -\frac{1}{k}\frac{dW(z)}{dz}\sin kx, \tag{2.93a}$$

$$v = 0. \tag{2.93b}$$

In case of rigid-rigid boundaries $k_c = 3.117$. The fact that k_c is close to the numerical value of π indicates that convective cells with almost square cross section are preferred. In convective cells with $k < k_c$ (hence $\lambda > \lambda_c$) the kinetic energy associated with the vertical motion is insufficient to overcome viscous dissipation. Conversely, cells with $k > k_c$ (hence $\lambda < \lambda_c$) are less likely to occur because heat conduction between up-flowing and down-flowing streams is strong and damps the available buoyancy.

Square cells are formed by superposition of two patterns of convection rolls with normal axes and are represented by the equation

$$w = W(z)(\cos kx + \cos ky), \tag{2.94}$$

or alternatively by

$$w = W(z)\cos(kx)\cos(ky). \tag{2.95}$$

In case the wave numbers in the x and y-directions are not equal one has rectangular cells represented by either one of the equations

$$w = W(z)(\cos k_1 x + \cos k_2 y), \tag{2.96}$$

$$w = W(z)\cos(k_1 x)\cos(k_2 y). \tag{2.97}$$

The critical wavenumber is in this case $k_c = (k_1^2 + k_2^2)^{1/2}$. In principle every combination of k_1 and k_2 in this formula is possible, thus even very flat square cells are possible, but in practice the fluid follows a rectangular pattern with aspect ratio close to 1.

Hexagonal cells can be represented by the equation

$$w = W(z)\left[\cos\left(\sqrt{3}kx\right)\cos(ky) + \frac{1}{2}\cos(2ky)\right] \tag{2.98}$$

which shows that they arise from the superposition of rectangular cells with rolls.

In polar coordinates (ρ, φ) Eq. (2.98) can be written as

$$w = W(z)\{\cos[k\rho\sin\varphi] + \cos[k\rho\sin(\varphi + \pi/3)] + \cos[k\rho\sin(\varphi + 2\pi/3)]\} \tag{2.99}$$

with $\rho = \sqrt{x^2 + y^2}$. Equation (2.99) shows that hexagonal cells arise from the superposition of three fields of convection rolls having axes displaced in orientation by 60°. Hexagonal cells can also be described mathematically by a combination of triangular cells.

In case of liquids, hexagonal cells have a rising stream in the core of the cells, and descending flows at the cells boundaries. The apposite occurs usually in gases. The reasons for this different behaviour are the temperature dependence of viscosity in the different fluids. In the atmosphere both flow patterns can be seen.

Ring cells are represented by the equation

$$w = W(z)J_0(kr) \tag{2.100}$$

where $J_0(kr)$ is the Bessel function of zero order.

Solutions of the linear theory corresponding to $n \geq 2$ are referred to as "higher modes" and they correspond to critical Rayleigh numbers, thus to critical temperature gradients, much larger than in normal mode $n = 1$. Formula (2.83) shows that Ra_c is proportional to n^4, and ΔT_c is proportional to Ra_c (see Eq. (2.85)). Higher modes, which correspond to multiple patterns, are difficult to realize experimentally because the fluid usually sets in a normal-mode pattern.

2.2 Applications to Molecular Dynamics

Following the analytical description of the Bénard-Rayleigh problem we turn now to the molecular dynamic approach, which, compared to the previous one, appears to by conceptually simpler. This simplicity is, however, only apparent because the correct application of the molecular dynamic requires both deep knowledge and computational experience.

In the following subsection we present the structure of the program, dubbed **SQUARE CAVITY**, to simulate the Bénard-Rayleigh experiments. The program actually consists in the simulation of three species, water vapour, air, and hydrogen and was at first written for simulating deflagration of hydrogen in a mixture with air and water vapour in case of an hypothetical accident in a pressurized water reactor. For this reason all fields allocated in the program are threefold fields, treating at the same times all species. In the application for the square cavity problem the fields treating water vapour and hydrogen have concentrations set to zero. Besides, the code can be run both in 2D and 3D options.

The program is too large for being proposed as a test case. Interested researchers wishing to receive the program may contact the publisher and the authors. However, it should be kept in mind that the code is far from being completed: it should be considered as a research tool for further investigations.

2.2.1 *Structure of the SQUARE CAVITY Program*

The code, written in FORTRAN IV and organized in 37 subroutines, consists of about 3500 statements including comments illustrating the organization of the program and block statements. The program is divided in a series of blocs, each corresponding to main features of the code. The main blocs are: B1—Initialization, B2—Time marching and data sampling, B3—Output data.

1. **Bloc B1: Initialization**
   (i) This bloc starts with a short description of subroutines and functions listed in alphabetical order.
   (ii) Follows a list of the main FORTRAN symbols, listed in alphabetical order.
   (iii) Read input data, subroutine RINPUT.

(iv) A flag called IREST is used to switch on a restart file. Default is IREST = 0; in case of IREST = 1 a previously written restart file is read. This feature allows for continuation of long calculations.
(v) Subroutine GEOMS is used for defining the geometrical data of the system.
(vi) Subroutine PHYSP defines the physical properties of materials in the system.
(vii) Subroutine INITF initializes flow of molecules, i.e. initial positions and velocity components in the full computational domain. Computes maxima and minima of velocity components.
(viii) Subroutine MOLIDO (MOLecole In DOmain) checks if all molecules are inside the computational domain.
(ix) Subroutine MIMXVE computes minimum and maximum of the velocity components of the molecules.
(x) Subroutine CHKMEC computes mean values of velocity components for all molecules.
(xi) Subroutine CHKMET computes instantaneous translational temperature.
(xii) Subroutine CDIAEF computes effective cross section of air, based on the Lennard-Jones potential.
(xiii) Subroutine ZEROFL zeroes counters and fields, only at the beginning of a calculation.

2. **Bloc B2: Time Marching and Data Sampling**

In a loop which runs over an index IT, from a minimum to a maximum, with a time step Δt, the following steps are made:

(xiv) Subroutine MOLIDO (MOLecole In DOmain) checks again that molecules are inside the computational domain.
(xv) Subroutine SCREEN checks that a molecule has not moved outside of the computational domain.
(xvi) Subroutine SPECDI computes normal velocity component and position of molecules after diffuse reflection from material surfaces or specular reflection from symmetry surfaces.
(xvii) Subroutine REPLACE replaces a molecule reflected out of the computational domain.
(xviii) Subroutine INDEXING fills arrays LC, LCR, ICEL for indexing of molecules in cells.
(xix) Subroutine COLMOD (COLlision MODel) computes velocity components of a pair of molecules which collide.
(xx) Arrays for data sampling are filled at regular intervals during the calculation.
(xxi) OUTRES (Restart) file is written at regular intervals during the calculation and at the last time step.

3. **Bloc B3: Output Data**

(xxii) Output data are written as final results.
(xxiii) OUTRES (Restart) file is written for the last time step.

The subsequent section presents results of the numerical simulations.

2.2.2 *Flow Heated from Below: Bénard Problem*

A qualitative verification of our codes for molecular dynamic calculations was made by simulating numerically the insurgence of instability in the Bénard problem, with a thin layer of fluid between infinite parallel plates heated from below. Starting from an isotropic and homogeneous state in the fluid, the temperature of the bottom plate was slowly increased, till a critical Rayleigh number was reached and dynamical instability occurred, manifested by the formation of convection rolls.

We simulated in a 2D (x, z) domain the insurgence of instability in air. Data about the domain and the calculations were as follows (Table 2.2):

In the final conditions, after increasing the temperature of the lower plate, the Rayleigh number was Ra = 2110.

Figure 2.5 shows the temperature distribution in the domain in the situation in which convection rolls are well developed and transport heat from the lower to the upper plate, but, at this point of the simulation, the Rayleigh number is not as high as to induce transition to turbulence and to chaotic behaviour in the fluid.

Figure 2.6 shows the velocity plot in the developed flow.

Developed counter-rotating cells are shown in Fig. 2.7. The spots highlighted in red denote streams directed in the right (positive) direction, those highlighted in blue denote streams directed to the left (negative) direction. The viewgraph shows the horizontal velocity component of the air in the phase of the full development of the convection cells in the fluid. The picture shows an anti-symmetric pattern with respect to a vertical axis through the middle of the domain, at a given instant of the numerical simulation. In reality, as time evolves, the convection rolls are not strictly fixed in space but oscillate around a mean equilibrium position.

Figure 2.8 shows the distribution of the z-component of the velocity vector. The spots highlighted in red denote ascending streams, and those highlighted in blue denote descending streams. The viewgraph shows the vertical velocity component of the air with fullly developed convection cells. The pattern is symmetric with respect to a vertical axis through the middle of the domain and shows fully developed cells in the middle of the domain with half-cells at the boundaries.

Figure 2.9 shows a diffuse distribution of the velocity vector.

Table 2.2 Data for the Bénard problem

Size of simulated domain: Xmax	= 0.0226 m
Zmax	= 0.0070 m
Aspect ratio (Xmax/Zmax)	= 3.223
Number of meshes (400× 80)	= 32,000
Size of meshes: Δx	= 56.5 μm
Δz	= 87.5 μm
Number of sample molecules, Nmol	= 320,000
Time step Δt	$= 1.0 \times 10^{-5}$ s
Temperature upper plate	= 303 K
Temperature of lower plate (end value)	= 383 K

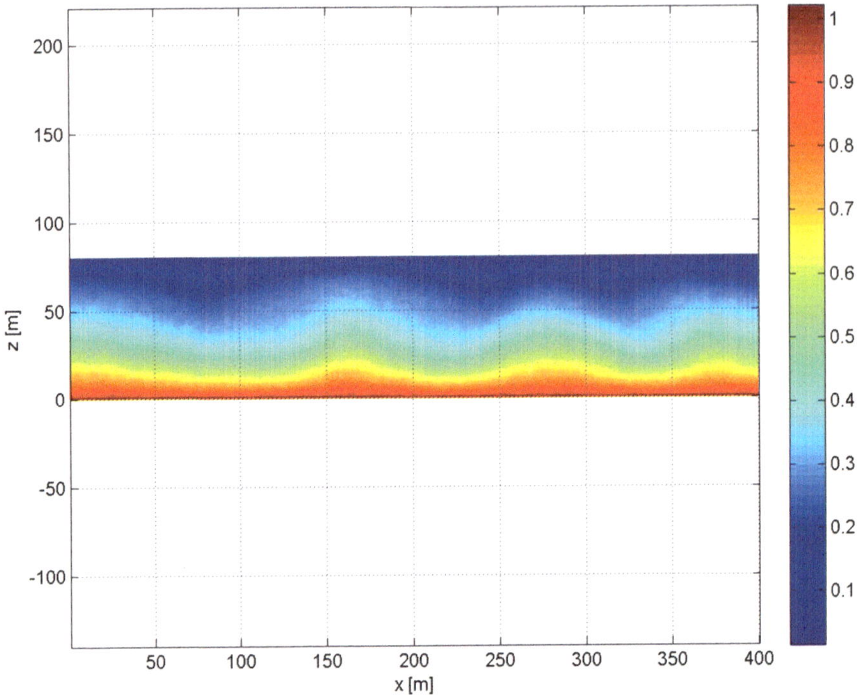

Fig. 2.5 *Temperature distribution in the domain*

Figures 2.10 and 2.11 show the normalized density distribution and the normalized temperature distribution, respectively.

The young scientists are encouraged to investigate the impact of the aspect ratio (Xmax/Zmax) on the size and distribution of the convective rolls in the domain and to investigate the frequency of the oscillations of the rolls.

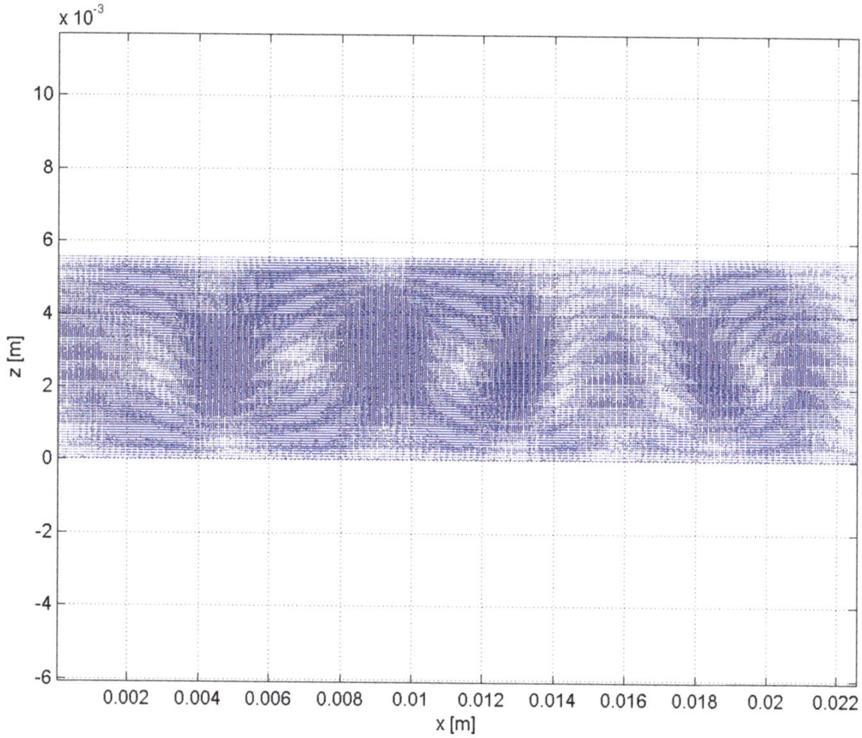

Fig. 2.6 *Velocity plot*

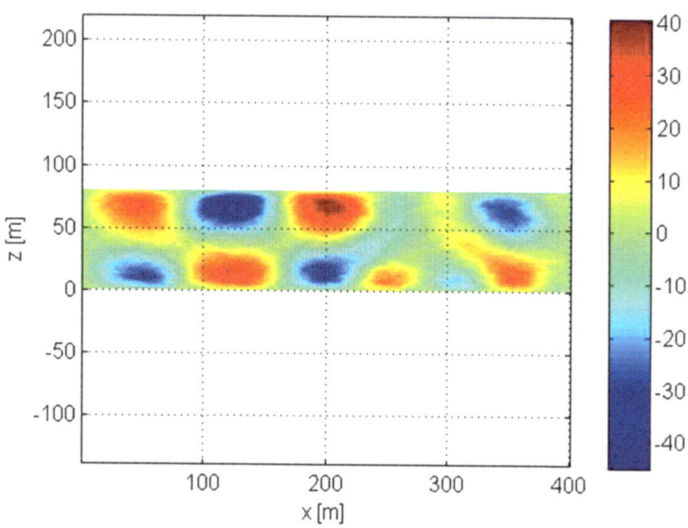

Fig. 2.7 *Distribution of the x-component of the velocity vector*

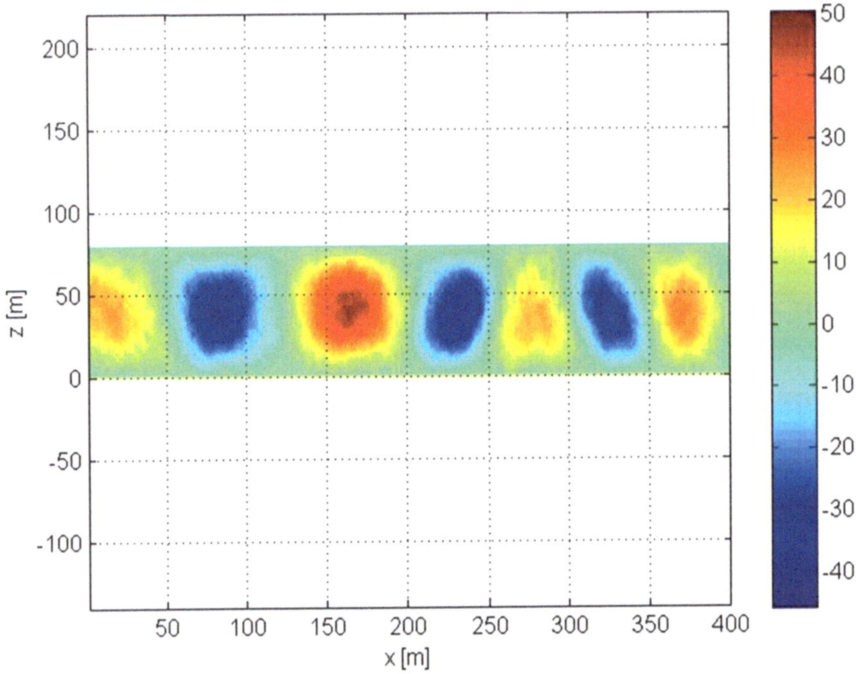

Fig. 2.8 *Distribution of the z-component of the velocity vector*

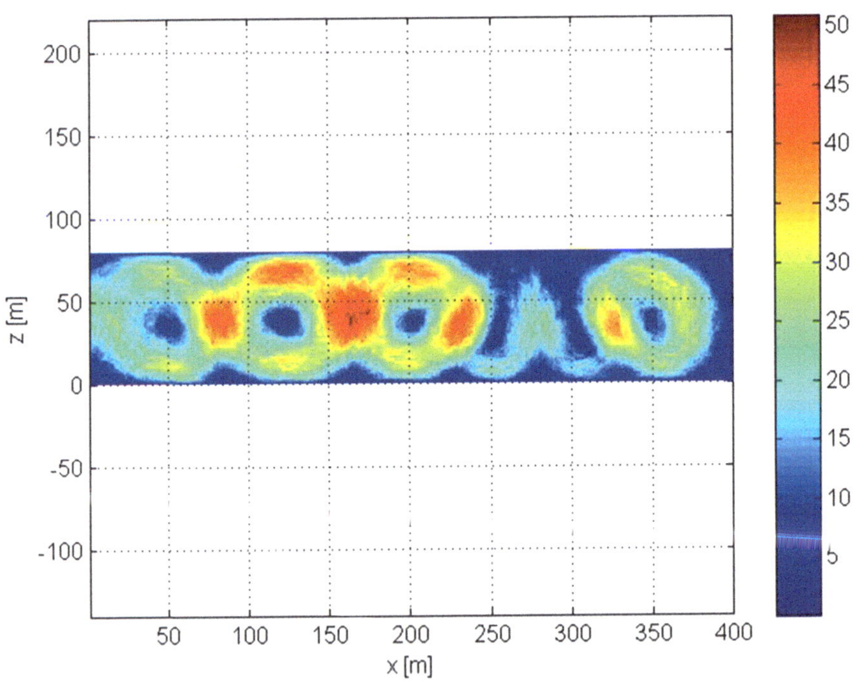

Fig. 2.9 *Diffuse distribution of the velocity vector*

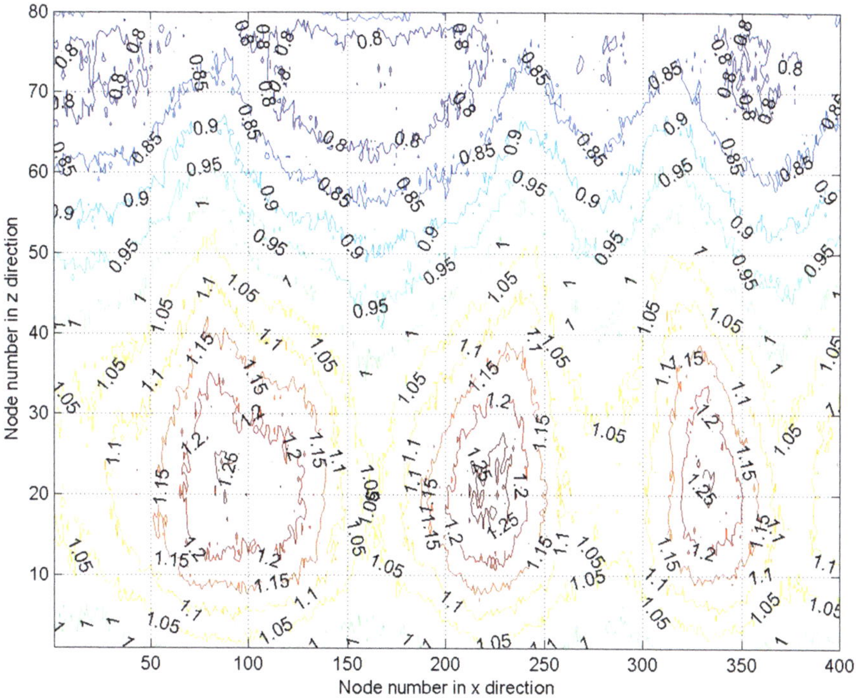

Fig. 2.10 *Normalized density distribution*

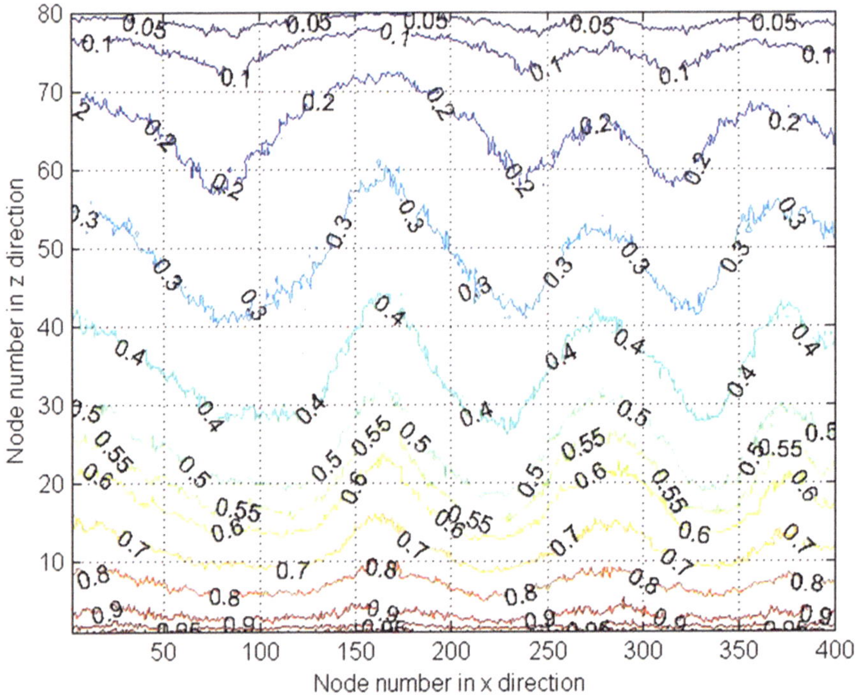

Fig. 2.11 *Normalized temperature distribution*

2.3 Bénard Problem Between Concentric Circles

The Bénard problem has been investigated numerically in the 2D domain between an inner circle of diameter D1 = 6 mm and an outer radius of diameter D2 = 20 mm. The surfaces of the circles were kept at T1 = 300 K and T2 = 240 K, respectively. Air molecules were initialized in the inner domain with initial temperature of 270 K. A transient calculation was then started from these initial conditions and carried on till 15,000 time steps of $\Delta t = 10^6$ s (Table 2.3).

The surface of the computational domain inside the outer circle and outside the inner circle is $S = \pi\left(R_2^2 - R_1^2\right) = \pi \cdot 91\ \text{mm}^2$ and corresponds to 10,291 elementary cells. The number of sample molecules is therefore about six per cell.

Figure 2.12 shows the normalized molecular density in the domain at the end of the calculation.

Figure 2.13 shows the vertical component *W* and the horizontal component *U* of molecular velocities after 9.000 time steps. The colour bar shows the velocity scale in m/s.

Figure 2.14 shows the vertical component *W* and the horizontal component *U* of molecular velocities at the end of the calculation, after 15.000 time steps.

Table 2.3 Data for the Bénard problem between concentric circles

Size of simulated domain: Xmax	= 20 mm
Zmax	= 20 mm
Number of meshes 120 × 120	= 14.400
Size of meshes: Δx	= 0.16667 mm
Δz	= 0.166667 mm
Number of sample molecules, Nmol	= 60,000
Time step Δt	$= 1.0 \times 10^{-6}$ s
Temperature of inner circle	= 300 K
Temperature of outer circle	= 240 K

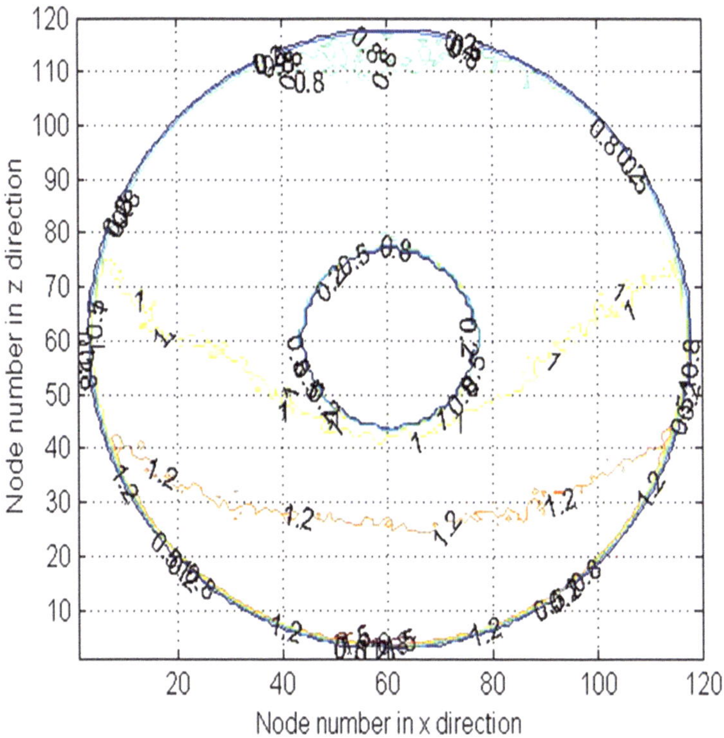

Fig. 2.12 *Normalized molecular density between inners and outer circular 2D domains*

Combining the information displayed in the above pictures one sees two main counter-rotating circulation zones, one counter-clockwise at the left part of the domain and another one clockwise at the right part.

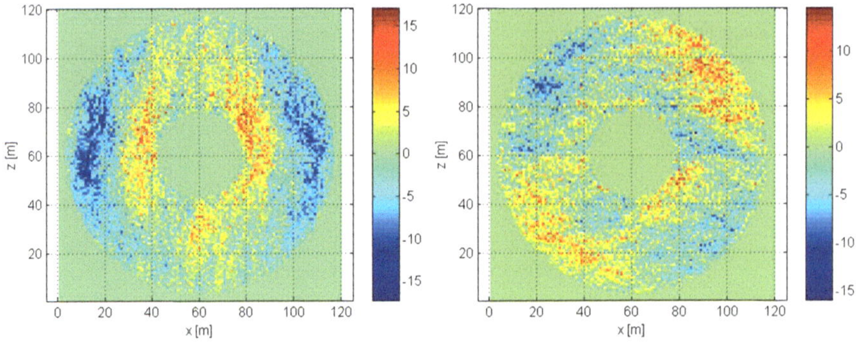

Fig. 2.13 *Vertical component W and horizontal component U of molecular velocity after 9.000 time steps*

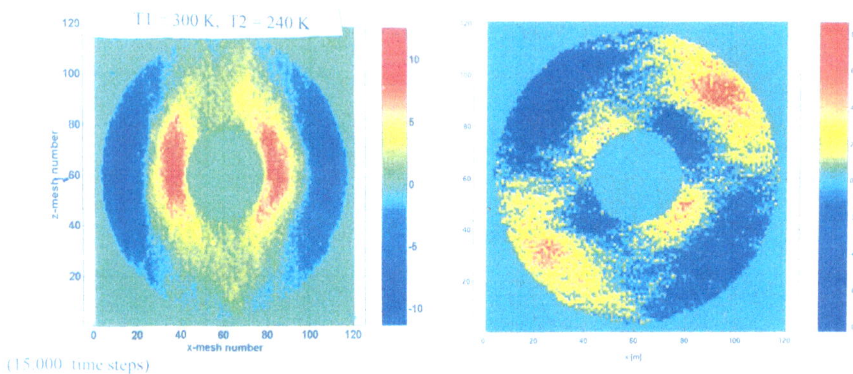

Fig. 2.14 *Vertical component W and horizontal component U of molecular velocity after at the end of the calculation after 15,000 time steps*

2.4 2D Flow Around a Circular Obstacle

Although our computational code allows for 3D applications we give a sample of a 2D calculation because the 3D case would display symmetry around a vertical axis through the midpoint of the simulated domain.

In the 2D case, the air flow directed in vertical z-direction into a square computational domain with given velocity at inlet has been simulated numerically to verify qualitatively the distributions of molecular density, kinetic temperature, and components of flow velocity around the obstacle. We show a few results obtained for a reference calculation, referred to (in our internal documentation) as Calculation D6, at high inlet velocity of the air flow. Data for this calculation are given in Table 2.1.

Figure 2.15 shows the normalized distribution of molecular density around the obstacle. Values of the normalized molecular density of up to three are computed in front of the obstacle where air molecules impinge on it and are diffusively reflected, while a rarefaction zone with values of the normalized molecular density in the range 0.2–0.5 is computed in the wake of the obstacle.

Table 2.4 Data for calculation D6

Radius R of obstacle	= 0.003 m
Size of simulated domain Xmax = Zmax	= 0.036 m
Number of meshes 120 × 120	= 14,400
Size of meshes $\Delta x = \Delta z$	= 300 μm
Mean velocity in z direction, W	= 500 m/s
Reynolds number, Re	= 266,000
Number of sample molecules, Nmol	= 60,000
Time step Δt	$= 1.0 \times 10^{-6}$ s
Temperature of air flux Ta	= 300 °K
Temperature of surface of obstacle Ts	= 240 °K

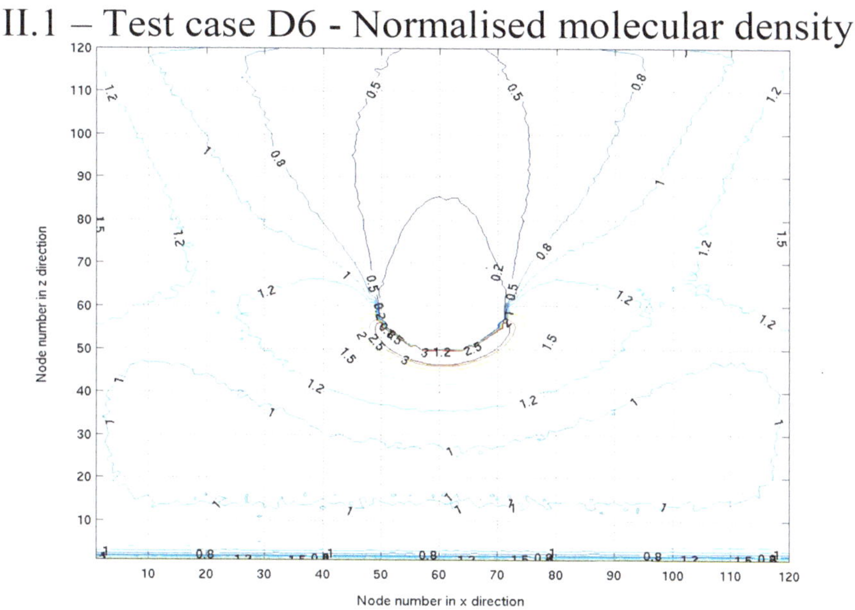

Fig. 2.15 *Normalized molecular density.* [The title of the figure refers to coding of internal reports]

Figure 2.16 shows the molecular temperature. The total energy of the molecules obtained as sum of the kinetic (translational) energy and of the internal (rotational) energy allows computing the molecular temperature. The largest values of the molecular temperature are obtained in front of the obstacle. The lowest values are at the surface of the impingement zone, where the molecules are diffusively reflected and in the wake where molecules are almost at rest.

Figure 2.17 shows the horizontal component *U* of the air velocity distribution around the obstacle. The pattern is anti-symmetric due to the deflection of the airflow from the obstacle and reattachment beyond the wake. Maximum values of the horizontal velocity components reach about 2/3 of the value of the mean flow velocity in the vertical direction.

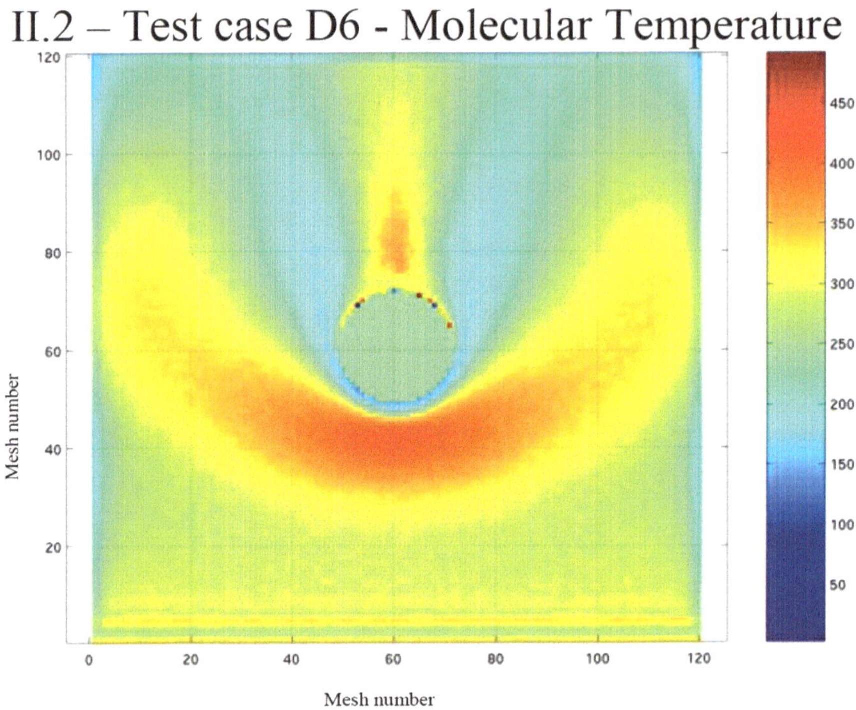

Fig. 2.16 *Molecular temperature*

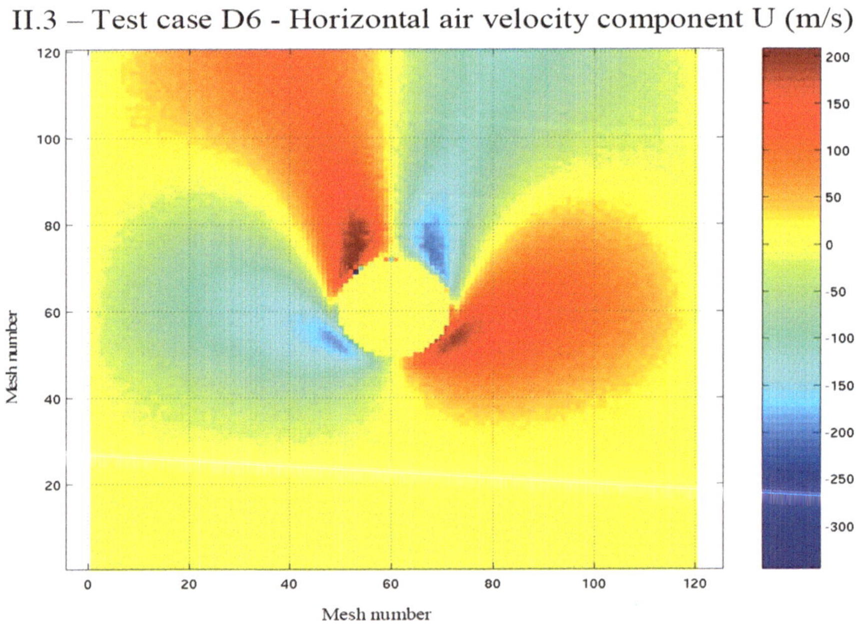

Fig. 2.17 *Horizontal air velocity component*

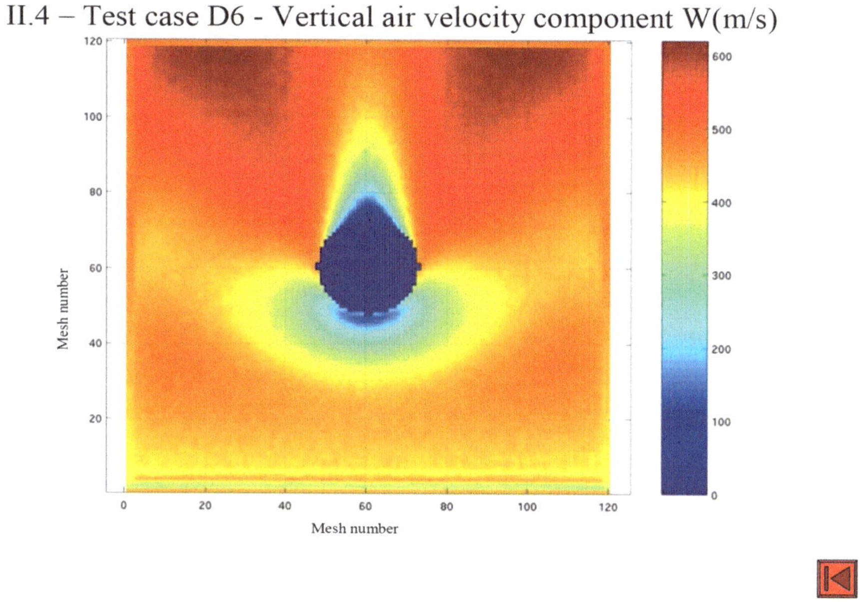

Fig. 2.18 *Vertical air velocity component*

Figure 2.18 shows the symmetric distribution of the vertical component *W* of the air flow velocity around the obstacle. Maximum values are about 20% larger than the mean air flow velocity, while minimum values, with molecules almost at rest, are obtained in the stagnation zones in front of the obstacle and in the wake.

2.5 Thermal Creep

Thermal creep is the phenomenon by which, in a fluid domain with a temperature gradient, a flow is induced, directed along the temperature gradient of the bounding surfaces. Micro-machine engineering exploits thermal creep-induced flows, because pumping effects can be achieved without any moving part of the system. Recently the thermal creep phenomenon has received increasing attention in a variety of fields ranging from nanotechnology to the physics of cosmic particles.

Numerical investigations of thermal creep are documented for instance in Sone et al. (1966), which report DSMC calculations aiming at quantifying velocity which can be achieved and pressure gradients which can be sustained by a one-dimensional flow. A state-of-the-art computational program has been released by the Bird's group (G.A.B. Consulting Pty Ltd, 1999), which includes thermal creep simulations among its basic test cases. Applications of thermal creep flows, in concomitance with thermophoretic forces, are being considered in the frame of micro-particle physics in outer space environment.

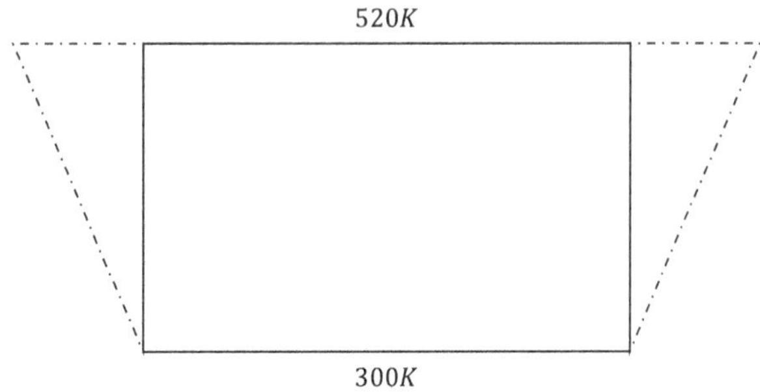

Fig. 2.19 *Domain for the thermal creep calculation*

Table 2.5 Conditions for the thermal creep calculation

Linear dimensions: xmax	= 0.0001 m
zmax	= 0.0005 m
Discretization of the domain: Imax	= 40 cells
Kmax	= 20 cells
Temperature: lower plate: T_1	= 300 K
Upper plate: T_2	= 520 K
Linear temperature profile along the vertical plates	
Number of sample molecules: 80,000 (100 mol/cell)	
F_N (number of real molecules represented by one sample molecule) = 2.5×10^{22}	
$\lambda = 52.3$ µm (mean free path)	
$K_n = 0.074$ (Knudsen number)	
$\rho = 1.2 \times 10^{-3}$ kg/m^3 (mean density)	
Time step: $\Delta t = 10^{-8}$s	

In this section we consider a simple application of the thermal creep in a two-dimensional domain bounded by material plates. With reference to a (x, z) coordinate system we consider two horizontal plates kept at different but constant temperature, connected by vertical plates with a linear temperature profile (Fig. 2.19).

Conditions for the calculations are given in Table 2.5:

The computation was continued till a steady state pattern was established. The most important result of this calculation is shown in Fig. 2.20 which displays two counter rotating whirls. Along the vertical plates thermal creep induces ascending flows while in the centre of the domain a strong descending flow develops. The maximum flow velocity in the whirls is about 6 m/s.

Figures 2.21 and 2.22 show the x- and z-components of the velocity vectors. The interpretation of the figures is immediate comparing them with the velocity distribution of Fig. 2.20.

In Fig. 2.22 the blue spot denotes strong negative components of the x-velocity vector; the red spot denotes positive components.

In Fig. 2.23 the blue spot denotes strong negative components of the y-velocity vector; the red spot denotes positive components.

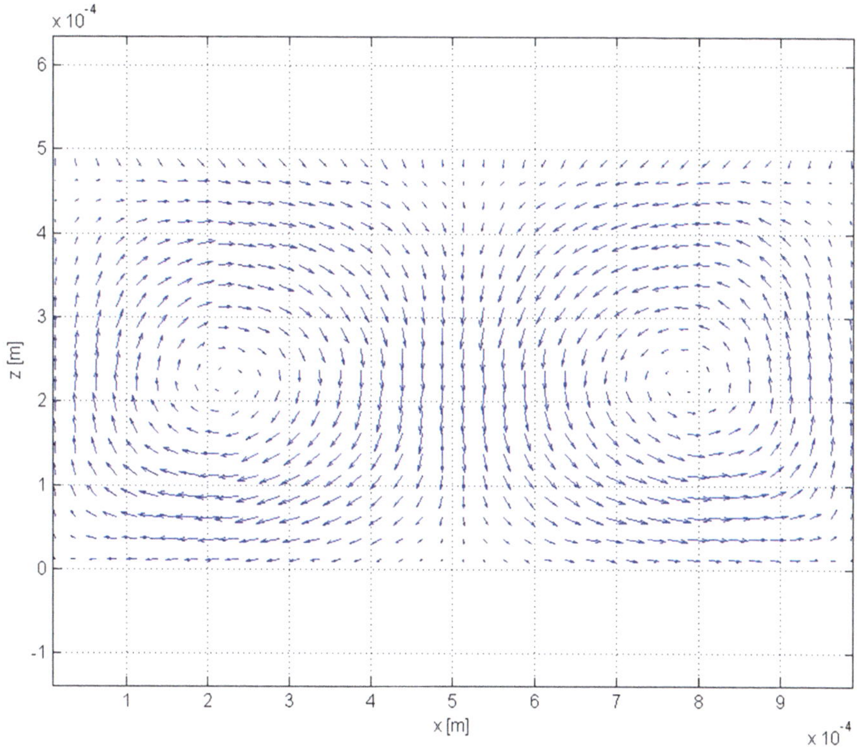

Fig. 2.20 *Velocity vectors in the thermal creep calculation*

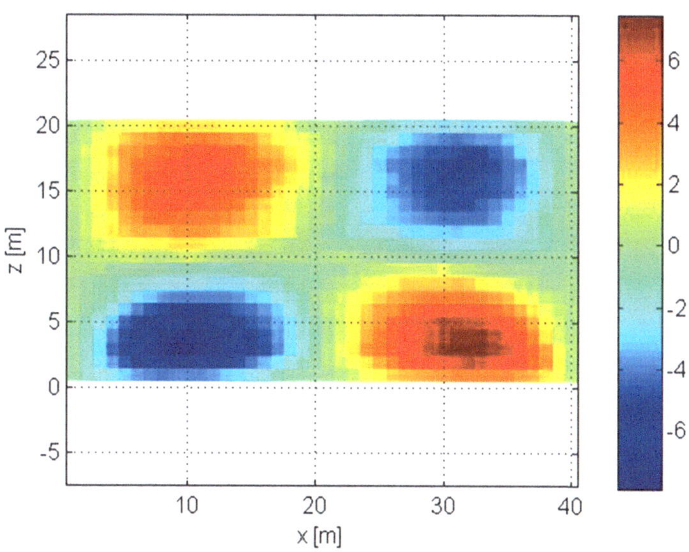

Fig. 2.21 *x-component of the velocity vector*

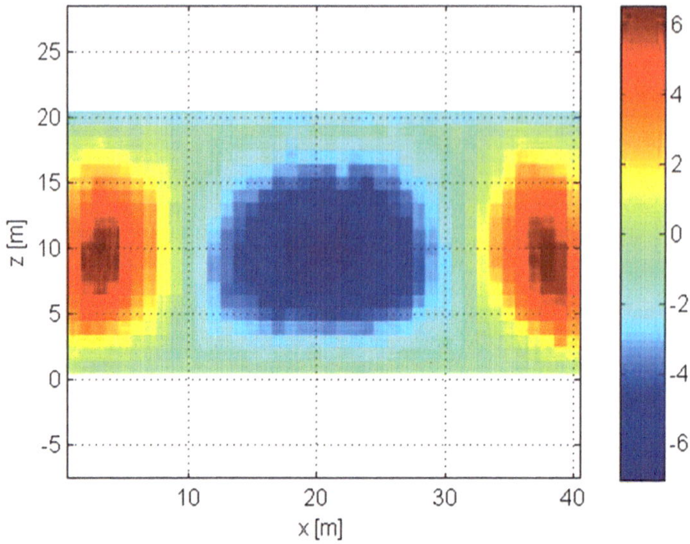

Fig. 2.22 *z-component of the velocity vector*

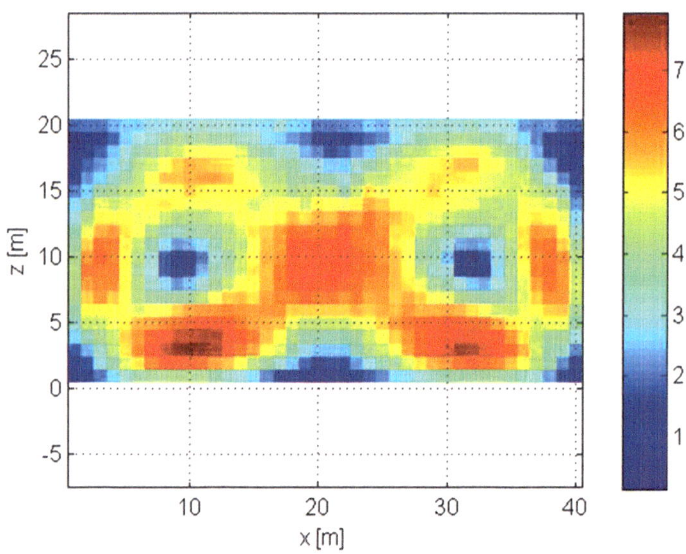

Fig. 2.23 *Normalized velocity distribution*

Figure 2.24 displays the normalized vector of the velocity distribution independent of its direction. The stream is almost at rest in the corners, in the upper and lower middle regions and in the two central whirls.

The normalized density and the normalized temperature distribution are shown in Fig. 2.24 and Fig. 2.25, respectively.

Physical properties of air are given in Table 2.6.

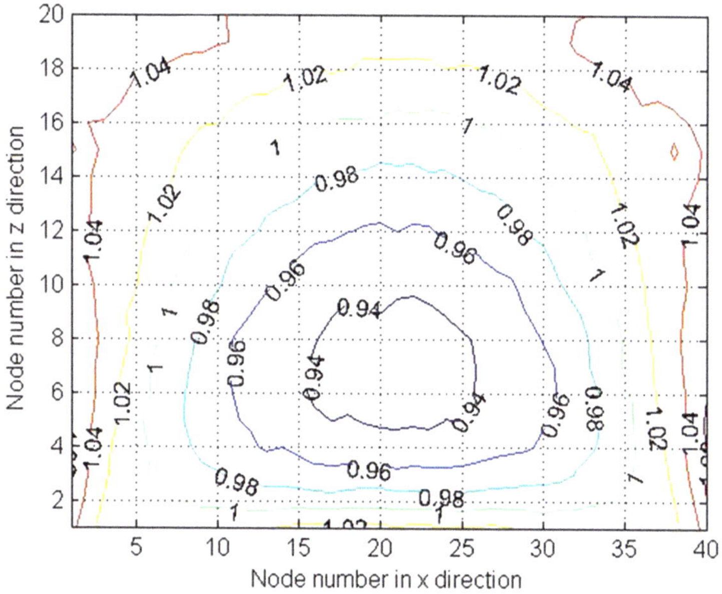

Fig. 2.24 *Normalized density*

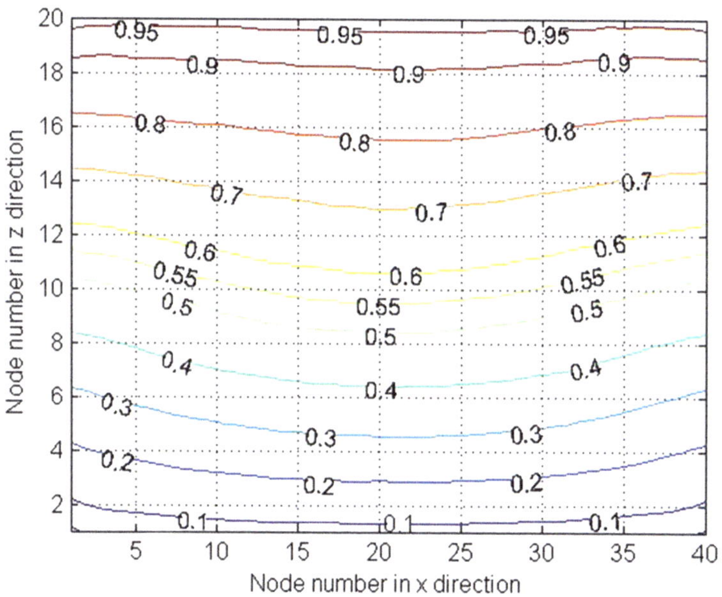

Fig. 2.25 *Normalized temperature distribution*

Table 2.6 *Physical properties of air*—[Data generated from the EES (Engineering Equation Solver)]

Properties of air at 1 atm pressure							
Temp. T(K)	Density (kg/m^3)	Specific heat cp (J/kg K)	Thermal conductivity (W/m K)	Thermal diffusivity (m^2/s)	Dynamic viscosity (kg/m s)	Kinematic viscosity (m^2/s)	Prandtl number Pr
−150	2.866	983	0.01171	4.158 $10*^{-6}$	8.636 $10*^{-6}$	3.013 $10*^{-6}$	0.7246
−100	2.038	966	0.01582	8.036 $10*^{-6}$	1.189 $10*^{-6}$	5.837 $10*^{-6}$	0.7263
−50	1.582	999	0.01979	1.252 $10*^{-5}$	1.474 $10*^{-5}$	9.319 $10*^{-6}$	0.7440
−40	1.514	1002	0.02057	1.356 $10*^{-5}$	1.527 $10*^{-5}$	1.008 $10*^{-5}$	0.7436
−30	1.451	1004	0.02134	1.465 $10*^{-5}$	1.579 $10*^{-5}$	1.087 $10*^{-5}$	0.7425
−20	1.394	1005	0.02211	1.578 $10*^{-5}$	1.630 $10*^{-5}$	1.169 $10*^{-5}$	0.7408
−10	1.341	1006	0.02288	1.696 $10*^{-5}$	1.680 $10*^{-5}$	1.252 $10*^{-5}$	0.7387
0	1.292	1006	0.02364	1.818 $10*^{-5}$	1.729 $10*^{-5}$	1.338 $10*^{-5}$	0.7362
5	1.269	1006	0.02401	1.880 $10*^{-5}$	1.754 $10*^{-5}$	1.382 $10*^{-5}$	0.7350
10	1.246	1006	0.02439	1.944 $10*^{-5}$	1.778 $10*^{-5}$	1.426 $10*^{-5}$	0.7336
15	1.225	1007	0.02476	2.009 $10*^{-5}$	1.802 $10*^{-5}$	1.470 $10*^{-5}$	0.7323
20	1.204	1007	0.02514	2.074 $10*^{-5}$	1.825 $10*^{-5}$	1.516 $10*^{-5}$	0.7309
25	1.184	1007	0.02551	2.141 $10*^{-5}$	1.849 $10*^{-5}$	1.562 $10*^{-5}$	0.7296
30	1.164	1007	0.02588	2.208 $10*^{-5}$	1.872 $10*^{-5}$	1.608 $10*^{-5}$	0.7282
35	1.145	1007	0.02625	2.277 $10*^{-5}$	1.895 $10*^{-5}$	1.655 $10*^{-5}$	0.7268
40	1.127	1007	0.02662	2.346 $10*^{-5}$	1.918 $10*^{-5}$	1.702 $10*^{-5}$	0.7255
45	1.109	1007	0.02699	2.416 $10*^{-5}$	1.941 $10*^{-5}$	1.750 $10*^{-5}$	0.7241
50	1.092	1007	0.02735	2.487 $10*^{-5}$	1.963 $10*^{-5}$	1.798 $10*^{-5}$	0.7228
60	1.059	1007	0.02808	2.632 $10*^{-5}$	2.008 $10*^{-5}$	1.896 $10*^{-5}$	0.7202
70	1.028	1007	0.02881	2.780 $10*^{-5}$	2.052 $10*^{-5}$	1.995 $10*^{-5}$	0.7177
80	0.9994	1008	0.02953	2.931 $10*^{-5}$	2.096 $10*^{-5}$	2.097 $10*^{-5}$	0.7154
90	0.9718	1008	0.03024	3.086 $10*^{-5}$	2.139 $10*^{-5}$	2.201 $10*^{-5}$	0.7132
100	0.9458	1009	0.03095	3.243 $10*^{-5}$	2.181 $10*^{-5}$	2.306 $10*^{-5}$	0.7111
120	0.8977	1011	0.03235	3.565 $10*^{-5}$	2.264 $10*^{-5}$	2.522 $10*^{-5}$	0.7073
140	0.8542	1013	0.03374	3.898 $10*^{-5}$	2.345 $10*^{-5}$	2.745 $10*^{-5}$	0.7041
160	0.8148	1016	0.03511	4.241 $10*^{-5}$	2.420 $10*^{-5}$	2.975 $10*^{-5}$	0.7014
180	0.7788	1019	0.03646	4.593 $10*^{-5}$	2.504 $10*^{-5}$	3.212 $10*^{-5}$	0.6992
200	0.7459	1023	0.03779	4.954 $10*^{-5}$	2.577 $10*^{-5}$	3.455 $10*^{-5}$	0.6974
250	0.6746	1033	0.04104	5.890 $10*^{-5}$	2.760 $10*^{-5}$	4.091 $10*^{-5}$	0.6946
300	0.6158	1044	0.04418	6.871 $10*^{-5}$	2.934 $10*^{-5}$	4.765 $10*^{-5}$	0.6935
350	0.5664	1056	0.04721	7.892 $10*^{-5}$	3.101 $10*^{-5}$	5.475 $10*^{-5}$	0.6937
400	0.5243	1069	0.05015	8.951 $10*^{-5}$	3.261 $10*^{-5}$	6.219 $10*^{-5}$	0.6948
450	0.4880	1081	0.05298	1.004 $10*^{-4}$	3.415 $10*^{-5}$	6.997 $10*^{-5}$	0.6965
500	0.4565	1093	0.05572	1.117 $10*^{-4}$	3.563 $10*^{-5}$	7.806 $10*^{-5}$	0.6986
600	0.4042	1115	0.06093	1.352 $10*^{-4}$	3.846 $10*^{-5}$	9.515 $10*^{-5}$	0.7037
700	0.3627	1135	0.06581	1.598 $10*^{-4}$	4.111 $10*^{-5}$	1.133 $10*^{-4}$	0.7092
800	0.3289	1153	0.07037	1.855 $10*^{-4}$	4.362 $10*^{-5}$	1.326 $10*^{-4}$	0.7149
900	0.3008	1169	0.07465	2.122 $10*^{-4}$	4.600 $10*^{-5}$	1.529 $10*^{-4}$	0.7206
1000	0.2772	1184	0.07868	2.398 $10*^{-4}$	4.826 $10*^{-5}$	1.741 $10*^{-4}$	0.7260
1500	0.1990	1234	0.09599	3.908 $10*^{-4}$	5.817 $10*^{-5}$	2.922 $10*^{-4}$	0.7478
2000	0.1553	1264	0.11113	5.664 $10*^{-4}$	6.630 $10*^{-5}$	4.270 $10*^{-4}$	0.7539

Table 2.7 *Dimensionless numbers used in thermal and fluid-dynamics problems*

Auxiliary definitions for this table:	
$\alpha^2 = \frac{\lambda}{\rho c_p}$	(m^2/s)
$\beta = -\frac{1}{\rho}\left(\frac{\partial \rho}{\partial T}\right)$	Volumetric expansion coefficient (K^{-1})
$\mu = \nu\rho$	Dynamic viscosity (kg/m s)
$\nu = \frac{\mu}{\rho}$	Kinematic viscosity (m^2/s)
$F = \frac{u}{\left(g\frac{\Delta\rho}{\rho}L\right)^{1/2}}$	Froude number
$Gr = \frac{L^3 g\rho^2\beta\Delta T}{\mu^2}$	Grashof number (for vertical plates)
$Gr = \frac{D^3 g\rho^2\beta\Delta T}{\mu^2}$	Grashof number (for pipes)
$Nu = \frac{\alpha D_h}{\lambda}$	Nusselt number
$Pe = \mathrm{Re}\cdot\mathrm{Pr} = \frac{wD_h}{\alpha^2}$	Peclet number
$\mathrm{Pr} = \frac{\nu}{\alpha^2} = \frac{\mu c_p}{\lambda}$	Prandtl number
$Ra = \frac{G\beta g H^4}{\nu\alpha^2} = \frac{\partial T}{\partial z}\frac{\beta g H^4}{\nu\alpha^2}$	Rayleigh number (H = distance between plates)
$\mathrm{Re} = \frac{wD_h}{\nu} = \frac{\rho\nu D_h}{\mu}$	Reynolds number
$Ri = -\frac{\lvert g\rvert}{\rho}\frac{\partial\rho/\partial z}{(\partial u/\partial z)^2}$	Gradient Richardson number
$Ri = -\frac{\Delta\rho}{\rho}\frac{gh}{u^2}$	Richardson number
$Sc = \frac{\nu}{D}$	Schmidt number (D = mass diffusion rate)
$St = \frac{Lf}{u}$	Strouhal number (f = frequency of vortex shedding)

Table 2.7 gives the definitions of useful dimensionless numbers used in thermal and fluid-dynamics problems.

References

Bénard, H. (1900). Les tourbillons cellulaires dans une nappe liquide. *Revue générale des Sciences pures et appliqués, 11*, 1261–1271. et 1309–28.

Bénard, H. (1901). Les tourbillons cellulaires dans une nappe liquide transportant de la chaleur par convection en régime permanent. *Annales de Chimie et de Physiques, 23*, 62–144.

Chandrasekhar, S. (1961). *Hydrodynamic and hydromagnetic stability*. Dover Publications.

G.A.B. Consulting Pty Ltd. (1999). *The DS2G program user's guide, Version 3.2.*

Koschmieder, E. L. (1993). *Bénard cells and Taylor vortices*. Cambridge University Press.

Rayleigh, L. (1916). On convective currents in a horizontal layer of fluid when the higher temperature is on the under side. *Philosophical Magazine, 32*, 529–546.

Sone, Y., Waniguchi, Y., & Aoki, K. (1966). One-way flow of a rarefied gas induced in a channel with a periodic temperature distribution. *Physics of Fluids, 8*, 8.

Chapter 3
Special Applications of the Discrete Random Deviation

Contents

This chapter is dedicated to special applications of techniques which could be called "discrete random deviations". This term implies the use of random deviations [real numbers in the open interval $(0 < r < 1)$] applied in the one- or n-dimensional case to simulate events occurring in n-dimensional space. Examples of application are given for several variants of Diffusion Limited Accretion (DLA), which consists in diffusion of small particles carried by a fluid and in formations of aggregates of particles upon impingement with the growing aggregates. The phenomenon occurs in soot formation, in clusters of aerosol particles in the atmosphere, in dendrites formed by mineralization, and the like. Numerical simulations show formation of

Text by M. Bottoni

M. Bottoni et al., *Molecular Dynamics - Theory and Applications*, Mechanical Engineering Series, https://doi.org/10.1007/978-3-031-84567-3_3

branches in a confined quadrant of the definition domain; grow of dendrites in an unbound domain and geodes growth un spherical geometry.

Another example of application is given for the numerical solution of the Langevin equation with the method of stochastic approach in the frame of the Fokker-Plank theory. Computer programs developed for these purposes are explained and numerical results are given.

The following appendices illustrate computer programs dedicated to variants of DLA:

Appendix A.1: Listing and explanation of code DIFFLA;
Appendix A.2: Listing and explanation of code DLA_1;
Appendix A.3: Listing and explanation of code DLA_2;
Appendix A.4: Listing and explanation of code CORR1D;
Appendix A.5: Listing and explanation of code CORR2D (With PROCESS);

The following appendix illustrate the computer programs dedicated to the numerical solution of the Langevin equation:

Appendix A.6: Listing and explanation of code LANGEVIN;

3.1 Diffusion Limited Accretion (DLA)

3.1.1 Diffusion Limited Aggregation (DLA) and Saffman-Taylor Finger

Diffusion of small particles and their aggregation to form clusters is a physical phenomenon which occurs frequently in the natural world as, for instance, in the formation of soot, clusters of aerosol particles, and dendrites formed by percolation of mineralizations in rock matrices. The rate-limiting step in the aggregation process is due to the finite diffusion speed of the particles; hence, the process is defined as Diffusion Limited Aggregation (DLA). The process can be simulated numerically partitioning the two- or three-dimensional definition domain in a square lattice, assuming in the lattice a seed (a single "sticky" particle) and following the random diffusion of a sample particle from far away till it reaches a node adjacent to the seed, where it is supposed to stick to it. Then a new particle is started and halted in a position adjacent to the forming aggregation. The forming aggregation displays scale invariance, and its density correlation functions is given by

$$\langle \rho(r+s)\rho(r) \rangle = s^{-A}. \tag{3.1}$$

where the exponent A is related to the space dimension d and to the Hausdorff or fractal dimension D by the relation $D = d - A$. In two dimensional space ($d = 2$) ensemble averaging over a large number of calculations yields $D \approx 1.71$, while, for instance, in the self-avoiding random walk, referred to in the following, one has $D \approx 1.33$. In a three-dimensional DLA the fractal dimension D is larger. Numerical simulations give results in the range 2.4–2.5 (Peitgen et al. 2004).

At the beginning of the 1980's Witten and Sander (Witten and Sander, (1981), (1983)) discovered that DLA strongly resembles the Saffman-Taylor finger (Saffman & Taylor, 1958) originated by infiltration of an inviscid fluid into a viscous one (f. i. air into oil) in a two-dimensional Hele-Shaw channel (Hele-Shaw, 1898; Shaw, 1898) with width W much larger than the thickness B of the channel, hence with large aspect ratio W/B. The formation of the Saffman-Taylor finger is a prototypical problem of moving interfaces, which has relevance in a variety of practical problems as, for instance, flame propagation in combustion processes. The problem can be treated in a closed analytical form (Crowdy & Tanveer, 2004). The Saffman-Taylor finger can be constructed numerically by solving the equations

$$\nabla \cdot \vec{u} = 0, \tag{3.2}$$

$$\nabla^2 \Phi = 0, \tag{3.3}$$

$$u_n = (\partial \Phi / \partial n)_s, \tag{3.4}$$

where $\vec{u}$ is the velocity of the infiltrating inviscid fluid, Φ is the velocity potential, and subscript s denoted the surface of the forming aggregation, which advances with normal speed u_n.

Recent work carried on by Sander and Somfai (2005) aims at establishing on a firmer basis the still open problem of the identification of averaged DLAs with the Saffmann-Taylor finger in wedge geometry with different angles. Cluster formation reported in the latter reference was obtained with typically 10^6 particles and averaging over an ensemble of 400 realizations.

3.1.2 Methodology of DLA

In this section we aim at investigating some of the basic approaches to the numerical solution of problems erasing in statistical mechanics, namely:

(i) The *Simple Sampling* (in which, shortly formulated, all probabilities of a series of events are equal);
(ii) *Biased Sampling* (with different probabilities of a series of events), and
(iii) *Importance Sampling* (better defined in the following).

An example of application of the Simple Sampling method is given by the *random walk* problem with some variants, like the *non-reversal random walk*. A second example of application of the Simple Sampling method is given by *Percolation*, which basically consists in the recognition of patterns.

The so-called *Self Avoiding Random Walk* can be classified as belonging to the Biased Sampling method. Its main concept consists in preventing continuation of chains of events in case of a return to a previously occupied point.

Importance Sampling becomes important when the method of Simple Sampling and of Biased Sampling are not suitable. This is the case when contributions to some critical events under investigation come from just a few islands of the state phase, rather than from all areas if the state phase. This is the typical case in the study of ferromagnetic materials in the frame of Ising models. A treatment of these cases is given in Binder and Heermann (1988).

In reference Bottoni (2022) the method of Simple Sampling is applied to test the quality of two different random number generators by applying the Einstein's law. This law states that in repeating a large number of N-steps walks and taking the mean values $\overline{R}_N^2$ of the displacements of the walks from the original points, in the limit for $N \to \infty$, $\overline{R}_N^2$ is proportional to the length of the walks: $\overline{R}_N^2 \propto N$.

In the following, some variants of the Diffusion Limited Aggregation (DLA) are examined and the results are compared with each other. A typical calculation proceeds in the following steps:

(i) Set up of the problem with choice of the lattice and of the size of the problem.
(ii) Choice of the variant of the calculation.

  (a) Variant DIFFLA: main subroutine in which particles going out of the boundaries are lost.
  (b) Variant DLA-1: particles going out of the domain are reflected specularly into the domain. Walkers (moving particles) are originated at the borders of the domain, but not in the corners.
  (c) Variant DLA-2: Like variant DLA-1, but walkers are originated within the domain in positions chosen at random and not belonging to the aggregation.

(iii) Computation of the density correlation by means of the subroutines DENSCORR and CORREL. The autocorrelation is computed for two data sets DATA1 and DATA2, each of length N (including user supplied zero padding). N must be a power of 2.

We recall that the correlation of two functions g, h, defined in time domain, is given by

$$Corr(g, h) = \int_{-\infty}^{\infty} g(\tau + t)h(\tau)d\tau, \tag{3.5}$$

If $g = h$ one has an autocorrelation.

The answer of the autocorrelation, given by the subroutine CORREL, is returned as the first N points in ANS, stored in wap-around order, i.e. correlations at increasingly negative lags are in ANS(N) on, down to ANS(N/2 + 1), while correlations at increasingly positive lags are in ANS (1) (zero lag) on, up to ANS(N/2).

Array ANS must be supplied by the calling program with length at least 2*N. Sign convention in this subroutine: if DATA1 lags DATA2, i.e. is shifted to the right of it, then ANS will show a peak at positive lags.

The theory involved applies the "Discrete Correlation Theorem" which states that the discrete correlation of two real functions g and h is symbolically related to their discrete Fourier Transforms by

$$Corr(g,h) <====> G\,H^{*}, \tag{3.6}$$

(* denotes complex conjugate). The functions G and H^* are defined in frequency domain and the arrow denotes correspondence between time domain and frequency domain.

Hence, for the autocorrelation one has

$$Corr(g,g) = G(f)H^{*}(f). \tag{3.7}$$

Equation (3.7) can be computed by the Wiener-Khinchin theorem

$$Corr(g,g) = |G(f)|^{2}. \tag{3.8}$$

Therefore the correlation is computed as follows:

- FFT the two data sets;
- Multiply one resulting transform by the complex conjugate of the other;
- Inverse transform the product.

The result, say $R(k)$, will be a complex vector of length N. However, it will have all its imaginary parts equal to zero, because the original datasets are real. The components of $R(k)$ are the values of the correlation at different lags. The correlation at lag zero is in $R(0)$, the first component; the correlation at lag 1 is in $R(1)$, the second component; the correlation at lag -1 is in $R(N-1)$, the last component, etc...

(iv) Post-processing of computed results, in particular computation of the density of the computed pattern.

Main auxiliary subroutines are the following:
CORREL: Drives the computation given by Eq. (3.7);
FOUR1: computes Fourier transform or anti-transform of one array of data;
REALTF: makes Fourier anti-transform of an array of complex numbers;
TWOFFT: makes Fourier transform of two real arrays of data;
FR (function): computes random numbers in the open interval (0–1).
SCREEN, SCREEN1: check that particles are inside the computational domain.

Examples of the application of autocorrelation in 1D domain and in 2D domains are given in Sects. 3.1.7 and 3.1.8.

Acknowledgment: Subroutines CORRL, FOUR1, REALTF, TWOFFT are from the Press Library (Press et al., 1986); function RF has been loaned from the Bird Library (Bird, 1994).

3.1.3 Variant DIFFLA

In the variant DIFFLA we describe a main subroutine in which particles going out of the boundaries of the definition domain are lost.

We discuss results on the basis of a simple 2D application made in a lattice of 100 × 100 cells. The scale is irrelevant but we may assume that the domain is normalized to $(0 < x < 1, 0 < y < 1)$.

The points of the lattice are classified and used as follows:

(i) A "**seed**" is set in the domain. In our case the seed is in position ($xi = 50$, $yj = 30$). The seed is the starting point of subsequent aggregations and is given an index = 3. Cells which are neighbours of the seed are set to an index = 2. These cells are called "**stick points**" and are potential aggregation points.

(ii) A so-called **walker** starts from a rand cell, called "**starting point**" (excluding the corners).

The walker moves at random with probability 1/4 for every coordinate direction. If the walker approaches the seed, it sticks to it and forms an "**aggregate**". Aggregates are excluded as new starting points. Cells which are on the path of the walker are turned to index = 1, so that the path of the walker can be followed till it goes out of the domain or it joins the aggregation.

(iii) Cells which join to the aggregation are set to an index = 3, and their neighbouring, as potential aggregates get index = 2.

When a walker joins the aggregate, or gets out of the domain, all cells with index = 1 are turned to index = 0, a new starting point is set and a new walk starts from a rand cell.

Figure 3.1 shows the aggregate at four subsequent time points of the calculation, namely after 9000/18,000/27,000/36,000 walks.

Figure 3.2 shows the aggregate at the end of the calculation, after 38,100 walks. The calculation ends when the aggregate grows to include the rand cells because there are not free rand points for starting a new walk. In this calculation there were 1216 aggregations while 36,884 walks ended with the walker going out of the domain.

At the beginning of the calculation, when a new starting point is set in a rand cell, the probability that the walker gets out of the domain at the first step is 1/4. The probability for the walker of going out of the domain at the beginning of the process is very high, but it decreases as soon as the aggregate becomes larger. This effect becomes evident from Fig. 3.3 which plots the percentage of cells in the aggregate versus the number of walks performed. At the end of the simulation made in this example the percentage of cells in the aggregate is 0.035.

An example of trajectories which join the aggregation is given in Fig. 3.4. It shows the ensemble of paths described by a walker at walk 20,000.

Appendix A.1 is a listing of the code DIFFLA. It gives detailed information about the organization of the lattices from the initial sticky point to the subsequent formation and growth of the aggregate.

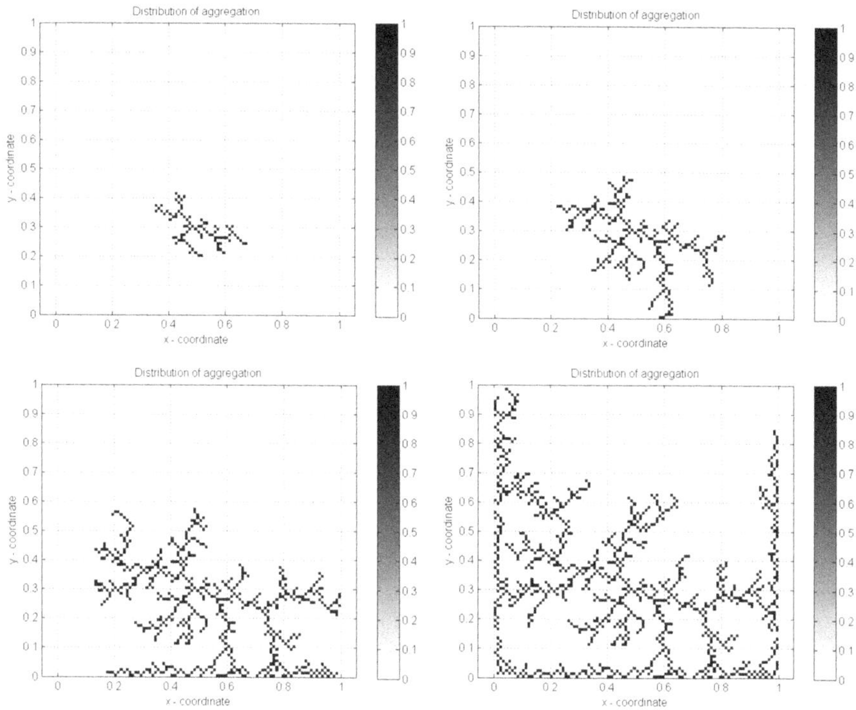

Fig. 3.1 Evolution of the aggregate at four subsequent time points of the calculation, namely at 9000/18,000/27,000/36,000 walks

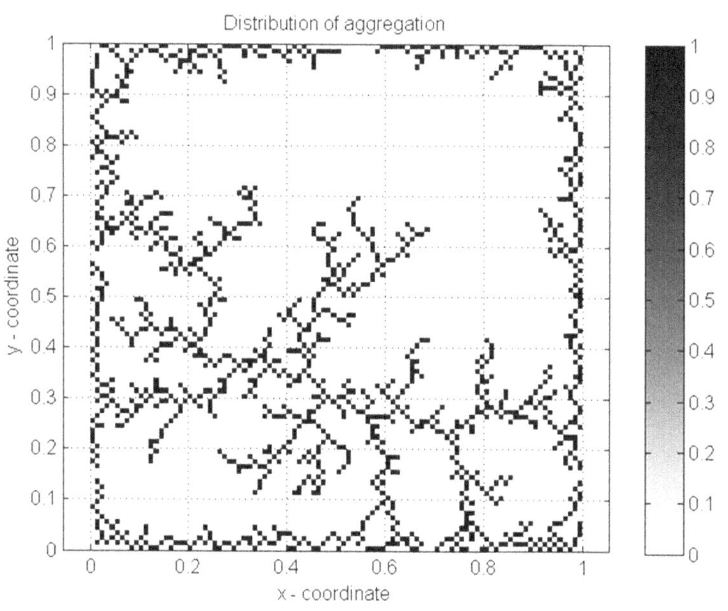

Fig. 3.2 Aggregate at the end of the calculation, after 38,100 walks

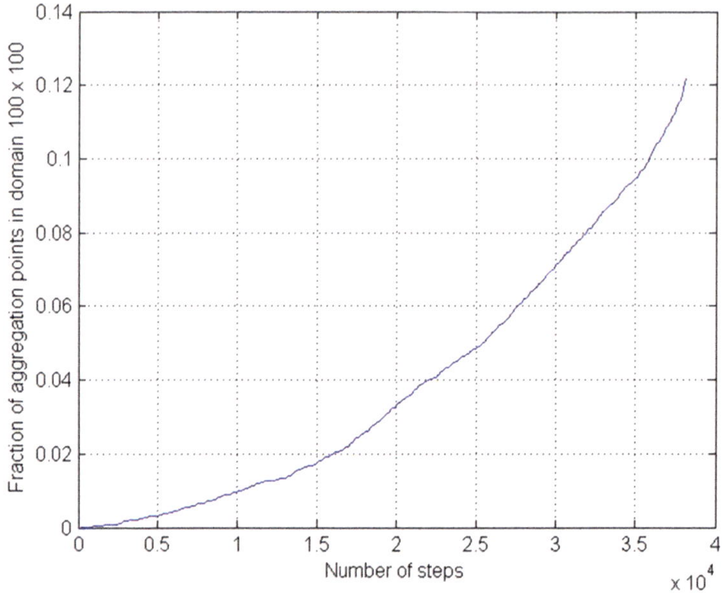

Fig. 3.3 Percentage of aggregation versus number of walks

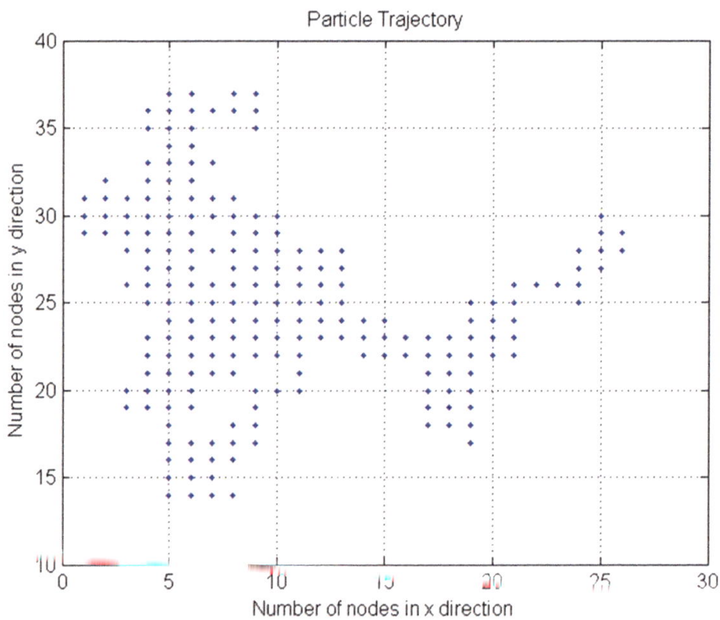

Fig. 3.4 Ensemble of trajectories described by a walker starting in mesh (1, 31) and joining the aggregate in mesh (25, 30)

3.1.4 Variant DLA_1

Variant DLA-1: particles going out of the domain are reflected specularly into the domain. Walkers (moving particles) are originated at the borders of the domain, but not in the corners.

The condition that a walker cannot leave the domain because it is reflected by the boundary is justified as a periodic boundary condition: if a walker goes out it is replaced by another walker moving in the opposite direction, as if it were reflected by the wall.

Figure 3.5 refers to a calculation with a lattice of 100 × 100 cells. It shows the aggregate at four subsequent time points of the calculation, namely after 250/500/750/1000 walks. In the case of the variant DLA-1, there are not losses of walkers beyond the boundaries of the domain and therefore the accretion of the aggregate is much faster than it was in the variant DIFFLA. The rate of accretion is in this case linear with the number of walks.

Figure 3.6 shows the aggregate at the end of the calculation, after 1140 walks.

Appendix A.2 is a listing of the code DLA_1. It gives essentially the same information as explained in Appendix A.1.

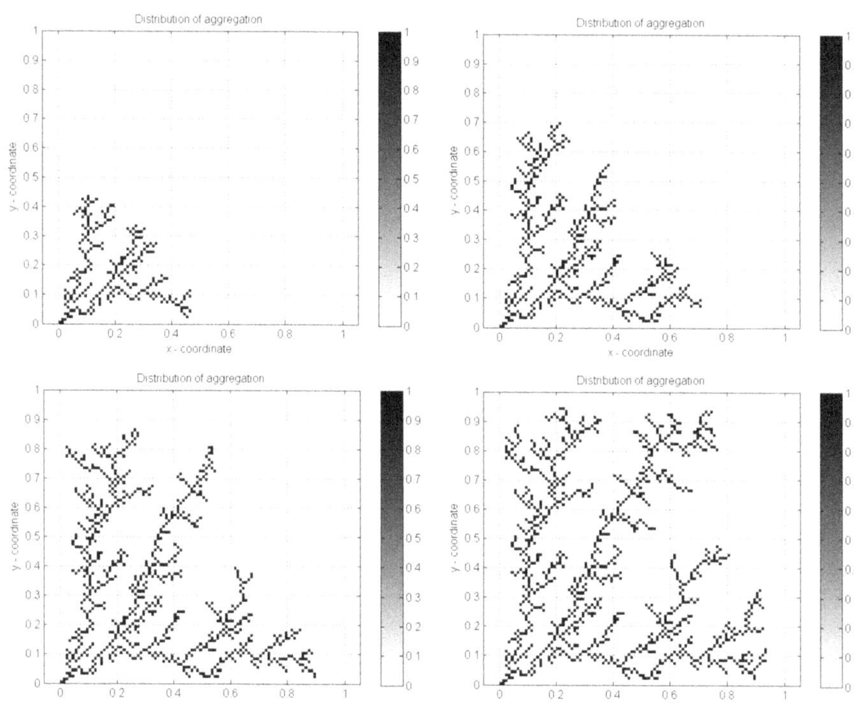

Fig. 3.5 Evolution of the aggregate at four subsequent time points of the calculation, namely at 250/500/750/1000 walks

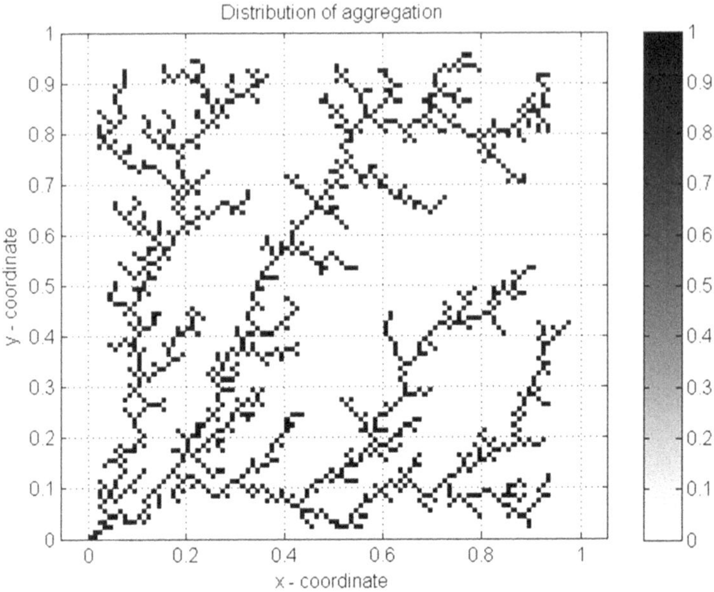

Fig. 3.6 Aggregate at the end of the calculation, after 1140 walks

3.1.5 *Variant DLA_2*

Variant DLA_2: Like variant DLA_1, but walkers are originated within the domain in positions chosen at random and not belonging to the aggregation.

Figure 3.7 refers to a calculation with a lattice of 100 × 100 cells and shows the aggregate at four subsequent time points of the calculation, namely after 250/500/750/1000 walks. As in the case of the variant DLA_1, there are not losses of walkers beyond the boundaries of the domain and therefore the accretion of the aggregate is much faster than it was in the variant DIFFLA.

Figure 3.8 shows the aggregate at the end of the calculation, after 1140 walks.

3.1.6 *Additional Realizations for Several Lattices*

Because of the aesthetic value of the figures we give a selection of a few realizations made with the code variant DLA-2 with several lattices. The reader may appreciate the increasing complexity of the aggregates with the increment of the computational domain.

(i) DLA-2_200—A realizations with a 200 × 200 lattice is shown in Fig. 3.9.
(ii) DLA-2_300—A realizations with a 300 × 300 lattice is shown in Fig. 3.10.
(iii) DLA-2_400—A realizations with a 400 × 400 lattice is shown in Fig. 3.11.
(iv) DLA-2_500—A realizations with a 500 × 500 lattice is shown in Fig. 3.12.
(v) DLA-2_1000—A realizations with a 1000 × 1000 lattice is shown in Fig. 3.13.

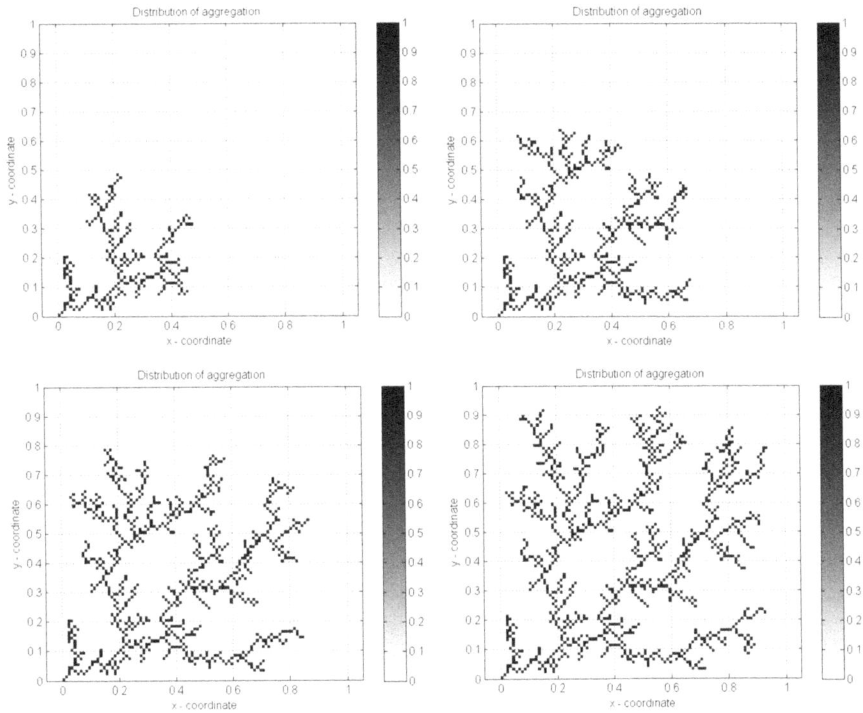

Fig. 3.7 Evolution of the aggregate at four subsequent time points of the calculation, namely at 250/500/750/1000 walks

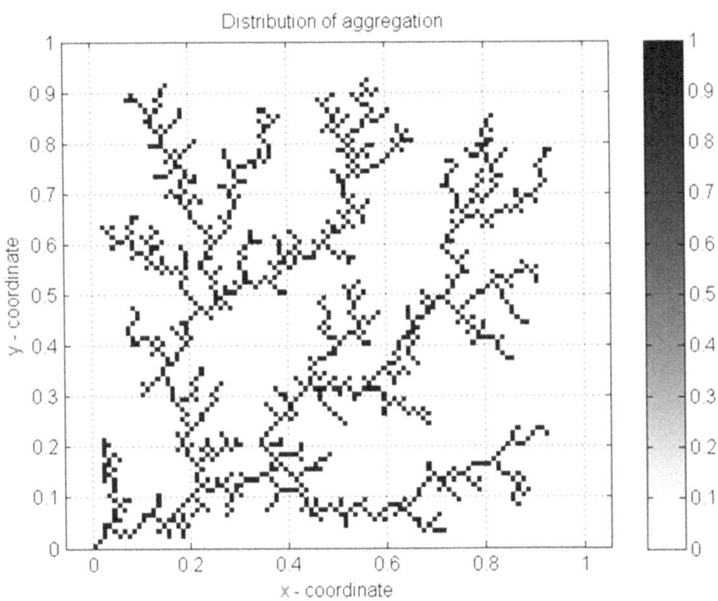

Fig. 3.8 Aggregate at the end of the calculation, after 1140 walks

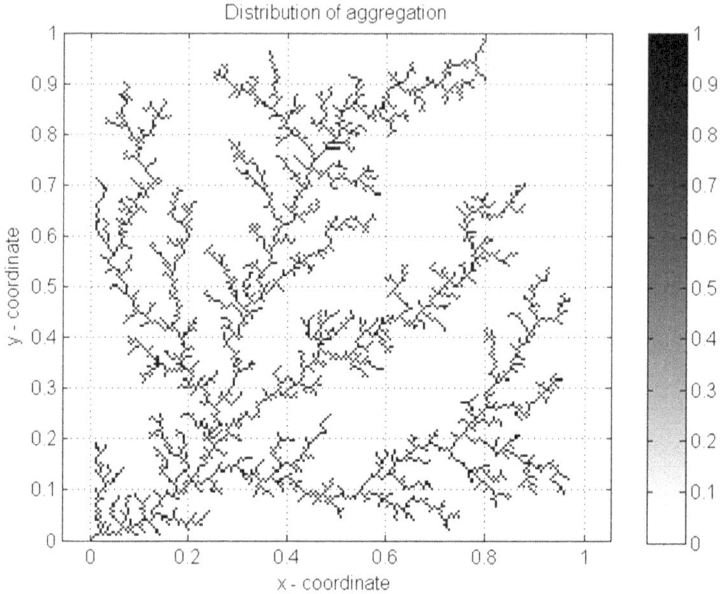

Fig. 3.9 Realizations with a 200 × 200 lattice

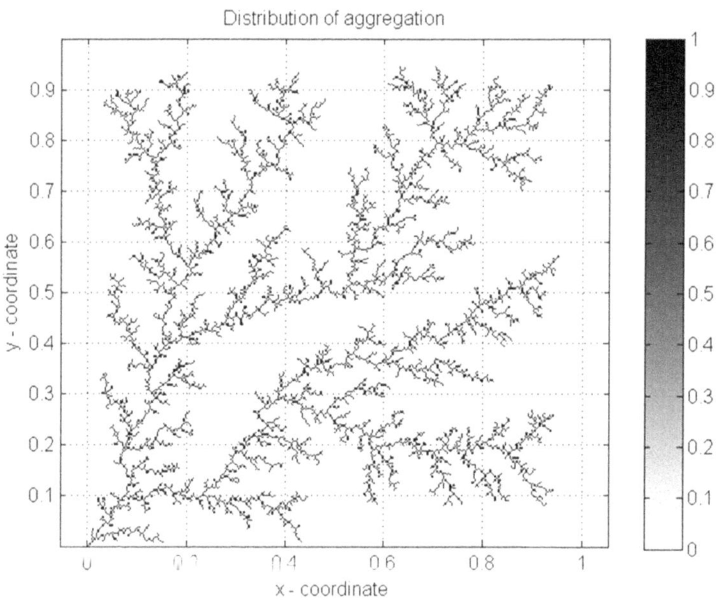

Fig. 3.10 Realizations with a 300 × 300 lattice

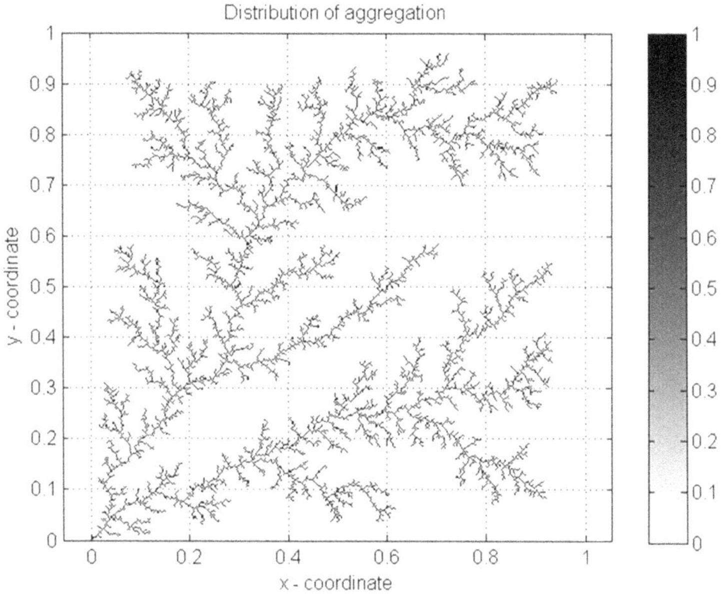

Fig. 3.11 Realizations with a 400 × 400 lattice

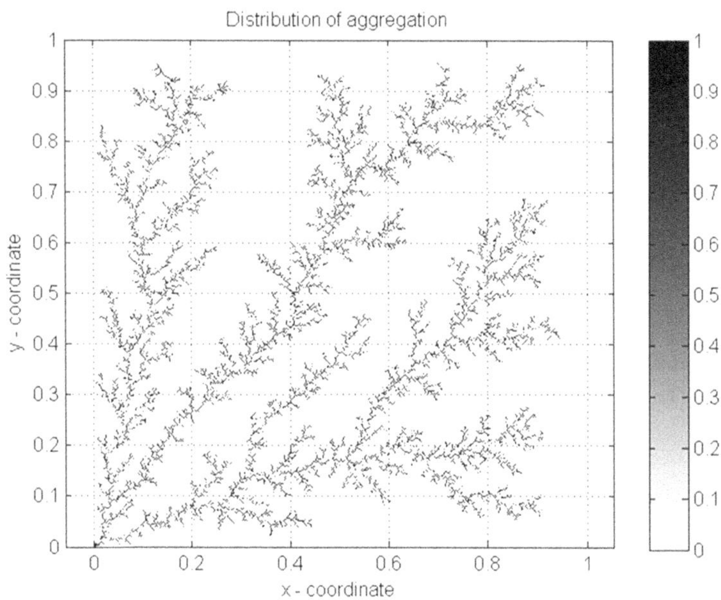

Fig. 3.12 Realizations with a 500 × 500 lattice

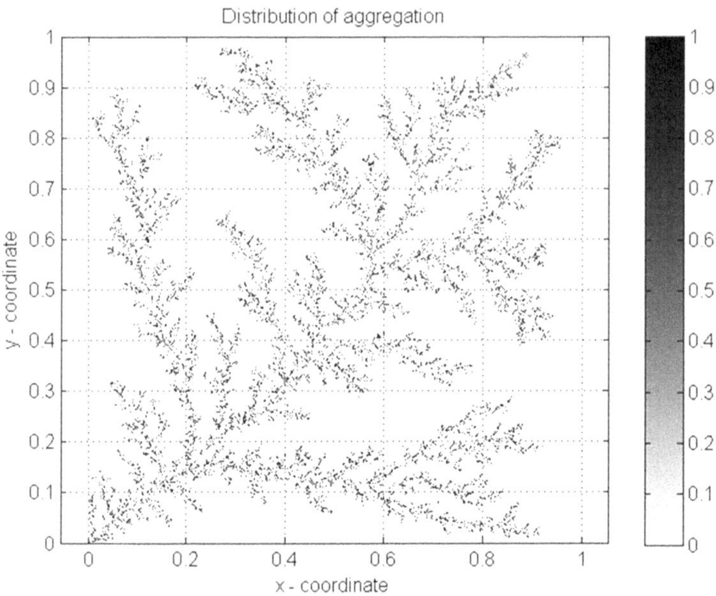

Fig. 3.13 Realizations with a 1000 × 1000 lattice

Appendix A.3 is a listing of the code DLA-2. It gives essentially the same information as explained in Appendix A.1.

3.1.7 Example of Application of Autocorrelation in 1D Domain

Appendix A.4 is a listing of the code **CORR1D**. One of the key points illustrated in the previous section involves the calculation of the density autocorrelation of the resulting two-dimensional aggregate. In Appendix A.4 a simple example of the application of the technique represented by Eq. (3.7) is given with the subroutine CORREL. The example treats an array of 64 data which are all zeros with exception of 7 data. The autocorrelation of these data can be computed by hand shifting beside each other two strips of the data representing the two arrays and counting the number of non-zero digits facing each other; but we apply the subroutine CORREL to learn its correct use.

Let us consider the following array of 64 data, referred to as DATA1, defined to be either 0 or 1. Data different from 0 are listed in the following Table 3.1;

Irrespective of the fact that the data can be interpreted as data in time domain of in space domain, the autocorrelation at lag 0 (we use the term "lag" as if data were in time domain) is 7. This result is evident if one considers a strip of the same data, say

Table 3.1 List of data to illustrate the autocorrelation

data1(1) = 1
data1(2) = 1
data1(3) = 1
data1(4) = 1
data1(5) = 1
data1(9) = 1
data1(10) = 1

Table 3.2 Results obtained with the numerical approach explained in Sect. 3.1.2. Left column: Leg; Right column: Correlation number at the given lag. For lags >9 all correlations are zero

0	0.70000E+01
1	0.50000E+01
2	0.30000E+01
3	0.20000E+01
4	0.20000E+01
5	0.20000E+01
6	0.20000E+01
7	0.20000E+01
8	0.20000E+01
9	0.10000E+01

DATA2, placed parallel to the strip of DATA1, and one considers the number of digits different from zero that are superimposed. With the same procedure, shifting the strip DATA2 by one position to the right, one sees thet the autocorrelation at lag 1 is 5; at lag 2 it is 3 and so on. The procedure made by hand is clearly difficult to follow. Using the numerical analysis explained in Sect. 3.1.2 one obtains the results shown in Table 3.2.

3.1.8 Example of Application of Autocorrelation in 2D Domain

In the case of a 2D domain, as represented by a lattice (i, j) ($i = 1$-imax, $j = 1$-jmax), we apply an "ad hoc" procedure which aims at establishing the uniformity of the distribution of non-zero digits in the lattice.

Let us consider the following matrix, shown in Fig. 3.14, obtained by applying the DLA-2 program to a 512 × 512 lattice, starting with a seed situated in the node (3, 3). The choice of a lattice with a linear dimension equal to a power of 2 simplifies the numerical treatment explained in the following.

Figure 3.15 shows the final aggregation at 25,000 steps.

In a subsequent code, donoted POSTPRO (POST PROcessing), we re-read a density matrix, called DENS, with the same dimensions as the lattice, and proceed to treat it is as follows:

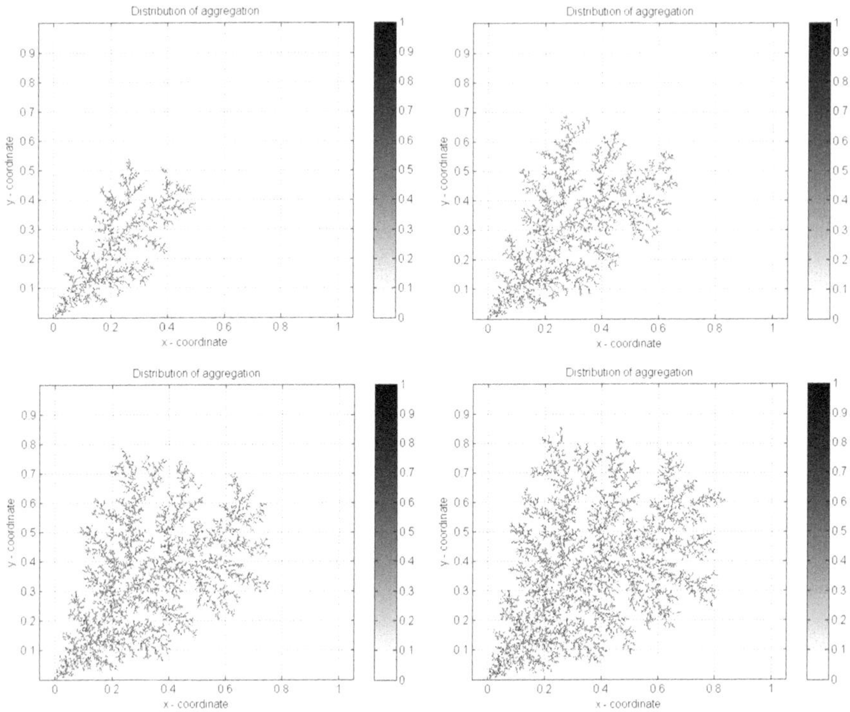

Fig. 3.14 Calculation made with the program DLA-2 for a 512 × 512 lattice. The four aggregations are shown at 5000, 10,000, 15,000, and 20,000 steps

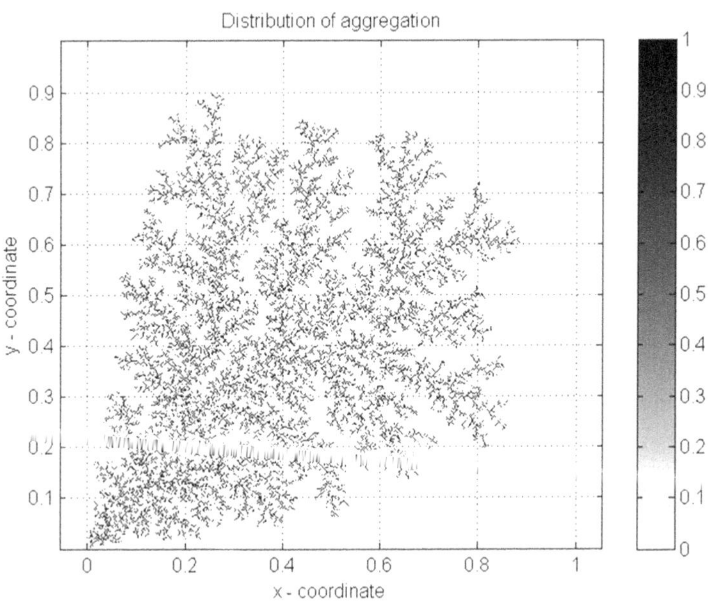

Fig. 3.15 Calculation made with the program DLA-2 for a 512 × 512 lattice. The final aggregation is shown at 25,000 steps

(i) The matrix DENS is used column by column and to every column is applied the code CORREL to compute the autocorrelation. The results of the autocorrelations are summed up and stored in a separate array (AUX).
(ii) The results stored in AUX are averaged and compared with the distribution given by (see Eq. 3.5).

$$g(\tau + t)g(\tau) = S^C, \tag{3.9}$$

where S is the leg distribution computed at step (i) and the coefficient $C = 0.31$ is estimated on the basis of optimum fitting.

The results of this pseudo-correlation give information about the non-uniformity of arrays shown in previous figures. In case of non-uniformity of the data, the result of the distribution of Eq. (3.9) shows picks. We used the program POSTPRO to obtain information about density of the aggregates, but with poor success. The topic is therefore open to further investigations.

Figure 3.16 shows a comparison between theoretical and computed autocorrelation.

Details of the application of the code POSTPRO are given in Appendix A.5.

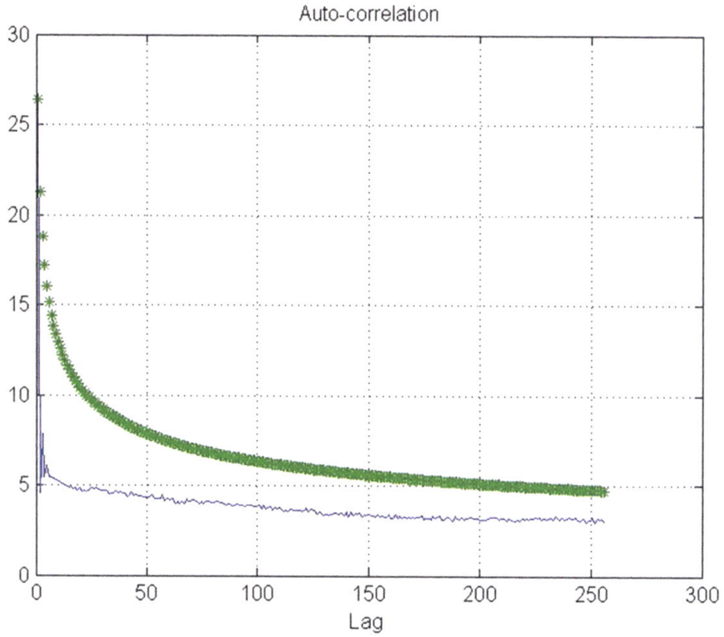

Fig. 3.16 Comparison between theoretical and computed autocorrelation for DLA

3.1.9 Dendrite Accretion in Unbound Domain

Experimental observation of aggregation were described by (Matsushita, 1989) in a Petri-dish aqueous solution of zinc sulphate with a layer of acetate. An electrochemical deposition of the zinc sulfate starts by applying a voltage of a few Volts to a carbon cathode and a ring-plate of zinc as anode.

In another branch of investigations, Diffusion-limited Growth (DLG) was reported, among others, by Matsushita and Fujikawa (1990) as emerging within a microbial colony on a solid substrate. Subsequent investigations modelled the experimental results. Results of this kind provided to the domain of molecular dynamics an additional research tool for the analysis of biological problems.

We present in this section numerical results showing the accretion of a dendrite in a theoretically unbound domain. In practice the domain, consisting of a square lattice of 500 × 500 cells, is large enough to prevent deformation of the results due to boundary effects. These results were obtained with the code DLA_2, starting from an initial stick point set in the centre of the definition domain.

Figure 3.17 shows the evolution of the dendrite at four subsequent time points of the calculation, namely at 2000/4000/6000/8000 walks.

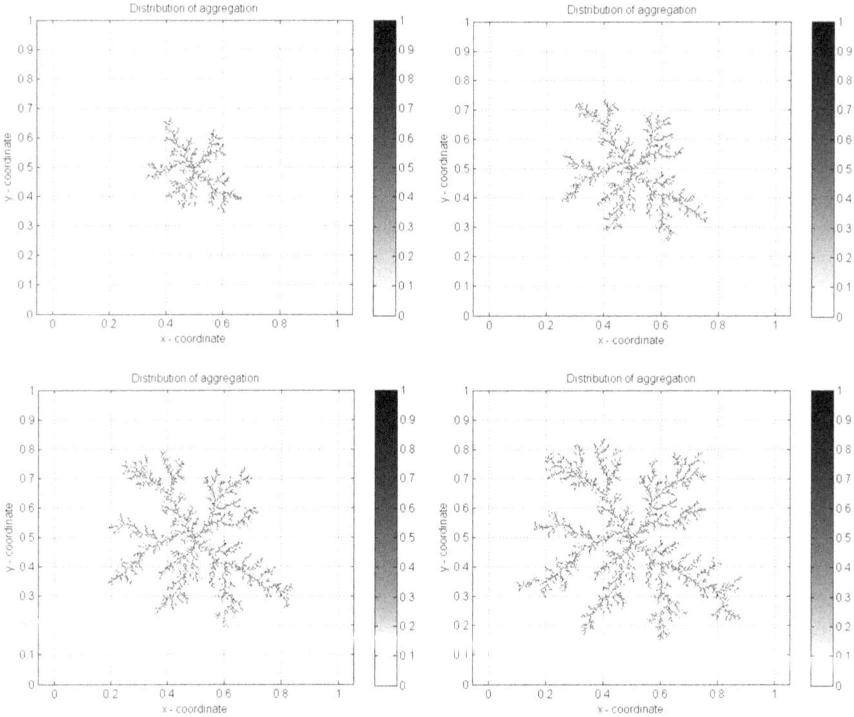

Fig. 3.17 Evolution of the dendrite at four subsequent time points of the calculation, namely at 2000/4000/6000/8000 walks

3.1.10 Geode Growth in Circular Domain (2D)

3.1.10.1 Taxonomy of Geodes Formation and Growth: Shape, Characteristics, and Structure

In this section we treat a test case for simulating the accretion of a geode in two-dimensional geometry, with the novel approach of molecular dynamics. Before explaining the application of this method we introduce the topic by examining different processes which can trigger the formation of proto-geodes and then their time evolution. Although the classification may be blurred and superpositions of processes are likely to occur, we attempt to classify formation and growth of geodes according to the following scheme:

(i) *Formation of geodic cavities in tubular shape*

One of the most common processes of formation of geodic cavities consists in cooling of hot basaltic magma originally in tubular shape, flowing on a rocky substrate. The process can be considered as a "*dry process*" because it does not involve hydrothermal fluids (to which is dedicated the next section). When the hot magma starts cooling, after the paroxystic event of its formation, an upper solidification front starts proceeding downwards and likewise a lower front forms proceeding upwards. Tubular, inward growing protuberances are also formed.

The morphology of the developing geodic cavity is controlled by several processes: degassing of magma by depressurization of volatile components present in the magma; bubble nucleation, following the depressurization; bubble coalescence to form larger interconnected voids.

This process of geodes formation has been simulated numerically (Proust & Fontaine, 2007) assuming an initial liquidus temperature of 1473 K, a solidus temperature of 1253 K, and taking into account the release of the latent heat of crystallization in the cooling process. Solidification of amethyst bearing basaltic flow in the form of micro-crystalline amethyst was predicted to occur in the time span of about 12 years.

Conclusions drown by the authors of the above quoted reference emphasize *that (1) the formation of geodes…can be entirely explained … by the exolution of magmatic gas from the supersaturated melt with no need for external surface water supply; (2) the particular tubular habit of the geodes results from the higher cooling rate of the magma in contact with the accumulating bubbles; and (3) the abnormal meter-sized geodes do not result from the diffusive/ decompressing bubble growth but from the coalescence of inward-progressing tubular cavities.*

This last quoted sentence will form the basis of the assumptions made in Sect. 3.1.10.3 for modelling the growth of geodic cavities with molecular dynamics approach.

(ii) *Hydrothermal origin (Generation of cavities by hydrothermal fluids)*

The technical term of "*ballowing*" is used to explain the growth of a cavity in a ductile medium, like basalt, under the pressure of water and its vapour. The size of a proto-geode depends on the water pressure, which is related to the thickness of the overlaying basalt, typically in the range of a few meters up to about 20 m. Ballowing occurs especially in ductile deformations of a magma layer and can be associated with a preliminary briddle process characterized by the formation of breccia. One can therefore assume that in many geodes generated from cavities by hydrothermal fluids there was an initial brittle fracture of the basalt, or andesite, followed by a ductile ballowing of the altered magma to form large cavities. Fragments of breccia are effectively found at the bottom of several geodes which developed upwardly. Both prolate and oblate cavities can be formed (Hartmann et al., 2015), depending on the stress during the initial formation process.

Long time studies of geode formation by hydrothermal fluids assumed that opening of cavities by ballowning occurred by degassing of lava at high temperature, about 1473 K. Recently however new models have been proposed (Hartmann et al., 2012) which explain cavity generation in basalt at low temperature (<423 K). In the quoted reference a numerical simulation is made, by means of the finite element method, of an axi-symmetric geode in a 2D representation in which the domain is reduced to a rectangle vertically oriented. One vertical late of the rectangle is the symmetry axis, the second vertical late is the boundary of the magma interested by the ballowing process. The floor of the cavity is considered as fixed while the ceiling of the cavity can be affected by the ballowing deformation. The main result obtained by the numerical simulation consists in the statement that the volume reached by the cavity depends on the Young's modulus $E = \sigma/\varepsilon$, where σ is the axial stress and ε is the axial strain.

The interpretation of geode genesis and growth explained in this section supersedes an older accepted scenario. According to this new research, formation of large voids does not require hydrothermal activity but can be related to degassing phenomena from a primary magma.

(iii) *Petrol geodes from the Warsaw formation*

A special mention is made of the geodes of the Warsaw formation in USA because of their peculiarity. Geodes of the lower Warsaw formation are found in the States of Iowa, Illinois, and Missouri, with a remarkable concentration in the Keokuk region of Iowa, which is considered the Type Locality (TL). It is generally accepted among scholars that original voids of fossils, for instance larger brachiopods, crinoid calices, and every kind of solution cavities in sedimentary rocks are at the origin of geode formations. Most geodes are ellipsoidal in form with the larger axes parallel to their bedding and the shorter axes normal to it.

The genesis of geodes is significantly different from that of concretions. Concretions grow outwardly by means of a sequence of depositions, while geodes grow inwardly from the wall of a cavity towards the center. The

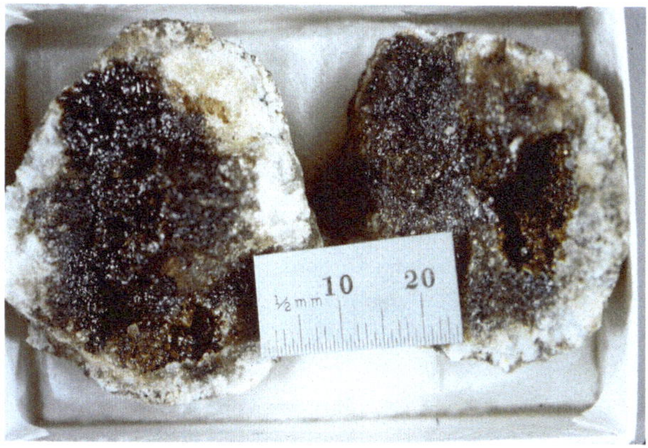

Fig. 3.18 Petrol geode from the Warsaw formation. (From the private collection of the author)

sequence of events leading to the formation of geodes can be summarized as follows (Hayes, 1964). Calcite or dolomite are the minerals at the origin of the process of geodes formation. Recrystallization of calcite generally occurs along with the replacement of calcite spherules with chalcedony. In this process the surface of the chalcedony is frequently coated with quartz crystals. Decay of organic matter may replay the original calcite generating bituminous matter which give some geodes the appearance and the name of "*petrol geodes*". See Fig. 3.18.

Similar conclusions about the genesis of geodes are drawn in (Chowns & Elkins, 1974) where an overview is given about characteristics of formations similar to those of the Keokuk region. These include the Woodbury area in Tennessee and the arid supratidal flats of the Persian Gulf, which were given the name of "*sabkhas*". In these areas, anhydrite, gypsum, and halite develop as precipitates within peritidal sediments because of high evaporation rates in marginal lagoons. Recrystallization of calcite and its replacement with chalcedony is confirmed in the above referenced work. Skeletal remains, mainly due to corals and sponges, but also including crinoids, blastoids, bryozoans and also foraminifers, are important in the process of calcite replacement with calcedony.

3.1.10.2 Modelling a Cavity in 2D Rectangular Geometry with Molecular Dynamics

In the numerical application of the molecular dynamics method presented in this section, the physical reference model consists of a domain filled with molten saturated silica in meta-stable condition at high temperature. The starting conditions is a pool the liquid magma inside a frame bounded by a thin crust. This preliminary condition, which is not modelled, refers to the thermodynamic status mentioned in item (i) above (Sect. 3.1.10.1), referring to the nucleation of bubbles in the proto-magma. As the molten silica cools slowly, first nucleations of tiny bubbles of about 50 μm diameter occurs. Spreading and growth of micro-bubbles depends on diffusion of gases dissolved in the magma ("*diffusion growth*") and in decompression of hydrostatic pressure when CO_2 and H_2O contained in the magma are released ("*decompression growth*"). Following decompression by release of volatile components, the magma, initially saturated at high pressure, becomes over-saturated.

Our numerical simulation starts at this point, when molecules of condensed silica migrate at random in the domain, till they deposit on a crust of solid silica assumed to be the boundary of the proto-magma. The model explained in the following is intended to simulate the growth of the aggregate and its characteristics in terms of volume and density.

In the following we compare experimental observations from the literature with results of our numerical simulations. Experimental observations are those of the quoted reference (Proust & Fontaine, 2007) in which solidification of an amethyst-bearing basaltic flow is presented with information about speed and duration of the solidification process. The basaltic flow consists of a column of 30 m thickness at the initial temperature of 1273 K. Cooling of the column proceeds from the upper surface downwards and from the bottom surface upwards till the two frozen surfaces join somewhere in the middle the column. The cooling time has been estimated from the experimenters in about 12 years. Figure 3.19 (reproduced from the above reference with kind permission of the editor) gives information about the progression in time of the upper and lower solidification fronts (Part a) and about the development of the flow structure (Part b) versus time. The solidification fronts merge at a height of 18.70 m. The upper solidification front proceeds faster than the lower one because the upper surface is a free surface.

We developed a variant of the DLA-2 program to simulate the basaltic column with a 300 × 300 lattice consisting of rectangular cells of 0.0333 and 0.1 m in the x- and y-directions, respectively. The domain is then normalized to 1 in either direction. The upper and lower boundaries of the domain are considered as sticking points where the formation of aggregates can start. Walkers are initialized as usual in the middle of the domain in positions chosen at random but not adjacent to the boundaries.

Figure 3.20 shows the evolution of the aggregate at five points of the calculation.

Our numerical simulation does not give information about the solidification time of the basaltic rock. This information is taken from the experiment and correspondingly the solidification time in our numerical results is set to 12 years.

Figure 3.21 shows the growth of the aggregate from above and from bottom. The lines are the boundaries between liquid (in the middle) and solid materials. The y-coordinate in the plot is given in terms of the number of cells used in the discretization of the domain. The upper and lower boundaries in our simulation meet exactly in the mid point ($y = 150$) of the domain. In the results of Fig. 3.19 the boundaries meet at 18.70 m, below the mid point of the domain. The discrepancy in our calculation depends on the fact that we did not simulate the cooling of the uppermost surface of the magmatic slab. This effect could be simulated assuming different probabilities of the walkers of moving in the upward and downward

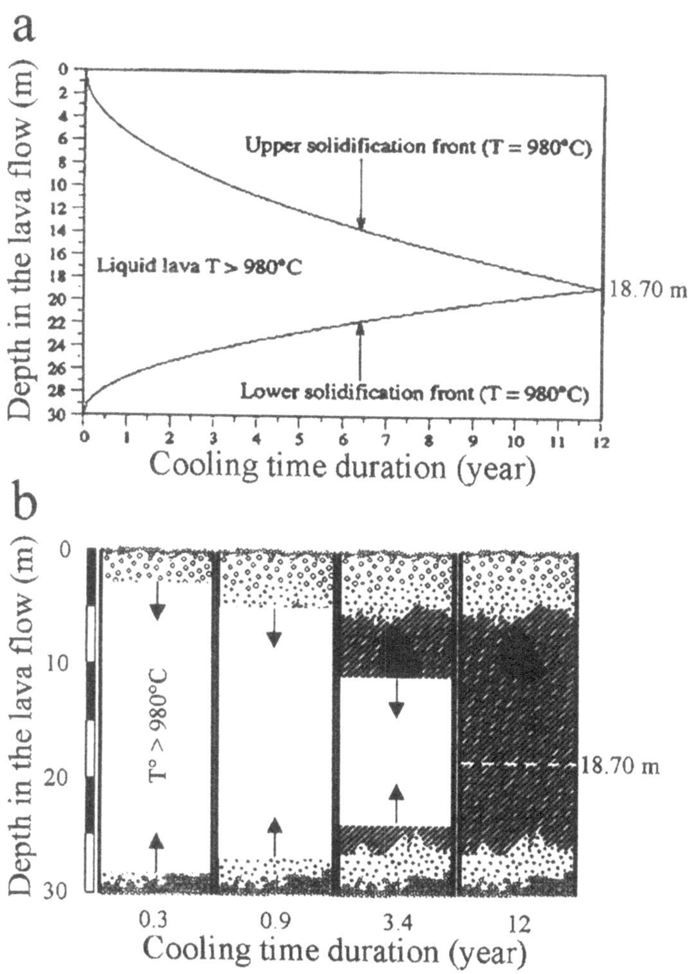

Fig. 3.19 Plot (**a**): Upper and lower solidification fronts versus time; Plot (**b**): Development of flow structure versus time

directions: a probability slightly larger than 1/4 for a move in the upward direction and a probability slightly smaller than 1/4 for a move in the downward direction. Thus the walkers would then move more frequently towards the upper freezing boundary. This investigation, however, has not been made thus far.

In our simulation, the boundaries between the solid and liquid domains do not display a continuous, but rather a fractal nature. The simulation does not terminate after about 16,000 walks but continues for 4000 walks more. The walkers released in this last span of the simulation provide a compaction of the already established boundaries. Whether this simulation has physical relevance is a topic which may be discussed among scholars.

Fig. 3.20 Evolution of the aggregate at four subsequent points of the calculation, namely after 4000/8000/12,000/16,000/20,000 walks

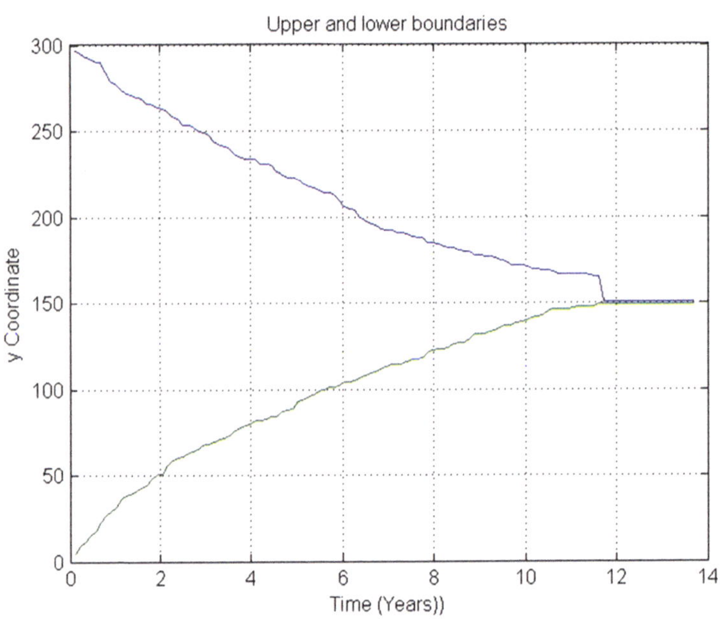

Fig. 3.21 Growth of the aggregate from above and from bottom

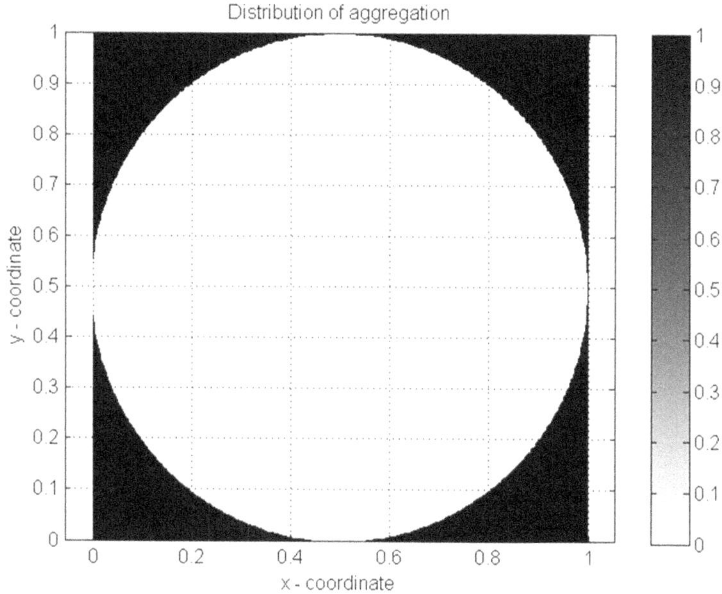

Fig. 3.22 Domain for the simulation of the growth of the geode

In some references, geodes were defined as sedimentary nodules with a drusy internal structure consisting of a coat of numerous small mineral crystals. Simulations of geodes accretion with fractal nature may therefore be a future challenging endeavour.

3.1.10.3 Modelling a Cavity in Circular 2D Geometry

We developed a variant of the DLA-2 program more suitable for the treatment of a circular domain bound by the circular frame of the crust. Some modifications with respect to the DLA-2 program were necessary because, for instance, the events of walkers migrating beyond the circular domain are excluded.

The starting condition is visualized by Fig. 3.22. In the following figures, the blackening of the domain outside of the circular frame is omitted.

The domain is modelled by a lattice of 500 × 500 (=250,000) cells, with 53,700 cells outside of the circular domain (excluded from the computations), 1476 cells on the circular crust and 194,824 cells inside the domain.

The simulation is initialized by a seed set adjacent to the circular crust. Subsequent "walkers" are generated at random in the inner domain.

Figure 3.23 shows the evolution of the aggregation at four subsequent points of the calculation, namely after 10,000/20,000/30,000/40,000 walks. The final aggregation computed at 50,000 steps is shown in Fig. 3.24.

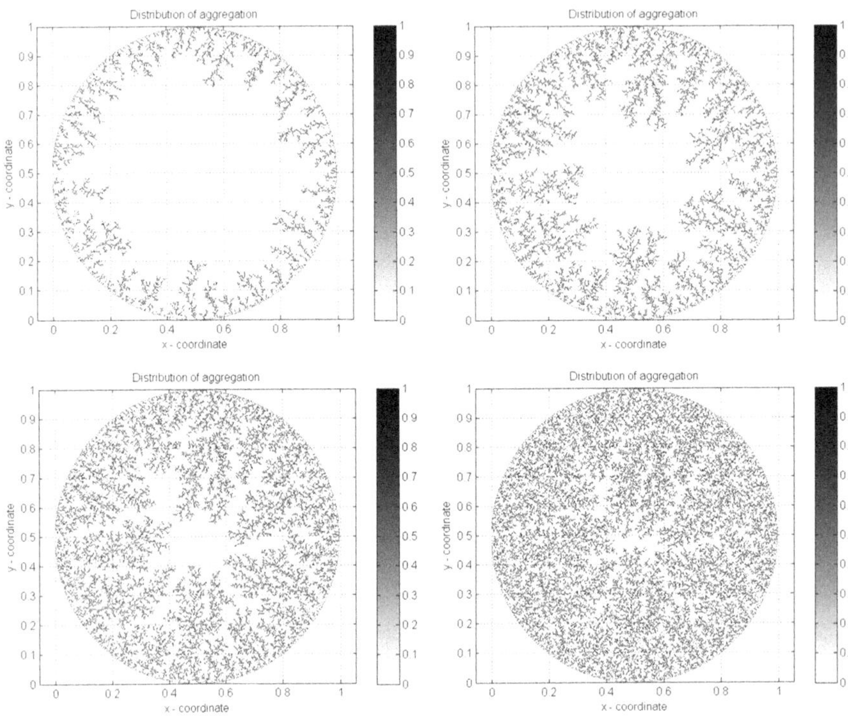

Fig. 3.23 Evolution of the aggregation at four subsequent points of the calculation, namely after 10,000/20,000/30,000/40,000 walks

An alternative calculation is made assuming that a condensation nucleus is initially formed within the domain thus forming an accretion that grows together with the crust, finally merging with it. The evolution of the aggregate after 10,000/20,000/30,000/40,000 walks is shown in Fig. 3.25.

The final result is shown in Fig. 3.26. It is not very different from the result of Fig. 3.24, but it shows an inner fracture resulting from the incomplete coalescing of the crust with the original inner nucleus.

It is difficult to estimate from the previous pictures the distribution and density of the aggregates. Another approach consists in the evaluation of the *Correlation Length* of the aggregate, defined as the *average distance between two cells in the aggregate*. In the geometry considered in this test case the correlation length varies between a maximum of 0.998 (= 499/500) (if the aggregate consists of two cells diametrically opposed to each other), and a minimum to be established theoretically. This investigation has not been made thus far.

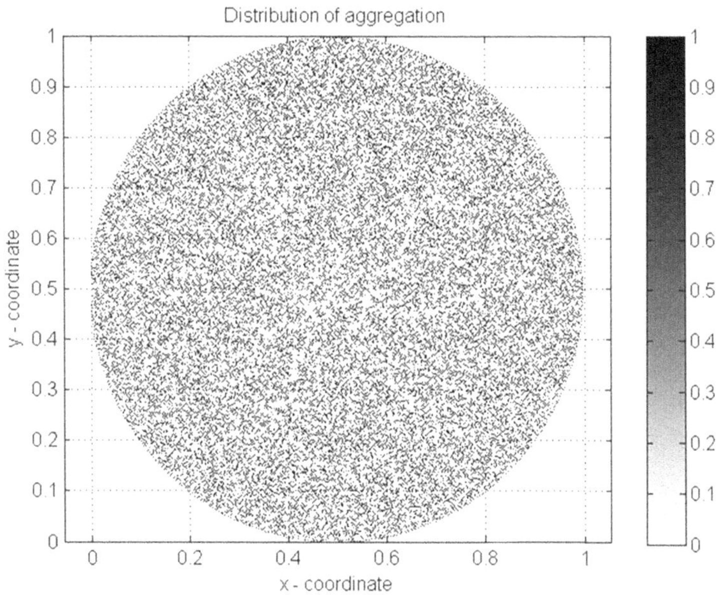

Fig. 3.24 Final aggregation computed after 50,000 walks

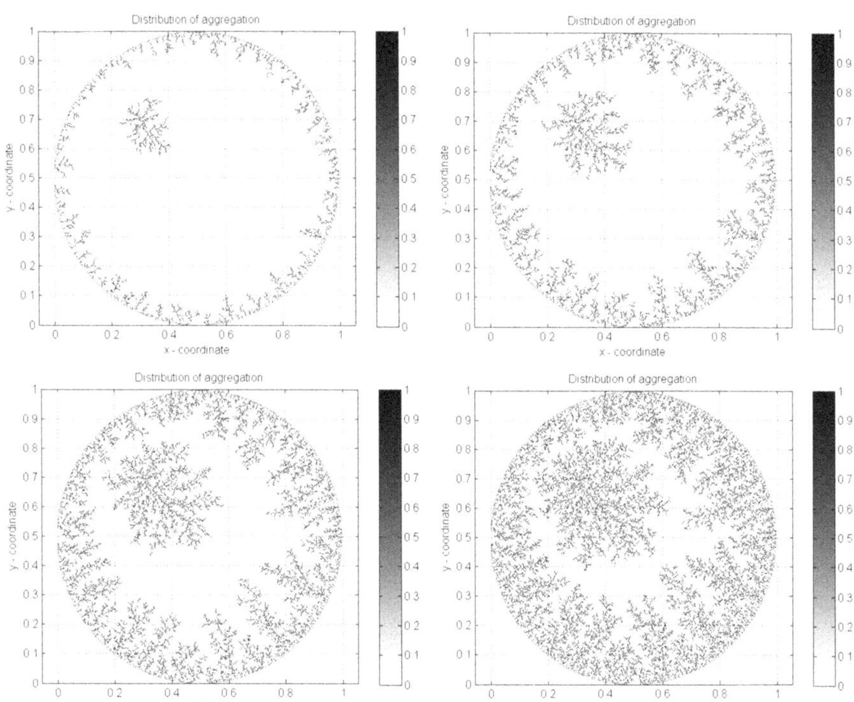

Fig. 3.25 Evolution of the aggregation with an initial nucleus in the inner domain at four subsequent points of the calculation, namely after 10,000/20,000/30,000/40,000 walks

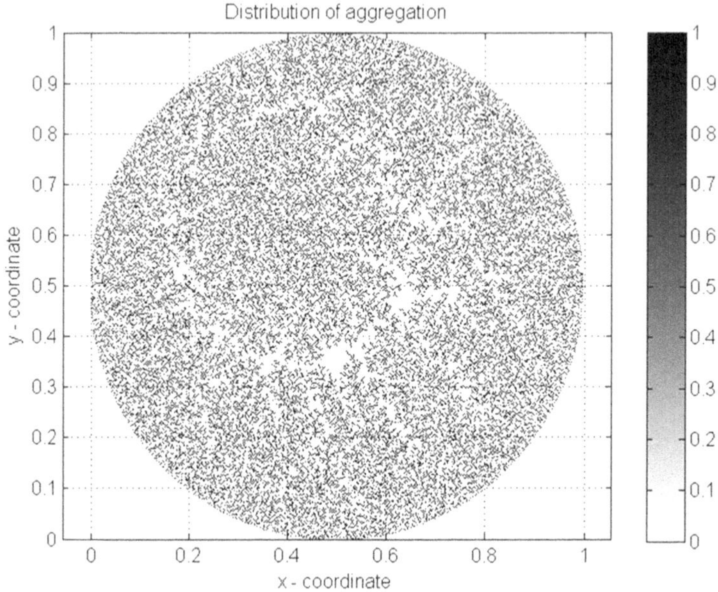

Fig. 3.26 Final aggregation with an initial nucleus in the inner domain, computed after 50,000 walks

3.2 Solution of Langevin Equation with the Molecular Dynamic Method

3.2.1 Introduction to the Topic

In this introduction to the topic, we follow the guideline given, with more details, in (Risken, 1989).

A particle with mass m, immersed in a fluid, experiences a friction (or damping force) F_f, given by the Stokes law, as being proportional to the velocity v in the neighbouring of the particle:

$$F_f = -\alpha v, \tag{3.10}$$

where the proportionality factor has dimensions $[\alpha] = [\text{kg/s}]$. The minus sign in Eq. (3.10) is justified by the fact of being a damping force. Writing the force as mass x acceleration, on derives the simplest expression of the equation of motion of the particle:

$$mv' + \alpha v = 0. \tag{3.11}$$

Dividing Eq. (3.11) by m, one obtains

$$v' + \gamma v = 0, \tag{3.12}$$

with the definition

$$\gamma = \alpha/m = \frac{1}{\tau}, \tag{3.13}$$

where τ has the dimension of time and is referred to as "relaxation time".

Integrating Eq. (3.13) with initial condition $v(t = 0) = v_0$ yields

$$v = v_0 e^{-\gamma t} = v_0 e^{t/\tau}. \tag{3.14}$$

Equation (3.12), however, is correct only in a first approximation if only the damping force F_f is considered. In reality the particle undergoes also a fluctuating force $F_f(t)$ of stochastic (or random) nature which must be added to the continuous force $F_c(t)$ thus giving a total force

$$F(t) = F_c(t) + F_f(t) = -\alpha v + F_f(t). \tag{3.15}$$

The contribution of the stochastic force $F_f(t)$ is considered by replacing Eq. (3.11) with

$$mv' + \alpha v - F_f(t) = 0, \tag{3.16}$$

$$v' + \frac{\alpha}{m} v - \frac{1}{m} F_f(t) = 0, \tag{3.17}$$

$$v' + \gamma v = \Gamma(t), \tag{3.18}$$

with the definitions of the "Langevin force"

$$\Gamma(t) = \frac{F_f(t)}{m}, \tag{3.19}$$

which represents the fluctuating force per unit of mass.

Equation (3.18) is called "Langevin equation" (Langevin, 1908); it is a stochastic differential equation because it contains the stochastic force $v' + \gamma v = \Gamma(t)$. In the following we make some basic assumptions about the characteristics of the Langevin stochastic force:

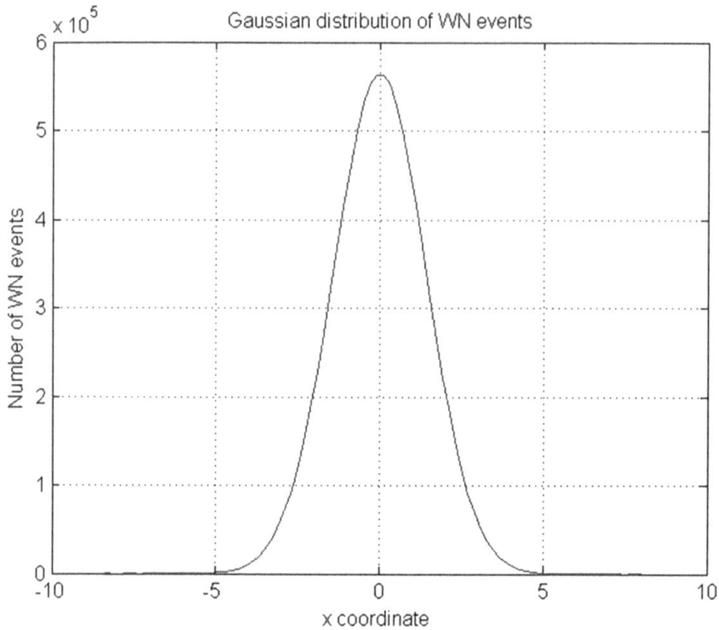

Fig. 3.27 Gaussian distribution computed with the Subroutine TEST_WN (see Appendix A.6)

(i) $\Gamma(t)$ is assumed to have a Gaussian distribution function with mean value equal to 0 (see Fig. 3.27):

$$<\Gamma(t)> = 0. \tag{3.20}$$

(ii) $\Gamma(t)$ has a correlation function proportional to the delta function

$$<\Gamma(t)\Gamma(t')> = q\delta(t-t'). \tag{3.21}$$

The proportionality function q is determined by the condition that the mean energy of the particle of mass m is given by the equipartition law as

$$\frac{m}{2} < v^2 > = \frac{1}{2}kT, \tag{3.22}$$

where k is the Boltzmann constant and T the absolute temperature.

From this condition one derives (Risken, 1989)

$$q = \frac{2\gamma}{m}kT, [\mathrm{W/kg}] \tag{3.23}$$

In numerical simulations of Brownian motion the Langevin force is due to the random impact of the fluid molecules on the particle. The impacts are almost instantaneous, much shorter than the relaxation time τ.

The Wiener-Khinchin theorem states that the spectral density $S(\omega)$ of the Langevin function is the Fourier transform of the correlation function (3.21) which is given by

$$S(\omega) = e\int_{-\infty}^{+\infty} e^{-\omega\tau} q\delta(\tau)d\tau = 2q \tag{3.24}$$

and is independent of the frequency ω.

Because the spectral density is independent of ω the noise force (3.14) is called a "white noise", in analogy with the white spectrum of the electromagnetic radiation. If the spectral density depends on ω, the noise is called "coloured noise".

3.2.2 *Treatment of the Langevin Equation*

A generalization of the methods of molecular dynamics has allowed numerical solution by computer simulation of the Langevin equation. Restricting ourselves, at first, to the one-dimensional case, let us consider the Langevin equation for one stochastic variable x:

$$dx/dt = h(x, t) + g(x, t)\Gamma(t), \tag{3.25}$$

with initial value $x(0) = x_0$. Equation (3.25) corresponds to the previously given Eq. (3.18) with $h(v) = \gamma v$ and $g = 1$. As stated in the previous section, the Langevin force $\Gamma(t)$ is assumed to be a Gaussian random variable with zero mean and a correlation function given by a δ Dirac function:

$$\langle\Gamma(t)\rangle = 0, \tag{3.26}$$

$$\langle\Gamma(t)\Gamma(t')\rangle = 2\delta(t - t'). \tag{3.27}$$

Integration from $t = 0$ to a given finite time T is obtained by partitioning the integration time in intervals of length $\Delta t = T/N$. Referring to the time points $t^n = n\Delta t$, the stochastic variable is computed at these time points by

$$x^{n+1} = x^n + D_1(x^n, t^n)\Delta t + \sqrt{D_2(x^n, t^n)\Delta t}\cdot W^n \ (n = 1, 2, \ldots N - 1) \tag{3.28}$$

where W^n are independent random variables with Gaussian distributions with zero mean and variance equal to 2:

$$\langle W^n \rangle = 0, \tag{3.29}$$

$$\langle W^n W^m \rangle = 2\delta_{nm}. \tag{3.30}$$

The coefficients $D_1(x^n, t^n)$ and $D_2(x^n, t^n)$ are the coefficients of the Kramers-Moyal expansion, given by

$$D_1(x^n, t^n) = h(x,t) + \frac{\partial g(x,t)}{\partial x} g(x,t), \tag{3.31}$$

$$D_2(x^n, t^n) = g^2(x,t). \tag{3.32}$$

A derivation of the coefficients of the Kramers-Moyal expansion is given in the above quoted text by Risken (1989).

A random number generator allows for constructing the sequence of Gaussian random variables W^n by

$$W^n = \sqrt{\frac{24}{M}} \sum_i^M (R_i - 1/2), \tag{3.33}$$

where R_i are random deviates in the open interval $(0 - 1)$ and M is a large number in the suggested order of magnitude around $M = 100$.

With this procedure one obtains a realization $x^N = x(T)$. By constructing a large number of such realizations and performing an ensemble averaging, one obtains the mean value $\langle x(T) \rangle$or any other functionally dependent expectation value $\langle f[x(T)] \rangle$.

3.2.3 Numerical Application of the Langevin Equation

In this section we present an application of the two-dimensional Langevin Equation under the assumption that we treat two independent Gaussian variables, each one under the constraint that they have zero mean value and variance equal to 2. The moving particle is an aerosol particle of 0.1 μm diameter and thus characteristic of the aerosols moving in the atmosphere. This problem is treated in more details in Chap. 5.

Data of the aerosol particle for the following calculation are given is Table 3.3.

The calculation starts in the program LANGEVIN performing two basilar tasks:

(i) The first task consists in testing the formula

$$WN = \text{DSQRT}(24/M)^* \ \text{SUM}(i = 1, \ldots M)[Ri - 1/2] \tag{3.34}$$

Table 3.3 Data for the calculation of the displacement of an aerosol particle with molecular dynamic method

Radius = 1.d-07 (m)
Specific heat = 2140.d0 (J/kg K)
Thermal conductivity = 0.15 (W/m K
Density = 890.d0 (kg/m^3)
Mass = 3.728×10^{-18} (kg)
Temperature = 300 (K)
Initial particle position = 0. (m)
Initial particle velocity = 0.001 (m/s)

Table 3.4 Mean values and variance computed with formula (3.34)

ISET	Mean value	Variance
1000	3.51 e−02	2.069
10,000	2.50 e−02	2.031
100,000	6.72 e−03	2.0068
1,000,000	2.41 e−03	2.0032
10,000,000	6.38 e−04	2.00036

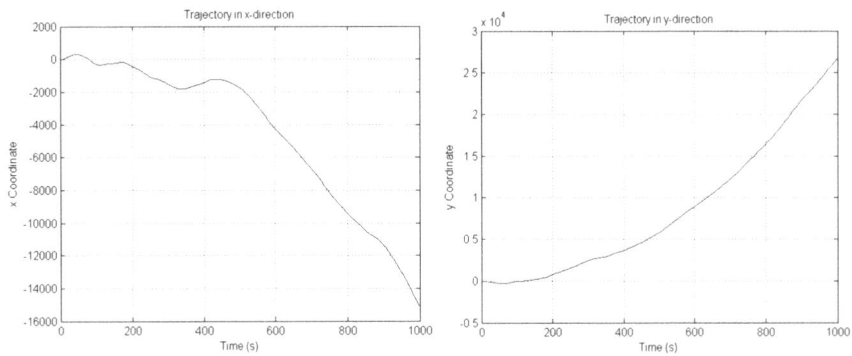

Fig. 3.28 Time-dependent displacement of particle in the x and y directions

where M = Large integer (100), Ri is a random deviate, WN is the random variable with Gaussian distribution which should turn out to have mean value = 0, variance = 2.

Results, using M = 100 and a number ISET of trials, are given in the following Table 3.4:

(ii) The second task consists in computing data for plotting the Gaussian distribution. The results are shown in Fig. 3.27.

In the subsequent parts of the programme LANGEVIN calculations of the displacement of the particle in the (x, y) plane are performed as explained in Sect. 3.2.2. The result obtained are given in Figs. 3.28, 3.29, 3.30, and 3.31.

Appendix A.6 gives a listing and explanations of the computational code LANGEVIN.

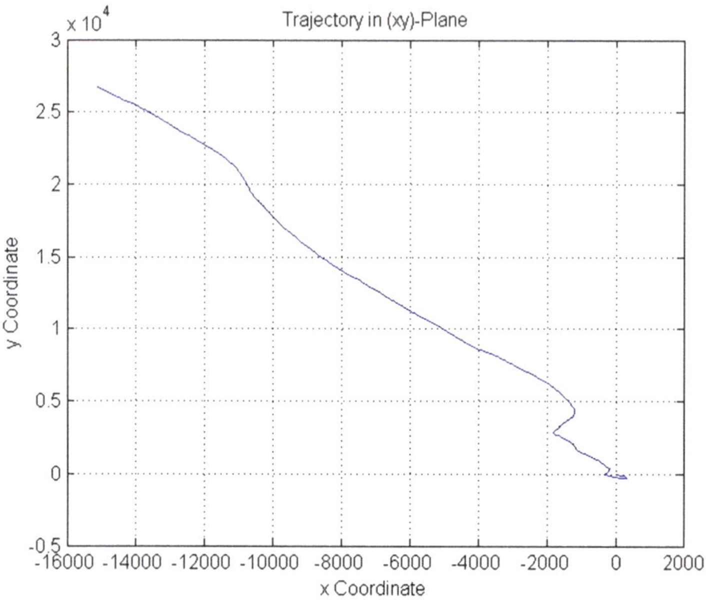

Fig. 3.29 Time-dependent displacement of particle in the (x, y) plane

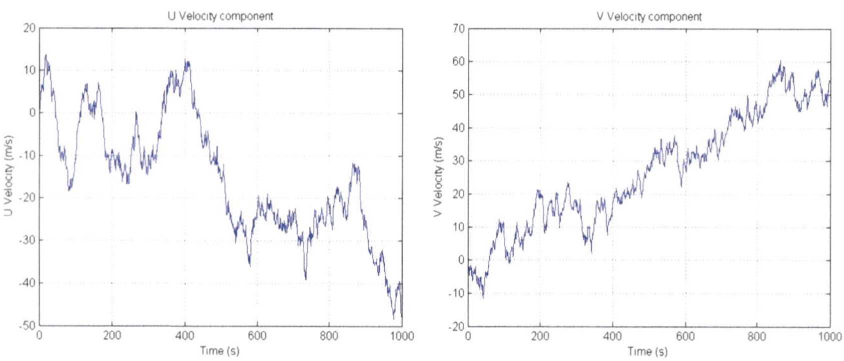

Fig. 3.30 Time-dependent velocity of particle in the x and y directions

We renounce to further investigations with applications of the Langevin equation because they would involve sample averages of numerous computed results, a task which exceeds the capacity of personal computer facilities.

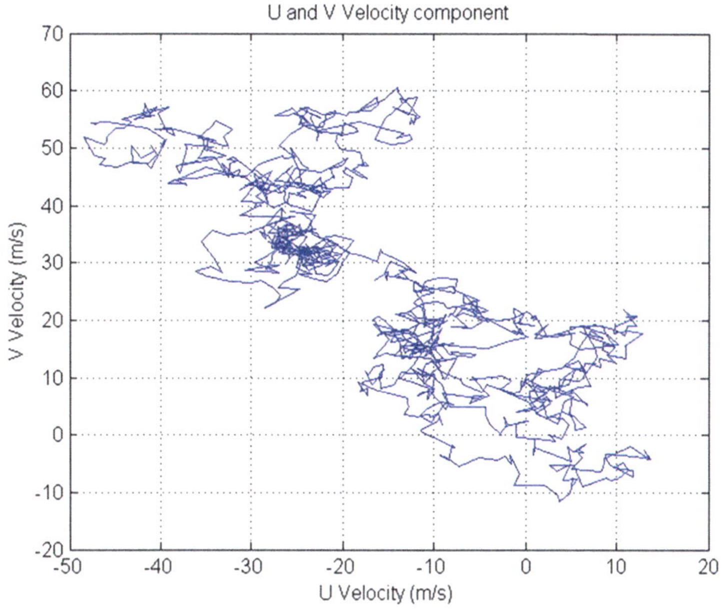

Fig. 3.31 *U* and *V* velocity components (with time as implicit parameter)

3.3 Concluding Remarks

Uhlenbeck and Ornstein (1930), already in the year 1930, subsequently Chandrasekhar (1943), treated a generalized Langevin equation for modelling Brownian motion of particles in a gas at low pressure. In spite of these theoretical advances, these pivotal ideas could not find application till the advent of fast computers. As remarked in Choi et al. (1966) in recent years Brownian motion has been investigated by numerical solutions of the Langevin equations for a variety of problems, for instance, wafer contamination in semiconductor processing, deposition of contaminant particles in turbulent channel flows. As long as parallel computation was not available, the approach to simulation of Brownian motion via numerical solution of the Langevin equation remained a difficult task for the need to construct large numbers of simulated trajectories of test particles in order to reduce statistical scattering. With the advent of parallel computers, a revival of the application of the Langevin equation to the simulation of Brownian motion occurred, as exemplified by the earlier quoted reference (Choi et al., 1966) with the appropriate title "*Massive parallel simulations of Brownian dynamic particle transport in low-pressure parallel plate reactors*". This reference explains that "*…particle trajectories … are calculated by solving the Langevin equation including particle inertia, external forces, and a fluctuating Brownian force. This Brownian Dynamics Simulation (BDS) is implemented on a massively parallel (MP) computer, which* *makes it feasible to perform the large number of particle-trajectory calculations needed to accurately predict particle transport*".

Since the famous article by Einstein (Einstein, 1905) about the explanation of the Brownian motion of small particles suspended in a fluid, observed and described by the Scottish botanist R. Brown in 1827, few domains of science have led to such a flourishing of studies as the interpretation of Brownian motion with the methods of statistical physics. In the year 2005, the centennial anniversary of Einstein's famous publication, a special issue (No. 15) of the Scientific Magazine CHAOS was especially dedicated to articles commemorating that important event in the history of science and summarizing most recent advances in the theoretical interpretation of Brownian motion and diffusion, ranging from stochastic processes to theoretical modeling of chaotic systems. The introductory article of the series of papers published in the above referenced magazine, with the significant title "*Introduction: 100 years of Brownian motion*" (Haenggi & Marchesoni, 2005), traces the summary of the body of theoretical advances made in the century elapsed since Einstein's publication. Subsequent articles of the series give all most modern advances from both viewpoints of theoretical and computational approaches.

References

Binder, K., & Heermann, D. W. (1988). *Monte Carlo simulation in statistical physics*. Springer Verlag.

Bird, G. A. (1994). *Molecular gas dynamics and the direct simulation of gas flows*. Clarendon Press.

Bottoni, M. (2022). *Physical modeling and computational techniques for thermal and fluid-dynamics*. Springer.

Chandrasekhar, S. (1943). Stochastic problems in physics and astronomy. *Reviews of Modern Physics, 15*, 1–83.

Choi, S. J., Rader, D. J., & Geller, A. S. (1966). Massively parallel simulation of Brownian dynamics particle transport in low pressure parallel-plate reactors. *Journal of Vacuum Science and Technology A, 14*(2), 660.

Chowns, T. M., & Elkins, J. E. (1974). The origin of quartz geodes and cauliflower cherts through the silicification of anhydrite nodules. *Journal of Sedimentary Petrology, 44*(3), 885–903.

Crowdy, D., & Tanveer, S. (2004). The effect of finiteness in the Saffman-Taylor viscous fingering problem. *Journal of Statistical Physics, 114*(5/6), 1501.

Einstein, A. (1905). Ueber die von der molekularkinetischen Theorie der Waerme geforderte Bewegung von in ruhenden Fluessigkeiten suspendierten Teilchen. *Annalen der Physik, 17*, 549.

Haenggi, P., & Marchesoni, F. (2005). Introduction: 100 years of Brownian motion. *Chaos, 15*, 026101.

Hartmann, L. A., Medeiros, J. T. N., & Petruzzellis, L. T. (2012). Numerical simulations of amethyst geode cavity formation by ballooning of altered Paraná volcanic rocks, South America. *Geofluids, 12*(2), 133–141.

Hartmann, L. A., Medeirosm, J. T. N., & Baggio, S. B. (2015). Controls on prolate and oblate geode geometries in the Veia Alta basalt flow, largest world producer of amethyst, Paraná volcanic province, Brazil. *Ore Geology Reviews, 66*, 243–251.

Hayes, J. B. (1964). Geodes and concretions from the Mississippian Warsaw formation, Keokuk region, Iowa, Illinois, Missouri. *Journal of Sedimntary Petrology, 34*(1), 123–133.

Hele-Shaw, H. S. (1898). The flow of water. *Nature, 58*(1489), 34–36.

Langevin, P. (1908). Comptes Rendus de Academie des. *Sciences (Paris), 146*, 530–533.

Matsushita, M. (1989). Experimental observation of aggregation. In A. Avnir (Ed.), *The fractal approach to heterogeneous chemistry: surfaces, colloids, polymers*. Wiley.

Matsushita, M., & Fujikawa, H. (1990). Diffusion-limited growth in bacterial colony formation. *Physica A: Statistical Mechanics and its Applications, 168*, 498–506.

Peitgen, H.-O., Jürgens, H., & Saupe, D. (2004). *Chaos and fractals* (2nd ed.). Springer Verlag.

Press, W. H., Flannery, B. P., Teukolsky, S. A., & Vetterling, W. T. (1986). *Numerical recipes*. Cambridge University Press.

Proust, D., & Fontaine, C. (2007). Amethyst-bearing lava flows in the Paraná basin (Rio Grande do Sul, Brazil): Cooling, vesiculation and formation of the geodic cavities. *Geological Magazine, 144*, 53.

Risken, H. (1989). *The Fokker plank equation* (2nd ed.). Springer Verlag.

Saffman P. G., & Taylor G. I. (1958). The penetration of a fluid in a porous medium of Hele-Shaw cell containing a more viscous fluid. *Proceedings of the Royal Society A: Mathematical, Physical, Engineering Sciences, 245*.

Sander, L. M., & Somfai, E. (2005). Random walks, diffusion limited aggregation in a wedge, and average conformal maps. *Chaos, 15*, 026109.

Shaw, S. H. (1898). *Investigation of the nature of surface resistance of water and of stream-line motion under certain experimental conditions*. Inst. N.A.

Uhlenbeck, E. G., & Ornstein, S. L. (1930). On the theory of the brownian motion. *Physical Review, 36*, 823.

Witten, T. A., & Sander, L. M. (1981). Diffusion-Limited Aggregation, a Kinetic Critical Phenomenon. *Physical Review Letters, 47*, 1400.

Witten, T. A., & Sander, L. M. (1983). Diffusion limited aggregation. *Physical Review B, 27*, 5686.

Chapter 4
Kaburaki, Nambu, and the Japanese School

Contents

The chapter opens with an excursion in the theoretical and numerical work performed by Kaburaki, Nambu, and their school, before applications were made for the safety analysis of nuclear reactors. In **Part A** of the chapter water-cooled thermal reactors are considered with particular reference to the safety analysis of the Chinese pressurized reactor at Dayawan.

The application of molecular dynamics to safety analysis of water reactor is an alternative to the whole reactor analysis made with integral codes like Gasflow

Text by M. Bottoni

M. Bottoni et al., *Molecular Dynamics - Theory and Applications*, Mechanical Engineering Series, https://doi.org/10.1007/978-3-031-84567-3_4

(Travis et al., 1998). Molecular dynamics allows a more detailed three-dimensional analysis in reactor cells where otherwise only global information (mean values over large volumes) would have been possible.

Part B is dedicated to safety issues of fast reactors with particular reference to the design of the German SNR 300 fast breeder reactor, which, however, never became operational. This part is preceded by a summary of the physical modelling and computational techniques for thermal hydraulic safety analysis of fast reactors. It includes a taxonomy of events and phase transitions in the course of a Hypothetical Core Disruption Accident (HCDA). This summary, together with the main description of the computer codes BLOW-3A and BACCHUS-3D/TP, is intended to stress the difference between the traditional approach of safety analysis based on macroscopic analysis of the single events and the microscopic representation of the same events.

One contribution to this topic concerns a detailed analysis of the sodium coolant behaviour in the case of hypothetical nuclear accident leading to boiling of the coolant with subsequent formation of dry-out regions, condensation of coolant in cooler reactor sections, formation of liquid slugs, and rewetting of hot sections of subassemblies. This sequence analysis is carried out coupled with a computer code based on the traditional macroscopic treatment of the coolant. The interaction between the traditional approach and the novel approach is examined in detail. One example of application of the methodology explained above is given for a 7-pin bundle with the geometry typical of the SNR reactor bundle.

One subsection of the chapter is dedicated to the special topic of fuel particle relaxation following the disruption of a fuel pin in case of uncontrolled hypothetical accident.

One further section concerns an application to thermal creep simulations in the 7-pin bundle treated above.

4.1 Introduction to the Novel Approach

The seminal book by Bird (1994) was for several years the paradigm of the Direct Simulation Monte Carlo (DSMC) method. Besides the thoughtful acquaintance with the book, in 1996 the author was initiated to the DSMC method through the interest of his colleague Prof. Hisashi Ninokata of the Tokyo Institute of Technology and through the mediation of his colleague Dr. Muramatsu Toshiharu of the thermal Hydraulic Research Section of the Power Reactor and Nuclear Fuel Development Corporation in O-arai, Japan.

Through the cooperation with Dr. Muramatsu, the author had the opportunity of contacting what I call the "Japanese School" of the DSMC method, which should not be interpreted as an alternative to the Bird method, but rather as a complement to it. Taking into account the development of thoughts experienced by the author in the years after 1996, it seems to be appropriate to recall the main achievements made by the Japanese School in the past decades.

4.1.1 Novel Approach

The first international conference on supercomputing in nuclear applications (Nuclear Energy Data Center, 1990) was held in Mito, Japan, on March 12–16, 1990. In this conference not only advanced supercomputing techniques were discussed but also a new approach was presented for the application of the Direct Simulation Monte Carlo (DSMC) method comparing the traditional and well established Bird method with a modified and novel Nambu method (Nambu, 1980a, b, c).

In the DSMC method one of the main paradigms is the assumption of the "*principle of uncoupling*" in which collisions between particles in a computational cell are considered for a short time interval as uncoupled from the particle motion. For a short time interval *Dt* the movement and the collisions of pairs of selected molecules are not affected by the presence of the flow in the cell considered.

In the traditional DSMC method the size of the collision cells is usually taken of the order of magnitude, or slightly smaller, than the mean free path of the molecules. In the modified treatment of the DSMC method cells initially defined are subdivided in finer cells if this is required by the condition that the cell size be smaller than the mean free path of the molecules. As a practical rule, it is suggested that the cell size be 2–5 times larger than the mean free path of the molecules. This numerical technique is referred to in the following as "*method of auto-generation*" of the collision sub-cells.

This is particularly important in some key problems in which the mean free path of the molecules in the flow varies largely with time. One typical example is the process of atomic vapour isotopic separation in laser technology where the atomic flow is generated bombarding an electron beam upon a support surface. As the atomic vapour expands into a vacuum space, the mean free path of the molecules varies largely, and correspondently the cells in which the process occurs must be adapted to the new situation.

Sampling of couples of molecules that collide within a cell is explained for the two different techniques, in the Bird method and in the novel Nambu method. Two numerical examples are given: (1) The lid driven cavity flow and (2) Gas expansion from a slit source into a vacuum space.

In test case (1) the feasibility of the automatic division of the original cells in more refined cells is assessed under conditions of extreme air rarefaction as it would be at 100 km height in the Earth atmosphere with initial temperature Ts = 183 K, pressure $p_s = 0.002$ pa, Knudsen number $K_n = 0.042$. Calculations made with both the Bird method and the Nambu method are compared with each other but a final assessment about the quality of the results is not possible because there were not experiments available in the conditions examined.

In the application (2) the method of auto-generation of the collision sub-cells is applied to the expansion of atomic vapour from a slit of 10 mm width. Atomic vapour is emitted from a source into a vacuum space. Pressure and temperature at the source are 250 K and 628 pa, respectively, with a Knudsen number at the source of

0.039. The analysis reported in the article show advantages of the cell system automatically adapted to the variation of the mean free path of the molecules in the expansion of the molecular stream into the vacuum space.

4.1.2 *Supporting the Novel Approach*

In a subsequent article published in the Proceedings of the Joint International Conference on Mathematical Methods on Supercomputing in Nuclear Applications (**M&C + SNA '93**) held in Karlsruhe, Germany, from April 19 to 23 , 1993, the method of auto-generation of collision sub-cells is further analyzed and compared with experimental results. The article presented at that conference Kaburaki and Yokokawa (1993) was introduced by the following summarizing worlds:

> *The Navier-Stokes equations are valid in the continuum region where the Knudsen number Kn is low. However, these equations fail as the Knudsen number increases above 0.1. Generally, the simulation of dilute gas flows is made through the use of the Boltzmann equation. The solution of this equation by the finite difference method is generally difficult to obtain. The Direct Simulation Monte Carlo (DSMC) method is one of the methods to integrate the Boltzmann equation for a wide range of Knudsen numbers. This method can in principle be applied to any regions of flows from continuum to rarefied gas. Generally, particles methods including the DMSC method have some advantages over numerical methods using discretized partial difference equations in terms of numerical errors and instabilities due to discretization. Moreover, boundary conditions at atomic level can be set easily in these methods. However, statistical averages have to be imposed to derive macroscopic properties.* [More precisely: Macroscopic quantities are obtained by sampling particles properties in sampling cells which are generally larger than the collision cells.] *To reduce the statistical errors, the number of particles used in the simulation has to be increased as the Knudsen number decreases. Therefore, a large number of simulated particles are required to simulate near continuum flows.*

The last sentence justifies the effort made by the authors in assessing the effect of the number of simulated particles on the quality of the computational results.

Numerical investigations are made for the two-dimensional square cavity flow. Initial conditions of temperature and pressure in the cavity are 183 K and 0.002 Pa, respectively, with a Knudsen number $Kn = 0.042$ in the initial conditions. The upper lid cavity is modelled to move at a constant speed corresponding to 0.6 times the average thermal speed. The number of the sampling cells in the cavity is 64 and the collision sub-cells are automatically generated in the original sampling cells as the calculation proceeds.

A standard case is examined with 2000 simulated particles per sampling cell and 1000 samplings. Computations with the Nambu method showed that results were not significantly affected by the number of simulated particles but were not good if the number of samplings was low. It was therefore inferred that the number of samplings is crucial and that the results are significantly improved when the number of sampling increases.

The above quoted reference reports computational analysis of evaporating spherical jets used in spectroscopic experiments to investigate atomic spectra. Atomic beams are usually produced by heating a metal contained in a crucible. A vapour stream arising from the evaporating surface undergoes a sharp change of the Knudsen number as the jet expands in the vacuum space. The authors conclude that with the introduction of the system for automatically generating sub-cells, a key feature of the novel Nambu methodology, "*the accurate calculations of the strongly evaporating free expanding jet is made possible.*"

4.1.3 Simulation of the Rayleigh-Bénard Instability

Having established by the work of Kaburaki and colleagues that large scale vortices can be observed in the DSMC method, provided the number of simulated particles is sufficiently large and the collision cells are larger than the mean free path of the molecules, a subsequent challenge was the simulation of the Rayleigh-Bénard system. When the temperature difference between the top and bottom walls exceeds a critical value, a transition occurs from conduction to convection, which is known as the Rayleigh-Bénard (RB) instability.

In the study made by Watanabe et al. (1994) a two-dimensional rectangular domain is considered, 11.3 mm in width and 5.6 mm in height, with aspect ratio of 2016. Initial temperature and pressure were 80 K and 20 Pa, respectively. Under these conditions the number density was $1.81 \times 10^{-22}/\text{m}^3$ and the mean free path of the molecules was 0.0981 m. The computational domain was filled with hard sphere particles with a diameter of 3.7×10^{-10} m and a mass of 4.8×10^{-26} kg.

The upper and lower surfaces are diffusively reflecting surfaces. Specularly reflection is assumed at the side walls: upon impingement of a molecule with the surface, the normal velocity of the molecule is reversed, while the tangential velocity remains unchanged.

Numerical calculation were made with 40×20 sampling cells, which were subsequently divided in 5×5 collision cells. The number of simulated particles was 400. Bifurcation between purely conductive and convective regimes was observed around the theoretical value of 1708, which was confirmed by experimental investigations. It was confirmed that the number of sample molecules had to be sufficiently high to reproduce well the Rayleigh-Bénard instability.

4.1.4 Link to the THEMIS Codes

Thus far we have been treating general purpose topics which constitute the background of the DSMC method applied the basic problems of fluid- and thermal-dynamics. In this section we introduce, as an important side step, a computer program called THEMIS which can be considered as the precursor of a different kind of applications, concerning nuclear reactor safety analysis. The THEMIS series of programs was developed by Dr. Toshiharu Muramatsu who worked at the Thermal Hydraulic Research Section of the Power Reactor and Fuel Development Corporation in O-arai, Ibaraki ken, Japan. The author had the opportunity of interacting with Dr. Muramatsu in the years following 1996 and establishing with him a fruitful cooperation which led to the applications explained in the following contributions of this chapter.

The reference (Muramatsu, 1995) presents an analysis of thermal striping phenomena characterized by stationary random temperature fluctuations in the region immediately above the core exit of a Liquid Metal-cooled Fast Reactor (LMFR) due to the interaction of cold sodium flowing out of a the control rod assembly and hot sodium flowing out of adjacent fuel assemblies. Components located in the core outlet region must be protected against stationary random processes which induce high-cycle fatigue. Because experimental analysis of these phenomena of thermal striping are both difficult and expensive to perform, the computer code THEMIS was developed and tested to assess its applicability for safety analysis of fast nuclear reactors.

A computational domain in which thermal striping must be investigated is partitioned in cells filled with sample molecules representative of some 10^{15} real molecules to which is assigned an initial $F(c)$ velocity distribution. The subsequent evaluation of the $F(c)$ function is determined by the application of the DSMC method with a set of procedures which compute positions and velocities of the simulated molecules as function of time.

Computations were made in a two-dimensional domain partitioned in $50 \times 50 = 2500$ cells with diffusive reflection conditions applied at the upper and lower walls and with specular reflections applied to the vertical walls. Results of the calculations made for the reference case "*confirmed that (1) fluid-structure interactions for momentum and energy can be described by the DSMC method and (2) the THEMIS code is applicable for the evolution of the non-stationary heat transfer phenomena in specific engineering applications such as for thermal striping.*"

The computer program THEMIS formed the basis for further development work performed by the author. A version of the code, referred to as THEMIS/3D.1 (Bottoni, 2004) models the geometrical configuration of a 60-degree sector of a 7-pin bundle to apply the DSMC method to safety analysis of fast breeder reactors under hypothetical accident conditions. This code version describes sodium evaporation, condensation, and rewetting of dried out fuel pins combining the microscopic treatment of the molecular dynamics, applied to sodium vapour, to a macroscopic description of the liquid film of sodium formed by condensation of the vapour in cold sections of the simulated bundle. A subsequent development, describing a full 7-pin bundle, has been documented in Bottoni et al. (2006).

4.1.5 Nambu's Theoretical Masterwork

Most computational codes applying molecular dynamics to fluid-dynamic problems rely upon the theoretical foundations posed by Nambu (1980a, b, c) who derived directly from the Boltzmann equation a simulation scheme which divides in three steps the task of determining position and velocities of molecules after a small time step. The first step consists in determining the maximum collision probability of a molecule in a fluid from basic principles, but without specifying at first its collision partner. The second step, comparing the maximum possible collision rate with the real collision rate, determines whether or not the chosen sample molecules are accepted to undergo a collision event. In case the second step determines that the collision event does occur, eventually the third step samples the probability density of the relative velocity of the molecules and thus determines the post collision velocity of the collision partners. The scheme developed by Nambu satisfies strictly the Boltzmann equation, provided the time step is small compared with the mean free time between collision of the molecules.

In the scheme developed by Nambu the often mentioned principle of uncoupling is applied by splitting the calculation in the collisionless motion of the molecules, which are assigned updated positions by $x_i^{n+1} = x_i^n + u_i^n \Delta t$, followed by the simulation of collision events among the molecules. The principle of uncoupling the collisionless motion over a time step Δt from the collision events, modelled at the end of the time step, is justified on the basis of the following argument. In the Boltzmann equation, governing time-dependent molecular motion, the convective operator $D = \overline{c}(\partial f / \partial \overline{x})$ can be split from the collision integral Jf with an approximation of the order of Δt^2. We generally use $\Delta t = 10^{-7} s$.

4.2 Safety Analysis of Thermal Nuclear Reactors

4.2.1 Distribution of Air, Water Vapour, and Hydrogen Within the Containment of a Pressurized Water Reactor (PWR) During Hypothetical Accident Conditions

4.2.1.1 Introduction

Analysis of Pressurized Water Reactors (PWR) were made with the whole reactor analysis computer program GASFLOW-II (Travis et al., 1998) both at the Los Alamos research centre in USA and at KfK in Germany. In these analyses the reactor containment is discretized in cells of relatively large dimensions (typically 350 m^3). Because in every cell only mean values of all relevant physical information are computed detailed information is lost. In many cells it is however necessary to retain more detailed physical information because critical parameters, like concentrations of hydrogen, could reach dangerous values and trigger deflagrations. For this purpose it is necessary to link the whole analysis code to another module in which the distributions of important parameters ca be evaluated in detail. It is therefore necessary to provide the output from the whole analysis code as input to a subsequent module in which detailed distributions are computed. This new module has been developed with the computer program THEMIS which simulates fluid flows from a microscopic viewpoint. This code, based on the Monte Carlo statistical approach, was first developed as a two-dimensional version at the Power Reactor and Fuel Development Corporation (PNC) (now Japan Nuclear Cycle, JNC) of the Japanese Research Center of O-Arai. Subsequently, starting from the two-dimensional version provided by PNC, a three-dimensional version was derived but limited to the simulation of a 60-degrees sector of a 7-pin bundle, with a hexagonal rod lattice, typical of a fast reactor.

Another three-dimensional code version was developed and applied to the simulation of condensation of steam inside a water reactor containment, simulating the conditions which would occur in case of a hypothetical loss of flow accident with loss of coolability of the reactor core. Preliminary work was made by the author at the INET (Institute of Nuclear Energy Technology) in Beijing, at the Tsinghua University, between October and December 1999, to establish a code simulating a reactor containment. Between February 2000 and July 2000 a new version, called TH3SL, was developed simulating condensation of steam, with or without not-condensable gases (air and hydrogen), in a prototypical cell of a parallelepiped form, in which the space domain adjacent to the containment surface could be discretized.

The main new features of the code are the following:

(i) Detailed simulation of physical conditions inside a partial domain of a water reactor containment;
(ii) Calculation of an effective collision cross section, using the Lennard-Jones potential;
(iii) Simulation of pure steam condensation, without not-condensable gases.
(iv) Simulation of steam condensation, with not-condensable gases.

4.2.1.2 Definition of the Geometrical Domain

The spherical surface of a containment is discretized in a spherical coordinate system in a set of cells by means of a net of meridians and parallels. The domains adjacent to the cells on the surface can be defined considering a given thickness in radial direction. Because condensation of steam on the surface of a containment occurs in a relatively thin boundary layer adjacent to the surface, the thickness of the computational domain can be of the same order of magnitude as the linear dimensions of the cells on the spherical surface. The computational domains can be approximated at first by means of parallelepipeds. One of the surfaces bounding the parallelepiped simulated the wall of the containment. The surface is treated in the Monte Carlo simulation of molecular motions as "diffuse reflecting" surface. This means that the direction of a molecule after impact with the surface is randomly distributed and its velocity has a Maxwellian distribution corresponding to the temperature of the wall. On this material wall condensation of impinging molecules can also occur. The other surfaces are ideal surfaces treated as "specularly reflecting" surfaces. This corresponds to the fact that, statistically, in equilibrium conditions molecules crossing the surface in one direction are replaced by other molecules crossing the surface in opposite direction.

In the code, the top and bottom surfaces bounding a control volume are labelled with indices 1 and 2, respectively. The lateral surfaces are labelled with indices 3–6 starting from the wall surface with index 3 and proceeding counterclockwise.

4.2.1.3 Calculation of Effective Collision Cross Section with the Lennard Jones Potential

For this topic see Sect. 1.3.3 (formulas 1.152–1.156) in Chap. 1.

4.2.1.4 Simulation of Conditions in Reactor Containment

In application to safety analysis of water reactors, the conditions inside a containment following a hypothetical accident with core melting, release of steam and production of hydrogen due to dissociation of steam molecules, are generally computed with three-dimensional safety analysis codes. The output of theses codes provide the initial and boundary conditions for the simulation of the computational domain in the TH3SL program.

In the program TH3SL it is assumed that steam is in saturation conditions at a given temperature in presence of air, and the partial pressure of hydrogen is also given. Optionally a relative humidity smaller than unity can be specified, thus simulating non-saturated steam conditions. The simulation of hydrogen is optional. It can be eliminated by setting its partial pressure to zero. A subroutine called PRERUN has been written to simulate the physical conditions of the three species (steam, air, and hydrogen) at the given temperature and partial pressure of hydrogen. Output of this subroutine are, besides other parameters, the mass fractions of the three species and the number of sample molecules which must be simulated in the code TH3SL to represent the desired conditions in the reactor containment. In this section we give some details about the calculations made in the subroutine PRERUN.

4.2.1.4.1 Steam

Given the temperature T_s of the steam-air-hydrogen mixture in the containment, the partial pressure of saturated steam is computed by

$$p_{vs} = 7.26 \cdot 10^{10} \exp(-5040/T_s) \tag{4.1}$$

and its density, at saturation, is given by

$$\rho_{vs}(T_s) = \frac{p_{vs}(T_s)}{R_v T_s}. \tag{4.2}$$

In Eq. (4.2) $R_v = \mathfrak{R}/M_v$ is the gas constant of steam, considered as a perfect gas. $\mathfrak{R}$ is the universal gas constant and M_v is the molecular weight of steam.

We take into account the relative humidity φ, defined by

$$\varphi = \frac{\rho_v}{\rho_s} = \frac{p_v}{p_s}. \tag{4.3}$$

The partial pressure of steam and its density are given by

$$p_v = \varphi p_{vs}, \tag{4.4}$$

$$\rho_v = \varphi \rho_{vs}. \tag{4.5}$$

The number density of steam is

$$n_v = \frac{\rho_v A}{M_v}, \tag{4.6}$$

where A is the Avogadro number.

The number of steam molecules in the computational domain simulated, of volume V, is

$$\widetilde{n} = n_v V. \tag{4.7}$$

Denoting m the mass of a steam molecule, the total mass of steam in the computational domain is

$$Ma_v = \widetilde{n} m_v. \tag{4.8}$$

We assume that for all species simulated, one sample molecule is representative of F real molecules. Therefore the number of sample molecules for the steam is

$$ns = \frac{n}{F_v}. \tag{4.9}$$

4.2.1.4.2 Air

The partial pressure of air at the temperature T_s is

$$p_a = \rho_a R_a T_s, \tag{4.10}$$

where R_a is the gas constant of air.

The number density and the number of molecules of air in the computational volume are

$$n_a = \rho_a A / M_a, \tag{4.11}$$

$$\widetilde{n} = n_a V. \tag{4.12}$$

The total mass of air is

$$Ma_a = \widehat{n} m_a, \tag{4.13}$$

where m_a is the mass of one air molecule, taken as weighted mean value of the masses of the constituent gases in the mixture.

The number of sample molecules of air is

$$ns_a = \, = \widehat{n}_a / F_v. \tag{4.14}$$

4.2.1.4.3 Hydrogen

Assuming that the partial pressure of hydrogen is given at the temperature T_s, it density is

$$\rho_h = \frac{p_h}{R_h T_s}. \tag{4.15}$$

The number density, the total number of molecules in the computational domain, the total mass and the number of sample molecules of hydrogen are given by the formulas identical with Eq. (4.11) through (4.14) , just replacing subscript a with h.

4.2.1.4.4 Mass Fractions

The mass fraction of the three species are given by

$$\omega = \frac{Ma_i}{Ma_T}, (i = v, a, h), \tag{4.16}$$

where $Ma_i = Ma_v + Ma_a + Ma_h$ is the total mass of the three species in the computational domain.

4.2.1.5 Simulation of Steam Condensation in the Presence of Air and Hydrogen

Simulation of pure steam condensation can be made in the TH3SL code, but we step to the treatment of the case with not-condensable gases. In the course of the code development a first assessment of the code capabilities was made by computing a steam, air and hydrogen mixture at the temperature of 400 K with 10,000 sample molecules for steam, 8000 for air and 2000 for hydrogen. The computation domain of lengths 1.0 m, 0.1 m, and 1.0 m in the x, y, and z coordinate directions was partitioned in 50, 1, and 30 meshes respectively. The coordinate $x = 0$ corresponded to the outer surface of the simulated containment. About 60,000 time steps with a time step length of 5×10^{-6}s were simulated for a total problem time of 0.3 s. The small time step is required by the fact that the most probable velocity of the hydrogen molecules is, at the given temperature, of about 1850 m/s and by the constraint that in a time step a molecule must not travel more than a mesh length.

Some representative results of this calculation are presented in the following. The physical conditions simulated inside the containment were:

- Temperature 390 K;
- Relative humidity equal to 1, hence saturated steam;
- Partial pressure of steam 1.772 bar, of air 2.295 bar, of hydrogen 1.530 bar;
- Mass fractions 0.322, 0.648, 0.030, respectively;

Figure 4.1 shows concentrations of steam, air, and hydrogen along the x coordinate axis at time $t = 0$, at the beginning of the simulated accident. The concentrations of all species of steam, air, and hydrogen and the total concentration are shown versus mesh number in the x-direction in Fig. 4.2 for the time point $t = 0.3$ s. The hydrogen is concentrated slightly in the middle of the x-direction.

Figures 4.3, 4.4, and 4.5 show the two-dimensional (x, z) distributions of the species.

The code THEMIS was eventually delivered to INET at the Tsinghua University and issued as internal (unpublished) report. Under the author's leadership, the code was later used for obtaining a master degree, documented in Gan et al. (2002)

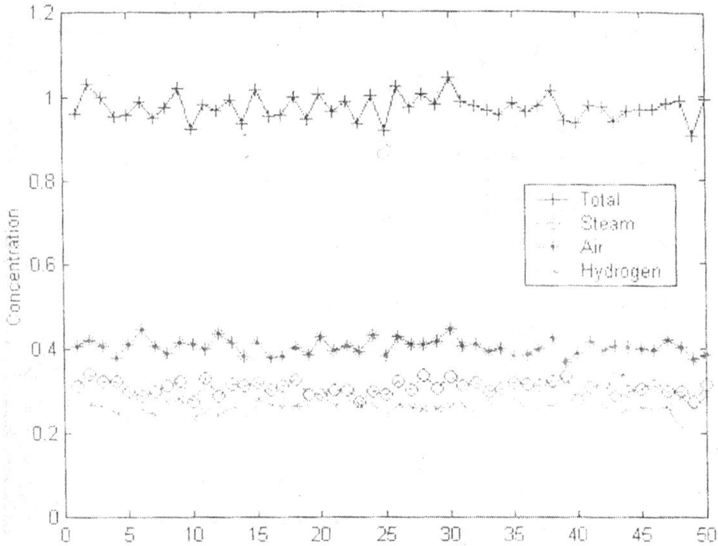

Fig. 4.1 Concentrations in the x-direction at $t = 0$ s

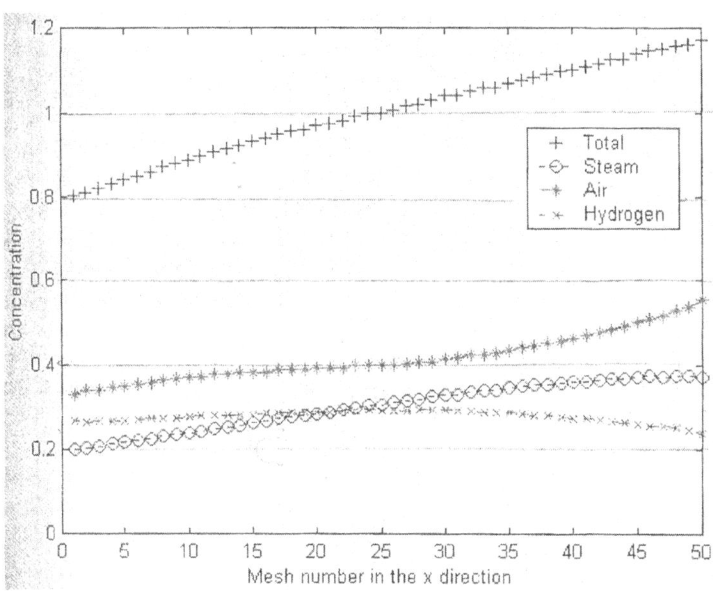

Fig. 4.2 Concentrations in the x-direction at $t = 0.3$ s

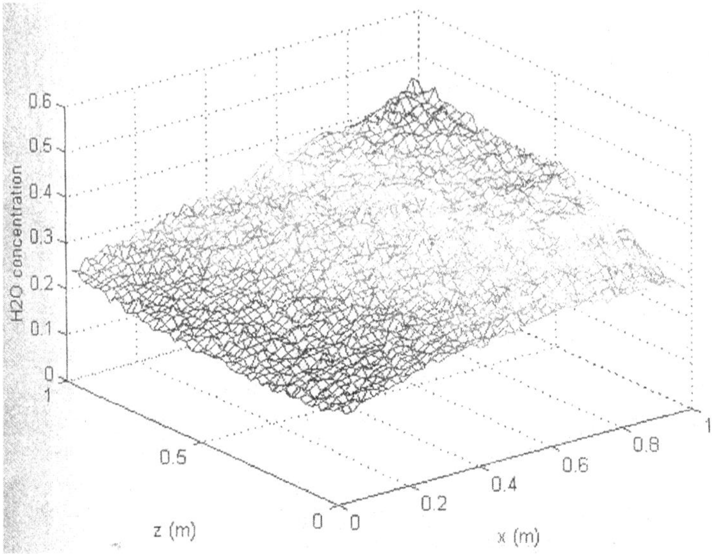

Fig. 4.3 Concentration of H_2O in x-direction at $t = 0.3$ s

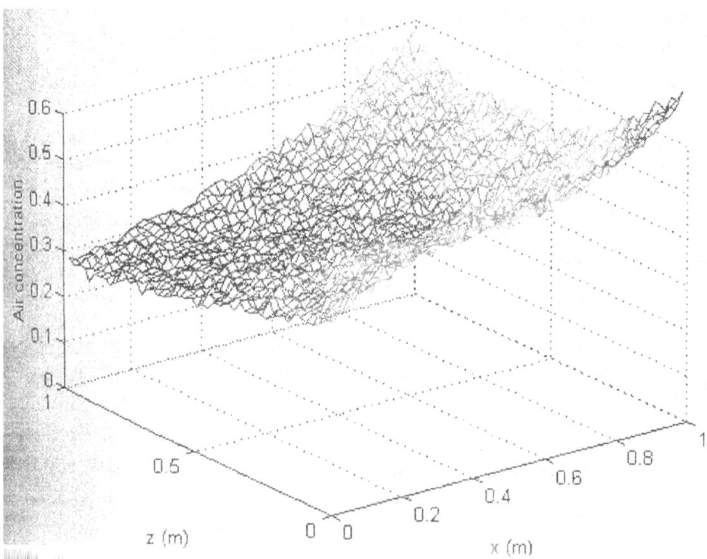

Fig. 4.4 Concentration of Air in x-direction at $t = 0.3$ s

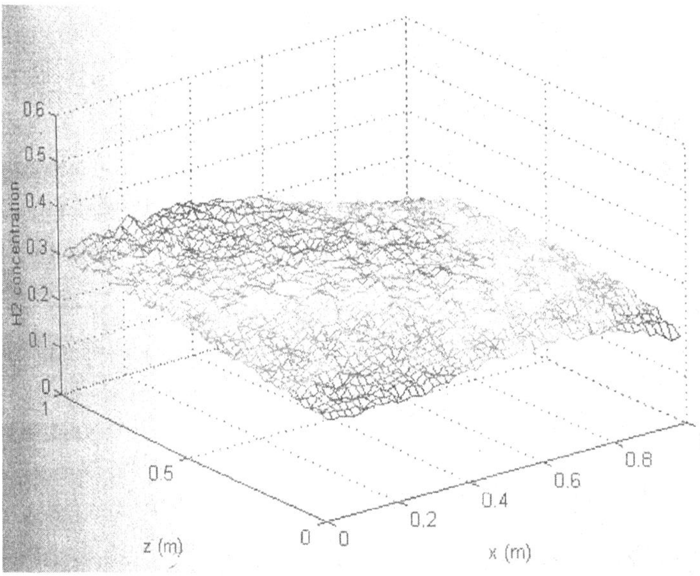

Fig. 4.5 Concentration of Hydrogen in x-direction at $t = 0.3$ s

4.3 Safety Analysis of Fast Nuclear Reactors

4.3.1 Introduction to Fast Reactor Safety Analysis: The Traditional Approach

4.3.1.1 Taxonomy of Events and Phase Transitions in the Course of a Hypothetical Core Disruption Accident (HCDA)

With regard to the evaluation of the consequences of low-probability hypothetical accidents in nuclear reactors, the main difference between Light Water Reactors (LWRs) and Fast Breeder Reactors (FBRs) consists in the core configuration. In LWRs, the core materials of fuel, cladding, coolant, and structures are arranged to maximize the effective criticality factor k_{eff}. Any material relocation from this optimum configuration tends to shut down the reactor neutronically. Unlike in LWRs, in FBRs material relocation, for instance, fuel compaction due to fuel melting and slumping, would increase at first the reactivity, although subsequent fuel dispersal would lead to reactivity loss.

Because of the absence of a moderator, boiling of coolant in a FBR would decrease the neutron absorption, and, more important, harden the spectrum thus leading to reactivity increase, at least for voiding in the core regions of a large reactor. The same event in an LWR would decrease the flux of thermal neutrons and lead to a negative reactivity feedback. In fast reactors the fraction of delayed neutrons ($\beta \approx 0.004$) is smaller than in an LWR ($\beta \approx 0.007$) and the neutron generation time ($\Lambda \approx 4 \times 10^{-7}$) is about two order of magnitudes smaller than in an LWR ($\Lambda \approx 2 \times 10^{-5}$). Therefore, the kinetic response of FBRs in case of major accidents is faster than in LWRs.

The major difference between the two reactor lines consists in the energy stored in the coolant which is very high in LWRs due to the pressurization of the system (about 70 bar in Boiling Water Reactors (BWRs) and about 160 bar in Pressurized Water reactors (PWRs). In Liquid Metal-cooled FBRs (LMFBRs) the coolant is subcooled by about 350 °C under normal operating conditions. Consequently, in case of a Loss of Coolant Accident (LOCA), following the blow down phase, it is a major concern to keep the core of an LWR covered with coolant, and it may be unavoidable to have a temporarily uncovered core. Conversely, in LMFBRs there is no coolant flashing and the coolant temperatures remain for some while (about 5–20 s) below saturation in loop type reactors, while in pool type reactors natural convection prevents altogether to reach saturation temperature in the coolant *(passive safety system)*.

For these reasons, the major topic involved in safety analysis of LMFBRs is neither a highly improbable large pipe break (large LOCA) nor a more probable small pipe break (small LOCA), as in LWRs. It is rather the concern about material relocation, and the associated recriticality, which led to the consideration of the two major accidents initiators for LMFBRs safety analysis:

(i) The Transient Over-Power (TOP) accident due to positive reactivity insertion;
(ii) The Transient Under Cooling (TUC) accident in which the coolant flow is reduced by constant reactor power. The Loss of Flow (LOF) accident is considered as a subcase of the TUC accident.

In order to investigate the accident scenarios and their consequences on the control and coolability of LMFBRs, a series of computer programs has been developed worldwide, mainly in the USA, Germany, France, Japan and United Kingdom, and verified against the results of in-pile and out of pile test facilities. In the following section we review briefly the characteristics of the main computer codes developed for LMFBR safety analysis and, within the general frame, determine the position occupied by the computer code BLOW-3A and of the code BACCHUS-3D/TP, both developed at KfK (now KIT).

The sequence of events following an initiating perturbation, TOP or TUC, of the normal reactor operation, leading to severe core damage and possibly to core disruption is generally referred to as Hypothetical Core Disruption Accident (HCDA). The sequence of phases involved in the course of a HCDA and transitions between the phases are shown in the simplified scenario of Fig. 4.6.

The case labelled with 0 refers to normal reactor operations. Perturbations of the normal conditions lead to the events labelled with subsequent digits from 1 (Initiating phase) up to 6 (Extended core material motion).

An *Early Neutron Shut Down* (Phase labelled 2) is likely to occur when the reactivity ramp following the initiating event is low (about a few cents per second).

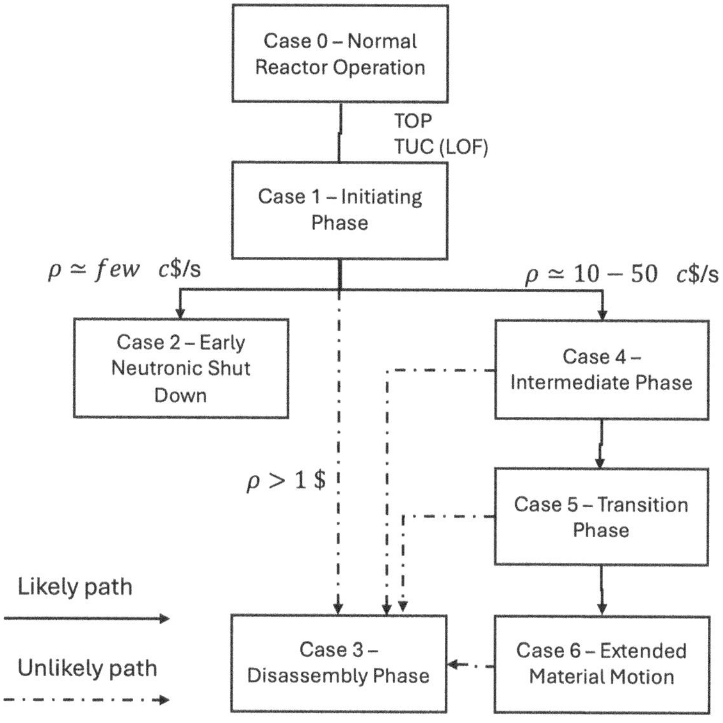

Fig. 4.6 Simplified scenario of phase transitions in the course of a HCDA (Hypothetic Core Disruptive Accident)

In this case fuel expansion or sweep out of fuel into the coolant channels leads rapidly to sub-criticality. In the later case, however, local fuel damage would occur.

The *Disassembly Phase*, also called phase of energetic core disassembly, is characterized by formation of fuel vapour and build up of high pressures in a compacted reactor core, which becomes prompt critical. This was the situation originally postulated by the Bethe-Tait analysis (Bethe & Tait, 1956) in 1956 in which an originally cylindrical reactor core was assumed to compact into a sphere of molten fuel without coolant and structural materials. Pressure build up and energetic disassembly would follow. In light of modern research a transition from the Initiating Phase to the Disassembly Phase is however highly improbable, because of negative Doppler reactivity insertion mitigating the reactivity ramp of the compacting core. Besides, a complete exclusion of other material than the fuel from the molten pool is unrealistic. The original Bethe-Tait scenario also neglected incoherency effects between different parts of the reactor core, which, according to recent results obtained with numerical simulations, have a significant impact on the steepness of the reactivity ramp.

The *Intermediate Phase* is characterized (a conventional definition) by progressive failure of fuel elements with still intact hexagonal canning structures of the subassemblies. Dominating mechanisms for this phase are: (i) Voiding dynamics within the subassemblies, with continuous release of positive void reactivity; (ii) Cladding melting and motion with possible blockage formation; (iii) Fuel melting and motion with possible formation of fuel crusts; (iv) Fission gas release from molten and slumping fuel; (v) Fuel-Coolant Interaction (FCI).

The *Transition Phase* is characterized, by conventional definition, by phenomena extending beyond the hexagonal structure of the single subassemblies. The dominant mechanisms are the same as listed above with the addition of (vi) Melting and slumping of structural materials of the hexagonal canning and thermal interaction with coolant, cladding, and fuel material. Recent research seems to support the scenario in which fission gas released from the fuel, steel vapour, and fuel vapour hinder the core compaction so that a prompt critical geometrical arrangement, with direct transition to the Disassembly Phase, cannot be reached. The generality of this conclusion is however questionable.

Ruling out, as unlikely, a direct transition to the Disassembly Phase, the most likely scenario is a path to the phase of *Extended Material Motion* in which a sub-prompt critical molten pool of steel and fuel interacts with the reactor vessel, slumps down into a core catcher initiating physical and chemical interaction with the concrete bulk. Water vapour is released by chemical reaction with the concrete. A transition of this molten pool mass to prompt-critical conditions, hence to the Disassembly Phase, is considered to be highly unlikely. Thus neutronic shut down occurs and the major remaining problems are thus related to the long-term coolability of the system.

The most likely path to the *Disassembly Phase* which occurs in case of intermediate reactivity ramp rates, of about some tents of cents per second, is from the Initiating Phase to the Transition Phase via an Intermediate Phase (often referred to as Transition to the Transition Phase). The Disassembly Phase is characterized, by definition, by a reactivity insertion which makes the reactor prompt critical, with extremely rapid power excursion. This leads to fuel vapour pressure build up and consequently to ejection of core material into upper and lower plena. The fuel redistribution finally leads to neutron shut down of the core.

4.3.1.2 Main Characteristics of Computer Codes Describing Phenomenological Events in the Course of an HCDA

Referring to the most likely successions of events sketched in the simplified scenario of Fig. 4.6, we distinguish three different classes of computer programs, each class consisting of codes dedicated to the numerical simulation of a specific phase, with overlapping, in some cases, of a subsequent phase. We refer to as Group-1 Codes those codes which describe the initiating phase and the intermediate phase in a deterministic way and provide initial conditions for programs simulating subsequent events. Likewise, Group-2 Codes are those simulating the phenomenology of the transition phase, with loss of integrity of the assembly geometry, and provide boundary conditions for the Group-3 Codes. The latter computer programs deal with the disassembly phase in which loss of geometrical integrity spreads to all subassemblies, and finally describe a totally collapsed core. In the following we summarize the characteristics of the main computer programs of these groups (or classes) mainly from the viewpoint of FBRs safety analysis. Major emphasis is given to the codes of the first group because to this group belong the two computer programs, BLOW-3A and BACCHUS-3D/TP, referred to above.

(I) Group-1 Codes

Codes of this class, starting from an initiating event like Transient Over Power (TOP) or Loss of Flow (LOF), and simulating a still intact subassembly geometry, describe in a deterministic way the phenomenology following the start of the accidental conditions. Individual phenomena like coolant boiling, expansion of two-phase flow region, dry-out and melting of cladding and fuel, material relocation, and fission gas release are dealt with in the most realistic possible way, compatible with to-date knowledge of the physics of the phenomena involved and with available computer facilities.

Codes of this group can in turn be classified in two classes: (i) (Class A) Codes based on the definition of an equivalent channel, representing a single or more subchannels of a subassembly, or representative of a single or more subassemblies of a reactor core; (ii) (Class B) Codes representing the full subassembly geometry where in turn, as explained later, a subchannel analysis concept or porous body approach can be followed.

A reactor code may be represented by several parallel equivalent channels, for instance representative of subassemblies located in radial direction from the center of a reactor core outwards. The parallel channels are coupled by the common inlet and outlet boundary conditions set in the simulated inlet and outlet plena of the reactor. The computation of several parallel channels allows for incoherency effects, i.e. the description of the occurrence of the same phenomenon in different regions of the reactor core at different times. This generally results in more realistic and milder reactivity feedback than in case the incoherency effects were neglected.

Codes of the class A, based on the definition of an equivalent channel, have been traditionally developed between 1960s and 1970s when the computer facilities did not yet allow a full subassembly representation. Codes of class B, describing the full subassembly with its real geometry, have been developed since the mid 1970.

Among the computer programs based on the definition of equivalent channel (Group-1, Class A) we mention the BLOW-3A code, the EAC (European Accident Code), and the codes of the SAS-series.

The BLOW-3A code (Bottoni & Struwe, 1982) was developed at the Kernforschungszentrum Karlsruhe (KfK, now KIT) between about 1970 and 1982. The basic methodology followed in the code consists in representing the two-phase flow of coolant in an equivalent channel representative of a reactor bundle (or part of a bundle), by means of a one-dimensional sequence of vapour bubbles separated by liquid coolant slugs. Between bubbles and claddings or structural surfaces a moving liquid film is simulated. The code describes the dynamics of this physical model which is representative, depending on the void fraction, of slug flow or annular flow. The model is referred to as a multi-bubble slug-ejection model because the dynamic of the vapour bubble expansion entails the ejection of coolant liquid slugs from the computational domain into a simulated upper or lower plenum.

The BLOW-3A code has provided the basis for the fluid-dynamic models integrated into the EAC (European Accident Code) (Reynen et al., 1989) developed since about 1980 at the Joint Research Center (JRC) of the European Community at Ispra, Italy. The EAC code provides for a strict coupling between the fluid-dynamic part and the neutronic diffusion calculations, with computation of reactivity feedbacks due to voiding expansion and Doppler effect.

The well-known series of the SAS codes was developed at the Argonne National Laboratory (ANL), USA, from about 1960 to about 1995 when ongoing research on the Integral Fast Reactor (IFR) design was terminated. The phenomenology described in the SAS codes is also based on the multi-bubble slug-ejection model and actually inspired the initiation and methodology followed in earlier versions of the BLOW code at KfK (Germany).

The latest released versions SAS-4A (Tentner et al., 1985; Cahalan et al., 1994) still represent a bundle by means of an equivalent channel with given inlet and outlet boundary conditions, replacing the modeling of the primary loop. However, in parallel to the development of these versions, also a new development was undertaken, which led to the completion of the version SASSYS (SYS for System, or primary loop simulation). The latest SASSYS-1 version (Dunn & Wei, 1998) presents two basic improvements with respect to earlier versions: (i) A bundle can be represented by a series of adjacent channels, each representative of a two-dimensional domain, bounded (in a hexagonal lattice) by planes through the pin axes. We refer to these domains as hexagonal rings. The rings are coupled to each other thermally, since heat exchange is simulated, but cross flow of coolant cannot occur; (ii) The full primary loop of a reactor can be simulated. The second improvement removes the need of specifying time-dependent boundary conditions at the inlet and outlet of the reactor core and allows for the simulation of natural convection in the primary loop, which would develop in case of loss of flow accident, as foreseen by the modern passive safety systems.

Among the codes of Group-1, Class B, we mention the BACCHUS-3D/TP (Two-Phase) code (Bottoni et al., 1992), the SABRE, and SABENA codes. The BACCHUS computer program, first developed at the Centre de Energie Nucleaire de Grenoble (CENG), France, as steady-state, two-dimensional, single-phase versions, was subsequently developed at the Kernforschungszentrum Karlsruhe (KfK) from about 1980 through 1990 and resulted in a three-dimensional transient single- and two-phase version. The bundle geometry is treated in this code with the porous body concept. The program BACCHUS-3D/TP provides two options for the two-phase flow model: (i) A slip model based on three equations for the fluid mixture, which contains as a submodel, in case of no slip between the phases, the Homogeneous Equilibrium Model (HEM), and referenced to as TSM (Improved Slip Model); (ii) A Separated Phases Model (SPM) based on five conservation equations: Two mass conservation and two momentum conservation equations for the separated phases and one energy equations for the mixture of the phases, assumed to be in thermodynamic equilibrium. The code user can choose to apply either model or, as a third option, an integrated code version in which a smooth transition from one model to the other occurs when the void fraction in a cell goes from 0.6 to 0.8.

The SABRE code (McDougall, 1978), based on the subchannel concept (an elementary cell is the triangular region between three adjacent pins in the hexagonal lattice), was developed at the United Kingdom Atomic Energy Authority (UKAEA) at Harwell. This code is especially dedicated to the simulation of blockages in bundle geometry.

The computer code SABENA (Ninokata & Okano, 1990), developed at the Power Reactor and Nuclear Fuel Development Corporation (PNC, now Japan Nuclear Cycle, JNC) of the O-Arai Engineering Center, Japan, is also based on the subchannel analysis concept. In the physical modeling and numerical treatment, this code is comparable with the separated phases version of the BACCHUS-3D/TP program.

(II) Group-2 Codes

Codes of this class, describing the damage propagation from subassembly to subassembly, are generally based on a strict coupling between the thermo-fluid-dynamic part and the neutronic calculations. The latter are generally based on the solution of the point kinetic equations with the calculation of space-dependent reactivity feedbacks due to Doppler broadening, voiding expansion, temperature changes, and relocation of materials (cladding and fuel slumping and possibly blockage formation).

Among the codes of this class we mention the KADIS program (Schmuck et al., 1977), developed at the Kernforschungszentrum Karlsruhe (KfK) since 1970 and the VENUS-II and VENUS-III (Weber & Ganapol, 1976) code versions developed at the Argonne National Laboratory (ANL). The latter is applicable from the transition phase till the disassembly phase and is therefore classified as disassembly code as well.

(III) Group- 3 Codes

Computer programs of this class are especially dedicated to the simulation of the phase of extended core material motion and to the disassembly phase. The initial conditions for these simulations are generally provided by the results of codes operating in the transition phase. Besides the already mentioned versions of the VENUS program, the most relevant codes of this class are those of the SIMMER series developed over 20 years, between about 1970 and 1990, at the Los Alamos National Laboratory (LANL), USA.

The prototype code of the SIMMER series, SIMMER-II (Smith & Sheheen, 1980), computes neutronic and fluid-dynamic behaviour in axis-symmetric (two-dimensional) cylindrical geometry. The code provides for the numerical solution of either the time-dependent multigroup neutron diffusion equation or the neutron transport equation. Core materials are grouped in field (solid, liquid, and vapour field) for which the fluid-dynamic equations are solved taking into account enthalpy, mass, and momentum transfer between the fields.

After a long hiatus in the development of the SIMMER code, a new tripartite cooperation has been started in about 1990 between the JNC (Japan Nuclear Cycle Development Institute, formerly PNC), CEA (Commissariat a l'Energie Atomique, France), and the Forschungszentrum Karlsruhe (FZK), Germany. This cooperation has led to the release of the SIMMER-III version (Kondo et al., 1999) (still two-dimensional) which has been also intensively verified. The main new features available in SIMMER-III are related to two-phase flow models and turbulence in presence of particles produced by fuel fragmentation.

4.3.1.3 Physical Models and Numerical Techniques Used in the Multi-bubble Slug-Ejection Computer Program BLOW-3A

The computer program BLOW-3A (Bottoni & Struwe, 1982) describes sodium boiling phenomena in fast reactor subassemblies as a consequence of different failure conditions like mass flow reduction, inlet blockages of single subassemblies, or reactivity transients. The code is based on description of two-phase flow by means of a multi-bubble slug ejection model: vapour bubbles develop between liquid slugs which can be ejected from the computational domain into simulated upper and lower plena of the reactor. Between bubbles and cladding or structure surfaces a moving liquid film is simulated. The two-phase flow model is representative for slug flow and annular flow with void fraction larger than about 0.6 and 0.8, respectively.

The code BLOW-3A is based upon the definition of equivalent channels (up to 10 channels can be simulated in the to-date code version) each one representative of a region of a reactor core or of a subassembly or of a part of a subassembly. The term "equivalent channel" implies that the pin diameter is adopted; the cross section of the region represented is transformed into a cylindrical gap, between two cylindrical surfaces, with the same hydraulic diameter. The typical configuration of an equivalent channel is shown in Fig. 4.7, with reference to the geometry of the German SNR 300 reactor with lower fission gas plenum. In a calculation with several channels, representative of different reactor regions or of different subassemblies, the channels are coupled thermal-hydraulically by the common inlet and outlet boundary conditions, but neither cross flow, nor heat exchange between the channels can be simulated.

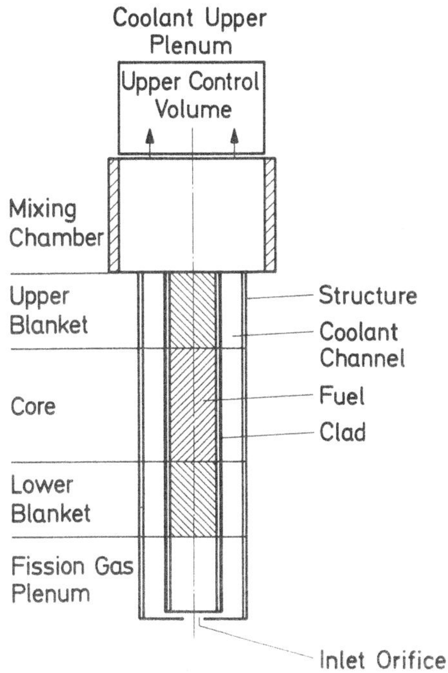

Fig. 4.7 Geometrical configuration of the test section modelled in the BLOW-3A programme

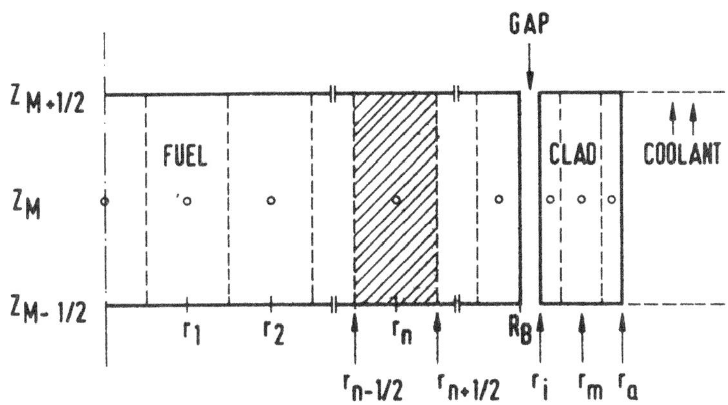

Fig. 4.8 Sketch showing the definition of radial meshes in fuel and clad

An equivalent channel can be partitioned in axial direction with up to 30 meshes. Up to 11 nodes are used for the discretization of the fuel pin; 3 nodes are used in the cladding, one for the coolant, and one for the structural material (see Fig. 4.8). In the pins the thermal gradient in the radial direction are much larger than in the axial direction, therefore only radial heat conduction is considered. The coupling between axial meshes of the pins occurs through the heat convected by the coolant.

The following equations are used for computing temperature distributions in fuel pins, cladding and structural materials:

(i) **Fuel**

The heat conduction equation in the fuel (subscript B for "Brennstoff") is

$$\left(\rho c_p\right)_B \frac{\partial T_B}{\partial t} = -\nabla \cdot \vec{q} + q_B \left[\mathrm{w/m^3}\right] \tag{4.17}$$

where q_B represents the power source and the heat flux is $\vec{q} = -\lambda_B \nabla T$. Assuming azimuthal symmetry around the pin axes, Eq. (4.17) becomes

$$\left(\rho c_p\right)_B \frac{\partial T}{\partial t} = \frac{\partial}{\partial r}\left(\lambda_B \frac{\partial T_B}{\partial r}\right) + \frac{1}{r}\lambda_B \frac{\partial T_B}{\partial r} + q_B \tag{4.18}$$

Equation (4.18) is solved with:

$$\text{Initial condition}: \; T_B(r,0) = T_{B0}(r) \tag{4.19a}$$

$$\text{boundary condition}: \; -\left(\lambda_B \frac{\partial T_B}{\partial r}\right)_{R_B} = \alpha_{BH}\left[T_B(t,R_B) - T_H(t,R_{Hi})\right] \tag{4.19b}$$

where T_H is the cladding temperature (subscript H for "Huelle").

Equation (4.18) has a singularity at $r = 0$. At this point it is replaced by

$$\left(\rho c_p\right)_B \frac{\partial T_B}{\partial t} = 2\lambda_B \frac{\partial^2 T_B}{\partial r^2} + q_{B,} \tag{4.20a}$$

with the symmetry condition

$$\left(\frac{\partial T_B}{\partial r}\right)_{r=0} = 0. \tag{4.20b}$$

(ii) **Cladding**

The heat conduction equation in the cladding is

$$\left(\rho c_p\right)_H \frac{\partial T_H}{\partial t} = \frac{\partial}{\partial r}\left(\lambda_H \frac{\partial T_H}{\partial r}\right) + \frac{1}{r}\left(\lambda_H \frac{\partial T_H}{\partial r}\right) + q_H, \tag{4.21}$$

with boundary conditions

$$\alpha_{BH}[T_B(t,R_B) - T_H(t,R_{Hi})] = -\lambda_H \left(\frac{\partial T_H}{\partial r}\right)_{R_{Hi}}, \tag{4.22a}$$

$$-\lambda\left(\frac{\partial T_H}{\partial r}\right)_{R_{Ha}} = \alpha_{Hk}[T_H(t,R_a) - T_k(t)], \tag{4.22b}$$

where T_k is the coolant temperature (subscript K for "Kuehlmittel") and α_{BH} and α_{Hk} are heat transfer coefficient fuel to cladding and cladding to coolant, respectively.

(iii) **Structure**

The thermal balance in the one-node structure material is;

$$(\rho c_p)_s \frac{\partial T_s}{\partial t} = \alpha_{ks}\frac{F_{Si}}{V_s}[T_k(t) - T_s(t)] + q_s - \alpha_{sw}\frac{F_{SW}}{V_s}[T_s(t) - T_w(t)]. \tag{4.23}$$

$T_w(t)$ is the temperature of the outer medium. F_{si}, F_{sw} are inner and outer surfaces of the structure per unit axial length.

From the viewpoint of the treatment of fuel, cladding, and structural materials, the code version BACCHUS-3D/TP is also based on the above equations. The difference between the standard solution methods adopted in these codes and the methods of molecular dynamics does not concern the structural materials but only the treatment of the coolant behaviour in the two-phase flow regions. The difference between the standard solutions and the molecular dynamics approach is illustrated in the following subsections.

4.3.1.4 Initialization of Vapour Bubbles, Coalescence, Growth, and Collapse

During the single-phase flow calculation, at the end of every time step the program checks whether the peak coolant temperature has exceeded the saturation temperature by a given (input specified) amount of superheat. If this is the case, a first vapour bubble is initialized at the lowermost boundary of the axial cell where the boiling criterion was satisfied. The initial bubble length is assumed to be 0.5 mm. The vapour bubble is assumed to fill instantaneously the cross section of the coolant channel apart from an input specified liquid film thickness on both sides, cladding and structure, of the equivalent channel. Experimental investigations suggest assuming the initial film thickness, in the case of sodium as a coolant, corresponding to about 0.8–0.85 void fraction. According to theoretical investigations (Schlechtendahl, 1970) the growth of a vapour bubble in spherical form up to the size which fill about 80% of the cross section occurs in such a short time (a few milliseconds) that modeling of the event can be skipped in a computer code.

The initial temperature of the vapour in the bubble is the coolant temperature in the mesh where initialization occurs. The pressure of vapour is the saturation pressure corresponding to its temperature. The initialization of a bubble with large superheat (typically 40–70 K) thus implies a large pressure peak in the node of the bubble formation. Therefore the bubble expands rapidly displacing the inlet and outlet liquid slugs. With the vapour expansion, the pressure peak broadens and drops in amplitude thus allowing a liquid film evaporation which sustains for a while the pressure inside the bubble. Boiling inception with a large amount of superheat may occur at the very first transition from metastable single- to two-phase flow conditions and is thus characterized by the formation of a large bubble which expands rapidly. Conversely, boiling inception without or with negligible superheat is characterized by the nearly simultaneous formation of several bubbles which move upwards with roughly the average liquid slug velocity and expand at moderate speed. Initialization of subsequent bubbles, after the first one, occurs with a newly specified superheat, normally in the range 0.5–2 K.

Bubbles displacement and growth or shrink is determined by the solution of the momentum equation for the liquid slugs between the bubbles and for the inlet and outlet slugs. The pressure of vapour at the phase interfaces with the slugs provide the boundary conditions for the solution of the momentum equation of the slugs. When the new axial position of these interfaces has been determined, as a result of the slug displacements, energy and momentum equations for the vapour inside the bubbles are solved.

A vapour bubble is considered to be short if it does not overlap three meshes; conversely, a bubble is consider long if it overlaps, at least partially, three meshes in the axial direction. For short bubbles only integral values of the vapour pressure and temperature (but not vapour velocity distribution) are computed, because axial discretization of the respective conservation equations is not possible in the frame of the meshing available. For long bubbles a rigorous calculation of the velocity and pressure distribution of vapour inside the bubbles is made. In case of simulation of hypothetical accidental conditions in bundles typical of fast reactor subassemblies the vapour velocity in the bubbles may attain values of 100–150 m/s in both directions, usually from the hottermost mesh, with the pressure peak, downwards and upwards. In case bubbles shrink from long to short, as it may be the case after power switch off in a simulated experiment, the program resumes the integral calculations of pressure and temperature.

Two adjacent bubbles coalesce when the length of the intermediate liquid slugs shrinks to <2 mm or when the pressure gradient in these slugs exceeds a given threshold. In the case of coalescence of bubbles, the mass of the vanishing liquid slug is distributed to the liquid film in the adjacent meshes. A bubble is modelled to vanish when its length shrinks below 0.4 mm.

The typical configuration modelled for a vapour bubble in the equivalent channel, with liquid film on the structure side, partially evaporated film on the cladding side, and with upper and lower liquid slugs, is shown in a schematic form in Fig. 4.9.

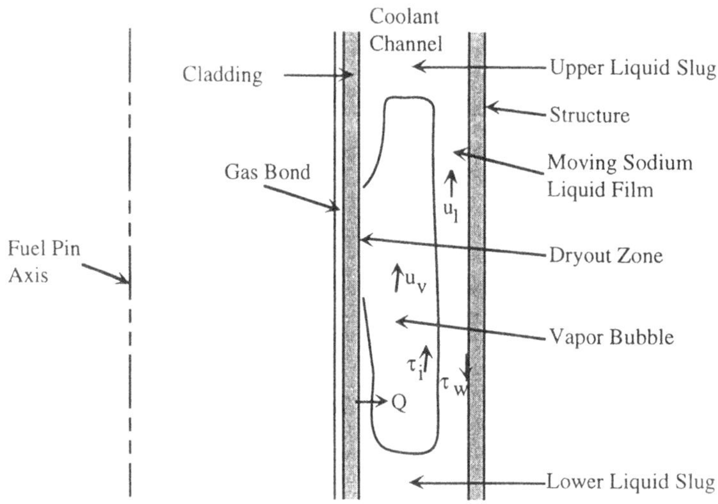

Fig. 4.9 Schematic diagram of annular two-phase flow region as modelled in BLOW-3A code

4.3.1.5 Calculation of Liquid Film Velocity

The concept of two-phase pressure drop multipliers, introduced by Lochart and Martinelli (1949), is used in a heterogeneous multi-bubble slug ejection model with an algorithm that simultaneously computes liquid/vapour interfacial shear stresses and mean velocity of the liquid film. The method used for calculating the liquid film velocity and the interfacial friction factor is based, in conformance with the Lockhart and Martinelli method, on the assumption that the two-phase flow can be represented by the "separate cylinders model". Further, we assume that in the multi-bubble domain, the pressure drop is due only to friction, i.e., acceleration forces are negligible.

Let us assume that we estimate a preliminary value of the liquid phase velocity $u_l = \beta u_v$ with $0 < \beta < 1$. Therefore, a first estimation of the vapour quality x and of the Lockhart-Martinelli parameter X_{LM} can be made. Results obtained by Lockhart and Martinelli have been approximated for the calculation of the two-phase pressure drop multipliers for the liquid phase as

$$
\begin{aligned}
\Phi_{tt} &= 4X_{LM}^{-0.590+0.0855\log_{10}X_{LM}}. && \text{for} X_{LM} \leq 1 \\
\Phi_{tt} &= 4X_{LM}^{-0.590+0.231\log_{10}X_{LM}.} && \text{for } 1 < X_{LM} \leq 10 \\
\Phi_{tt} &= 1.75 && \text{for } 10 < X_{LM}
\end{aligned}
\tag{4.24}
$$

for turbulent regimes in both liquid film and vapour and as

$$\begin{aligned} \Phi_{u} &= 2.66X_{LM}^{-0.516+0.142\log_{10}X_{LM}} && \text{for } X_{LM}\le 1 \\ \Phi_{u} &= 2.66X_{LM}^{-0.516+0.276\log_{10}X_{LM}} && \text{for } 1<X_{LM}\le 10 \\ \Phi_{tt} &= 1.53 && \text{for } 10<X_{LM} \end{aligned} \tag{4.25}$$

for laminar flow regimes in both phases. If only one phase is laminar, it has been assumed that

$$\Phi_{tl}=\Phi_{lt}=\frac{1}{2}(\Phi_{tt}+\Phi_{ll}). \tag{4.26}$$

The theoretical relationship between the two-phase multipliers, referred to the vapour and liquid phases, is

$$\left(\frac{1}{\Phi_v^2}\right)^{1/n}+\left(\frac{1}{\Phi_l^2}\right)^{1/n}=1. \tag{4.27}$$

with $n\approx 2.5$–4.0. Wallis (1969) suggested taking $n=3.5$ as a satisfactory approximation for all flow regimes. Given Φ_l by Eqs. (4.24)–(4.26), Φ_v can be obtained from Eq. (4.27). Calculating dp/dz (the real pressure drop) and $(dp/dz)_v$ (the pressure drop if only the vapour were flowing in the channel) by the single-phase flow theory, and utilizing the definition of two-phase multipliers yields

$$\psi_i=\psi_v\Phi_v^2\alpha^{5/2}, \tag{4.28}$$

where ψ_v is a known function of the vapour Reynolds number

$$\psi_v=f(\mathrm{Re}_v)=a\,\mathrm{Re}_v^{-b}, \tag{4.29}$$

with $a=0.3$ and $b=0.25$. The interfacial shear stress is then given by

$$\tau_i=\frac{\psi_i}{2}\rho_v(u_v-u_i)^2, \tag{4.30}$$

where u_i is the liquid film velocity at the vapor/liquid interface. Following the notation of the boundary layer theory, let

$$u_f^*=\sqrt{\frac{\tau_i}{\rho_l}} \tag{4.31}$$

be the friction velocity, and

$$\delta^{+} = \delta u_f^{*} / \nu_f \tag{4.32}$$

and

$$u_f^{+} = u_f / u_f^{*}, \tag{4.33}$$

be the dimensionless liquid film thickness and velocity, respectively. The mass flow of the liquid phase is given by

$$W_f = \int_{A_f} \rho_l u_f dA = \int_{R_{f1}}^{R_{f2}} 2\pi r \rho_l u_f dr = \rho_l \overline{u}_f A_f, \tag{4.34}$$

where $A_f = \alpha_f A$ is the area of cross-section A occupied by the film, $\overline{u}_f$ represents the mean value of the film velocity over area A_f, and R_{f1} and R_{f2} are the radii of the cylindrical surfaces that bound the liquid film.

We now distinguish between the liquid film wetting the cladding surface and the outer film on the structural surface and use the subscripts "c" and "w" for cladding and wall, respectively. The relative mass flows are hence denoted as W_{fc} and W_{fw}, respectively. The dimensionless mass flows are defined by

$$W_{fc}^{+} = \frac{W_{fc}}{2\pi R_c \mu_f} = \frac{1}{R_c \mu_f} \int_{R_c}^{R_{fc}} \rho_l u_f r dr \tag{4.35}$$

and

$$W_{fw}^{+} = \frac{W_{fw}}{2\pi R_w \mu_f} = \frac{1}{R_w \mu_f} \int_{A_{fw}} \rho_l u_f dA_{fw}. \tag{4.36}$$

Let us define a local coordinate system across the liquid film on the cladding, $R_c \le x \le R_c + \delta_c$, and a coordinate system directed inwardly across the film on the wall, $R_w - \delta_w \le y \le R_w$. The dimensionless coordinates are denoted as x^{+} and y^{+}, respectively. The Reynolds numbers of the liquid films are defined by

$$\mathrm{Re}_{\beta}^{*} = \frac{R_{\beta} u_f^{*}}{\nu_f} \; (\beta = c, w) \tag{4.37}$$

Using these definitions and expressing the integrals in Eqs. (4.35) and (4.36) in terms of the local coordinate systems, we obtain

$$W_{fc}^{+} = \int_{0}^{\delta^{+}} u_f^{+} \left(1 + \frac{x^{+}}{\mathrm{Re}_c^{*}}\right) dx^{+} \tag{4.38}$$

and

$$W_{fw}^{+} = \int_{0}^{\delta^{+}} u_{f}^{+}\left(1 - \frac{y^{+}}{\mathrm{Re}_{w}^{*}}\right) dy^{+}. \tag{4.39}$$

We assume that the dimensionless velocity distribution over the film thickness is given in both cases by the "universal velocity profile"

$$\begin{aligned} u_f^+ &= x^+, && x^+ < 5 \\ u_f^+ &= -3.05 + 5\ln x^+, && 5 \leq x^+ < 30 \\ u_f^+ &= 5.5 + 2.5\ln x^+. && 30 \leq x^+ \end{aligned} \tag{4.40}$$

Introducing Eqs. (4.40) in (4.34) and integrating, we derive the following expressions or the dimensionless mass flow in the laminar region ($\delta^+ < 5$), in the transition zone ($5 \leq \delta^+ < 30$), and in the turbulent layer $30 \leq x^+$:

$$\begin{aligned} W_{fc}^{+} &= \frac{\delta^{+2}}{2} + \frac{\delta^{+3}}{3\mathrm{Re}_c^*} && \delta^+ < 5 \\ W_{fc}^{+} &= 12.51 - 8.05\delta^+ + 5\delta^+\ln\delta^+ + \frac{10.45}{\mathrm{Re}_c^*} - \frac{2.775}{\mathrm{Re}_c^*}\delta^{+2} + \frac{2.5}{\mathrm{Re}_c^*}\delta^{+2}\ln\delta^+, && 5 \leq \delta^+ < 30 \\ W_{fc}^{+} &= -51.39 + 3\delta^+ + 2.5\delta^+\ln\delta^+ + \frac{10{,}918.93}{\mathrm{Re}_c^*} + \frac{2.125}{\mathrm{Re}_c^*}\delta^{+2} + \frac{1.25}{\mathrm{Re}_c^*}\delta^{+2}\ln\delta^+. && 30 \leq \delta^+ \end{aligned} \tag{4.41}$$

Similarly, introducing Eq. (4.40) into Eq. (4.39), we can derive the following expressions for the liquid film on the structural surfaces:

$$\begin{aligned} W_{fw}^{+} &= \frac{\delta^{+2}}{2} - \frac{\delta^{+3}}{3\mathrm{Re}_w^*} && \delta^+ < 5 \\ W_{fw}^{+} &= 12.51 - 8.05\delta^+ + 5\delta^+\ln\delta^+ - \frac{10.45}{\mathrm{Re}_w^*} + \frac{2.775}{\mathrm{Re}_w^*}\delta^{+2} - \frac{2.5}{\mathrm{Re}_w^*}\delta^{+2}\ln\delta^+, && 5 \leq \delta^+ < 30 \\ W_{fw}^{+} &= -63.89 + 3\delta^+ + 2.5\delta^+\ln\delta^+ + \frac{573.2}{\mathrm{Re}_w^*} - \frac{2.125}{\mathrm{Re}_w^*}\delta^{+2} - \frac{1.25}{\mathrm{Re}_w^*}\delta^{+2}\ln\delta^+. && 30 \leq \delta^+ \end{aligned} \tag{4.42}$$

From Eq. (4.34) and from the definitions given in Eqs. (4.35) and (4.36), we derive the mean values of the liquid film velocities as

$$\overline{u}_{f\beta} = \frac{W_{f\beta}^{+}\nu_f}{\vartheta_\beta}, \qquad (\beta = c, w) \tag{4.43}$$

having used

$$A_{f\beta} \approx 2\pi R_\beta \delta_\beta \qquad (\beta = c, w) \tag{4.44}$$

The value of the $\overline{u}_{fc}$ (or $\overline{u}_{fw}$) given by Eq. (4.43) replaces the initial guess $u_f = u_l \approx \beta u_v$ and the calculation can be repeated from Eq. (4.24) through Eq. (4.43). Convergence, based on the requirement of a relative error $\left|u_f^{r+1} - u_f^r\right|/u_f^r \leq 10^{-6}$, where r is an iteration index, is usually achieved in 10–20 sweeps.

4.3.1.6 Physical Models Used in the Computer Program BACCHUD-3D/TP

In the following we refer to real problems treated at the Nuclear Research Centre of Karlsruhe (Kernforschungszentrum Karlsruhe, KfK), Germany, in the framework of safety analysis of Liquid Metal Fast Breeder Reactors (LMFBRs). In this framework KfK has sponsored the development of a computer program describing the multi-dimensional thermal-hydraulic behaviour of a fuel rod bundle under nominal and accident conditions. The simulation of the coolant behaviour in the reactor core, including incoherencies (time shifts) of boiling initiation of the coolant in case of hypothetical accidents, requires the correct representation of the coolant channels and of the geometrical configuration of the fuel rods. Therefore a correct three-dimensional representation of the fuel bundle is mandatory. Computer programs dedicated to these analysis must be able to describe not only single-phase flow conditions, but, in case of core accidents, also two-phase flow conditions (liquid and vapour) in which the coolant of the reactor (liquid sodium in case of fast reactors) boils. At KfK the computer program BACCHUS-3D/TP (three-dimensional, two-phases) (Bottoni et al., 1992) has been developed over several years. In the following a simplified description is given of the geometrical model used in this code to represent correctly a typical fuel bundle mainly consisting of rods of fuel, coolant flowing along the rods, and an outer hexagonal canning.

The hexagonal arrangement of fuel rods in bundles typical of LMFBRs is shown in Fig. 4.10 for the half of a 37 pin bundle, although in the reactors much larger bundles with up to 397 fuel rods are used. Control cells for mass and enthalpy balances in the coolant are bounded in axial direction by planes perpendicular to the z-axis of the bundle and Δz apart; in radial directions the control cells are bounded by planes through the pin axes and Δr apart; in the azimuthal direction by planes passing through the bundle axis with azimuthal aperture of $\Delta\theta = 2\pi/NTH$ (NTH is the number of azimuthal cells used).

The cells in which a bundle is partitioned are labelled in the axial direction with an index JC = 1, 2,…,MC; in the radial direction with an index IC = 1, 2,…,NC; in the azimuthal direction with index IT = 1, 2,…,NTH. Therefore the full bundle is partitioned in a total number NP = MC * NC * NTH of cells. In the case of a 37-pin bundle, there are 4 radial cells. In case of the large bundle with 397-pin rods there are 12 radial cells. In radial direction the number of cells is suggested by the number of

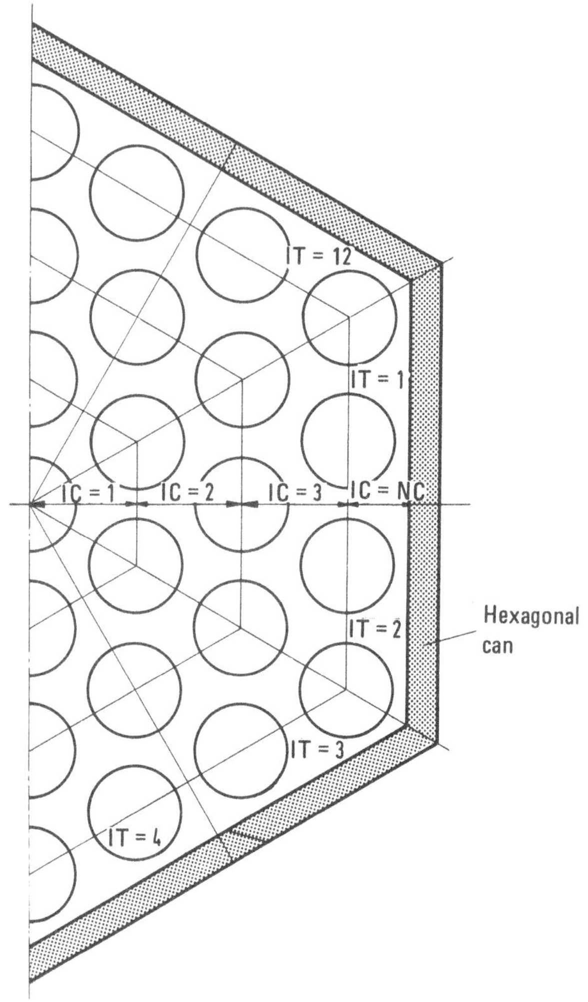

Fig. 4.10 Discretization of a 37-pin bundle with indexing of control cells in radial and azimuthal directions

hexagons (in jargon referred to as "hexagonal rings") of the bundles. A finer discretization would be difficult, though possible. The same holds for the azimuthal discretization. The axial discretization is not bound by the geometrical configuration of the bundle.

The main dependent variables describing the coolant behaviour (mass, pressure and enthalpy) are described by conservation equations integrated over the cell volumes and therefore representative of the mean values in the cells, for this reason also called "control volumes".

We refer to the domain shown in Fig. 4.10 as "pseudo"-cylindrical domain, where the term "pseudo" refers to the hexagonal geometry with cells bounded in the azimuthal directions by planes passing through the z axis. The radial, axial, and azimuthal coordinate directions are labelled with indices, r, z and s, respectively. The simulated domain may contain structures. For these reasons computational cells are divided in a fluid region and in a solid region and have a volume porosity, defined as the fraction occupied by the fluid. Bounding surfaces are characterized by a surface permeability, defined as the fraction of the surface open for the flow. The sub-case of a continuum is recovered by setting all volume porosities and surface permeabilities to one.

Centred control cells are denoted with indices (i, j, k) for the radial, axial, and azimuthal directions, respectively. These indices actually refer to the central node of a cell and bounding surfaces are denoted by $(i \pm 1/2, j \pm 1/2, k \pm 1/2)$. Cells displaced (staggered) in the three coordinate directions have centred point labelled as $(i \pm 1/2, j, k)$, $(i, j \pm 1/2, k)$ and $(i, j, k \pm 1/2)$, respectively.

Space discretization of the conservation equations is done with reference to the staggered meshes. Scalar quantities, like coolant pressure, enthalpy, and other physical properties of the fluid, are defined at the center point (i, j, k) of a control volume. Velocity components of the coolant (u, w, v for the r, z, s directions, respectively) are defined at the mid points of the boundary faces.

All cells are characterized by a total volume V, a volume occupied by the fluid V_f, by the areas of the lateral faces, S_t, S_b (top, bottom, perpendicular to the z axis of the bundle), S_i, S_e (internal, external, perpendicular to the radial coordinate r), S_m, S_p (bounding the cell in the azimuthal direction, where the subscripts m (minus), p (plus) denote the sequence considered in the positive clockwise direction). These geometrical elements are used to define volumetric porosities and surface permeabilities for every cell.

Let $S_{f_t}, S_{f_b}, S_{f_i}, S_{f_e}, S_{f_m}, S_{f_p}$ be the cross flow areas of the bounding faces. For every cell we define surface permeabilities ζ, Ψ, ξ for the axial, radial, and azimuthal directions, respectively, as the ratio of the respective cross flow area to total area, for instance $\Psi_i = S_{f_i}/S_i$ and $\zeta_b = S_{f_b}/S_b$, etc. The volume porosity of a cell is defined as the ratio of the volume occupied by the fluid to the total cell volume, i.e.: $\varepsilon = V_f/V$. In an undisturbed geometry the volume porosity is equal to the surface permeabilities in the axial direction.

4.3.2 *Traditional Approach Coupled with Monte Carlo Simulation of Coolant Behaviour*

In the frame of safety analysis of Liquid Metal Fast Breeder Reactors (LMFBRs) under hypothetical Unprotected Loss-of-Flow (ULOF) conditions, two-phase flow of sodium is simulated in a reactor core. Traditional approaches used in safety analysis codes to simulate sodium vapour condensation and vaporization rely upon application of macroscopic semi-empirical correlations for heat transfer and vapour condensation or evaporation rates. As an alternative to this macroscopic approach, the author developed a microscopic methodology based upon the application of the basic laws of the kinetic theory for the determination of the evaporation and condensation rates of vapour in a reactor bundle.

This microscopic approach is based upon a Monte Carlo simulation of the molecular trajectories, collision rates between vapour molecules and of the molecules with the surfaces of the claddings of the pins of a reactor bundle. The pins surfaces are treated in the Monte Carlo simulation as diffusely reflecting surfaces. Scattering of sodium particles is simulated with the "hard sphere" collision model. The "step splitting" technique is applied, which consists in separating the collisions dynamic calculation from collisionless paths of the molecules. Vapour particles are assumed to condense on the surfaces of the pins when, after diffuse reflection, their velocity would be less than one third of the most probable velocity corresponding to the wall temperature. Rewetting of dried out regions of the cladding surfaces is simulated with a dynamic film model that computes the velocity distribution of the liquid across the film thickness and then the mean liquid film velocity.

Evaporation of sodium molecules from the film yields a source of molecule which re-enters into the Monte Carlo calculation of the molecular dynamic approach. The coupling of the micro- and macroscopic models has been applied to the numerical simulation of an out-of-pile sodium boiling experiment run at the Nuclear Research Center of Karlsruhe (KfK), Germany.

4.3.2.1 Preliminary Investigations

Numerical simulation of condensation of sodium vapour and of rewetting of dried out regions in a simulated bundle of a fast breeder reactor during hypothetical Unprotected Loss-of-Flow (ULOF) conditions has been made with the application of the Monte Carlo technique using the computer program THEMIS (Bottoni, 1998a, b). The computer code THEMIS (Muramatsu, 1995) first developed as a two dimensional program at the O-Arai Engineering Center, Japan Nuclear Cycle Development Institute (JNC), Ibaraki-ken, Japan, was applied to a mock up of a reactor bundle consisting of an equivalent annular domain between two co-axial cylindrical surfaces. The inner cylindrical surface was equivalent to the heated surfaces of the powered pins of a bundle while the outer cylindrical surface simulated the cold hexagonal canning. Because of the azimuthal symmetry of the simulated bundle the problem was treated in an axis-symmetric (r; z) geometrical representation.

More recent developments rely upon a three-dimensional representation of the geometry of a reactor bundle which enables simulation of incoherency effects in the bundle, like for instance occurrence of dry-out spots in different locations at different times. For a detailed numerical simulation of the microscopic phenomena occurring in the two-phase flow domain of a bundle, made with the approach of the kinetic theory, a three-dimensional representation of the bundle is necessary.

For this reason a three-dimensional version of the THEMIS program, referred hereunder as THEMIS/3D.1 (Bottoni, 2004), has been developed and applied to a 60-degrees sector of a 7-pin bundle. The limited sector, bounded in the azimuthal direction by symmetry planes, and the small size bundle were chosen at first to develop and test the necessary numerical technique.

The basic numerical techniques have been pioneered by the theoretical work of (Nambu, 1980a, b, c), among others, and consists in applying the "principle of uncoupling," separating the collisionless molecular motion over the length of a time step from the simulation of the collision events among randomly chosen couples of the molecules within a computational cell.

4.3.2.2 Geometrical Configuration

Figure 4.11 shows the cross section of a 60-degrees sector of a 7-pin bundle with hexagonal lattice. The geometrical configuration of the cross section is fully defined giving the pin diameter R, the pitch P and the distance of the hexagonal canning from the origin of the main coordinate system (x, y, z) defined as follows: Origin on the axis of the central pin at the simulated inlet plane of the bundle; z axis coincident with the axis of the central pin and directed upwardly; y-axis normal to the hexagonal canning and oriented outwardly; x axis normal to the (y, z) plane and oriented as to form a right-handed system. The centre points of the three pins are referred to as O (0,0), C1[$P\cos(\pi/3)$, $P\sin(\pi/3)$], C2[$-P\cos(\pi/3)$, $P\sin(\pi/3)$]. The 60-degrees azimuthal sector is bounded in the azimuthal direction by two symmetry planes defined by the angles $\vartheta_{\min} = \pi/3$ and $\vartheta_{\max} = 2\pi/3$ which they form with the x axis.

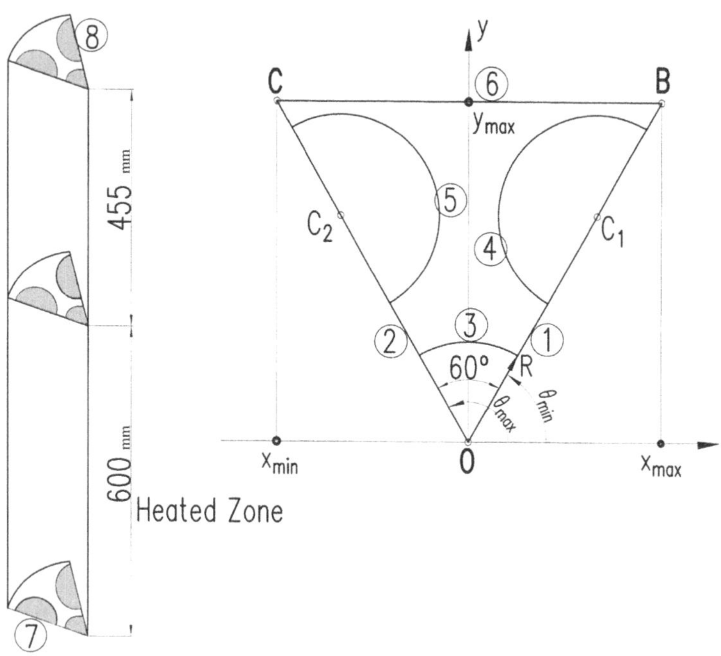

Fig. 4.11 Perspective view and cross section of a 60-degrees sector of a 7-pin bundle with main Cartesian coordinate system and indexing of surfaces bounding the computational domain

For computational purposes, the cross section of the bundle is considered to be included in the rectangle defined by $[x_{\min} = -x_{\max} < x < x_{\max} = y_{\max} \cot \vartheta_{\min}]$, $[y_{\min} = 0 < y < y_{\max}]$. In the axial direction the coordinates of the inlet and outlet planes are given ($z_{\min} = 0, z_{\max}$). Besides the main Cartesian coordinate system (x, y, z), we consider in the cross section a polar (ϑ, r) system. In this system the computational domain is within the sector S defined by $S = [(\vartheta_{\min} < \vartheta < \vartheta_{\max}), (R < r < R_{\max})]$ with $R_{\max} = \text{OB}$. Part of this circular sector does not belong to the computational domain, namely the calotte beyond the hexagonal canning and the regions inside the upper right and upper left pins. The sector S is partitioned, in this polar system, in cells of radial dimension $\Delta r = (R_{\max} - R)/\text{NI}$ and of azimuthal aperture $\Delta\vartheta = (\vartheta_{\max} - \vartheta_{\min})/\text{NJ}$, where NI and NJ are the number of cells in the respective coordinate directions. The computer program takes into account which ones of the cells or which fraction of the cells defined in sector S does really belong to the computational domain, i.e. lies outside of the pins and inside the hexagonal canning. In the axial direction the sector S is partitioned in cells of length $\Delta z = (z_{\max} - z_{\min})/\text{NK}$.

The computational domain is bounded by eight surfaces namely: Two symmetry planes (OB and OC, respectively, shown in Fig. 4.11 with circled numbers 1 and 2); three pin surfaces (central, upper right, and upper left pin, identified with numbers 3, 4, and 5, respectively); surface of the hexagonal canning (number 6); inlet and outlet planes (numbers 7 and 8 in the perspective view of the sector of the bundle).

4.3.2.3 Inlet Flow

In this section, there is some overlapping with the treatment given in Chap. 1, Sects. 1.4 and 1.5. This overlapping allows a clearer description of the geometrical configuration to which the Figs. 4.11, 4.12, 4.13, 4.14 and 4.15 refer.

(i) **Inlet Flow Rate**

We simulate with the Monte Carlo method a stream of monoatomic sodium vapour molecules flowing through the inlet section of the bundle and we assume that the mean stream velocity vector, $\vec{c}_0$, is parallel to the z axis. From the definition of the instantaneous molecular velocity

$$\vec{c} = \vec{c}_0 + \vec{c'} \tag{4.45}$$

as sum of the mean value $\vec{c}_0$ and of the fluctuating component $\vec{c'}$ we derive

$$u = u', \quad v = v', \quad w = w' + c_0 = w' + w_0. \tag{4.46}$$

At the inlet plane of the computational domain the inward velocity flux is (Bird, 1994)

$$N_i = n\overline{w} = n\int_{-\infty}^{+\infty}\int_{-\infty}^{+\infty}\int_{0}^{+\infty} wf_0 dudvdw. \tag{4.47}$$

In Eq. (4.47) the third integral is taken over only positive velocity components w, therefore only for a flow entering into the computational domain; n is the number density of the molecules and f_o is the equilibrium or Maxwellian distribution function, which describes the distribution of the molecular speed c' and is given by

$$f_0 = \frac{\beta^3}{\pi^{3/2}} \exp\left(-\beta^2 c'^2\right) \tag{4.48}$$

$\beta = 1/c'_m$ is the inverse of the most probable thermal speed in the velocity distribution of Eq. (4.48). Inserting Eq. (4.48) into (4.47) one obtains

$$N_i = n\overline{w} = \frac{n\beta^3}{\pi^{3/2}}\int_{\infty}^{\infty}\int_{\infty}^{\infty}\int_{0}^{\infty} \exp\left(-\beta^2 c'^2\right) dudvdw \tag{4.49}$$

Using Eq. (4.46) and changing the integration variables from u, v, w to u', v', w' one obtains

$$N_i = \frac{n\beta^3}{\pi^{3/2}}\int_{\infty}^{\infty}\int_{\infty}^{\infty}\int_{-c_0}^{\infty} (w' + c_0)\exp\left[-\beta^2\left(u'^2 + v'^2 + w'^2\right)\right] du'dv'dw'. \tag{4.50}$$

The variables in Eq. (4.50) can be separated to give

$$\begin{aligned} N_i = \frac{n\beta^3}{\pi^{3/2}} & \int_{\infty}^{\infty} \exp\left(-\beta'^2 u'^2\right) du' \int_{\infty}^{\infty} \exp\left(-\beta'^{2\prime} v'^2\right) dv' \\ & \times \int_{c_o}^{\infty} (w' + c_o)\exp\left(-\beta'^2 w'^2\right) dw'. \end{aligned} \tag{4.51}$$

Using the definition of the error function

$$erf(a) = \frac{\pi}{2}\int_0^a e^{-x^2} dx \tag{4.52}$$

and the integrals

$$\int_{\infty}^{\infty} \exp\left(-\beta'^2 u'^2\right) du' = \int_{\infty}^{\infty} \exp\left(-\beta'^2 v'^2\right) dv' = \frac{\sqrt{\pi}}{2\beta}, \tag{4.53}$$

$$\int_{\infty}^{\infty} w' \exp\left(-\beta'^2 w'^2\right) dw' = \frac{1}{2\beta^2}. \tag{4.54}$$

Equation (4.51) becomes

$$N_i = \frac{n}{2\beta\sqrt{\pi}} \left\{ \exp(-c_0^2\beta^2) + \sqrt{\pi} c_0 \beta (1 + erf(c_0\beta)) \right\}. \tag{4.55}$$

Let

$$s_n = c_0\beta = \frac{c_0}{c_m} = \frac{c_0}{\sqrt{2RT}} \tag{4.56}$$

be the molecular speed ratio, or ratio of the mean stream velocity c_o to the most probable molecular velocity c_m. Equation (4.55) can therefore be written

$$N_i = \frac{n}{2\beta\sqrt{\pi}} \exp(-s_n^2) + \sqrt{\pi} s_n [1 + erf(s_n)] = \frac{n\sqrt{2RT}}{2\sqrt{\pi}} K(s_n), \tag{4.57}$$

with the definition

$$K(s_n) = \exp(-s_n^2) + \sqrt{\pi} s_n [1 + erf(s_n)]. \tag{4.58}$$

(ii) **Inlet Flow Velocity**

The distribution function of the product of β with the thermal velocity component w' of the inlet flow is (Bird, 1994)

$$f(\beta w') = f(w'/c_m') = \beta w \exp\left(-\beta^2 w'^2\right) = (\beta w' + s_n) \exp\left(-\beta^2 w'^2\right). \tag{4.59}$$

Equating the derivative $\partial f_{\beta w'}/\partial w'$ to zero one finds that the distribution has a maximum for

$$\beta w' = \frac{1}{2}\left(-s_n \pm \sqrt{s_n^2 + 2}\right) \tag{4.60}$$

and the maximum is

$$f_{\beta w',\max} = \frac{1}{2}\left(\sqrt{s_n^2+2}+s_n\right) \times \exp\left[-\left(\frac{1}{2}+\frac{s_n^2}{2}-\frac{s_n}{2}\sqrt{s_n^2+2}\right)\right]. \quad (4.61)$$

Hence we compute the ratio of the probability P of a molecule entering trough the inlet surface with thermal velocity w' to the maximum of its distribution $P_{\max} = \left[f_{\beta w'}\right]_{\max}$ as

$$\frac{P}{P_{\max}} = \frac{f_{\beta w'}}{f_{\beta w',\max}} = \frac{2(\beta w' + s_n)}{s_n + \sqrt{s_n^2+2}} \exp\left(\frac{1}{2}+\frac{s_n}{2}\left(s_n - \sqrt{s_n^2+2}\right) - \beta^2 w'^2\right). \quad (4.62)$$

Equation (4.62) is used in the following way: Let R1 and R2 be two uniform deviates. Let us assume that the limits of the distributions of $\beta w'$ are $\pm(\beta w')_{\max}$, replacing infinity. A random value

$$\beta w' = -(\beta w')_{\max} + 2(\beta w')_{\max} R_1 \quad (4.63)$$

is generated, with uniform distribution in the interval $\pm(\beta w')_{\max}$. Then the ratio $P = P_{\max}$ is computed from Eq. (4.62). The value $\beta w'$ from Eq. (4.63) is accepted if

$$R_2 \leq P/P_{\max} \quad (4.64)$$

otherwise $\beta w'$ is rejected. Finally

$$w = \frac{\beta w' + s_n}{\beta} = (\beta w' + s_n)c_m{}' \quad (4.65)$$

is computed. This procedure generates inlet molecule velocities under the condition that $\beta w = \beta w' + s_n \geq 0$, hence that the velocity component w is directed into the computational domain.

4.3.2.4 Wall Reflection

In this section we compute the components of particle position and velocity after diffuse reflection from the material surfaces and after specular reflection from the symmetry surfaces. Diffuse reflection from material surfaces is justified by the fact that impinging molecules are trapped by the surfaces and, with a time delay negligible for the DSMC computations, are released in a direction that preserves no memory of the direction of incidence. Specular reflection from the symmetry surfaces is justified by the assumption of statistical equilibrium, which implies that for every molecule crossing a symmetry boundary in one direction there is another one crossing in the opposite direction.

(i) **Diffuse Reflection from the Surface of a Pin**

Let us consider a pin of radius R represented in Fig. 4.12 by the shadowed cylindrical surface with its circular projection to the plane $z = 0$. The axis of the pin intersects the $z = 0$ plane in the centre O_p with coordinates (x_p, y_p). Let P_1 (x_1, y_1, z_1) be the position of a particle at time point t_n, and P_2 (x_2, y_2, z_2) its position at time t_{n+1}. If, during the time interval $\Delta t = t_{n+1} - t_n$ the particle has hit the surface of the pin at a point P_c (x_c, y_c, z_c) with projection P_{c0} in the $z = 0$ plane, before being deflected, the coordinates of the point P_c must be computed first and then the coordinates of P_2 can be computed.

Let $P_2^{\sim}\left(x_2^{\sim}, y_2^{\sim}, z_2^{\sim}\right)$ with projection $P_{20}^{\sim}$ be the point, inside the surface of the pin, where the particle would arrive at time point t_{n+1}, if it were not reflected by the pin surface. Combining the equation of the line $L = P_1 P_2^{\sim}$ with the equation of the cylindrical surface of the cladding one obtains the coordinates (x_c, y_c, z_c) of the collision point P_c. In practice we determine first the intersection (A or B in Fig. 4.12) of the line ℓ with the circle C, which is the projection of P_c on the plane $z = 0$, and then compute the coordinate z_c. After determining the collision point, the position and velocity of a particle after diffuse reflection from the surface of a pin can be computed.

(ii) **Diffuse Reflection from the Surface of the Hexagonal Canning**

As in case of reflection from the surface of a pin, we must first compute the coordinates of the collision point of the particle with the hexagonal canning. From the equation of the trajectory of the particle, and knowing that the collision point has coordinate $y_c = y_{\max}$, the coordinates of the collision point $P_c(x_c, y_c, z_c)$ are determined easily.

We now consider a local coordinate system (x_ℓ, y_ℓ, z_ℓ) with origin in P_c and defined as follows: Axis is horizontal and oriented in the opposite direction of the x-axis; axis y_ℓ is normal to the wall and oriented inwardly; the z_ℓ axis vertical. We also consider a spherical coordinate system $(\vartheta \Phi)$ with the same origin in P_c and defined as follows: Given a vector OP and its projection OA on the (x_ℓ, y_ℓ) plane, ϑ is the angle between the vector OP and the z_ℓ axis; Φ is the angle between the projection OA and the x_ℓ axis.

The velocity components in the local coordinate system after reflection are computed by

$$v_{2l} = c'_m \sqrt{-\ln R_1}, \quad u_{2l} = c'_m r \cos \zeta, \quad w_{2l} = c'_m r \sin \zeta \tag{4.66}$$

with the definitions

$$\zeta = R_2 2\pi, \tag{4.67a}$$

$$r = \sqrt{-\ln R_3}, \tag{4.67b}$$

where R_1, R_2, R_3 are uniform deviates. Notice that $v_{2\ell}$ is always positive while the other two components, $u_{2\ell}$ and $w_{2\ell}$, may be positive or negative. The

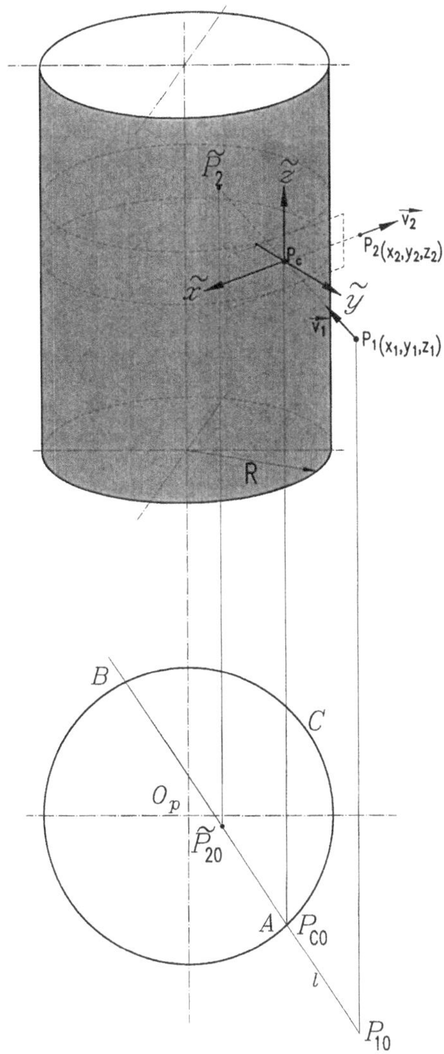

Fig. 4.12 Reference system used to determine the collision point of a particle with the surface of a pin

transformation from the local system to the main (x_c, y_c, z_c) system yields $u_2 = -u_{2\ell}$, $v_2 = -v_{2\ell}$, $w_2 = -w_{2\ell}$. The component v_2 is always negative.

The coordinates of the particle after diffuse reflection in the local system and in the main Cartesian system are

$$\begin{cases} x_{2l} = x_c - x_2 = R_{P2} \sin\theta \cos\varphi \\ y_{2l} = y_c - y_2 = R_{P2} \sin\theta \sin\varphi \\ z_{2l} = -z_c + z_2 = R_{P2} \cos\theta \end{cases} \tag{4.68}$$

(iii) **Specular Reflection from the Symmetry Planes**

The sector of the bundle simulated in the code is bounded by two symmetry planes inclined by angles $\alpha = \pi/3$ and $\alpha = 2(\pi/3/)$, with respect to the x-axis. These symmetry planes are treated as perfectly reflecting surfaces. If the point $P_1(x_1, y_1, z_1)$ defines the position of a particle before reflection from a symmetry plane, the position of the particle after reflection, $P_2(x_2, y_2, z_2)$, is given by

$$\begin{cases} x_2 = x_1 \cos 2\alpha + y_1 \sin 2\alpha \\ y_2 = x_1 \sin 2\alpha - y_1 \cos 2\alpha \\ z_2 = z_1 + w_1 \Delta t \end{cases} \tag{4.69}$$

Likewise for the velocity components one has

$$\begin{cases} u_2 = u_1 \cos 2\alpha + v_1 \sin 2\alpha \\ v_2 = u_1 \sin 2\alpha - v_1 \cos 2\alpha \\ w_2 = w_1 \end{cases} \tag{4.70}$$

4.3.2.5 Collision Model

As mentioned earlier, in the numerical simulation the principle of uncoupling is applied, by splitting the calculation in the collisionless motion of the molecules, which are assigned positions updated by $x_1^{n+1} = x_1^n + u_i^n \Delta t$, followed by the simulation of collision events among the molecules. The principle of uncoupling the collisionless motion over a time step Δt, from the collisional events, modelled at the end of the time step, is justified in (Nambu, 1980a) on the basis of the argument given at the end of Sect. 4.1.5.

(i) **Collision Frequency**

Let us consider a cell of the computational domain with volume V_c containing a number $N = nV_c/F_N$ of sample molecules, where F_N is the number of real molecules represented by every sample molecule. The collision number between all molecule pairs in the cell is (Bird, 1994)

$$C_N = \frac{N^2}{2} F_N \frac{\sigma_T c_r \Delta t}{V_c} \tag{4.71}$$

where σ_T is the collision cross section and c_r is the relative velocity of the molecules. The maximum theoretical collision number $C_{N,\max}$ is given by the same formula (4.71) taking the maximum theoretical value $(\sigma_T c_r)_{\max}$ of the product between the collision cross section and the relative velocity of the molecules. In the numerical computations the maximum number $C_{N,\max}$ of collisions is considered and, among this theoretical number of collisions events, only those are accepted which satisfy the condition

$$R_1 \leq \frac{C_N}{C_{N,\max}} = \frac{\sigma_T c_r}{(\sigma_T c_r)_{\max}} \tag{4.72}$$

where R_1 is a random deviate.

(ii) **Collision Dynamics**

Within a computational cell, collision pairs are chosen randomly among the sample molecules. Let indices 1 and 2 refer to the two molecules of a random chosen couple and let $\vec{c}_r = \vec{c}_1 - \vec{c}_2$ be the relative velocity before collision. For molecules of equal mass the centre of mass velocity is $\vec{c}_m = \left(\vec{c}_1 + \vec{c}_2\right)/2$ Let us denote with a star (*) as superscript the post-collision velocities. For the collision dynamics we use the isotropic "hard sphere" collision model, an acceptable approximation for mono-atomic molecules, in which the post-collision relative velocity has the same magnitude as the pre-collision relative velocity: $\left|\vec{c}_r^{\,*}\right| = \left|\vec{c}_r\right|$. In an isotropic scattering model the components of the post-collision relative velocity can be computed by multiplying a unit vector generated with the Marsaglia technique (Bird, 1994) with the module of the relative velocity:

$$\begin{cases} u_r^* = Sc_r \cos\beta \\ v_r^* = Sc_r \sin\beta \\ w_r^* = C_S c_r \end{cases} \tag{4.73}$$

using two uniform deviates R_1 and R_2 and computing

$$\beta = 2\pi R_1, \;\; C_S = 1 - 2R_2, S = \sqrt{1 - C_S^2}. \tag{4.74}$$

Because the components $u_{m,i}$ (using index "i" for the three coordinate components) of the centre of mass velocity are known from the pre-collision velocity components, the components of the post-collision velocities of the two molecules are given by

$$u_{1,i}^* = u_{m,i} + u_r^*/2 \tag{4.75}$$

$$u_{2,i}^* = u_{m,i} - u_r^*/2 + .$$

4.3.2.6 Liquid Film Model

(i) **Condensation Rate and Thickness of Condensed Film**

The collision rate of the molecules with the pin surfaces is computed following the trajectories of the sample molecules. In the upper unheated part of the test section, condensation may occur. The thickness of the sodium liquid film formed by condensation in the upper cold section of the bundle is computed as follows. Let N_{cond} be the number of sample molecules computed to condense during a time step Δt. Because every simulated molecule of mass m represents F_N real molecules, N_{cond} molecules correspond to a condensation rate given by

$$G = (N_{cond} m F_N)/\Delta t. \tag{4.76}$$

The volumetric condensation rate is then

$$Q = \frac{G}{\rho_\ell}. \tag{4.77}$$

The thickness of liquid film forming in the unit of time is

$$\delta' = Q/A_{cond} \tag{4.78}$$

where A_{cond} represents the area of the surface of the pin where condensation occurs. The increment of the liquid film thickness in one time step is then

$$\Delta\delta = \delta' \Delta t. \tag{4.79}$$

(ii) **Distribution of Liquid Film Thickness and Velocity**

The distribution of the liquid film thickness on the surface of the claddings is obtained, in a one-dimensional time-dependent simulation, by solving the continuity equation

$$\frac{\partial \delta}{\partial t} + \frac{\partial (\overline{w}_f \delta)}{\partial z} = \delta' \tag{4.80}$$

where $\overline{w}_f$ is the mean film velocity. Equation (4.80) allows to compute the updated value of the liquid film thickness.

Condensation of vapour molecules on a material surface is assumed when the particle velocity, after diffuse reflection, is less than one third of the most probable velocity corresponding to the temperature of the reflecting surface. Conversely, evaporation of sample molecules representing the liquid film is assumed when their thermal velocity is larger than three times the most probable velocity of the molecules at the temperature of the liquid film.

In most computational codes simulating numerically heat transfer rates and vapour condensation, condensation rates are computed on the basis of the Reynolds analogy of heat and mass transfer. The heat and mass transfer analogy is justified by the fact that the governing equations for enthalpy and mass transfer are formally similar. It holds however only under the condition that the Levis number be approximately equal to one, which is most often not satisfied in engineering systems. The alternative to the macroscopic approach, offered by the microscopic representation of the phenomena involved, implies tackling the following tasks. When test particles collide with a surface bounding the computational domain, they must be assigned a new velocity determined by sampling a probability distribution function, which takes into account the physical conditions at the material surface. Scattering experiments of molecular beams on material surfaces show that the molecular impact on the surface cannot be simulated by an isotropic scattering model.

The search for an adequate surface scattering model involves in turn two kinds of problems: (a) Simulation of intermolecular forces between the impinging molecules and the molecules of the material surface; (b) Evaluation of a distribution function for the velocities of re-emerging molecules after the impact. The distribution function must take into account the interaction of impacting molecules with the superficial layersof the material surface. In case all impinging molecules are assumed to re-emerge from the surface, after some absorption time, the velocity distribution function is normalized to one. Otherwise, molecules are trapped by the surface either diffusing in it or condensing on it. In the case of evaporation of a liquid wetting a surface, one should simulate the interaction between molecules of a liquid by means of a short-range potential acting between the molecules and evaluate the likelihood of a molecule to be trapped by or released from the potential well of the neighbouring molecules.

4.3.2.7 Numerical Applications

(i) **Geometry and Characteristics of a 7-Pin Bundle**

In the framework of liquid metal fast breeder reactor (LMFBR) safety analysis a number of sodium boiling experiments has been performed at KfK in the NSK (Natrium Sieden Kreislauf) test loop (Aberle et al., 1976) with a bundle of electrically heated pins within hexagonal wrappers, simulating a fast reactor subassembly. The experiments of the series were intended to investigate the boiling behaviour and the cladding temperature excursions, in case of dry-out, following a simulated failure of the reactor to trip.

Liquid film evaporation was simulated in the hot part of the bundle up to dry-out conditions, but rewetting of the cladding surfaces could not be simulated. With the code versions available in the scientific community around the year 2000, the temperature excursions of the cladding surfaces were overpredicted. This modelling deficiency could be eliminated by linking the

Table 4.1 Experimental conditions of run 7-2/24 of the NSK series

Geometry	
Pin diameter [m]	0.0060
Pitch [m]	0.0079
Heated section [m]	0.600
Upper unheated section [m]	0.455
Average steady-state coolant conditions	
Heat flux [W/m^2]	99.3×10^4
Inlet/outlet pressure [N/m^2] $\times 10^{-5}$	2.03/1.05
Inlet/outlet temperature [K]	826/988
Coolant velocity [m/s]	2.15
Experimental conditions at voiding initiation ($t = 6.5s$)	
Coolant velocity [m/s]	0.33
Superheat [K]	68

available state-of-the art codes with the simulation of evaporation, condensation, and rewetting.

In the following we refer to the experimental conditions of the 7-2/24 run of the NSK series, characterized by the data of Table 4.1.

The experiment 7-2/24, with the largest superheat observed in the NSK experimental series, has been chosen for the simulation because it is characterized by a very rapid transient following boiling inception and large voiding forming in the coolant channel. Under these conditions there is a quick excursion of the cladding temperature following the formation of large dry-out zones. The traditional macroscopic approach, based on semi-empirical correlations for heat transfer coefficients under dry-out conditions, was not suitable to predict sodium condensation rates, hence rewetting of the dried-out zones, and was replaced by the microscopic description based on the molecular dynamics.

(ii) **Initial and Boundary Conditions**

In the numerical calculations of the experiment described above a length of 1000 mm (600 mm of the heated section and 400 mm of the upper cold section) was simulated.

The total simulated length has been partitioned in 100 axial meshes of 10 mm length.

The abscissa axis extends from the beginning of the heated section to the top end of the simulated domain. The temperature is assumed constant, 1220 K, between axial meshes 1 and 40, is assumed to increase by 120 K between meshes 41 and 60, to simulate an incipient temperature increase following dry-out, and a colder section at 1000 K is simulated between meshes 61 and 100. This temperature distribution corresponds to a smoothed temperature distribution at dry-out inception, obtained from experimental information.

(iii) **Discussion of Computed Results**

Monte Carlo calculations of molecular dynamics were made using 120,000 sample particles. Having used 10 cells in radial direction, 10 in the azimuthal direction and 100 in the axial direction, for a total of 10,000 cells in the full computational domain, there was an average of 12 simulated molecules per cell. This density of sample molecules is just sufficient to justify the statistical calculations of molecular collisions. The computational time step used (10^{-7} s) was small enough to ensure that even a molecule with high thermal velocity would not cross a computational mesh during the time step.

Figure 4.13 shows the axial distribution of the liquid film thickness forming by condensation in the unit of time upon the pin surfaces in the uppermost cold section of the bundle from $z = 0.6$ m to $z = 1.0$ m. Numbering and locations of these pins are shown in Fig. 4.11. Comparison with experimental data is not possible, because there were not experimental facilities to record liquid film thicknesses.

The strongest condensation occurs at the lowermost boundary of the cold section, within the first ten centimeters, then declines only slowly proceeding upwardly. These condensation rates have been evaluated after 0.5 s of problem time of the Monte Carlo calculation, when it was reasonable to assume that a quasi steady-state condition had been attained.

The liquid film flows downwards by gravity into the hot test section. The extent of the rewetted zone is limited by subsequent evaporation of the liquid in the hot part of

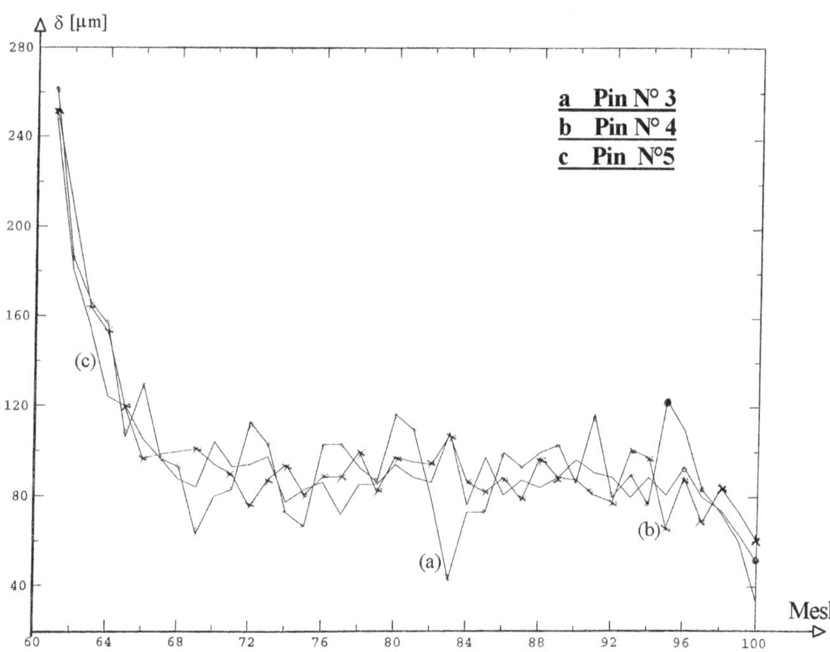

Fig. 4.13 Film thickness forming by condensation per unit of time on the surface of the pins in the uppermost cold section, versus mesh number

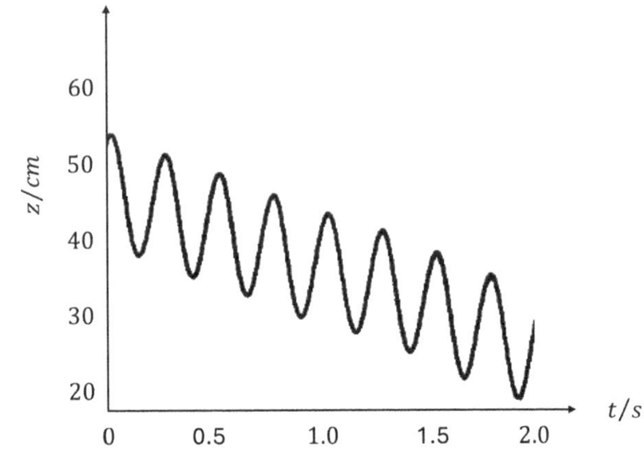

Fig. 4.14 Time evolution of the lower boundary of the liquid film along the axis of the innermost pin

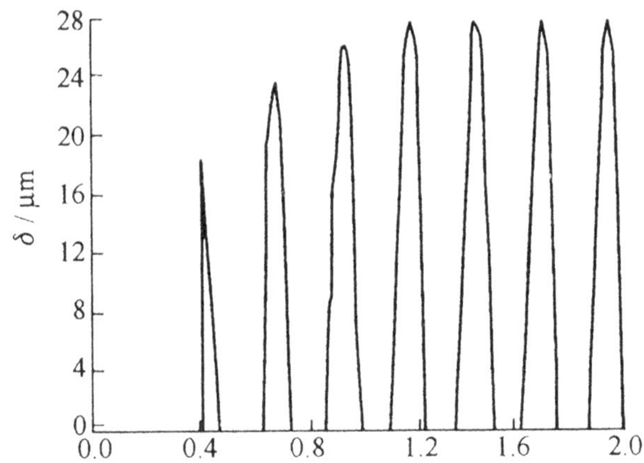

Fig. 4.15 Film thickness in mesh No. 40 of the innermost pin versus time

the test section. The time evolution of the lower boundary of the liquid film on pin 3 is shown in Fig. 4.14.

A horizontal section of the results shown by the plot of Fig. 4.14 at the axial level of mesh No. 40 displays the time variation of the liquid film thickness in this mesh. These results are shown in Fig. 4.15. After about 2 s into the simulated transient, following the initialization of the liquid film model, a steady-state situation, independent of the initialization itself, has been practically attained.

However, the thermal coupling between the outer surface of the pin pellet and the inner surface of the cladding has been simulated by assuming different heat transfer coefficients in case of dry-out and in case of wetted surfaces. In presence of the liquid film the heat transfer coefficient was computed by the ratio $h = \lambda_f/\delta$ of the liquid film conductivity and thickness. In case of dry-out the heat transfer coefficient was set to 5000 [W/m^2·K], which, for a thin liquid film, is more than one order of magnitude smaller than the heat transfer coefficient for wetted surfaces.

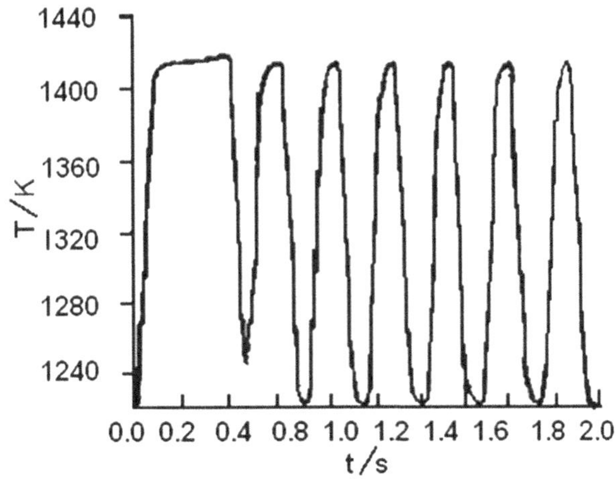

Fig. 4.16 Computed temperature distribution on the surface of the innermost pin at axial mesh No. 40, vs. time

The time variation of the temperature of the surface of the cladding of the innermost pin is shown in Fig. 4.16 for the axial mesh number 40, located at $z = 400$ mm, which was alternatively dried-out and rewetted. During the dry-out sequences the superficial cladding temperature rises to about 1420 K, while it drops to the initial value of about 1220 K during the time intervals of rewetting.

4.3.2.8 Final Assessment

The direct simulation Monte Carlo (DSMC) method turns out to be well applicable to the sodium vapour flow of monoatomic molecules into a sector of a 7-pin bundle boiling experiment by using the splitting technique of separating collisionless flow of molecules from the calculation of collision events. Sodium condensation rates, depending on the collision rate of sodium molecules with cold surfaces of the unheated section of the pins can therefore be successfully computed. These condensation rates provide the source term for a macroscopic film model flowing by gravity and rewetting the surfaces of the dry-out region of the pins. Molecules evaporating from the surface of the liquid film re-enter the Monte Carlo calculation in the domain between the pins.

The approach followed of coupling the simulation of microscopic phenomena, like condensation and evaporation, computed with the kinetic model, to macroscopic behaviour, represented by the liquid film model, has proved viable. Difficulties encountered in the simulations made with the one-dimensional transient computer code BLOW-3A and subsequently with the three-dimensional transient code BACCHUS-3D/TP arose from the absence in these codes of a rewetting or quencing model. Consequently, in those simulations the temperature of the surfaces of the cladding was consistently overestimated after dry-out inception. The model and algorithms proposed for computing dry-out and rewetting sequences eliminate an important deficiency of computer programs of the earlier generation.

The work summarized in this section has been the first attempt to combine traditional fluid dynamics and heat transfer methods with the novel approach of molecular dynamics in the safety analysis if a small reactor bundle.

4.3.3 7-Pin Bundle—Molecular Dynamic Simulation in 7-Pin LMFBR Bundle Under Hypothetical Accident Conditions

The transition between the previous section and the present one is a strait forward procedure which consists in removing the limiting as assumption of the 60-degrees domain and extending the domain to the full representation of a small bundle with seven pins. The further extension to a larger bundle does not imply any further difficulty apart from the increase of computational time for a full simulation of the accident conditions.

4.3.3.1 Geometrical Model

From the previous treatments it becomes evident that for a detailed numerical simulation of the microscopic phenomena occurring in the two-phase flow domain of a bundle a three-dimensional representation of the bundle is necessary.

The computer program described in Sect. 4.3.2 was further developed by building up the version THEMIS/3D-B7, a version of THEMIS/3D.1 (Bottoni, 2004) which simulates a small 7-pin bundle but in full size.

The DSMC method applied to the molecular dynamics inside a small size reactor bundle established the applicability of the model to reactor safety analysis, in view of incorporation of the methodology in established safety analysis computer codes. The results obtained for the temperature of the fuel pins, when taking into account the rewetting due to falling film of sodium, arising from condensation in the uppermost unheated section of the bundle, show that advantages, due to more realistic modelling of temperature distributions in the pins, would be obtained in computer codes which at present have not a rewetting model.

Figure 4.17 shows the trajectory of one sample molecule moving in upward direction from the entrance to the uppermost boundary, where it exits, being replaced by another sample molecule with the same index number. Periodic boundary conditions are assumed. In the time span of 30 ms the sample molecule exited a few times from the uppermost cross section of the bundle, being replaced by an entering molecule with the same index. The trajectories which cross the pins are spurious tracks due to these replacements of the sample molecule. In the following

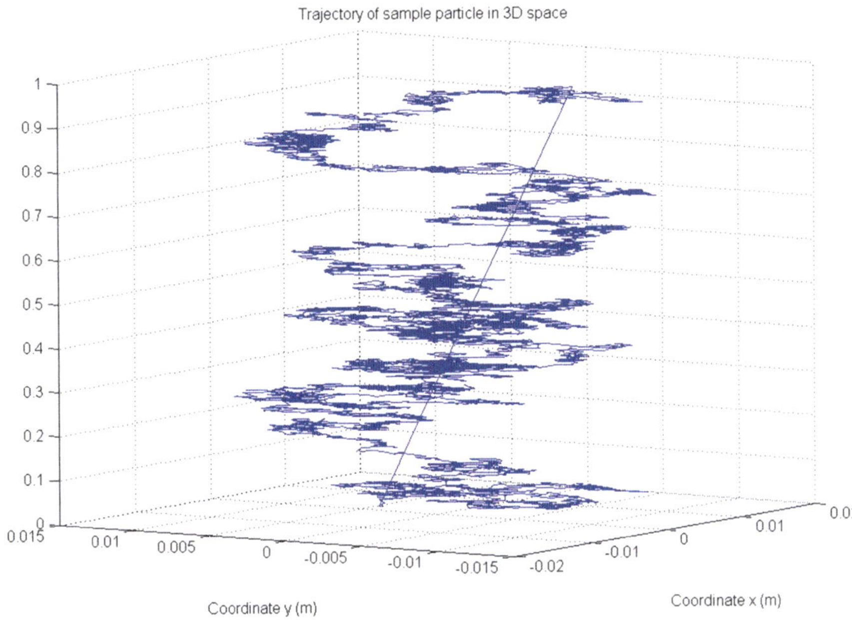

Fig. 4.17 Trajectory described by one simple particle

we refer to indexes of the pins from 1 to 6 starting from the uppermost right pin in the cross section and proceeding counterclockwise, while pin No 7 is the centre pin.

Figure 4.18 shows the envelope of fuel pins made by sample molecules during the molecular dynamic calculations. All the resulting trajectories form a full envelope around the pins and finally leads to the formation of a Fig. like the cross section shown in Fig. 4.19.

The geometrical configuration of the cross section of a 7 pin bundle is completely defined by the pin diameter D, pitch P and by the distance between the axis of the centre pin and the vertex of the hexagonal canning. The height of the bundle is called Z_{max}. We assumed a main reference Cartesian coordinate system with origin at the axis of the centre pin on the inlet plane ($z = 0$) of the bundle. The orientation of the x and y axes is implicitly shown in Fig. 4.19, while the z axis is vertical. Figure 4.17 was obtained by plotting the trajectory of a sample molecule during the first 50,000 steps, corresponding to 50 ms of problem time, of one molecular dynamic simulation. Changes in direction of the trajectories are due to the intermolecular collisions and to the diffuse reflections from the surfaces of the pin or from the hexagonal canning. In case of diffuse reflections the coordinates of the collision points with the surfaces were recorded so that the ensemble of the trajectories gives the envelop of these surfaces.

This procedure of drawing a cross section of the bundle as the ensemble of the envelopes, might be the most expensive one, but it allows for checking the

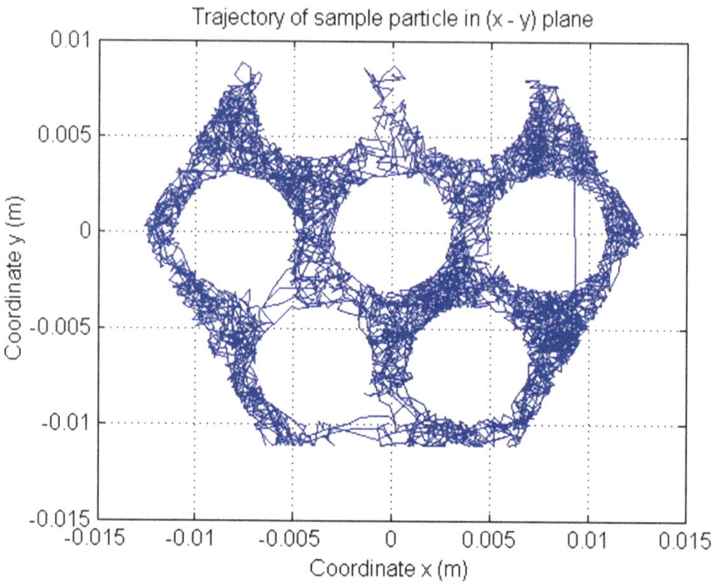

Fig. 4.18 Envelope of fuel pins made by sample molecules during the molecular dynamic calculations

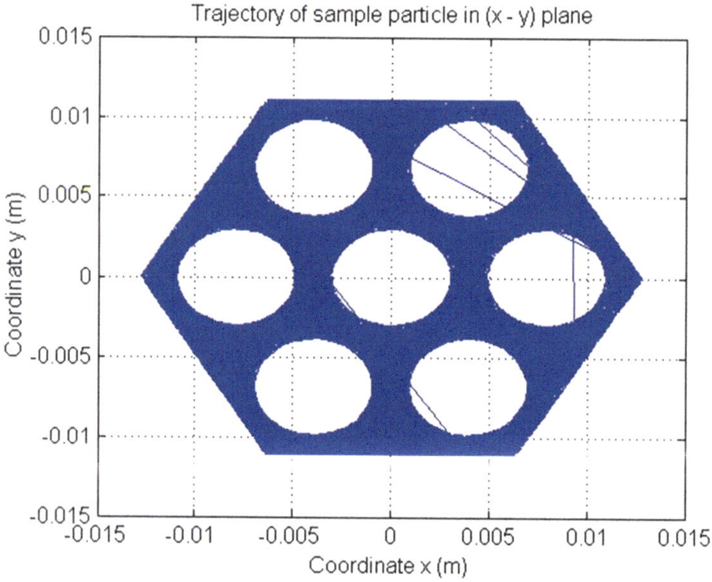

Fig. 4.19 Trajectories of a sample particle described during 50,000 time steps (0–50 ms)

correctness of the calculation of the trajectories of the molecules following the multiple reflections.

We assume the full bundle to be included in a square domain of late $2X_{\max}$ with centre in the origin of the (x,y) coordinate system. The domain in divided in a number $I_{\max}\ J_{\max}$ of computational cells of size $\Delta x = 2X_{\max}/I_{\max}$ and $\Delta y = 2X_{\max}/J_{\max}$. The length of the cells in the axial direction is $\Delta z = Z_{\max}/K_{\max}$. The computer program determines the cells which fall outside of the hexagonal canning. These cells are excluded from the molecular dynamic calculations. The program also identifies the cells which partially overlap the domain of definition, being cut by the hexagonal canning or by the surface of the pins. For these cells the program determines, with a Mont Carlo integration, the areas and volumes of the cell fractions which belong to the domain.

As in the previous section, we simulated the following bundle geometrical conditions: Pin diameter $D = 0.006$ m (radius of fuel pellet of uranium dioxide 0.0024 m, thickness of cladding 0.0006 m), pitch $P = 0.0079$ m, lower heated section $\Delta z_x = 0.600$ m, upper unheated section $\Delta z_u = 0.400$ m. Temperature and pressure boundary conditions in steady state, before the simulation of transient accident conditions: Inlet/outlet pressure: $(2.03/1.05) \times 10^5$ (N/m^2); inlet / outlet coolant temperature: 826/988 (K), coolant velocity: 2.15 (m/s).

4.3.3.2 Temperature Distribution

Calculations were made with 900,000 sample molecules in the full domain.

Figure 4.20 displays the steady-state temperature distribution in vertical cross section of the bundle. The figure clearly shows the boundary between the heated and unheated sections of the bundle. In the region belonging to the coolant, the temperature distribution is computed with the molecular dynamics, while in the other regions it is computed by solving the hear diffusion equations.

Calculations of temperature distributions in the pins were made in axis symmetric (r, z) geometry.

Boundary conditions for the temperature distributions in the pins are obtained by averaging in azimuthal direction the sodium coolant temperatures obtained with the molecular dynamic calculations. The heat transfer coefficients are determined, as specified later, by the presence or absence of a liquid film of sodium coolant wetting the pin surfaces.

We assumed an axial specific power distribution in the pins given, in the heated section, by

$$Q(z) = Q_{peak}(0.8 + 0.2\sin(\pi z/\Delta z_h), \tag{4.81}$$

with $Q_{peak} = 10^6$ (W/m^3). In the unheated uppermost sections of the pins we assumed a specific power of 3 per cent of the peak power, to simulate heat released by gamma radiation. Besides, a multiplying skew factor k_s was applied for the pins as

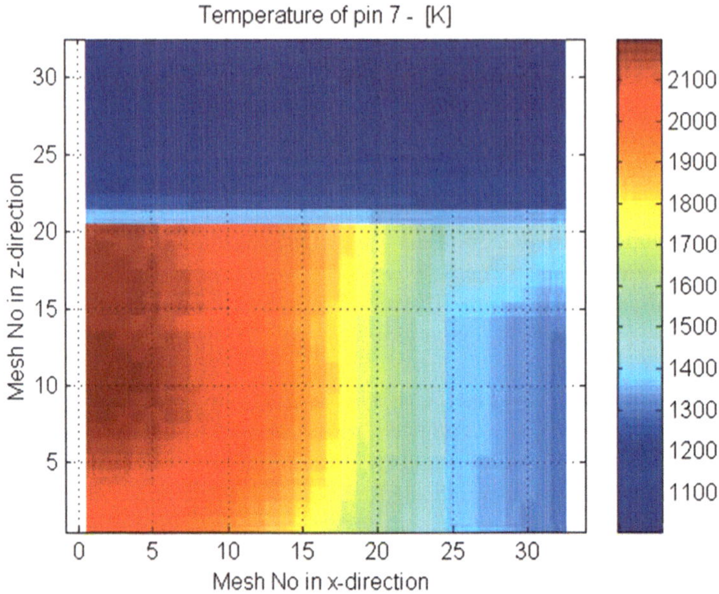

Fig. 4.20 Steady-state temperature distribution in vertical cross section of the bundle

follows: Pins No 1 and 2 $k_s = 0.9$, pins No 3 and 6 $k_s = 1.0$, pins No 4 and 5 $k_s = 1.1$, and pin No 7 $k_s = 1.2$. This resulted in total pin power ranging from 9,438 W for pins 1 and 2 to 12,584 W for the central pin No 7. The total bundle power was to 75,506 W. Figure 4.21 shows the skeweness of the temperature distribution in a horizontal cross section of the bundle. Figure 4.22 displays the temperature distribution in a vertical cross section of the bundle.

The full domain was divided in 32 meshes in all three coordinate directions. We assumed 32 radial meshes for the computation of the temperature distribution in the pins, out of which 25 nodes were used for the fuel, 6 nodes for the cladding, and one node for the boundary condition provided by the mean coolant temperature around the pins.

Following a simulated loss of coolant accident, sodium boiling started in the real experiment at $t = 6.5$ s into the transient (from the beginning of simulated pump run-down transient at $t = 0$ s). In typical voiding conditions sodium vapour flows with a speed of 50–150 m/s. We assumed a value of 128.5 m/s.

The Monte Carlo simulation is assumed to start when the dry-out region spreads to the unheated part of the test section by expulsion of the upper liquid coolant slug. Because of the high value of superheat (68 K) observed in this experiment at boiling inception there is a violent upper slug ejection with voiding and dry-out spreading upwards in a fraction of a second. The Monte Carlo simulation is set to start at this point at about 7.2 s into the transient.

Figure 4.23 shows the temperature distribution in a vertical cross section of the bundle which consists of four domains in the coolant region and of three domains in

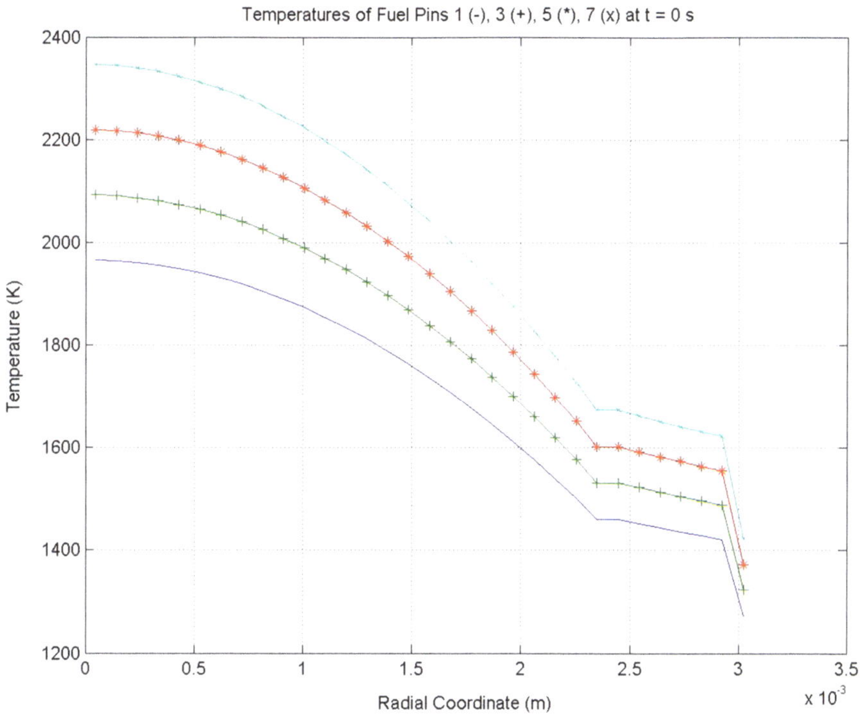

Fig. 4.21 Temperature distribution at $t = 0$ along the radial direction for selected pins

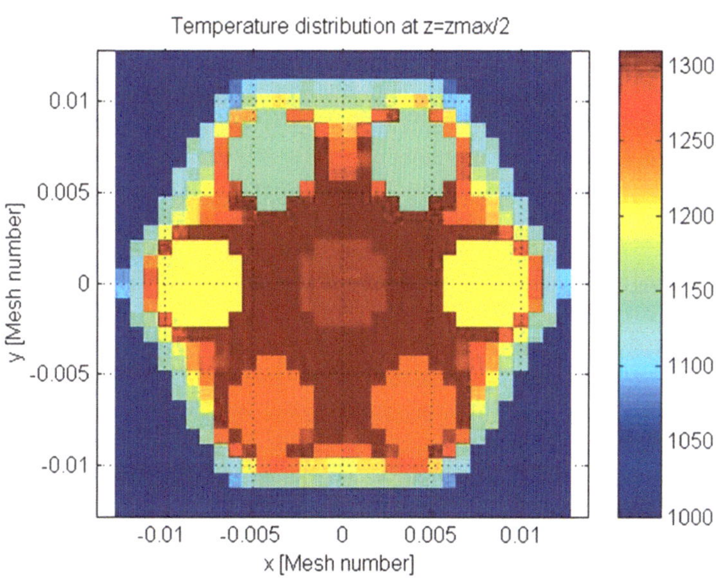

Fig. 4.22 Temperature distribution in a horizontal cross section of the bundle

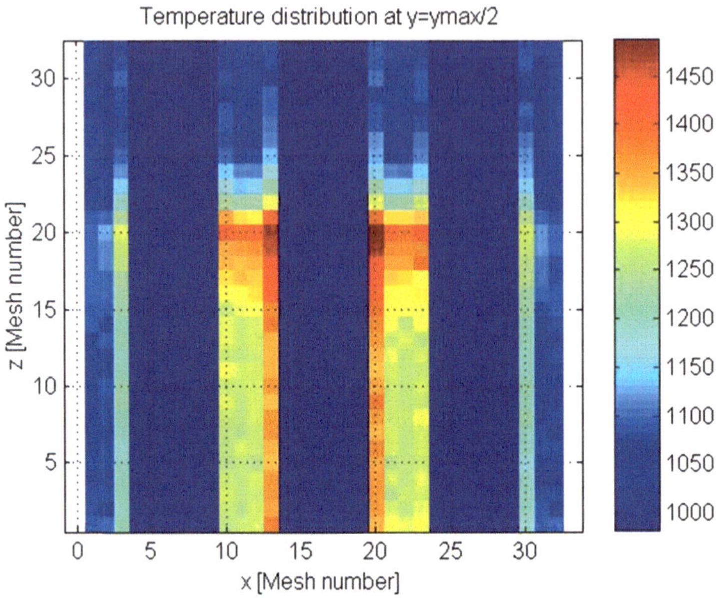

Fig. 4.23 Temperature distribution in a vertical cross section of the bundle

the Pins (Pin 3, Pin 7, and Pin 6). The temperature distribution shown in this figure is computed with the molecular dynamic and refers only to the coolant domain. The temperatures shown in the regions of the pins are meaningless.

4.3.3.3 Discussion of Computed Results

(i) **Film Thickness**

Figure 4.24 shows the time behaviour of the liquid film wetting four fuel pins at mesh level 13. Following a short transient of about four seconds the film thickness stabilizes at about 500 μm on all pins. The plots show that the film thickness is not wavy, nor displays strong oscillations in the amplitude. According to the criterion by Kapitza (1941) the waves are stable and are not likely to brake. A short summary of the theory of liquid films is given in (Fulford, 1964). Therefore, under the given conditions, there is good cooling of the surfaces wetted in the annular flow regime.

(ii) **Film Velocity**

Figure 4.25 Shows the distribution of liquid film velocity in four pins at time point $t = 20$ s into the transient. The distributions are very similar to each other, but it will be noticed that in the colder Pin 1 the film descends till mesh 5, while in the hottest Pin 7 the liquid film stops at mesh 8

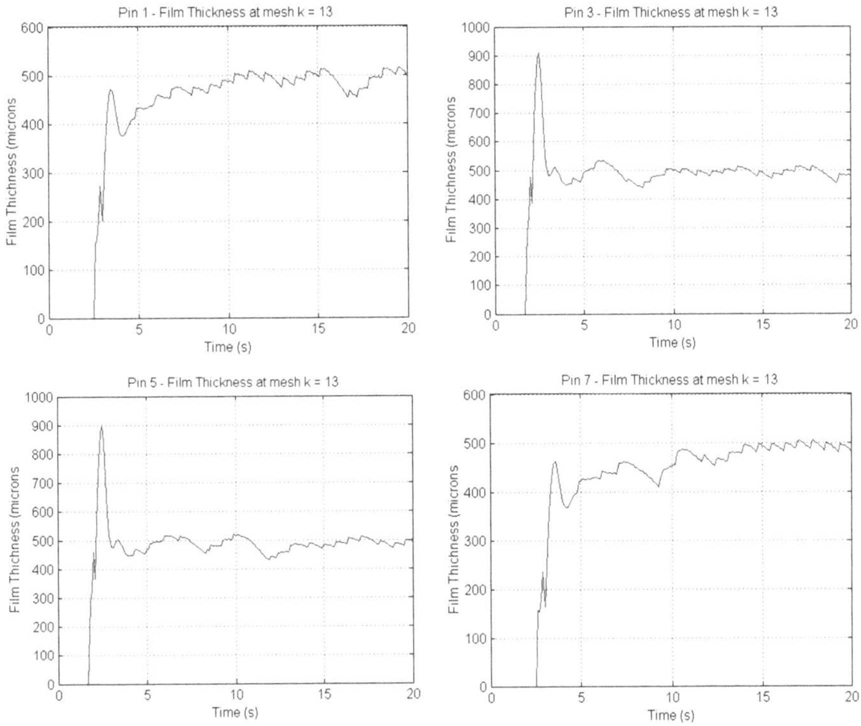

Fig. 4.24 Liquid film thickness on four Pins at mesh number 13, versus time

(iii) **Lowermost Wetted Meshes**

Figure 4.26 shows the lowermost wetted meshes in Pin 1, Pin 3, and Pin 7. Following the transient of 4–5 s, the plots stabilize in the range between the meshes 6–10.

4.3.3.4 Conclusions

The direct simulation Monte Carlo (DSMC) method can be successfully applied to the sodium vapour flow of monoatomic molecules in a 7-pin bundle simulating a boiling experiment by using the splitting technique of separating collisionless flow of molecules from the calculation of collision events. Sodium condensation rates, depending on the collision rate of sodium molecules with cold surfaces of the unheated section of the pins, computed with the molecular dynamic approach, provide the source term for a macroscopic film model flowing by gravity and rewetting the surfaces of the dry-out regions of the pins. Molecules evaporating from the surface of the liquid film re-enter the Monte Carlo calculation in the domain between the pins.

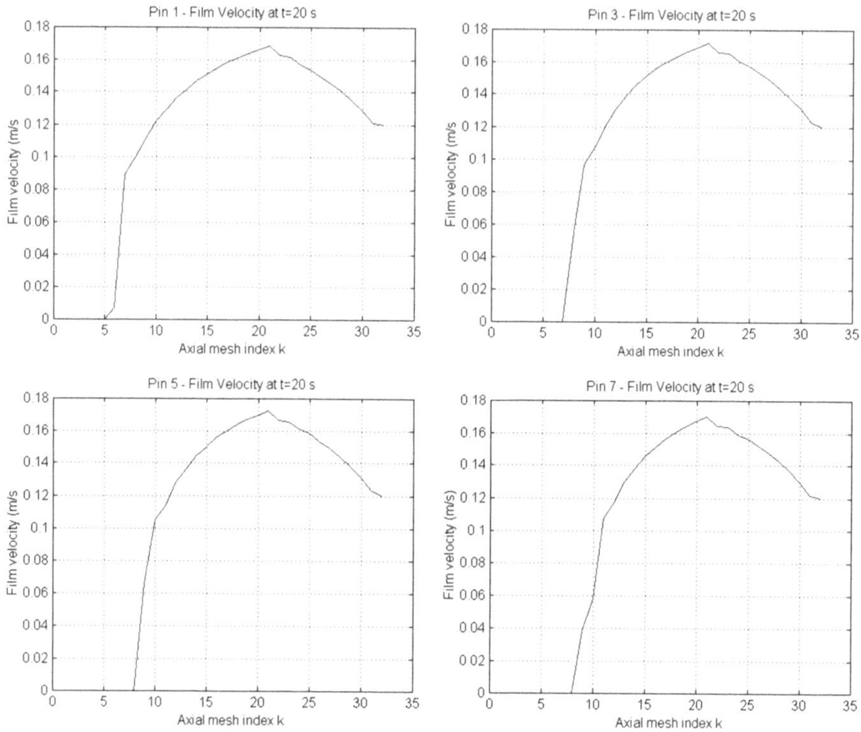

Fig. 4.25 Distribution of liquid film velocity in four pins at time point $t = 20$ s into the transient

The approach followed of coupling the simulation of microscopic phenomena, like condensation and evaporation, computed with the kinetic model, to macroscopic behaviour, represented by the liquid film model and pin model, has proved viable and results obtained are qualitatively consistent. Difficulties encountered in the simulations of fuel bundle experiments made in past times with reactor safety analysis codes in the pre-disassembly phase mainly arose from the unavailability of fuel pins rewetting or quencing models. Consequently, in those simulations, the temperature of the surfaces of the cladding was consistently overestimated after dry-out inception. As remarked in the previous section, the models proposed for computing dry-out and rewetting sequences eliminate important flows of earlier computer programs and provide better tools for reactor safety analysis.

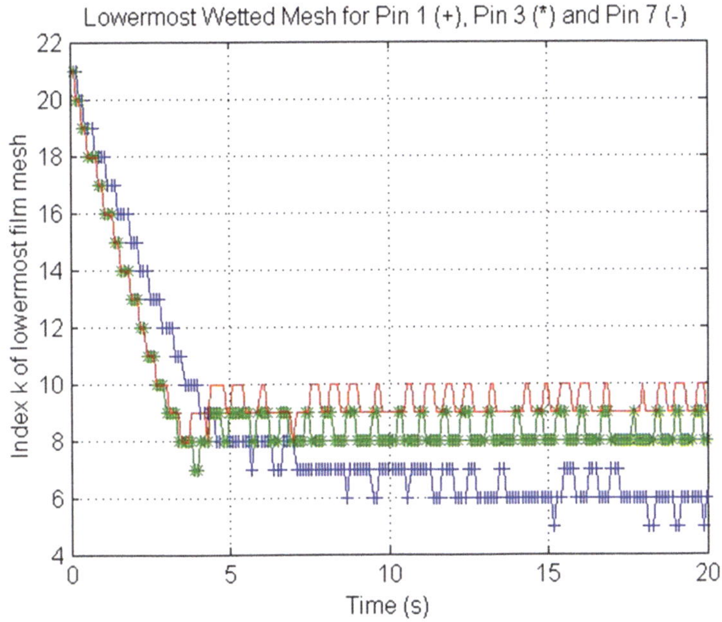

Fig. 4.26 Location of the lowermost wetted meshes in Pin 1, Pin 3 and Pin 7

4.3.4 Direct Monte Carlo Simulation of Fuel Particles Relocation in Nuclear Reactor Bundle During Hypothetic Core Disruptive Accident

In the safety analysis of fast reactors we referred to the evolution of phenomena from the so-called "initiating phase" through the "intermediate phase" to the "transition phase", as sketched in Fig. 4.6. The dominant phenomena were the progressive failure of fuel elements with melt down of cladding and molten fuel motion by still intact hexagonal canning. In the following section, we remove the assumption of still intact structural elements and examine the consequences of fuel motion involving subassemblies adjacent to those where the initiating failure was assumed to occur.

With reference to the simplified scenario of Fig. 4.6 we therefore refer to sequences of events occurring in the "transition phase", characterized by melting and slumping of structural elements mixing with coolant and fission products and eventually interacting with them in the frame of FCI (Fuel Coolant Interaction).

This section is organized as follows: We fist describe the condition that leads to the initiating failure and then examine the progression of events discussing the consequences of the assumption made. Because several options are available and consequently several scenarios of events are possible, we discuss the impact of the different choices of events upon the final results.

Starting from a simple preliminary approach to the description of the sequence of events, we proceed to improve the phenomenological treatment adding new possibilities to the previous description. In other words, we assume to proceed with a learning attitude from a preliminary stage to a more sophisticated one where more detailed phenomena can be taken into consideration. This might be the approach to be followed by young researchers willing to learn how to progress when confronted with new problems.

(*In the following part of this section we denote with cursive character improvements of the model which could be undertaken in the future.*)

4.3.4.1 Geometrical Model

The geometrical configuration is as shown in Figs. 4.27 and 4.28, which refer to temperature distributions obtained in steady-state calculations made before a transient pump run down experiment was started.

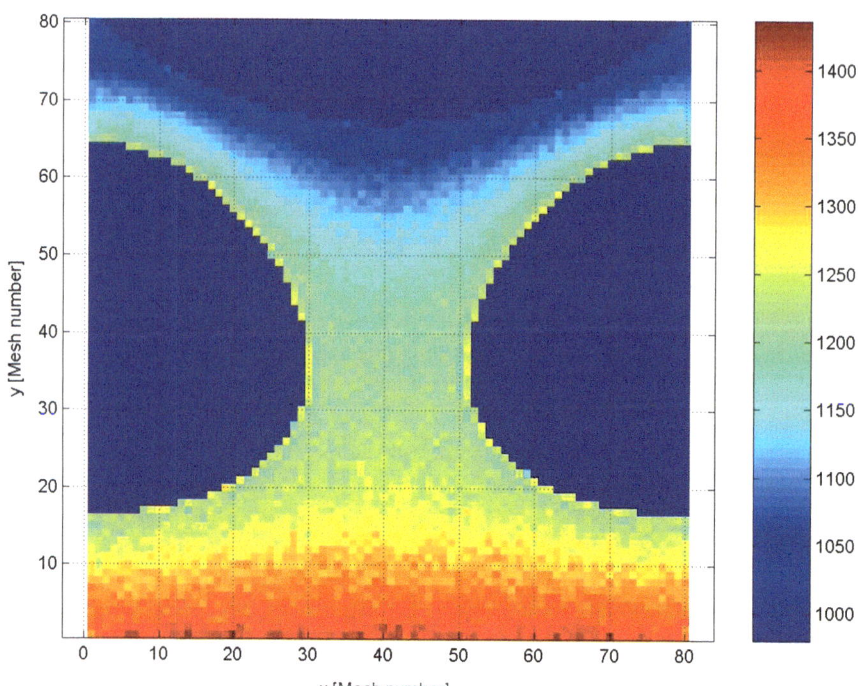

Fig. 4.27 Diffuse representation of the temperature distribution in a horizontal cross section of the 60-degrees sector of the hexagonal assembly

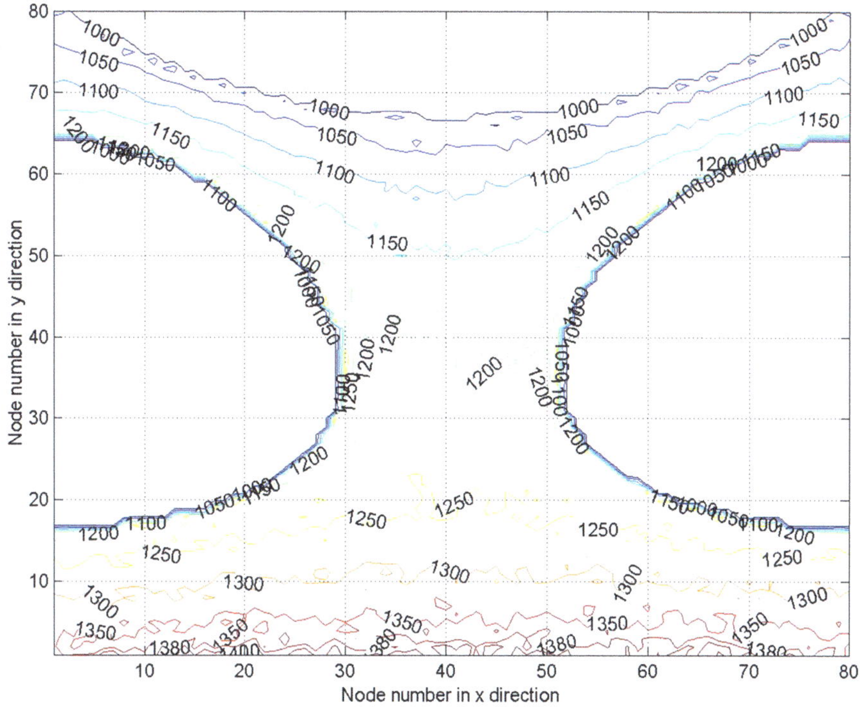

Fig. 4.28 Isotherms in a horizontal cross section of the 60-degrees sector of the hexagonal assembly

4.3.4.2 Conduction of the Experiment

A failure analysis experiment was conducted in the same way as explained in Sect. 4.3.3 with the only difference that not the full bundle, but a section of it, was simulated. A failure to the pump power supply leads to overheating of the fuel pins and, without switching off the heat generation in the pins, eventually leads to cladding failure and sputtering of fuel debris and fission gases into the residual coolant flow.

We assumed at first an input of 10^4 fuel particles injected into the coolant. This is however an issue which can be considered for improvements in future work.

(A model can be introduced for simulating a time-dependent injection of fuel particles into the coolant. A further improvement can be made simulating injection of particles of variables sizes according to a given probability distribution.)

4.3.4.3 Film Thickness Distribution After Pin Failure

Liquid film thickness formed by condensation per unit of time after pin failure in the uppermost cold unheated test section was computed with the method explained in Sect. 4.3.2.7.

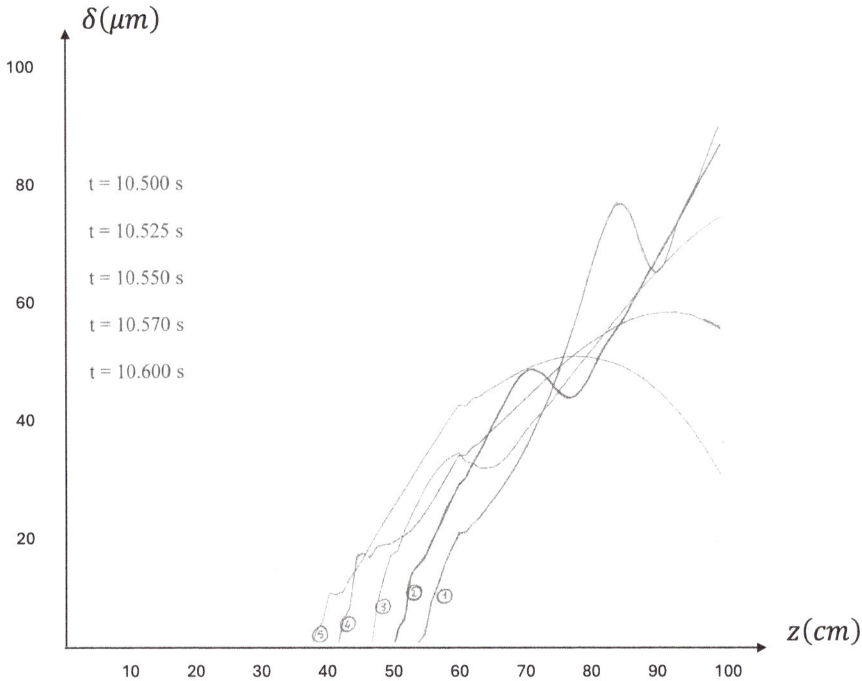

Fig. 4.29 Axial distribution of liquid film thickness in the central pin for selected time points ($z =$ 0–60 cm: Heated section; $z =$ 60–100 cm: Unheated section)

The time behaviour of the liquid film distribution is shown in Fig. 4.29 for the central pin for the time span between 10.5 and 10.6 s into the transient. In this time the lower boundary of liquid film thickness moves from mesh 53 to mesh 39 with a speed of about 150 cm/s.

4.3.4.4 Simulation of Thermophoretic Velocities

The thermal conditions in the sector of the 7-pin bundle are determined by the temperature gradient between the inner pin, surface 3, and the hexagonal canning, surface 6, which is 11.1 mm apart. Keeping the temperature of the hexagonal canning constant at 1000 K, the temperature of the inner pin was varied between 1200 and 1400 K.

Computational values of thermophoretic velocities were obtained by simulating a reduced section of the bundle of 0.01 m in height, without thermal gradients in the vertical direction and without gravitational forces. In a grid of 20×20 meshes in a horizontal plane there are 248 meshes belonging to the computational domain. We simulated 20 axial meshes for a total of 4960 computational cells. In this reduced section of the bundle 70,000 sample molecules of sodium were simulated at the pressure of 1.50 bar and temperature distribution resulting from the assigned gradient between the inner pin and the hexagonal canning. The temperatures of the surfaces of the intermediate pins were kept at 1200 K. Under these conditions, each one of the sample molecules corresponds to 3.31×10^{13} real molecules.

The DSMC method was applied to simulate the molecular dynamic in the given geometrical configuration. The symmetry surfaces 1 and 2 of the lateral planes were treated as specularly reflecting surfaces while the material surfaces of the pins and of the hexagonal canning, surfaces 3 through 6, were treated as diffusely reflecting surfaces.

In the numerical investigations thermophoretic velocities result from the different momenta given to a particle by molecules with different velocities diffusing from the surfaces between which the temperature gradient is applied. Because of the high dispersion of molecular velocities ranging from zero to infinity (a cut-off at twice the most probable velocity being applied) mean values of the velocities of molecules impinging and reflected from sample particles were made over a time interval of 0.00 s, computed with a time step of 10^{-7} s . In the temperature range shown by the abscissas in Fig. 4.27 we made several calculations providing the distributions of mean thermophoretic velocities shown in the plot.

From the viewpoint of the centre of a sample particle we refer to "north" and "south" directions as those from the hexagonal canning and from the inner pin, respectively. Molecules incoming from north have negative y-velocity (v velocity) components, those incoming from south have positive v component. After starting a calculation and computing collisions of molecules with the particle we first allow for 2 ms for the vapour molecules in the computational domain to adjust to the imposed boundary conditions. At this time point, till 5 ms, we compute both mean values of the velocities of the molecules impinging with the particle from north and from south and take their difference. This difference, which determines the thermophoretic velocity, is referred in the ordinate axis of Fig. 4.30 with the acronym AYDIFFNS (Average for the Y direction of the DIFFerence of the mean velocity of molecules impinging from North and from South) . The mean value of the plot in the interval 2–5 ms yields the ordinate of a point for Fig. 4.31.

In the low temperature range of the abscissas of Fig. 4.31 slightly negative values of the thermophoretic velocities were predicted, implying transport of particles along the temperature gradient. This result is explained by non-uniform distributions of molecular densities, due to the particular geometry and temperature distribution.

In the numerical applications referred to in the following, we used the previous results in the following manner. Because of the immense computational effort which would be required by simulating thermophoretic velocities acting on every one of the particles of fuel debris released in case of hypothetical accident, in the order of 10^4, we applied thermophoretic velocities to the debris particles sampling the computed plot of Fig. 4.31 at the relevant temperature and applying a Gaussian distribution with the mean value given by the ordinate of the plot.

(*Future applications made with massively parallel computers may allow simulating the thermophoretic forces acting upon every one of the debris particles without referring to the artifice represented by* Fig. 4.31 *but computing directly the thermophoretic forces for every particle of debris injected into the system.*)

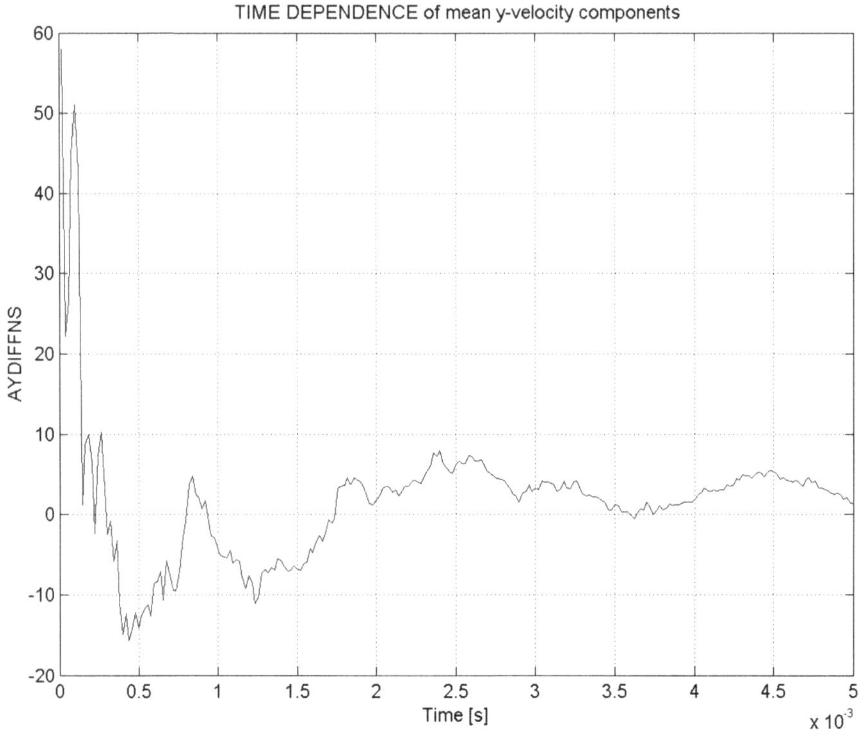

Fig. 4.30 Time dependence of mean velocity components (m/s) of molecules impinging on a sample particle from the side of the hexagonal canning ("north") and from the side of the central pin ("south"). [The acronym AYDIFFNS is explained in the text]

4.3.4.5 Temperature Distribution in a Fuel Particle

We consider a typical fuel particle, assumed to consist of uranium dioxide, as a sphere of given radius r_p and with known physical properties and compute the transient, three-dimensional temperature distribution in the sphere. We start from an initial distribution depending on the temperature gradient in the sodium vapour adjacent to the particle and compute the transient temperature distribution taking into account the kinetic energy transferred to the particle from the incident air molecules and taken from the particle from the reflected molecules. We refer to a spherical coordinate system (r, ϑ, Φ) with origin in the centre of the particle. ϑ is the co-latitude and Φ the azimuth.

The heat diffusion equation for the temperature T of a particle of density ρ and thermal conductivity λ, is, in the spherical system is

$$\rho c_p \frac{\partial T}{\partial t} = \nabla \cdot (\lambda \nabla T) - hS(T_S - T_g), \tag{4.82}$$

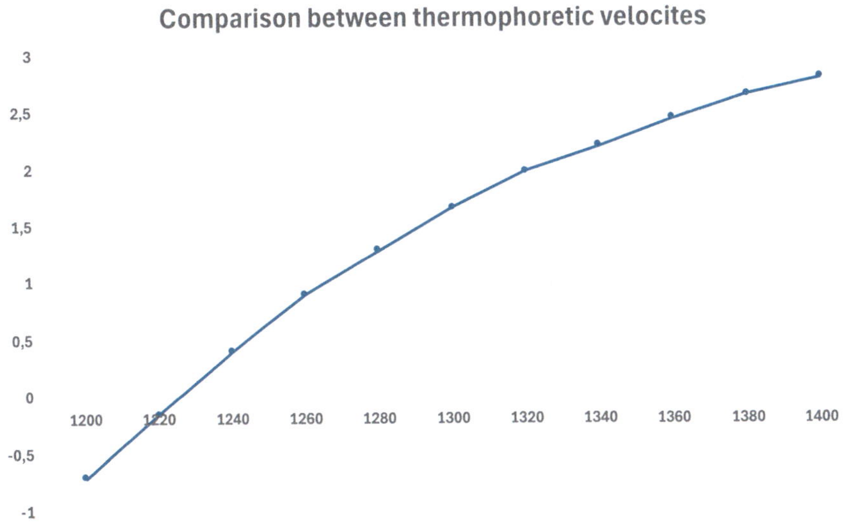

Fig. 4.31 Thermophoretic velocities (m/s · 10^4) computed with the DSMC method. [Abscissa: Temperature of inner pin (K)]

where S denotes the surface of the particle. The subscripts in the last term of Eq. (4.82) refer to the surface of the particle and to the surrounding gas or vapour, respectively. h is a heat transfer coefficient per unit volume.

Integrating Eq. (4.82) over the volume V_c of a computational cell and applying the Gauss theorem one replaces the volume integral of the divergence of the vector $\lambda \nabla T$ with the flux of the vector through the bounding surfaces, thus obtaining

$$\int_{Vc} \rho c_p \frac{\partial T}{\partial t} dV = \int_{S} \lambda \nabla T \cdot \vec{n}\, dS - \int_{Vc} hS(T_S - T_g) dV \tag{4.83}$$

where $\vec{n}$ is the outwards oriented unit vector, normal to the surface.

Equation (4.83) is discretized in the computer program and solved with the iterative Jacobi scheme. The resulting surface temperatures determine the energy released from the particles to impinging molecules and therefore are taken into account in the evaluation of thermophoretic forces.

(An improvement of the computational speed can be attained by replacing the numerical solution of Eq. (4.83), now made with the iterative Jacobi scheme, with a direct method as, for instance, the Cholesky matrix decomposition technique.)

4.3.4.6 Set-Up of the Numerical Simulations

For the numerical applications we set the channel height at $z_{\max} = 1.00$ m and partitioned the full domain in 20 cells in each coordinate direction. The temperature distribution of the surface of the inner pin was assumed to be a sinus function with a peak value of 1400 K at $z = z_{max}/2$ and minima of 1200 K at the upper and lower boundaries. The temperatures of the other two pin surfaces were uniform and set to 1200 K and the temperature of the hexagonal canning was set to 1000 K. During accident conditions, the coolant channel is chocked by the upper and lower liquid slugs and oscillating with a frequency of a few cycles per second. Because the mean flow velocity in chocked condition is zero we did not assumed a mean flow velocity, but allowed for a random components of particles velocities in axial direction to simulate local perturbations due to turbulent whirls in the vapour. This assumption does not affect the calculation of thermophoretic forces acting in horizontal direction but affects the dispersion of particles in axial direction.

(*Further developments if the model would imply integration of the simulated pin in a sub-module of a larger accident analysis code. In such a case the boundary conditions of the axial flow would be delivered by the calling program.*)

A typical calculation starts from a uniform temperature distribution, equal to the temperature of the lateral pins, in the simulated vapour coolant formed by sodium molecules. From this initial condition the calculation proceeds for 0.005 s, with a time step of 10^{-7} s, to allow for a thermodynamic quasi-equilibrium to set in the sodium vapour, in which the temperature adjusts to the simulated temperature gradients between the solid surfaces. Following this preliminary phase, it is assumed that failure of the cladding occurs in correspondence of the peak temperature of the centre pin. Sputtering of fuel debris of an axial section of a pin pellet injects into the computational domain a cloud of fuel particles with random size distribution within a given interval. In the calculation here documented we assumed that one sixth of the volume of a 5 mm height of fuel (a volume of 4.71×10^{-8} m^3) was sputtered into particles of sizes chosen randomly in the interval 30–100 μm. These assumptions lead to model a disperse cloud of about 26,000 particles, injected into the computational domain in the first discretization mesh adjacent to the surface of the inner pin.

(*Interactions between particles, with possible coalescence, but also possible fragmentation of particles, should be considered for model improvements*).

Upon injection into the computational domain, the thermophoretic velocity distribution of Fig. 4.31 is sampled to assign to each particle its velocity component.

(*In advanced simulations, with mean fluid velocity different from zero, can be considered a model of fuel particle entrainment based of the equation* $m_p \frac{dV_p}{dt} = 6\pi\tilde{\mu}_g\left(V_g - V_p\right)$, *where* m_p *is the mass of the particle,* V_g, V_p *are the fluid and particle velocities, respectively,* t *is time,* $\tilde{\mu}_g$ *is the effective (turbulent) dynamic viscosity of the gas. Additional terms may be considered at the right hand side of the above equation to account for fluid and particle accelerations.*)

4.3.4.7 Computational Results

Figure 4.32 shows the temperature distribution in a cross section of the bundle at axial level $z_{max}/2$, in correspondence of the peak temperature of the surface of the

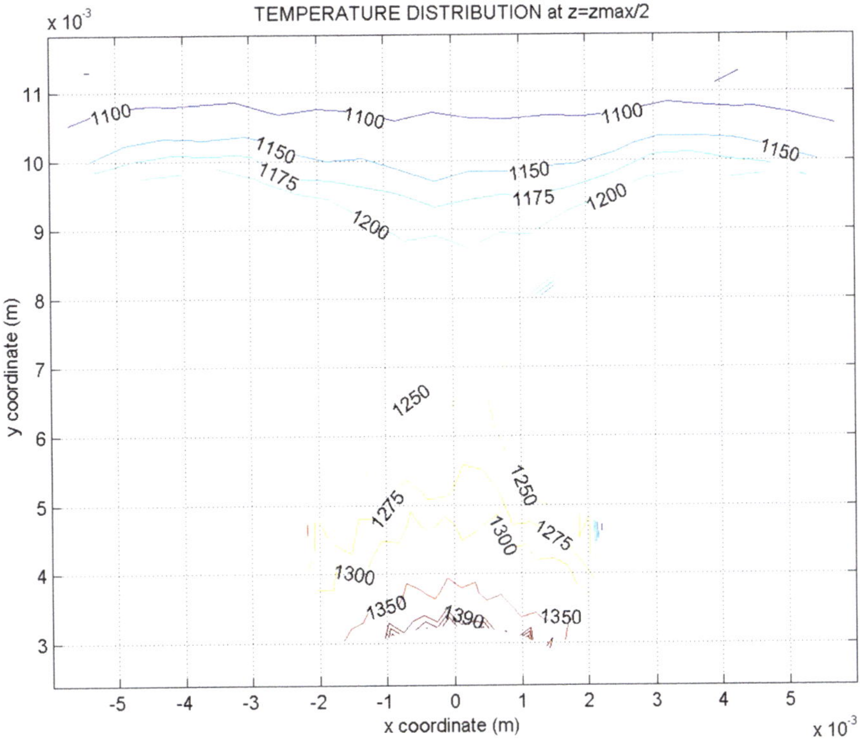

Fig. 4.32 Temperature distribution in the sector of the bundle at $z = 0.5$ m, at the time of particles injection

centre pin. The strongest temperature gradients are in the domain between the central and the lateral pins, whereas in the domain between the lateral pins and in the area between these and the hexagonal canning the gradients are smaller. Therefore the thermophoretic forces are not uniform in all space between the inner pin and the hexagonal canning. This implies that sampling thermophoretic velocities as we do from the distribution shown in Fig. 4.31 is actually only a preliminary approach which, as remarked above, should be replaced by a continuous calculation of thermophoretic velocities following the particles trajectories.

A snapshot of the particle distribution a $t = 3.3$ s after the injection of the particles is given in Fig. 4.33. The bulk of the particles moved 1.3 mm in the positive y-direction, from the cells adjacent to the centre pin towards the hexagonal canning. This displacement corresponds to a mean thermophoretic velocity of 310 μm/s (Fig. 4.34).

As particles approach the lateral pins and collide with their surface, the computational program assumes deposition of these particles on the pin surfaces with formation of a crust of fuel. The population of particles in the computational domain thus wanes as particles are depleted.

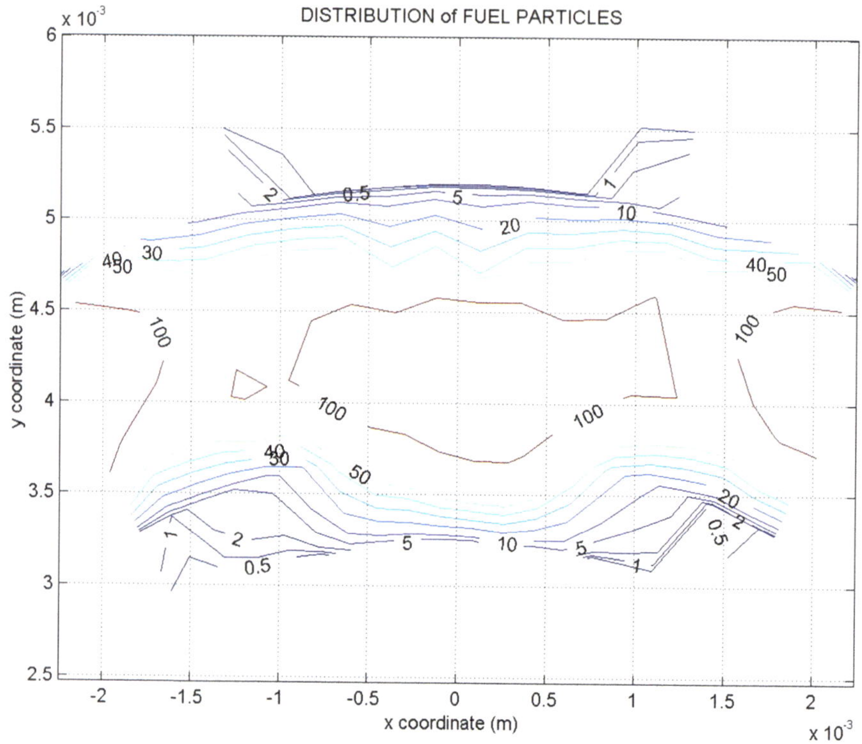

Fig. 4.33 Density distribution of fuel debris at $t = 3.3$ s since the ejection of particles

The distribution of residual particles in the computational domain at $t = 9$ s since the injection is shown in Fig. 4.34 where only a few per cent of the particles were left. Only a minority of the particles reaches and deposits upon the surface of the hexagonal canning. Most of the debris remain in the subchannel of the bundle where injection occurred and only a small percent of the debris propagate to other subchannels.

The density of the crusts formed upon the pin surfaces or on the surface of the hexagonal canning is smaller than the theoretical density of the fuel. Taking into account that the shape of the particles is in reality not spherical, but rather irregular, we assumed a crust porosity of 0.4 for calculating the volume of the depositions. For a visualization of crust formation, we assumed a local polar coordinate system with origin in the centre of the lateral pins. In the case of the upper right half-pin, the azimuthal angle is calculated counterclockwise from the right symmetry plane (surface 1). The periphery of the pin was partitioned in 20 meshes in this counter-clockwise direction. For the upper left half-pin, we numbered the cells along the periphery in the clockwise direction.

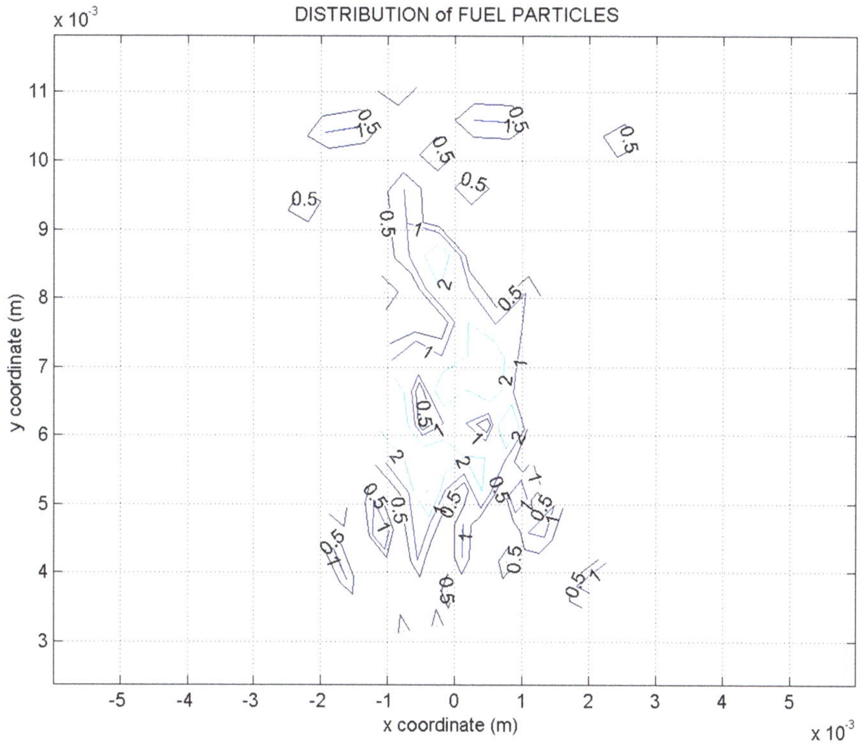

Fig. 4.34 Residual density distribution of fuel debris at $t = 9$ s since the ejection of particles

The thickness of fuel crusts forming along the periphery of the two pins at the axial level $z = 0.475$ m is shown in Fig. 4.35, where the abscissa is the index referring to the cells along the periphery of the pins. In our calculation deposition of fuel particles occurred in meshes 11 through 20. The distributions clearly show a maximum corresponding to the cell indexed 20, where the coolant gap is minimum.

(Computer programs based on neutron diffusion calculations would be invoked at this stage to evaluate the consequences of crust formation upon the reactivity of the reactor core.)

4.3.5 *Molecular Dynamic Simulation of Thermal Creep in a Sector of a 7-Pin LMFBR Bundle Under Hypothetical Accident Conditions*

In this section the direct simulation Monte Carlo (DSMC) method is applied to vapour of sodium monoatomic molecules in a 60-degrees sector of a 7-pin bundle

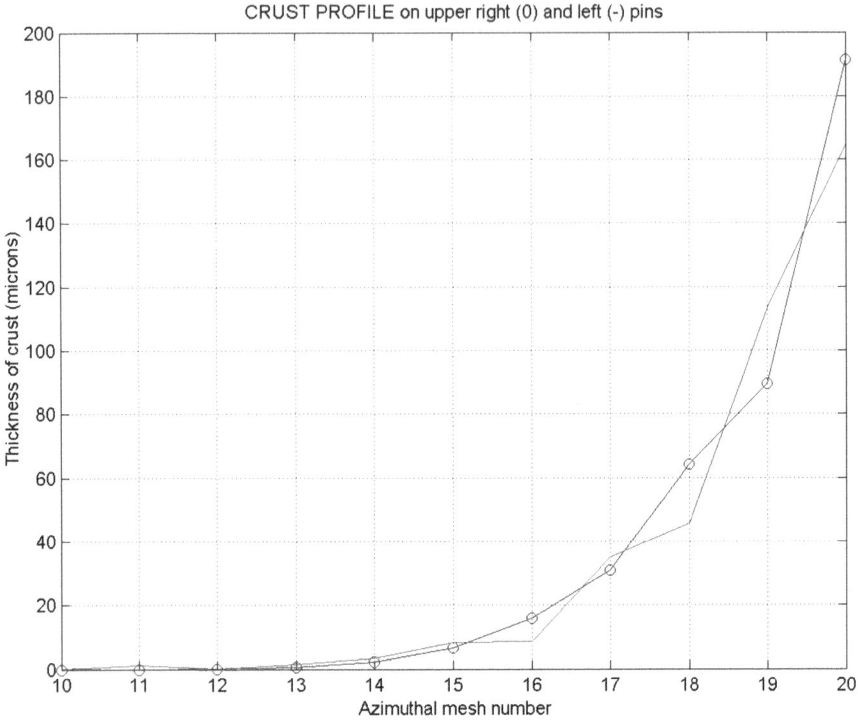

Fig. 4.35 Profiles of fuel crusts forming at axial level $z = 0.475$ m on the upper right (dotted plot) and left pins versus index of meshes along the periphery of the pins

under conditions which would be obtained in the course of an hypothetical accident leading to sodium coolant vaporization and growth of the voided region in a subchannel chocked by upper and lower liquid slugs.

Unlike in the previous sections of this chapter, in the following treatment we also consider the impact of stream induced by thermal creep, in addition to the flow inducted by thermophoretic forces.

It is assumed that under chocked channel conditions failure of the hottest pin at the location of peak temperature injects particles of fuel debris into the coolant channel. Thermophoretic velocities, arising from unbalance of momenta transferred from the vapour molecules to sample fuel particles, are simulated in a parametric way and used to calculate particle migrations and deposition on adjacent pin surfaces. Thermally induced creep velocities, due to the large temperature gradients along the pins surfaces in axial direction, are simulated and used to compute the entrainment of fuel debris in the axial direction of the bundle. Fuel debris are spread both in the cross flow planes in radial and azimuthal directions and in axial direction by the creep velocities. Deposition of fuel debris upon the surfaces of lateral pins leads to the formation of fuel crust much thinner than in the case the particles were diffusing in the cross section planes without the impact of the creep velocities. In thus far, consideration of these creep velocities is important in the evaluation of fuel recriticality under hypothetical accident conditions.

Some excerpts of this section are taken from Bottoni (2007), and the following figures are reproduced from this publication with kind permission of the editor.

4.3.5.1 Theoretical Background

Sample molecules of vapour of the sodium coolant are initialized in the computational domain with velocities obtained by sampling a Mawellian probability distribution

$$f = \frac{1}{(2\pi RT)^{3/2}} e^{-c'^2/(2RT_f)}. \tag{4.84}$$

In Eq. (4.84) c' is the thermal velocity, defined as difference between the instantaneous and the mean stream velocities, R is the gas constant and T_f is the temperature of the fluid, as set by input data. The Maxwellian distribution satisfies in equilibrium condition the Boltzmann equation for mono-atomic hard sphere molecules (Cercignani, 1988).

According to the DSMC method, molecules are moved collisionless for a time step Δt, assigning them new positions according to the formula $x_i^{n+1} = x_i^n + c_i^n \Delta t$ where the subscript i refers to the three coordinate directions and superscript n refers to time discretization. The collisionless motion of the molecules is followed at the end of the time step by the simulation of the intermolecular collisions. The frequency of the intermolecular collisions, hence the collision number to be simulated per time step, is obtained as explained in Bird (1994). Collisional pairs of molecules are chosen at random among the molecules located within a computational cell. Let subscripts 1 and 2 denote a pair of molecules colliding and an asterisk as superscript denote post-collision velocities. According to the isotropic hard-sphere model the post-collision velocities are given by

$$\vec{c}_1^{\,*} = \vec{c}_1 - \vec{n}\left[\vec{n} \cdot \left(\vec{c}_1 - \vec{c}_2\right)\right], \tag{4.85a}$$

$$\vec{c}_2^{\,*} = \vec{c}_2 + \vec{n}\left[\vec{n} \cdot \left(\vec{c}_1 - \vec{c}_2\right)\right], \tag{4.85b}$$

where $\vec{n}$ is a unit vector directed along the line which bisects the directions of $\vec{c}_1^{\,*}$ and $-\vec{c}_1$.

Molecules colliding with the pin surfaces and the surface of the hexagonal canning are assumed to be reflected diffusely, so that the direction of reflection does not preserve memory of the incident direction and is isotropically distributed in the outer half-space defined by a plane tangent to the surface in the point of impact.

Let T_w be the temperature of the surface in the impact point of a molecule. Diffused molecules are assigned velocities obtained by sampling a half-space Maxwellian distribution (Maxwell, 1965) given by

$$f_w = \frac{1}{2\pi (RT_w)^2} e^{-c l^2/(2RT_w)}. \tag{4.86}$$

Following the calculations of a time step, the particles positions and velocities are used to evaluate the distributions of density ρ, mean velocity $\overline{c}$, translational temperature T_{tr}, and pressure p according to the following formulas:

$$\rho = mn, \tag{4.87}$$

$$\overline{c} = \frac{1}{N} \sum_{i=1}^{N} c_i, \tag{4.88}$$

$$T_{tr} = \frac{1}{3\rho R} \sum_{i=1}^{N} (c_i - \overline{c})^2 \tag{4.89}$$

$$p = \rho RT, \tag{4.90}$$

where n is the molecular density and N denotes the total number of sample molecules.

4.3.5.2 Thermophoretic Velocities

The DSMC method is applied to simulate molecular dynamics in the same geometrical configuration as in Fig. 4.11 of Sect. 4.3.2.2. The symmetry surfaces 1 and 2 of the lateral planes are treated as specularly reflecting surfaces while the material surfaces of the pins and of the hexagonal canning, surfaces 3 through 6, are treated as diffusely reflecting surfaces. For the following simulation we assume a voided and chocked channel, bounded by oscillating upper and lower slugs but with zero mean flow of the vapour coolant. The inlet and outlet surfaces are also treated as specularly reflecting surfaces, without net inlet of molecules of the coolant.

The computational domain, with $z_{\max} = 1.0$ m, was divided in a grid of $60 \times 60 \times 60$ computational cells, with reference to the radial, azimuthal, and axial coordinate directions in the polar-axial reference system explained in Sect. 4.3.2.2. In a plane normal to the z-axis the real number of cells belonging to the fluid-dynamic computational domain is 1995, resulting in a total of 119,700 computational cells for the full sector of the bundle. Computations were made with 1,200,000 sample molecules, of sodium at the pressure of 1.50 bar and temperature distribution resulting from the assigned gradients between the pins and the hexagonal canning. We therefore simulated an average of about 10 sample molecules per cell, each one corresponding to 2.1×10^{14} real molecules.

Table 4.2 Discretization of the computational domain and physical conditions

Computational domain: $[x_{min} = -x_{max} < x < x_{max} = y_{max} \cot \vartheta_{min}]$
$[y_{min} = 0 < y < y_{max}]$
Number of computational meshes: 60 × 60 × 60
Number of computational meshes in coolant domain: 1995 (in a (x, y) plane)
119,700 (in full domain)
Number of sample molecules: 1,200,000
Number of real molecules corresponding to 1 sample molecule: $F_N = 2.1 \times 10^{14}$
Mean pressure of sodium coolant: 1.5 bar

The main information concerning the discretization of the computational domain and calculation conditions is summarized in Table 4.2.

The temperature distribution of the outer surface of the cladding in the inner pin is assumed as shown in Fig. 4.36. The temperatures of the other two pin surfaces had the same profile but were 80 K lower. The temperature of the hexagonal canning was set uniform at 1100 K. A typical calculation starts from a uniform temperature distribution, equal to the mean value of the temperatures of the pins, in the simulated vapour coolant formed by sodium molecules. From this initial condition the calculation proceeds for 0.005 s, with a time step of 10^{-7} s, to allow for a thermodynamic quasi-equilibrium to set in the sodium vapour, in which the temperature adjusts to the simulated temperature gradients between the solid surfaces. Following this preliminary phase it is assumed that failure of the cladding occurs in correspondence of the peak temperature of the centre pin. Because of rewetting sequences of a sodium liquid film formed by vapour condensed in the upper cold section of the bundle and flowing into the uppermost part of the heated section of the pins, computed as explained in detail in Bottoni (2004), the highest temperatures in the pins are not in the top heated cell but below that, at cell No. 30 in the axial direction (axial level $z = 0.492$ m).

Failure of the cladding is followed by sputtering of fuel into the computational domain as a cloud of particles with random size distribution within a given interval. We assumed that one sixth of the volume of a 2 mm height of fuel (a volume of 9.42×10^{-9} m^3) is sputtered into particles of radii chosen randomly in the interval 30–100 μm. These assumptions lead to model a disperse cloud of about 40,000 particles, injected into the computational domain in the first discretization mesh adjacent to the surface of the inner pin, at the average distance of 0.135 mm from the surface. The simulated problem time is reset to zero at the time point of the particles injection.

Upon injection into the computational domain, a thermophoretic velocity distribution is applied to the cloud of particles with the method explained in Sect. 4.3.4.4. Besides the thermophoretic forces, the fuel debris experience entrainment in the axial direction due to the creep velocities induced in the coolant sodium vapour by the temperature gradients along the pins, mainly along the hottest central pin.

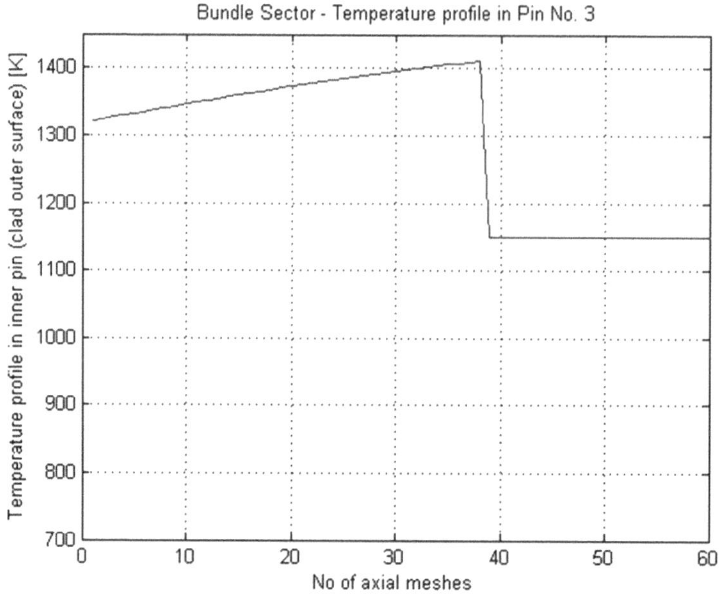

Fig. 4.36 Temperature distribution in Pin No. 3 at the outermost cladding surface at $t = 0$ s

4.3.5.3 Creep Velocity

To assess the importance of creep velocities in the complex geometrical configuration of the bundle is an arduous computational problem, mainly because the mean free path of the molecules of sodium under the conditions investigated is as small as 19 micrometres. A rigorous application of the DSMC method would require discretization of the computational domain in meshes smaller than the mean free path of the simulated molecules and simulation of a larger number of sample particles of vapour coolant per cell. Calculations were made with uniform mesh size in axial direction of 16.7 mm and of 0.135 mm in radial direction. The maximum mesh size in azimuthal direction is 0.019 mm adjacent to the hexagonal canning.

We present in Fig. 4.37 the mean value of the creep velocities computed in the inner cells of the heated section (meshes 1–36) of the bundle. The plot displays a sharp increase of the creep velocities in the uppermost heated section of the bundle.

Because of the rather strong creep velocities of fuel debris in the inner cells of the bundle, particles are entrained upwardly even in case of chocked subchannels with zero mean value of coolant mass transport. The entrainment of the particles in the radial direction is negligible, but in this direction, particles migrate under the effect of thermophoretic forces. The computed distribution of fuel debris in the axial direction is shown in Fig. 4.38 at four different time points, namely $t = 2$ s, $t = 5$ s, $t = 8$ s, and $t = 11$ s.

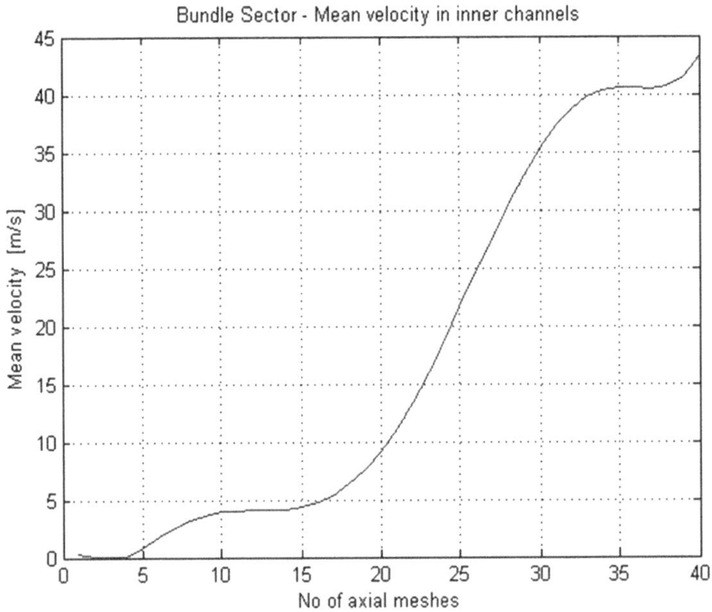

Fig. 4.37 Mean creep velocities computed in the inner meshes of the heated section of the bundle

The most flattered plot in Fig. 4.38, referring to the time point of 11 s since particle injection, shows that a considerable amount of fuel debris has already been depleted by deposition upon the surfaces of the lateral pins No 4 and No 5. In fact the percentages of debris mass deposited at this time point on these two pins are 38.0 and 37.5, respectively, while 24 per cent of the mass still remains in the fluid domain. According to our simulation only a few particles leaves the innermost cells of the bundle and are deposited on the outer hexagonal canning.

Particle concentration in a cross flow plane at axial level $z = 0.542$ m (axial mesh 31), is presented in Fig. 4.39, where the distribution of the number of particles per cell is plotted at 2 s after particle injection. At this time point the cloud of particles is still relatively compact and dense. Because of the spreading of the fuel debris both in the cross flow planes and in the axial direction, the concentration of fuel particles sharply decreases with time. Figure 4.40 gives an example of particle density distribution at $t = 11$ s after particles injection at axial level $z = 0.792$ (axial mesh 48) and shows clearly that the number of particles per cell reduces to just a few ones.

The thickness of a crust forming upon the fuel pins is thus much smaller than it was modelled in previous simulations, without taking into account the impact of thermally induced creep velocities in the coolant vapour. This is shown in Fig. 4.41 which compares the thickness of the crust of accumulated debris at axial level $z = 0.792$ without taking into account the creep velocities (left plot) and modelling the creep velocities (right plot).

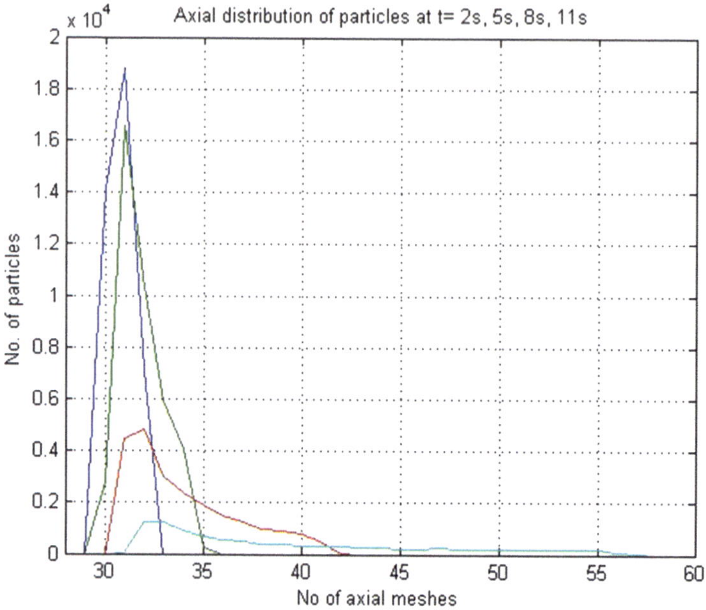

Fig. 4.38 Axial distribution of particles at four time points after particle injection at $t = 0$ s. (As time proceeds distribution flattens and moves to the right)

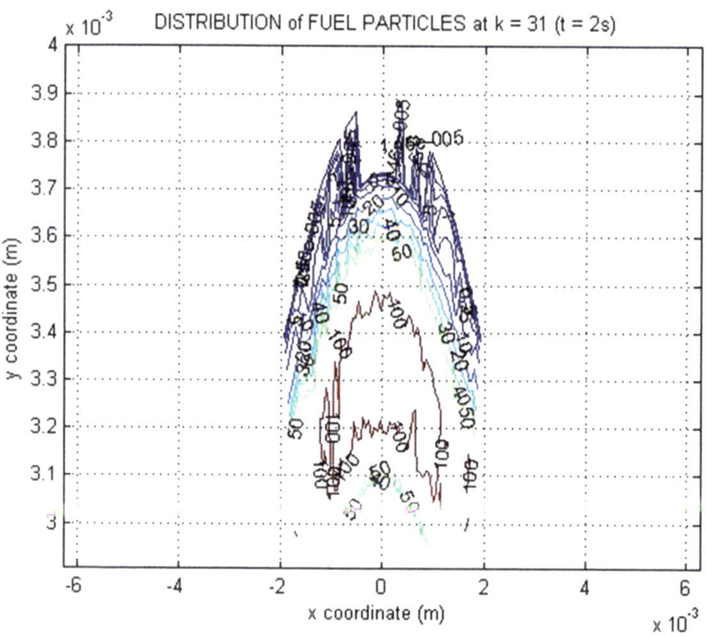

Fig. 4.39 Distribution of fuel particles in the cross flow plane at axial mesh No. 33 ($z = 0.542$ m) at 2 s after particles injection

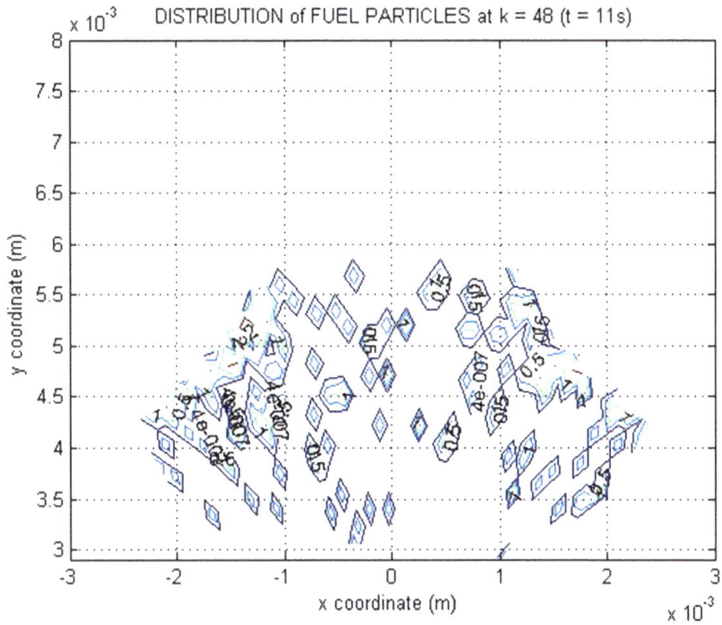

Fig. 4.40 Particle density distribution at mesh $k = 48$ ($z = 0.792$ m) at $t = 11$ s after particle injection

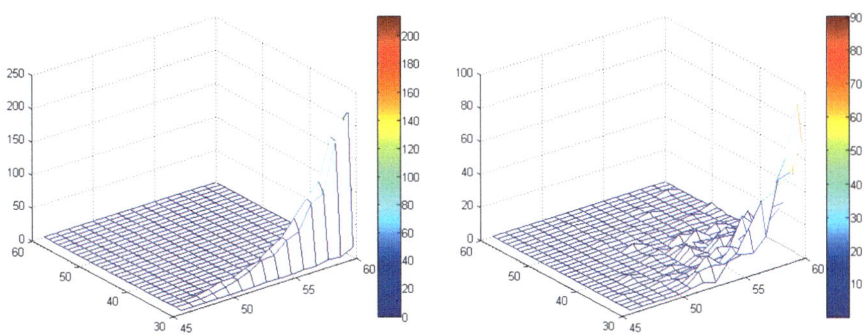

Fig. 4.41 Comparison of thickness of the crust of debris at axial level $z = 0.792$ computed without taking into account the creep velocities (left plot, peak value 200 μm) and modelling the creep velocities (right plot, peak value 90 μm)

The impact of fuel debris accumulation upon the possibility of re-criticality of a reactor under real accident condition could be evaluated only by inserting the computational models into a core-simulation computer program, like the codes of the SIMMER series (Kondo et al., 1999).

Subsequent work with the stand-alone computer program should be aimed at assessing the impact of the entrainment coefficient upon the velocities of the transported particles, eventually taking into account that the fuel debris have not a spherical shape but rather display a fractal structure, and modelling dual collisions and aggregation between the particles.

References

Aberle, J., Brook, A., Peppler, W., Rohrbacher, H., & Schleisiek, K. (1976). *Sodium boiling experiments in a 7-pin bundle under flow run down conditions*. Kernforschungszentrum Karlsruhe. KfK 2378.

Bethe, H. A., & Tait, J. H. (1956). An estimate of the order of magnitude of the vigorous interaction expected should the core of a fast reactor collapse. *UKAEA-RHM, 56*, 113. UKAEA Reactor Group.

Bird, G. A. (1994). *Molecular gas dynamics and the direct simulation of gas flow*. Clarendon Press.

Bottoni, M. (1998a). Monte Carlo calculation of sodium vapour condensation and rewetting simulation in equivalent LMFBR bundle under hypothetical accident conditions. In *Spring Meeting of the Japanese Atomic Power Society, Osaka*. (In Japanese).

Bottoni, M. (1998b). *Monte Carlo calculation of sodium vapour condensation and rewetting simulation in equivalent LMFBR bundle under hypothetical accident conditions*. Annual Meeting of the Energy Soc. Japan. (in Japanese).

Bottoni, M. (2004). Molecular approach to sodium vapour condensation, rewetting and vaporization in LMFBR bundle under hypothetical accident conditions. *Journal of Nuclear Science and Technology, 41*(5), 579–593.

Bottoni, M. (2007). Molecular dynamic simulation of thermal creep in 7-pin LMFBR bundle under hypothetical accident conditions. In *15th International Conference on Nuclear Engineering, (ICONE15), Nagoya, Japan, April 22–26, 2007*.

Bottoni, M., & Struwe, D. (1982). *BLOW-3A: A theoretical model to describe transient two-phase conditions in LMFBR coolant channels*. Kernforschungszentrum Karlsruhe GmbH (Germany). KfK-3317.

Bottoni, M., Dorr, B., Homann, C. (1992). The tree-dimensional transient two-phase flow computer programme BACCHUS-3D/TP. KfK 4760.

Bottoni, M., Bottoni, C., & Scanu, J. (2006). Molecular dynamic simulation of sodium in 7-pin LMFBR bundle under hypothetical accident conditions. In *14th International Conference on Nuclear Engineering, (ICONE14), Miami, USA*.

Cahalan, J. E., Tentner, A. M., & Morris, E. E. (1994). Advanced LMR safety analysis capabilities in the SASSYS-1 and SAS4A computer codes. In *Proceedings of the International Meeting on Advanced Reactor Safety, Pittsburg, Pennsilvania, April 17–21, 1994*.

Cercignani, C. (1988). *The Boltzmann equation and its applications*. Springer-Verlag.

Dunn, F. E., & Wei, T. Y. C. (1998). The role of SASSYS-1 in LMR safety analysis. In *Proceedings of the International Topical Meeting on safety of next generation power reactors, Seattle, Washington, May 1–5, 1988*.

Fulford, G. D. (1964). *Advances in chemical engineering* (The flow of liquids in thin films) (Vol. 5, pp. 151–236).

Gan, X., Bottoni, M., & Zhou, Z. (2002). TH3SL: 水堆安全壳内气体分布模拟计算。(In Chinese). (TH3SL: Simulation of gas distribution inside a water reactor containment). *Atomic Energy Science and Technology, 36*(1), 70–75.

Kaburaki, H., & Yokokawa, M. (1993). Numerical simulation of near continuum gas flow by the direct simulation Monte Carlo method. In *Mathematical methods and supercomputing in nuclear applications, M&C + SNA'93, Karlsruhe* (Vol. 1).

Kapitsa, P. (1941). The study of heat transport in Helium II. *Journal of Physics USSR, 4*.

Kondo, S., Tobita, Y., Morita, K., Brear, D. J., Kamiyama, K., Yamano, H., Fujita, S., Mashek, W., Fischer, E. A., Kiefhaber, E., Buckel, G., Coste, P., & Pigny, S. (1999). Current status and validation of the SIMMER-III LMFR safety analysis code. In *7th International Conference on Nuclear Engineering, Tokyo, Japan, April 19–23, 1999; ICONE-7249*.

Lochart, R. W., & Martinelli, R. C. (1949). Proposed correlation of data for isothermal two-phase, two-componwent flow in pipes. *Chemical Engineering Progress, 45*(1), 39–48.

Maxwell, J. C. (1965). 1879, Phil. Trans. Royal Soc., I (Appendix). Reprinted in *The scientific papers of J. C. Maxwell*. New York: Dover Publications.

McDougall, J. C. (1978). *SABRE-2: A computer program for the calculation of transient three-dimensional flows in rod clusters*. AEEW-R-1104, UKAEA.

Muramatsu, T. (1995). A direct simulation Monte Carlo code THEMIS for the evaluation of nonstationary heat transfer phenomena in LMFR thermohydraulics. In *8th IAHR Working Group meeting on advanced nuclear reactor thermal hydraulics, Nuclear Research Institute Rezplc, Czech Republic*.

Nambu, K. (1980a). Direct simulation scheme derived from the Boltzmann equation. I. Monocomponent gases. *Journal of the Physical Society of Japan, 49*(5), 2042–2049.

Nambu, K. (1980b). Direct simulation scheme derived from the Boltzmann equation. II. Multicomponent gas mixtures. *Journal of the Physical Society of Japan, 49*(5), 2050–2054.

Nambu, K. (1980c). Direct simulation scheme derived from the Boltzmann equation. III. Rough sphere gases. *Journal of the Physical Society of Japan, 49*(5), 2055–2058.

Ninokata, H., & Okano, T. (1990). SABENA: Subassembly boiling evolution numerical analysis. *Nuclear Engineering and Design, 120*(2–3), 349–367.

Nuclear Energy Data Center (1990); (SNA '90), Tokai Mura, Ibaraki ken (Japan); 1. In *International conference on supercomputing in nuclear applications*, 1990.

Reynen, J. et al. (1989). EAC-European Accident Code. In *Proceedings of the 10th LMBWG meeting, Karlsruhe, 1989*.

Schlechtendahl, E. G. (1970). Theoretical investigations on sodium boiling in fast reactors. *Nuclear Science and Engineering, 41*, 99–114.

Schmuck, P., Jacobs G., & Arnecke G., (1977). KADIS: A program to analyze the disassembly phase of hypothetical accidents in LMFBRs. KfK 2497.

Smith L. L., & Sheheen N. N (1980). *SIMMER-II: A computer program for LMFBR disassembly core analysis*. NUREG/CR-0453; LA-7515-M.

Tentner, A. M., et al. (1985). The SAS-4A LMFBR whole core accident analysis code. In *Proceedings of the International Topical Meeting on Fast Reactor Safety, Vol. 2, Knoxville, Tennessee, April 21–25*.

Travis J. R., Spore J. W., & Royl P. (1998). *GASFLOW-II: A three-dimensional finite-volume fluid-dynamics code for calculating the transport, mixing and combustion of flammable gases and aerosols in geometrically complex domains, Vol. 1: Theory and computational model*. Vol. 2: User's manual. NUREG/CR-6570, LA-13357-MS, FZKA-5994.

Wallis, G. B. (1969). *One-dimensional two-phase flow*. McGraw Hill.

Watanabe, T., Kaburaki, H., & Yokokawa, M. (1994). Simulation of two-dimensional Rayleigh-Bénard system using the direct simulation Monte-Carlo method. *Physical Review E, 49*(5), 4060–4064.

Weber, D. P., & Ganapol, B. D. (1976). Multi-field hydrodynamics disassembly calculations with space-time kinetics: FX2/VENUS-III. In *Proceedings of the international meeting on fast reactor safety and related physics, Conf-761001, Chicago, IL* (Vol. III).

Chapter 5
Aerosols and Atmospheric Physics

Contents

The original version of the chapter has been revised. A correction to this chapter can be found at https://doi.org/10.1007/978-3-031-84567-3_8

Text by M. Bottoni, S. Mantovani and G. Zanghirati

M. Bottoni et al., *Molecular Dynamics - Theory and Applications*, Mechanical Engineering Series, https://doi.org/10.1007/978-3-031-84567-3_5

This chapter is divided into six main sections.

The first section illustrates the relevance of thermophoretic calculations in research problems and technical domains like the simulation of non-Newtonian fluids, to name one among many possible applications.

The second section provides a detailed explanation of the concepts and methods used in the simulation of thermophoretic forces acting in a domain subjected to a temperature gradient. In the course of this explanation, details are also given concerning the organization of the computer program THEPHOR, developed and used for the simulation of thermophoretic forces.

The third section refers to the work made by the author for the simulation of thermophoretic forces in the frame of European cooperation, which led to the first application, at least in the European contest, of a successful thermophoretic simulation.

The fourth section explains the development of an advanced version of the computer program THEPHOR, which allows for substantial improvements in computational time, without sacrificing the precision of computational results. The improvement achieved is based on the following concept. When the simulated physical domain is partitioned in small cells, with linear dimensions of some micrometers, as fast molecules have large velocities, time steps used for the simulations in pristine—not dimensionless—quantities need to be as small as 10^{-8}s. This small time step ensures that one sample molecule does not overcome a computational cell during the time step. We relaxed this constraint by scaling down molecular velocities by a chosen factor, typically in the range of 10–100, and increased the momentum carried by the sample molecules. Kinetic temperatures, related to molecular velocities, are also scaled correspondingly. The feasibility of this approach, which allows for larger time steps, hence smaller computational times, has been tested by proving that physical quantities related to equilibrium gas properties are preserved when using the scaled variables.

The fifth section illustrates the open problem of negative thermophoresis with a brief discussion of the simulation conditions that should be further investigated.

The sixth and last section is dedicated to the first attempt to develop a parallel version of the THEPHOR program. It summarizes the preliminary work made with the supervision of Professor Gaetano Zanghirati of the Mathematical Department of the University of Ferrara, Italy.

The chapter is self-consistent, provided it is integrated with Sect. 1.2.1 from Chap. 1.

5.1 The Relevance of Thermophoretic Calculations in the Contest of Molecular Dynamics

One of the most important inheritances of modern science, transmitted over centuries from ancient science, is the atomistic conception of matter, which is usually attributed to Democritus of Abdera (460–370 A.C.) who conceived a chaotic vision of the world. According to Democritus, in the primordial universe, atoms of different species were floating in space with directions and velocities governed by the laws of chaos. From the aggregation of these atoms, the structure of matter was progressively formed into its current state. Modern science recognizes that, from a conceptual viewpoint, given the initial distributions of positions and velocities of the atoms, and knowing the laws that govern their interactions and their aggregations, the full evolution of the universe could be computed.

The eminent scholar Rapaport (Rapaport, 2004), in his book devoted to the art of numerical simulations of molecular dynamics, points out that the possibility of a deterministic solution to the so-called multibody problem had already been noticed by Pierre Simon de Laplace (1749–1827; De Laplace, 1951), who wrote the quotation reported in the Introduction (I).

Great precursors of modern science had foreseen a deterministic approach to the solution of the problems of the evolution of atoms and particles. Nevertheless, progress in many scientific fields relies now upon concepts of probabilistic theories and statistical methods. Mathematical calculation, which in past centuries has been conceived as a collection of interesting but useless intellectual delights, is now of great interest thanks to the advent of electronic computers and the development of the new domain of numerical analysis. In this category of new calculations made possible by the progress of the numerical disciplines also belongs simulations of molecular dynamics based on the laws that govern molecular collisions and aggregations of molecules.

There are however two important constraints inherent in the calculations of molecular dynamics that imply difficulties in two of the most important research fields developed in the past century—the Theory of Relativity and quantum mechanics. According to the Theory of Relativity, information between particles cannot be transferred at a speed greater than the speed of light in a vacuum. In numerical simulation of molecular dynamics, this difficulty is bypassed simply by accepting that information is transmitted instantaneously between particles. In quantum mechanics, the Indetermination Principle implies that it is impossible to precisely determine both the position and velocity of a particle in a given instant. It is rather the probability of finding a particle in a given position that can be computed with the Schroedinger equation. Nevertheless, in simulations of molecular dynamics, one accepts that the position and velocity of a particle can be determined in a given instant, thus violating one of the most important canons of modern science. In some particular cases, however, relativistic concepts can be applied also in the context of simulations of molecular dynamics (Bird, 1994).

The Navier-Stokes equations, which govern the space and time distribution of the velocity components of a fluid and its pressure, were for decades during the Twentieth Century, the paradigm for the analysis of fluid-dynamic problems. In most cases, the numerical solution of the Navier-Stokes equations is obtained by integrating the continuity equation and the momentum conservation equations over volume elements (cells) in which the definition domain of the problem is divided.

From these equations, averaged over the volume elements, an algebraic equation is obtained for every cell. The set of these algebraic equations forms a system in which the mean pressure of the cells is the unknown that must be computed. The solution of the algebraic system determines the pressure distribution in the fluid. The velocity distribution in the fluid cells is obtained by again applying the momentum conservation equations.

If the definition domain of the problem is not regular, but, for instance, limited by curved surfaces, not orthogonal to the coordinate axes, there are irregular cells that are responsible for a numerical error called *numerical diffusion*, which can be minimized but not completely suppressed.

The validity of the Navier-Stokes equations assumes furthermore the continuity of the fluid in which the motion must be simulated and is therefore informed in the case of rarefied fluids, especially rarefied gases. The macroscopic method is infirmed in rarefied gases in which the order of magnitude of the phenomena that must be simulated, for instance, vortices in the fluid, is comparable to or smaller than the mean free path of the molecules.

The rarefaction of a gas is expressed by the Knudsen number, $K_n = \lambda/L = 0.1$, where λ is the mean free path of the molecules and L is the scale of the phenomena which must be simulated—for instance, the size of vortices in the fluid. Very rarefied fluids characterized by large Knudsen numbers, such as, for instance, those that occur in cosmic space, are practically exempt from intermolecular collisions and cannot be simulated with the traditional methods based on the application of the Navier-Stokes equations.

Both of these problems, simulations of fluids in domains bounded by irregular cells and simulations of rarefied gases that cannot be treated with the macroscopic method, are eliminated with the microscopic method of molecular dynamics based on the Direct Simulation Monte Carlo (DSMC) method.

The DSMC method is based on the strict application of the fundamental laws of kinetic and molecular dynamics, as described in detail in the fundamental text of Vincenti and Kruger (Vincenti & Kruger, 1965). Complications arising from the geometric configuration are overcome because cells in the spatial domain are defined with the sole purpose of computing the frequency of intermolecular collisions in the same cells without affecting the accuracy of the calculation. Gases with every degree of rarefaction can be simulated.

The methods of molecular dynamics can be applied to neutral gases and also to ionizes gases. In the first case are simulated intermolecular forces with a short action range, equal or smaller than the mean distance of the molecules; in the second case of the ionized gases long-range forces are simulated and every particle undergoes the action of the full medium, more than the interaction of the particles in the immediate neighboring.

The physical problem is further complicated by the structure of monoatomic and polyatomic molecules. Monoatomic molecules can be modeled as small rigid spheres with the diameter of their atom. More complex molecules, such as oxygen O_2 and nitrogen N_2, the main constituents of air, do not have a spherical structure, and only with some approximation can they be modeled as rigid spheres. Moreover, multiatomic molecules, besides kinetic energy, have two other forms of internal energy—rotational energy, due to the rotation of the molecule upon itself, and vibration energy, due to the vibration of the atoms constituent in the molecule about their mean position. In case of intermolecular collisions, these forms of energy can be exchanged from one molecule to another one and energy can also be exchanged within the same molecule. The analytical expressions that govern the exchange of energy among molecules are complex and can be used in numerical simulations only relying on models of statistical physics.

Simulation of thermophoretic forces acting on a particle requires therefore the simulation of a complex of physical phenomena, the most important among which are:

(i) Impact of the molecules against the walls kept at a fixed temperature and reflection of the molecules from the walls.
(ii) Computation of the frequency of intermolecular collisions.
(iii) Intermolecular exchange of energy in the collisions.
(iv) Computation of the trajectories of the molecules before the collisions and of the new trajectories after the collisions.
(v) Simulation of the impact of the molecules against a particle and of their reflection from the surface of the particle.

In our computations, praxis calculations were made partitioning a two-dimensional spatial domain in square cells typically of dimensions $10 \times 10\ \mu m$ and simulating the progress of time in intervals Δt from t_n to $t_{n+1}=t_n + \Delta t$ ($n = 1, 2, 3, \ldots$). One of the fundamental canons in the fluid-dynamic simulations states that during a time step Δt, one fluid particle, or molecule, can drift from one cell to an adjacent cell but not to more distant cells. In other words, the information carried on by a molecule or by a particle, relative to its energy or its momentum, cannot overcome a cell. Because some molecules have a speed of the order of magnitude of 1000 m/s, the canon stated above implies that the time step should be smaller than 10_{-8} s. With this time step, the simulation of 1 s of a time problem requires 100 million time steps, which may give an idea of the computational time necessary in case one can not afford a parallel computer.

In atmospheric physics, which we will consider later on, the behavior of aerosol particles suspended in the atmosphere has an important role in the formation and growth of condensation nuclei within clouds and therefore in the formation of hydrometeors. Diffusion and transport of aerosol particles in the atmosphere are strictly connected with the behavior of the atmosphere in its complex and are not only due to convective or turbulent motions but also to the impact of thermophoretic forces due to temperature gradients.

Experimental investigations are long and expensive and now can be integrated by numerical simulations of thermophoretic forces acting on aerosol particles in conditions of microgravity in the higher regions of the atmosphere. Numerical investigations have a dual purpose: i) Verification of experimental results with calculations and, vice versa, verification of the consistency of calculations with good experimental results; and ii) application to conditions not yet investigated experimentally, as soon as a computational program has been verified and has reached a good degree of confidence. For instance, it can be applied to simulations of interactions between molecules and cosmic dust particles in rarefied conditions like those occurring in interplanetary space.

5.2 Concepts and Methods Used for the Simulation of Thermophoretic Force—Documentation of Code THEPHOR

5.2.1 The Early Approach

This chapter documents the computer program THEPHOR, which describes the displacement due to thermophoresis of a tiny particle in a gas domain subjected to a temperature gradient. This code has been developed starting from a variant of the THEMIS code. The precursor of this family of codes was written at the nuclear research center of O-Arai, Mito, in Japan (Muramatsu, 1995). The first version of the code modeled thermal transport from a flux of liquid sodium between two parallel infinite plates. The code was intended to study the problem of thermal striping, i.e., thermal oscillations of the material forming the walls, along with their impact upon the resistance of the material to fatigue. The flux of liquid sodium and the thermal transport to the walls were modeled with the statistical methods of molecular dynamics, based on the *Direct Simulation Monte Carlo (DSMC)* method.

The above-referenced version built the basis for a three-dimensional, one-component, two-phase flow version describing the flux of sodium vapor in the presence of liquid, in a seven-pin assembly of a fast nuclear reactor under hypothetical accident conditions. This code version (Bottoni, 2004), referred to as THEMIS/3D.1, allows simulation of sodium vapor condensation upon the cladding of the pins and the formation of a liquid film of sodium, which eventually may evaporate again. The purpose of this version was to predict the temperature distribution on the surface of the bars.

Another three-dimensional version, referred to as TH3SL (Gan et al., 2002, describing three components, namely water vapor, air, and hydrogen, was built to investigate the hydrogen concentration in the mixture of these components in a typical cell, in which the inner volume of the containment of a pressurized water reactor can be partitioned for thermal and fluid-dynamic analysis. This code version also describes water vapor condensation upon the inner walls of the containment.

Modules of the above code versions were used and, whenever necessary, modified to build up the code version documented in this chapter. At present, the code consists of a driving subroutine called THEPHOR and 51 subroutines or functions, out of which only the most important ones have been inserted in the flow chart of Fig. 5.2. The code consists of two main parts.

In the first part, which consists of the initialization of the code, the subroutine GEOMZE defines a two-dimensional computational domain, consisting of a square or rectangular array delimited from two upper and lower material surfaces, as well as by two lateral ideal symmetry surfaces. The upper and lower surfaces are kept at constant but different temperatures during the calculation and therefore give the temperature gradient from which the forces due to thermophoresis arise. By convention, the lower plate is kept at the lowest temperature. These two material surfaces reflect air molecules inside the domain as *diffusive reflecting* surfaces. The lateral symmetry surfaces reflect air molecules as *specularly reflecting* surfaces. The implication of these definitions upon the scattering of air molecules will be illustrated in the following. In the center of the computational domain is set an aerosol particle, assumed to be spherical and treated as a three-dimensional body with a given size and physical properties. The initial velocity of the aerosol particle is set to 0. A chosen number of sample air molecules with statistically determined initial positions and velocities are initialized (Subroutine INITF) inside the computational domain.

The second part of the code describes the molecular dynamics of the air sample molecules and the collisions among them with the walls and the aerosol particles. The difference of the momenta transferred from the air molecules to the aerosol particle in the upper half sphere of the particle and the lower half sphere determines the displacement of the particle due to thermophoresis. The problem time is split into time steps that should be sufficiently small to prevent fast air molecules from crossing a computational mesh during the time step. Because the size of the computation meshes is chosen to be of the order of magnitude of about ten times the particle radius, the time step results in an order of magnitude of 10^{-8} seconds. Because the aerosol particle moves during the calculation, at every time step the geometrical configuration of the computational domain must be redefined (Subroutine GEOMTR).

In Sects. 5.2.1, 5.2.2, 5.2.3, 5.2.4, 5.2.5, 5.2.6, and 5.2.7 we document the most important physical models used. In Sect. 5.2.8, the structure of the computational program is outlined and a compact synopsis of the code is given.

5.2.2 Preliminaries and Basic Concepts

The theoretical and computational backgrounds of the chapter are based on analytical and numerical information given previously that should be kept in mind for understanding the topics treated in the following. We recall in this section these basic preliminary concepts:

(i) Basic theoretical background is given in Sect. 1.2.1 (Chap. 1).
(ii) The Lennard-Jones potential is used to determine the equivalent cross-section of interacting molecules. This topic is treated in Sect. 1.3.3 (Chap. 1).

5.2.3 *Interaction Between Air Molecules and Aerosol Particles*

5.2.3.1 Kinetic Energy Incident upon and Reflected from the Aerosol Particle

We compute the kinetic energy transferred from an air molecule to the aerosol particle upon impact, the kinetic energy reflected, associated with a molecule emerging from the particle, along with their difference, which is the net energy transferred from the air molecule to the particle.

The kinetic energy of a sample molecule incident with velocity c_i upon the particle is

$$e_{kin,i} = \frac{1}{2} m c_i^2. \tag{5.1}$$

Let F_N be the number of real molecules represented by a sample air molecule. The real kinetic energy associated with a sample molecule is

$$E_{kin,i} = F_N e_{kin,i} = \frac{1}{2} m F_N c_i^2. \tag{5.2}$$

Similarly, the kinetic energy associated with a molecule reflected from the particle (subscript r) with velocity c_r is

$$E_{kin,r} = F_N e_{kin,r} = \frac{1}{2} m F_N c_r^2. \tag{5.3}$$

In the code, the components u_r, $w_r$of the velocity $c_r$are computed by sampling a Maxwellian velocity distribution corresponding to the temperature of the impact point.

5.2.3.2 Momentum Exchange

Let $\vec{C}_p$and M_p denote velocity and mass of the aerosol particle, and indices 1 and 2 refer to the velocities of an air molecule and the particle before and after a collision. The momentum conservation implies that

$$m\vec{c}_1 + M_p\vec{C}_{p1} = m\vec{c}_2 + M_p\vec{C}_{p2} \tag{5.4}$$

From Eq. 5.4, the post-collision velocity components of the particle are derived as

$$W_{p2} = W_{p1} + \frac{m}{M_p}(w_1 - w_2), \tag{5.5}$$

$$U_{p2} = U_{p1} + \frac{m}{M_p}(u_1 - u_2). \tag{5.6}$$

5.2.4 *Wall Reflection*

As explained earlier, the computational domain is bounded by two material surfaces (top and bottom) treated as diffusive reflecting surfaces, and by two symmetry surfaces (lateral surfaces) treated as specularly reflecting surfaces. Besides that, there is the surface of the particle that also behaves as a diffusive reflecting surface. In the following part of this section, we explain the kind of interaction an air molecule has with these surfaces.

5.2.4.1 Diffuse Reflection from Material Walls

5.2.4.1.1 Diffuse Reflection from the Upper and Lower Material Surfaces

We must first of all determine the coordinates of the collision point of an air molecule with the upper material surface located at $z = z_{\max}$. Let $P_1(x_1, z_1)$ be the point of a molecule before the collision, and let $\widetilde{P}_2(\widetilde{x}_2, \widetilde{z}_2)$ be the point that would be attained by the molecule if there were no collisions, which means beyond the surface. The equation of the trajectory of the particle determined by these two points is

$$\frac{x - x_1}{\widetilde{x}_2 - x_1} = \frac{z - z_1}{\widetilde{z}_2 - z_1}. \tag{5.7}$$

Letting

$$m_z = \frac{z_{\max} - z_1}{\widetilde{z}_2 - z_1}, \tag{5.8}$$

one derives from Eq. 5.7 that the x and z coordinates of the collision point $P_c(x_c, z_c)$ are given by

$$\begin{aligned} x_c &= x_1 + m_z(\widetilde{x}_2 - x_1), \\ z_c &= z_1 + m_z(\widetilde{z}_2 - z_1) = z_{\max}. \end{aligned} \tag{5.9}$$

We now consider a local coordinate system (x_ℓ, z_ℓ) with origin in P_c, defined as follows: Axis x_ℓ is horizontal and oriented in the opposite direction of the x axis; and axis z_ℓ is normal to the wall and oriented inwardly. We also consider a local polar system (ρ, ϑ) with the same origin in P_c and defined as follows: Given a vector $\rho =$ OP, ϑ is the angle between the vector OP and the z_ℓ axis.

The velocity components in the local coordinate system after reflection are computed by

$$\begin{aligned} w_{2\ell} &= c'_m \sqrt{-\ln R_1} \\ u_{2\ell} &= c'_m r \cos \zeta \end{aligned} \tag{5.10}$$

where

$$\begin{aligned} \zeta &= R_2 2\pi \\ r &= \sqrt{-\ln R_3} \end{aligned} \tag{5.11}$$

and R_1, R_2, R_3 are uniform deviates. The variable c'_m represents the most probable thermal velocity corresponding to the Maxwellian distribution of molecular velocities at the temperature of the surface. Notice that $w_{2\ell}$ is always positive, as it should be, while the component $u_{2\ell}$ may be positive or negative. The transformation from the local system to the main (x,z) system yields $u_2 = -u_{2\ell}$; $w_2 = -w_{2\ell}$ The component w_2 is always negative.

The velocity of the particle after reflection is then

$$V_2^{n+1} = \sqrt{u_2^2 + w_2^2}. \tag{5.12}$$

The particle travels, after collision with the surface, a distance

$$R_{P2} = \Delta t_{ac} V_2^{n+1}, \tag{5.13}$$

where Δt_{ac} is the fraction of the time interval after the collision. The orientation of the vector R_{P2} is determined, in the local polar system, randomly by

$$\vartheta = R_1 \pi. \tag{5.14}$$

The coordinates of the particle after diffuse reflection in the local system are

$$\begin{aligned} x_{2\ell} &= R_{P2} \cos \vartheta, \\ z_{2\ell} &= R_{P2} \sin \vartheta. \end{aligned} \tag{5.15}$$

The coordinates of the particle after diffuse reflection in the main coordinate system are

$$
\begin{aligned}
x_2 &= x_c - x_{2\ell} \\
z_2 &= z_c - z_{2\ell}.
\end{aligned}
\tag{5.16}
$$

Similar considerations hold for the collision of an air molecule with the lower material surface.

5.2.4.1.2 Diffuse Reflection from the Surface of the Aerosol Particle

Before computing the components of the position and velocity of a molecule after collision with the surface of the aerosol particle, it is necessary to compute the coordinates of the collision point. Let $P_1(x_1, z_1)$ be the position of a molecule at time point t_n, and $P_2(x_2, z_2)$ its position at time t_{n+1}. If, during the time interval $\Delta t = t_{n+1} - t_n$, the molecule has hit the surface of the aerosol particle at a point $P_c(x_c, z_c)$ before being deflected, the coordinates of the point P_c must be computed first and then the coordinates of P_2 can be computed.

Let $\widetilde{P}_2\,(\widetilde{x}_2, \widetilde{z}_2)$ be the point where the molecule would arrive at time point t_{n+1}, if it were not deflected by the surface of the aerosol particle. The equation of the line $L = P_1\widetilde{P}_2$ is

$$
\frac{x - x_1}{\widetilde{x}_2 - x_1} = \frac{z - z_1}{\widetilde{z}_2 - z_1}.
\tag{5.17}
$$

or

$$
z = z_1 + (x - x_1)\frac{\widetilde{z}_2 - z_1}{\widetilde{x}_2 - x_1} = z_1 + (x - x_1)tg\alpha,
\tag{5.18}
$$

with

$$
tg\alpha = \frac{\widetilde{z}_2 - z_1}{\widetilde{x}_2 - x_1}.
\tag{5.19}
$$

Eq. 5.18 can be written in the normal form

$$
ax + bz + c = 0,
\tag{5.20}
$$

with

$$
\begin{aligned}
a &= tg\alpha \\
b &= -1 \\
c &= z_1 - x_1 tg\alpha
\end{aligned}
\tag{5.21}
$$

The distance $d(L)$ of the line $L = P_1\widetilde{P}_2$ from the center (x_p, z_p) of the aerosol particle is

$$d(L) = \left| \frac{ax_p + bz_p + c}{\sqrt{a^2 + b^2}} \right| = \left| \frac{tg\alpha(x_p - x_1) + z_1 - z_p}{\sqrt{1 + tg^2\alpha}} \right|. \tag{5.22}$$

The equation of the circle C, a cross-section of the aerosol particle of radius R with the (x,z) plane is

$$(x - x_p)^2 + (z - z_p)^2 = R^2. \tag{5.23}$$

Combining Eq. 5.18 with Eq. 4.288, one obtains

$$(x - x_p)^2 + (z_1 + (x - x_1)tg\alpha - z_p)^2 = R^2. \tag{5.24}$$

Eq. 5.24 can be written

$$\overline{\alpha}x^2 + \beta z^2 + \gamma = 0, \tag{5.25}$$

with

$$\overline{\alpha} = 1 + tg^2\alpha, \tag{5.26a}$$

$$\beta = -x_p - x_1 tg^2\alpha + z_1 tg\alpha - z_p tg\alpha, \tag{5.26b}$$

$$\gamma = x_p^2 + z_p^2 + z_1^2 + x_1^2 tg^2\alpha - 2x_1 z_1 tg\alpha - 2z_1 z_p + 2x_1 z_p tg\alpha - R^2. \tag{5.26c}$$

Eq. 5.25 has the two solutions

$$X_1 = \frac{-\beta + \sqrt{\beta^2 - \overline{\alpha}\gamma}}{\overline{\alpha}}, \tag{5.27a}$$

$$X_2 = \frac{-\beta - \sqrt{\beta^2 - \overline{\alpha}\gamma}}{\overline{\alpha}}, \tag{5.27b}$$

which correspond to the z coordinates, from Eq. 5.18,

$$Z_1 = z_1 + (X_1 - x_1)tg\alpha, \tag{5.28a}$$

$$Z_2 = z_1 + (X_2 - x_1)tg\alpha. \tag{5.28b}$$

Thus, the two intersections between line L and the circle C are the points A and B with coordinates A $= (X_1, Z_1)$ and B $= (X_2, Z_2)$. To decide which one of these two points is the collision point P_c, we compute the distances

$$d_1 = AP_1 + A\widetilde{P}_2 = \sqrt{(x_1 - X_1)^2 + (z_1 - Z_1)^2} + \sqrt{(\widetilde{x}_2 - X_1)^2 + (\widetilde{z}_2 - Z_1)^2}, \tag{5.29a}$$

$$d_2 = BP_1 + B\widetilde{P}_2 = \sqrt{(x_1 - X_2)^2 + (z_1 - Z_2)^2} + \sqrt{(\widetilde{x}_2 - X_2)^2 + (\widetilde{z}_2 - Z_2)^2}. \tag{5.29b}$$

If $d_1 < d_2$, A is the collision point ($P_c = \text{A}$), otherwise B is the collision point ($P_c = \text{B}$).

The coordinates (x_c, z_c) of the collision point P_c are computed in the subroutine INTCIR.

Next, one has to compute the position and velocity of the molecule after diffuse reflection from the surface of the aerosol particle. We consider first a coordinate system $(\widetilde{x}, \widetilde{z})$ with origin in the collision point P_c, defined as follows: the $\widetilde{x}$ axis is tangent to the circle C and directed in a clockwise direction, and the $\widetilde{z}$ axis is directed along the line from the center of the circle C to the collision point. A second coordinate system referred to as local (index ℓ) system (x_ℓ, z_ℓ), is defined as follows: Axis x_ℓ is coincident with the $\widetilde{z}$ axis; and the z_ℓ axis is coincident with the $\widetilde{x}$ axis but oriented in the opposite direction. Finally, the coordinate system (x, y, z), with $y = y_p = c$, is the main Cartesian coordinate system in which the main code calculations are made.

Let

$$\alpha = ar\cos\left[(x_c - x_p)/R\right] \tag{5.30}$$

be the angle between the x_ℓ and the x axis.

The components of the molecule velocity and the components of its position vector after diffuse reflection are computed in the following steps:

Step 1: Calculation of velocity components in the local system (x_ℓ, z_ℓ).

The velocity components in the local coordinate system are computed by

$$\begin{aligned} u_{2\ell} &= c'_m \sqrt{-\ln R_1} \\ w_{2\ell} &= c'_m r \cos\zeta \end{aligned} \tag{5.31}$$

where

$$\begin{aligned} \zeta &= R_2 2\pi \\ r &= \sqrt{-\ln R_3} \end{aligned} \tag{5.32}$$

and R_1, R_2, R_3 are uniform deviates. The variable c'_m represents the most probable velocity corresponding to the temperature of the surface of the aerosol particle at the point of impact. Notice that $u_{2\ell}$ is always positive, as it should be, while $w_{2\ell}$ may be positive or negative.

Step 2: Calculation of velocity components in the main system (x,z).

A coordinate transformation between the local system (x_ℓ, z_ℓ) and the main Cartesian system (x,z) yields:

$$u_2 = u_{2\ell}\cos\alpha - w_{2\ell}\sin\alpha$$
$$w_2 = u_{2\ell}\sin\alpha + w_{2\ell}\cos\alpha \tag{5.33}$$

The velocity of the particle after reflection is then

$$V_2^{n+1} = \sqrt{u_2^2 + w_2^2}. \tag{5.34}$$

Step 3: Calculation of the components of the position of a particle in the $(\widetilde{x}, \widetilde{z})$ system

Let $P^{n+1} = P_2(x_2, z_2)$ be the position of a particle after diffuse reflection. The length of the vector $R_{P2} = P_cP^{n+1}$ is computed by

$$R_{P2} = P_cP^{n+1} = V_2^{n+1}\Delta t\left(P_c\widetilde{P}_2/P_1\widetilde{P}_2\right). \tag{5.35}$$

$\Delta t\left(P_c\widetilde{P}_2/P_1\widetilde{P}_2\right)$ is the time interval, or fraction of the time step, after the collision. Hence

$$\widetilde{x} = R_{P2}\cos\Phi,$$
$$\widetilde{z} = R_{P2}\sin\Phi. \tag{5.36}$$

with $\Phi = R_1\pi$, where R_1 is a random deviation.

Step 4: Calculation of the components of the position of the particle in the local coordinate system (x_ℓ, z_ℓ).

From the definition of coordinate systems, it is evident that

$$x_{2\ell} = \widetilde{y},$$
$$z_{2\ell} = -\widetilde{x}, \tag{5.37}$$

Step 5: Calculation of the components of the position of the particle in the main coordinate system (x, z).

The coordinate transformation between the local system (x_ℓ, z_ℓ) and the main system (x, z) yields

$$x_2 = x_c + x_{2\ell}\cos\alpha - y_2\sin\alpha,$$
$$z_2 = z_c + x_{2\ell}\sin\alpha + z_2\cos\alpha. \tag{5.38}$$

5.2.4.2 Specular Reflection from the Symmetry Planes

The computational domain is bounded by two vertical symmetry planes, i.e., inclined by an angle $\alpha = (\pi/2)$ with respect to the x axis. These symmetry planes are treated as perfectly reflecting surfaces. If the point $P_1(x_1, z_1)$ defines the position of a particle before reflection from a symmetry plane [$P_1(x_1, z_1)$ being beyond the boundary plane], the position of the particle after reflection, $P_2(x_2, z_2)$, is given by

$$\begin{aligned} x_2 &= x_1 \cos 2\alpha + z_1 \sin 2\alpha, \\ z_2 &= x_1 \sin 2\alpha - z_1 \cos 2\alpha. \end{aligned} \tag{5.39}$$

Likewise, for the velocity components, one has

$$\begin{aligned} u_2 &= u_1 \cos 2\alpha + w_1 \sin 2\alpha, \\ w_2 &= u_1 \sin 2\alpha - w_1 \cos 2\alpha, \end{aligned} \tag{5.40}$$

For $\alpha = (\pi/2)$, the above formulae are simplified to

$$\begin{aligned} x_2 &= x_1 \cos 2\alpha = -x_1 \\ z_2 &= -z_1 \cos 2\alpha. = z_1 \end{aligned} \tag{5.41}$$

and

$$\begin{aligned} u_2 &= u_1 \cos 2\alpha = -u_1 \\ w_2 &= -w_1 \cos 2\alpha = w_1 \end{aligned} \tag{5.42}$$

5.2.5 *Collisionless Free Motion and Collision Model*

In the numerical simulation, the principle of uncoupling is applied by splitting the calculation in the collisionless motion of the molecules, followed by the simulation of collision events among them.

5.2.5.1 Collisionless Free Motion

Following the collisionless motion of the molecules, their position coordinates are updated by

$$\begin{aligned} x^{n+1} &= x^n + u\Delta t, \\ y^{n+1} &= y^n + v\Delta t, \\ z^{n+1} &= z^n + w\Delta t. \end{aligned} \tag{5.43}$$

5.2.5.2 Collision Frequency

Let us consider a cell of the computational domain with a volume V_c containing a number $N = nV_c/F_N$ of sample molecules, where F_N is the number of real molecules represented by every sample molecule. The collision number between all molecule pairs in the cell is

$$C_N = \frac{N^2}{2} F_N \frac{\sigma_T c_r \Delta t}{V_c} \tag{5.44}$$

where σ_T is the collision cross section and c_r is the relative velocity of the molecules. The maximum theoretical collision number $C_{N,\max}$ is given by the same formula (5.44), taking the maximum theoretical value $(\sigma_T c_r)_{\max}$ of the product between the collision cross-section and the relative velocity of the molecules. In numerical computations, the maximum number $C_{N,\max}$ of collisions is considered and, among this theoretical number of collision events, only accept those that satisfy the condition

$$R_1 \le \frac{C_N}{C_{N,\max}} = \frac{\sigma_T c_r}{(\sigma_T c_r)_{\max}} \tag{5.45}$$

where R_1 is a random deviation.

5.2.5.3 Collision Dynamics

Within a computational cell, collision pairs are chosen randomly among the sample molecules. Let indices 1 and 2 refer to the two molecules of a randomly chosen couple, and let $\vec{c}_r = \vec{c}_1 - \vec{c}_2$ be the relative velocity before the collision. For molecules of equal mass, the center of mass velocity is $\vec{c}_m = \left(\vec{c}_1 + \vec{c}_2\right)/2$. Let us denote with a star (*) as a superscript the post-collision velocities. For the collision dynamics, we use the isotropic *hard-sphere* collision model, in which the post-collision relative velocity has the same magnitude as the pre-collision relative velocity–$\left|\vec{c}_r^{\,*}\right| = \left|\vec{c}_r\right|$. In an isotropic-scattering model, the components of the post-collision relative velocity can be computed by multiplying a unit vector generated with the Marsaglia technique (4), with the module of the relative velocity:

$$
\begin{aligned}
u_r^* &= S c_r \cos\beta, \\
v_r^* &= S c_r \sin\beta, \\
w_r^* &= C_s c_r,
\end{aligned}
\tag{5.46}
$$

using two uniform deviates R_1 and R_2 and computing

$$
\begin{aligned}
\beta &= 2\pi R_1, \\
C_s &= 1 - 2R_2, \\
S &= \sqrt{1 - C_s^2}.
\end{aligned}
\tag{5.47}
$$

Because the components u_m and w_m of the center of mass velocity are known from the pre-collision velocity components, the components of the post-collision velocities of the two molecules are given by

$$
\begin{aligned}
u_1^* &= u_m + \frac{|c_r^*|}{2}, \\
u_2^* &= u_m - \frac{|c_r^*|}{2},
\end{aligned}
\tag{5.48}
$$

$$
\begin{aligned}
w_1^* &= w_m + \frac{|c_r^*|}{2}, \\
w_2^* &= w_m - \frac{|c_r^*|}{2}.
\end{aligned}
\tag{5.49}
$$

5.2.6 *Temperature Distribution in the Aerosol Particle*

We consider a particle as a sphere of a given radius R with known physical properties and compute the transient, three-dimensional temperature distribution in the sphere. We start from an initial distribution depending on the temperature gradient in the air adjacent to the particle and compute the transient temperature distribution, taking into account the kinetic energy transferred to the particle from the incident air molecules and taken from the particle from the reflected molecules.

We refer to a spherical coordinate system (r, ϑ, Φ) with origin in the center of a particle (Point O). ϑ is the colatitude, and Φ the azimuth. The radial, latitudinal, and azimuthal coordinates are partitioned in the computer program in NSER, NSTH, and NSPH meshes, respectively. Concerning a computational cell shown in Fig. 5.1, we have the following relationships:

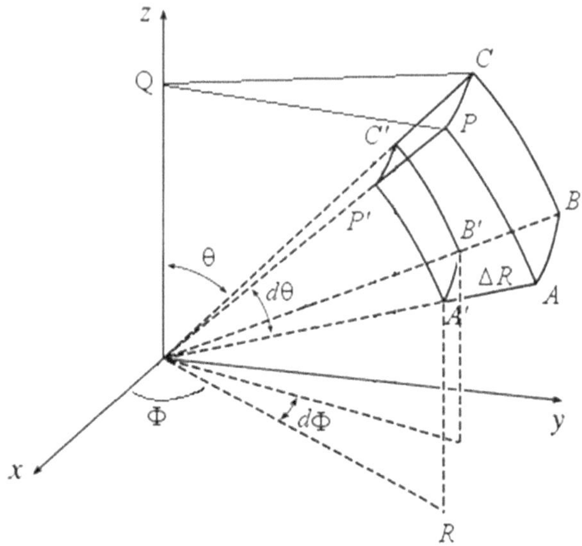

Fig. 5.1 Spherical coordinate system used for calculation of temperature distribution in the aerosol particle

$$AA' = CC' = \Delta R, \tag{5.50}$$

$$OQ = R \cos\vartheta, \tag{5.51}$$

$$PQ = R \sin\vartheta, \tag{5.52}$$

$$PC = PQ\ d\Phi = R \sin\vartheta\ d\Phi, \tag{5.53}$$

$$PA = CB = R\ d\vartheta, \tag{5.54}$$

$$dA(r) = PA \bullet PC = R^2 \sin\vartheta d\vartheta d\Phi, \tag{5.55}$$

$$dA(\vartheta) = PC \bullet P\,P' = R\ \Delta R\ \sin\vartheta d\Phi, \tag{5.56}$$

$$dA(\Phi) = AA' \bullet AP = R\ \Delta R\ d\vartheta. \tag{5.57}$$

In these equations $\Delta R = R / NSER$, $d\vartheta = \pi / NSTH$ and $d\Phi = 2\pi/NSPH$.

In Eqs. 5.55 through 5.57, the symbols dA(r), dA(ϑ), and dA(Φ) denote the areas of the surfaces of the outermost computational cell normal to the respective coordinate directions. Similar formulas hold for the inner computational cells, just replacing the radius *R* with the coordinate *r*.

The volume element in the spherical coordinate system is given by

$$dV = r^2 \sin\vartheta d\vartheta d\Phi dr. \tag{5.58}$$

The heat diffusion equation in the spherical system is

$$\rho c_p \frac{\partial T}{\partial t} = \nabla \cdot (\lambda \nabla T) + Q - \widetilde{h} S (T_S - T_a) \tag{5.59}$$

The subscripts in the last term of Eq. 5.59 refer to the surface of the particle and air, respectively. $\widetilde{h}\left[W/m^5K\right]$ is a heat-transfer coefficient per unit volume. The last term of Eq. 5.59 is considered only for the outermost cells of the discretized spherical volume.

Integrating Eq. 5.59 over the volume V_c of a computational cell, one obtains

$$\int_{Vc} \rho c_p \frac{\partial T}{\partial t} dV = \int_{Vc} \nabla \cdot (\lambda \nabla T) dV + \int_{Vc} Q dV - \int_{Vc} \widetilde{h} S (T_S - T_a) dV \tag{5.60}$$

Applying the Gauss theorem, one replaces the volume integral of the divergence of the vector $\lambda \nabla T$ with the flux of the vector through the bounding surfaces, thus obtaining

$$\int_{Vc} \rho c_p \frac{\partial T}{\partial t} dV = \int_{S} \lambda \nabla T \bullet \vec{n}\, dS + \int_{Vc} Q dV - \int_{Vc} \widetilde{h} S (T_S - T_a) dV \tag{5.61}$$

where $\vec{n}$ is the outwards-oriented unit vector, normal to the surface.

Equation 5.61 is discretized in the computer program by replacing the integrals over the volume of a cell with the mean volumes of the integrand over the cell, multiplied by the cell volume. Thus one obtains

$$\langle \rho c_p \frac{\partial T}{\partial t} \rangle V_c = \sum_{\alpha} \int_{S\alpha} \lambda \nabla T dS + \langle Q \rangle V_c - \langle \widetilde{h} S \rangle (T_S - T_a) V_c \tag{5.62}$$

The symbols $\langle\rangle$ in the above equation denote mean values over the cell volume. The first term at the right-hand side of Eq. 5.62 is the sum of the fluxes over the surfaces S_α bounding the control cell. In the following, the surfaces of the cell normal to the radial coordinate r are denoted with S_1 (innermost) and S_2 (outermost), respectively. The surfaces normal to the latitudinal coordinate are denoted by S_3 and S_4, and those normal to the azimuthal coordinate are denoted by S_5 and S_6.

We denote with indices (i, j, k) the center node of a cell, with $(i \pm 1/2, j, k)$ the center node of the surfaces normal to the r coordinate direction and likewise $(i, j \pm 1/2, k)$ and $(i, j, k \pm 1/2)$ for the other coordinate directions. Thus, for instance, the notation S_1 is equivalent to $S_{i-1/2, j, k}$. Using this notation, superscript n, to refer to the time point t_n, and referring to mesh lengths Δr, Δm and Δp in the radial, meridian, and azimuthal directions respectively, Eq. 5.62 can be discretized as follows:

$$
\begin{aligned}
&\left(\rho c_p V_c\right)_{i,j,k} \frac{1}{\Delta t}\left(T_{i,j,k}^{n+1}-T_{i,j,k}^{n}\right)= \\
&-\left[\lambda_{1-1/2,j,k} \frac{1}{\Delta r}\left(T_{i,j,k}^{n+1}-T_{i-1,j,k}^{n+1}\right)\right] S_1+\left[\lambda_{1+1/2,j,k} \frac{1}{\Delta r}\left(T_{i+1,j,k}^{n+1}-T_{i,j,k}^{n+1}\right)\right] S_2 \\
&-\left[\lambda_{1,j-1/2,k} \frac{1}{\Delta m}\left(T_{i,j,k}^{n+1}-T_{i,j-1,k}^{n+1}\right)\right] S_3+\left[\lambda_{1,j+1/2,k} \frac{1}{\Delta m}\left(T_{i,j+1,k}^{n+1}-T_{i,j,k}^{n+1}\right)\right] S_4 \\
&-\left[\lambda_{1,j,k-1/2} \frac{1}{\Delta p}\left(T_{i,j,k}^{n+1}-T_{i,j,k-1}^{n+1}\right)\right] S_5+\left[\lambda_{1,j,k+1/2} \frac{1}{\Delta p}\left(T_{i,j,k+1}^{n+1}-T_{i,j,k}^{n+1}\right)\right] S_6 \\
&+\left(Q V_c\right)_{i,j,k}+h S_2\left(T_{S,i+1/2,j,k}^{n}-T_a^{n}\right).
\end{aligned} \tag{5.63}
$$

In Eq. 5.63, we set $h=\widetilde{h}\mathrm{V}_{\mathrm{c}}$. The unknown in Eq. 5.63 is the temperature $T_{i,j,k}^{n+1}$. Solving the equation with respect to $T_{i,j,k}^{n+1}$, one obtains

$$
\begin{aligned}
& T_{i,j,k}^{n+1}\left[\left(\rho c_p V_c\right)_{i,j,k} \frac{1}{\Delta t}+\lambda_{i-1/2,j,k} \frac{S_1}{\Delta r}+\lambda_{i+1/2,j,k} \frac{S_2}{\Delta r}+\lambda_{i,j-1/2,k} \frac{S_3}{\Delta m}\right. \\
& \qquad \left.+\lambda_{i,j+1/2,k} \frac{S_4}{\Delta m}+\lambda_{i,j,k-1/2} \frac{S_5}{\Delta p}+\lambda_{i,j,k+1/2} \frac{S_6}{\Delta r}+h S_2\right] \\
& -T_{i-1,j,k}^{n+1}\left[\lambda_{i-1/2,j,k} \frac{S_1}{\Delta r}\right]-T_{i+1,j,k}^{n+1}\left[\lambda_{i+1/2,j,k} \frac{S_2}{\Delta r}\right] \\
& -T_{i,j-1,k}^{n+1}\left[\lambda_{i,j-1/2,k} \frac{S_3}{\Delta m}\right]-T_{i,j+1,k}^{n+1}\left[\lambda_{i,j+1/2,k} \frac{S_4}{\Delta m}\right] \\
& -T_{i,j,k-1}^{n+1}\left[\lambda_{i,j,k-1/2} \frac{S_5}{\Delta p}\right]-T_{i,j,k+1}^{n+1}\left[\lambda_{i,j,k+1/2} \frac{S_6}{\Delta p}\right] \\
& =T_{i,j,k}^{n}\left(\rho c_p V_c\right)_{i,j,k} \frac{1}{\Delta t}+\left(Q V_c\right)_{i,j,k}+h S_2\left(T_{S,i+1/2,j,k}^{n}-T_a^{n}\right).
\end{aligned} \tag{5.64}
$$

Equation 5.64 can be written in the compact form:

$$
\begin{aligned}
& \mathrm{A} T_{i,j,k}^{n+1}-\mathrm{B} T_{i-1,j,k}^{n+1}-\mathrm{C} T_{i+1,j,k}^{n+1}-\mathrm{D} T_{i,j-1,k}^{n+1}-\mathrm{E} T_{i,j+1,k}^{n+1}-\mathrm{ATM}\, T_{i,j,k-1}^{n+1} \\
& \qquad -\mathrm{ATP} T_{i,j,k+1}^{n+1}=G_{i,j,k}^{n}
\end{aligned} \tag{5.65}
$$

with the following definition of the coefficients and of the right-hand side:

$$
\begin{aligned}
\mathrm{A} & =\left(\rho c_p V_c\right)_{i,j,k} \frac{1}{\Delta t}+\lambda_{i-1/2,j,k} \frac{S_1}{\Delta r}+\lambda_{i+1/2,j,k} \frac{S_2}{\Delta r}+\lambda_{i,j-1/2,k} \frac{S_3}{\Delta m} \\
& \quad +\lambda_{i,j+1/2,k} \frac{S_4}{\Delta m}+\lambda_{i,j,k-1/2} \frac{S_5}{\Delta p}+\lambda_{i,j,k+1/2} \frac{S_6}{\Delta p} \\
& =\left(\rho c_p V_c\right)_{i,j,k} \frac{1}{\Delta t}+\mathrm{B}+\mathrm{C}+\mathrm{D}+\mathrm{E}+\mathrm{ATM}+\mathrm{ATP},
\end{aligned} \tag{5.66}
$$

$$\mathrm{B} = \lambda_{i-1/2,j,k} \frac{S_1}{\Delta r}, \tag{5.67}$$

$$\mathrm{C} = \lambda_{i+1/2,j,k} \frac{S_2}{\Delta r}, \tag{5.68}$$

$$\mathrm{D} = \lambda_{i,j-1/2,k} \frac{S_3}{\Delta m}, \tag{5.69}$$

$$\mathrm{E} = \lambda_{i,j+1/2,k} \frac{S_4}{\Delta m}, \tag{5.70}$$

$$\mathrm{ATM} = \lambda_{i,j,k-1/2} \frac{S_5}{\Delta p}, \tag{5.71}$$

$$\mathrm{ATP} = \lambda_{i,j,k+1/2} \frac{S_6}{\Delta p}, \tag{5.72}$$

$$G^n_{i,j,k} = T^n_{i,j,k}\left(\rho c_p V_c\right)_{i,j,k} \frac{1}{\Delta t} + \left(QV_c\right)_{i,j,k} + hS_2\left(T^n_{S,i+1/2,j,k} - T^n_a\right). \tag{5.73}$$

In the computer program, the system of Eq. 5.65, written for all computational cells in the particle, can be solved with an iterative Jacobi or Gauss scheme, or by applying the ADI (Alternating Direction Implicit) method. Improvement of the numerical solution, by minimizing the residuals, could be obtained by applying a direct method as, for instance, the Cholesky matrix decomposition technique.

5.2.7 *Convergence Criterion*

In thermophoresis calculations, because of the displacement of the particle, the calculation never reaches a steady state. Starting from a random initial distribution of kinetic temperatures of the air molecules, in the course of the calculation, the molecular velocities and kinetic temperatures reach a pretty stable situation. It is therefore useful to monitor a steady-state criterion that gives information about the stability of the molecular configuration attained. A convergence criterion has been introduced in the following way. Let $V = (u, v = 0, w)$ be the two-dimensional field of the sample air molecules simulated in the full computational domain, and let us consider the two N-dimensional vectors $U_i^{(1)}$ and $W_i^{(1)}$ $(i = 1, 2, 3, \ldots N)$, formed with the velocity components of the molecules along the x and z coordinates at time t_1, and the vectors $U_i^{(2)}$ and $W_i^{(2)}$ referring to time point $t_2 = t_1 + n\,\Delta t$ ($n = 10$, for instance). Starting from the beginning of the calculation, these vectors, sampled every n time steps, are averaged over the number of samplings made. Using the averaged vectors, we then compute the Euclidian distance between the couple of vectors $U_i^{(1)}$ and $U_i^{(2)}$ and the couple $W_i^{(1)}$ and $W_i^{(2)}$. For any two vectors a_i and b_i, the Euclidian distance is defined by $\mathrm{d(a,b)} = \left(\sum_i (a_i - b_i)^2\right)^{1/2}$. When the maximum Euclidian distances of the two couples of vectors $U_i^{(1)}$, $U_i^{(2)}$, and $W_i^{(1)}$, $W_i^{(2)}$, becomes smaller than a given threshold, a quasi-steady-state configuration is reached. Experience showed that a suitable value of this threshold is given by S = $N_{mol} \times 10^{-4}$, where N_{mol} is the number of sample molecules.

5.2.8 *Structure of the Computational Code*

The code THEPHOR is written in FORTRAN 77 and consists of about 5000 statements (including comments), divided into 51 subroutines or functions. The to-date stage of development of the code is summarized in the simplified flow chart of Fig. 5.2:

The functions of the main subroutines shown in the block diagram are summarized in the same sequence in which the subroutines are called:

Subroutine THEPHOR: Main program.

Subroutine RINPUT: Reads and writes the input file.

Subroutine OUTRES: In case of a continuation run, the restart file OUTRES is read, and files previously used for storing data for post-processing are reopened.

Subroutine GEOMZE: This subroutine is called at time t = 0, at the beginning of a new calculation, and it performs the following tasks:

(i) Computation of geometrical data (mesh lengths, cell areas, volumes, etc.) of the computational domain and the aerosol particle. As far as the computational meshes are concerned, two cases are possible: (a) meshes with dx = dz < $R_p/2$, where R_p is the particle radius; (b) meshes with ds = dz > $2R_p$. The first case, with very small meshes, may be used for a small domain around the particle, and the second case with meshes larger than the particle is used for simulating a larger domain. A flag defined for all meshes and called INDDOM gives information about the relative position of meshes and particles:

INDDOM(i, j, k) = 0, if the mesh (i, j, k) is completely inside the particle;
INDDOM(i, j, k) = 1, if the mesh (i, j, k) overlaps the particle;
INDDOM(i, j, k) = 2, if the mesh (i, j, k) is outside of the particle.

(ii) The sphere simulating the aerosol particle is partitioned in meshes along the radial coordinate, the meridians, and the parallels. In the three spherical coordinate directions, NSER, NSTH, and NSPH meshes, respectively, are defined.

Subroutine INITF: Initializes the flow of the molecules, i.e., initial positions and velocity components in the full computational domain. The most probable velocity of a molecule is computed taking into account the z coordinate of the molecule and the temperature gradient assigned between the top and bottom boundaries of the domain. This Subroutine INITF also initializes the temperature distribution in the aerosol particle.

Before subroutine SCREEN, positions and velocities of the particle are updated.

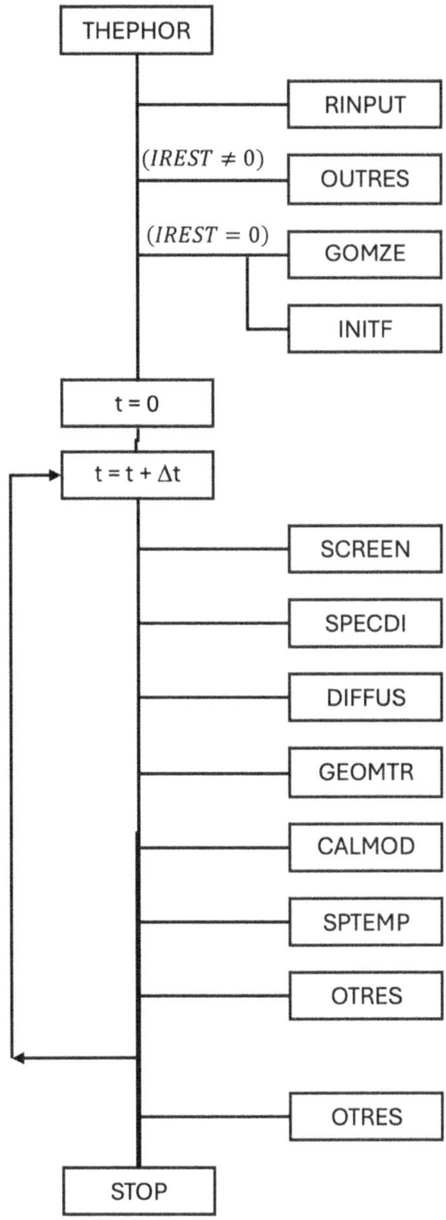

Fig. 5.2 Simplified block diagram of the THEPHOR main routines

Subroutine SCREEN: Given the Cartesian coordinates of a molecule, the subroutine returns an Index = 0 in case the molecule lies in the computational domain. Returns an index different from zero, suitably coded, in case the particle is outside of the computational domain.

Subroutine SPECDI: Computes diffusive reflection from upper or lower boundaries and specular reflection from all other surfaces bounding the computational domain.

Subroutine DIFFUS: Computes velocity components and coordinates of a molecule after diffuse reflection from the surface of the aerosol particle. Knowing the position of the molecule before the impact with the particle and knowing its velocity, the coordinates of the impact point are computed.

Subroutine GEOMTR: This subroutine is called at every time step during the transient calculation and performs the same task described for the subroutine GEOMZE, but it takes into account the displacement of the aerosol particle from its initial position.

Subroutine COLMOD: For every couple of collision molecules, with known pre-collision velocities, compute the post-collision velocities. The equations used are given in Sect. 1.3.

Subroutine SPTEMP: Computes the temperature distribution in the aerosol particle.

Subroutine OUTRES: The restart file is updated.

In case the final simulation time has not been attained, the time step is incremented, and a new loop starts from the subroutine SCREEN.

Before the end of the calculation, the restart file OUTRES is again updated.

5.3 Simulation of Thermophoretic Forces in the Frame of the European Cooperation

5.3.1 Experimental Conditions

As mentioned earlier, the physics of aerosol particles suspended in the atmosphere plays an important role in the formation and growth of liquid and ice nuclei within clouds and therefore in the formation and accretion of hydrometeors. Diffusion of aerosol particles is due not only to fluid-dynamic transport and turbulent behavior of the atmosphere but also to their migration under the impact of thermophoretic forces due to temperature gradients in the atmosphere. These forces, which act against the temperature gradients, may have an important impact on the scavenging of aerosol particles in the clouds.

Experimental investigations of thermophoretic forces and velocities of aerosol particles under microgravity conditions, both in drop tower facilities and in parabolic flights, are being carried out under cooperation between the Institute of Atmospheric Sciences and Climate (ISAC) of the National Research Council (CNR) of Italy, the Department of Physics of the University of Ferrara, Italy, and the Microgravity Research Centre of the Université Libre de Bruxelles, Belgium.

Experimental investigations were paralleled by a computational effort aimed at simulating thermophoretic forces and migration of aerosol particles in zero-gravity conditions. The purpose of these investigations was to develop a computational tool to simulate thermophoretic forces in conditions not yet investigated experimentally—for instance, at intermediate Knudsen numbers. The computational approach followed in the numerical investigations is based upon the application of the Direct Simulation Monte Carlo (DSMC) method in a bidimensional (x, z) computational domain with the aerosol particle initially set at the center of the domain and the thermophoretic forces acting in the negative z direction, i.e., with the temperature gradient directed upward.

The basic numerical techniques have been pioneered by the theoretical work of Nambu (1980), among others, and consist of applying the principle of uncoupling by splitting the calculation of molecular dynamics in the collisionless motion of the molecules, followed by the simulation of collision events among randomly chosen couples of molecules within a computational cell. For about 30 years, a number of theoretical works have been performed by Kaburaki and his coworkers at the Japan Atomic Energy Institute (JAERI) of Tokaimura (Kaburaki et al., 1990), aiming at establishing the basis and validation of the DSMC in classical problems of fluid dynamics. Among these are studies of vapor expansion in a vacuum space (Kaburaki & Yokokawa, 1993), of a jet into a vacuum, simulations of two-dimensional Rayleigh-Benard systems (Watanabe et al., 1994), and of other fundamental problems of thermal and fluid dynamics.

Simulations of thermophoretic forces in microgravity conditions were made with the computer code THEPHOR, as explained in detail in the previous Sect. 5.2.

5.3.2 *Numerical Simulations*

Numerical simulations of thermophoretic forces and the relative displacements were made for an aerosol particle of 0.5-μm radius initially set with zero velocity in the center of the domain, bounded by an upper and lower material plate and by lateral geometrical surfaces. In the simulations, the vertical two-dimensional cross-section (x, z) of this domain was considered. Table 5.1 summarizes the physical characteristics of the particle and data about the computational domain and boundary conditions. These are data of a reference calculation made with a simulated computational domain, narrow in the horizontal x direction and, consequently, with the minimum number of sample molecules and computational effort.

Other calculations, made in wider domains and with a larger number of sample molecules were made to assess possible qualitative and quantitative differences with the reference calculation. We performed numerical simulations from the time point $t = 0$, in which the aerosol particle was initialized in the midpoint of the computational domain.

Instantaneous values of the horizontal U and vertical W velocity components of the aerosol particle were sampled at intervals of 500 time steps. Figure 5.3 shows the recorded data for these two velocity components. The vertical velocity component, always negative, as expected, reaches a value of -11×10^{-5} (m/s). The horizontal velocity component oscillates around the zero value.

Table 5.1 Physical characteristics of the aerosol particle and data about computational domain, boundary conditions, and numerical simulation

Physical properties of the aerosol particle:
Radius [μm] 0.5
Density [kg/m3] 890.0
Specific heat [J/kg K] 2140.0
Thermal conductivity [W/m K] 0.15
Physical conditions of air in the definition domain:
Mean free path λ [m] 4.86×10^{-8}
Knudsen number Kn 2.5×10^{-4}
Temperature ($t = 0$) [K] 315.30
Density ρ [kg/m3] 1.2
Dynamic viscosity [kg/(m s)] 1.45×10^{-5}
Kinematic viscosity [m2/s] 1.13×10^{-5}
Boundary conditions and temperature gradient:
Temperature of top surface [K] 335.05
Temperature of bottom surface [K] 295.55
ΔT [K] 39.5
$\nabla T = \Delta T/(zmax - zmin)$ [K/m] 5.642×10^{-3}
$\nabla T/T$ [m − 1] 17.89
Domain (x, z), discretization, and calculation conditions:
$xmin-$ $xmax$ [μm] 3490–3510
$zmin-$ $zmax$ [μm] 0–7000
Δx [μm] 10
Δz [μm] 10
Total Nr. of cells: 1400
Nr. of simulated molecules: 21,000
The mean number of molecules per cell: 15
Time step [s]: 10^{-8}

The large amplitudes of the oscillations suggest that the number of simulated air molecules should be increased to reduce statistical scatter. A test calculation made in the same geometry, with 50 molecules per cell instead of 15, effectively displayed reduced scattering and damped amplitudes of the oscillations, but could not be carried on for a sufficiently large problem time due to the increased computational effort, which is proportional to the collision frequency and hence to the square of the number of sample molecules per cell.

The simulation of a narrow domain, however, has a spurious and unphysical effect on the numerical results. The mean value of the velocity of the descending molecules, with a negative z-velocity component after diffuse reflection from the upper material plate at the uppermost temperature, is an absolute value larger than the mean velocity component of the ascending molecules, reflected from the lower plate at the lowermost temperature. The consequence is that when statistical fluctuations are to some extent smoothed out, the mean flux of descending molecules becomes lower than the mean flux of ascending molecules. Therefore, the number of impacts of descending molecules with the aerosol particle also becomes smaller than the number of impacts due to ascending molecules. This effect compensates for the larger momentum associated with the descending faster molecules and acts against the thermophoretic force. The effect is however only of a numerical nature, and in

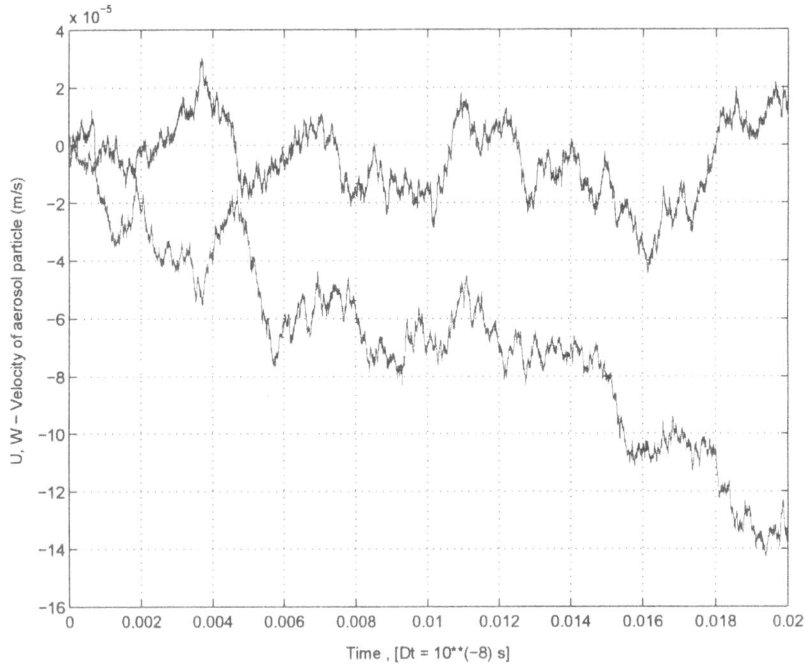

Fig. 5.3 Horizontal (U) and vertical (W) velocity components of the aerosol particle, versus time

effect disappears when a wider computational domain is simulated, because in that case there is a stronger mixing of molecules in the region of the aerosol particle with molecules coming from adjacent regions. When the narrow domain of 20 μm is replaced by a wider domain of 100 μm, with 10 meshes in the horizontal direction instead of only 2, the aforementioned numerical effect almost vanishes. The expense is however a roughly five-fold computational time, which in turn requires more powerful processors.

For reasons investigated by the author but never clarified, the tripartite cooperation mentioned above between the Institute of Atmospheric Sciences and Climate (ISAC) of the National Research Council (CNR) of Italy, the Department of Physics of the University of Ferrara, Italy, where the author was provisionally appointed, and the Microgravity Research Centre of the Universite Libre de Bruxelles, Belgium, was terminated in the course of the year 2005.

Meanwhile, it had become apparent that the approach followed thus far required very large computational times on a personal computer, of the order of magnitude of weeks for every tenth of a second of problem time, which made it impracticable for production runs. These difficulties motivated the search for an improvement of computational efficiency, which was attained after several attempts in the course of the year 2008, as documented in the following section.

5.4 Advanced Version of the Code THEPHOR—Thermophoretic Calculations with Pristine and Scaled Variables

The field of numerical simulations of particle dynamics, including the particular topic of thermophoretic problems, relies upon the massive use of parallel computers (Choi et al., 1996). However, even in this case, the key point of the simulations is the requirement that a sample gas molecule does not cross a computational cell completely within a one time step. In other words, information carried by a molecule cannot jump over a mesh cell during a time step. This fundamental paradigm of numerical computations implies a constraint on every single processor, even in the case of massive parallel calculations. The accuracy of calculations requires that the mesh size be at most one order of magnitude larger than the size of the particle subjected to thermophoretic forces. In most cases, we used a mesh size of 10 μm. Because fast gas molecules in the tail of a Maxwellian velocity distribution may have a speed larger than 1000 m/s, the time step should not exceed 10^{-8}s, which is a very strong constraint for the speed of a numerical simulation.

Starting from the background provided by the computer program THEPHOR, as documented in the previous section (Sect. 5.3), we investigated the possibility of overcoming the strong constraint of a very small time step. The improvements of this code version to the previous version are described in the following.

In the following calculations, we used the physical quantities summarized in Table 5.2:

5.4.1 *Equilibrium Gas Properties*

Let us consider the thermal velocity vector $\vec{c}'$ represented in velocity space (Fig. 5.4) with reference to a spherical coordinate system with azimuthal angle Φ and colatitude ϑ. A surface element in this coordinate system is

Table 5.2 Summary of physical quantities used in the numerical simulations

Air gas constant	$R = 0.28827x10^3[J/kg \cdot K]$
Molecular weight of air [averaged over the constituents O_2 (21%) and N_2 (79%)]	$0.21 \times 32 + 0.79 \times 28 = 28.84$ [kg/kmole]
Mass of air molecule	$m = 4{,}78902 \times 10^{-26}$ [kg]
Diameter of air molecule	$d = 4.15\ x\,10^{-10}$ [m]
Air density [at $T = 273$ K, $p = 10^5[N/m^2]$]	$\rho_0 = mn_0 = 4.78902x10^{-26}\ x2.68699\ x10^{25} = 1.286804\ [kg/m^3]$
Geometrical cross section of air molecules	$\sigma_T = \pi d^2 = 54.106x10^{-20}\ [m^2]$

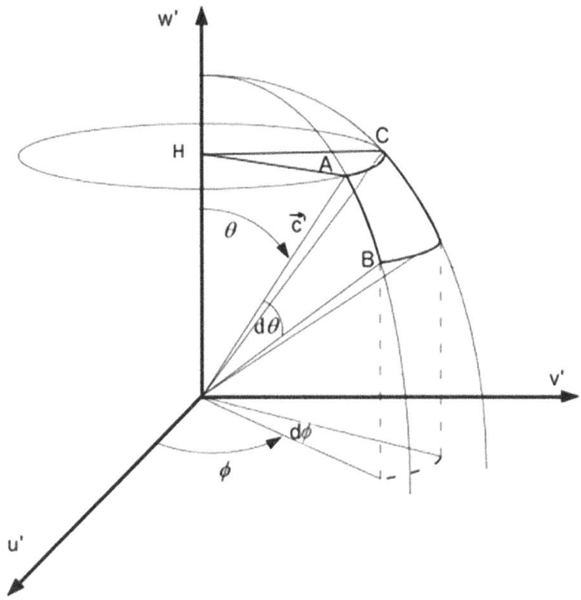

Fig. 5.4 Representation of thermal velocity vector $\vec{c}'$ in a spherical coordinate system

given by $dS = c'^2 \sin \vartheta \, d\vartheta \, d\Phi$ and a volume element is given by $d\bar{c}' = dS\, dc' = c'^2 \sin \vartheta \, d\vartheta \, d\Phi \, dc'$. The fraction of molecules with speed in the range $(\vec{c}', \vec{c}' + d\vec{c}')$, crossing the elementary surface dS = AB · AC, represented in the figure, is $dn = nf_0 dc'$, where

$$f_0 = \frac{\beta^3}{\pi^{3/2}} e^{-\beta^2 c'^2} \tag{5.74}$$

is the equilibrium distribution function with $\beta = 1/\sqrt{2RT}$. Hence we obtain

$$\frac{dn}{n} = f_0 d\bar{c}' = \frac{\beta^3}{\pi^{3/2}} c'^2 e^{-\beta^2 c'^2} \sin \vartheta \, d\vartheta \, d\Phi \, dc' \tag{5.75}$$

The fraction of molecules with speed in the range $(\bar{c}', \bar{c}' + d\bar{c}')$, irrespective of their direction, hence crossing any point of the sphere of radius $\vec{c}'$, is obtained by integrating Eq. 5.75 over the full solid angle. Using $\int_0^\pi \sin \vartheta \, d\vartheta = 2$ and $\int_0^{2\pi} d\Phi = 2\pi$, the result is

$$\frac{dn}{n} = \frac{\beta^3}{\pi^{3/2}} \int_0^\pi \int_0^{2\pi} c'^2 e^{-\beta^2 c'^2} \sin \vartheta \, d\vartheta \, d\Phi \, dc' = \frac{4\beta^3}{\sqrt{\pi}} c'^2 e^{-\beta^2 c'^2} dc' = f_{c'} \, dc' \tag{5.76}$$

with the definition of the distribution function

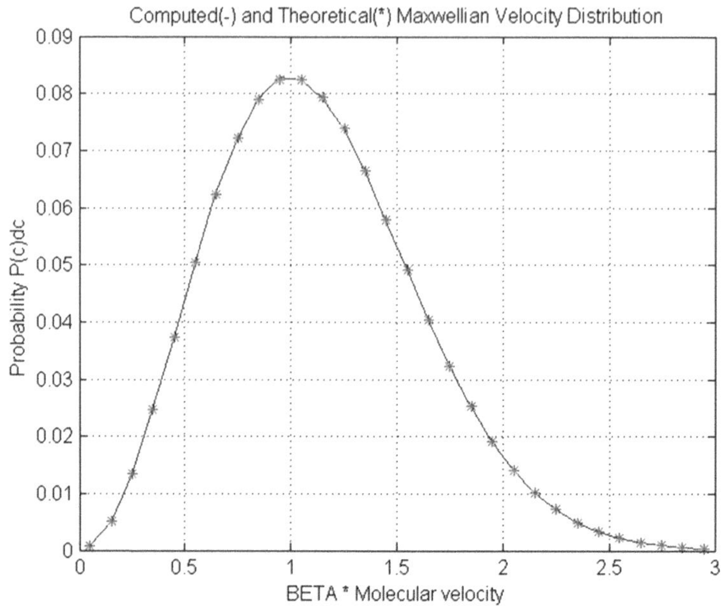

Fig. 5.5 Comparison between theoretical Maxwellian velocity distribution and the distribution obtained by sampling the distribution function

$$f_{c'} = \frac{4\beta^3}{\sqrt{\pi}} c'^2 e^{-\beta^2 c'^2} \tag{5.77}$$

The integral of the distribution function satisfies the normalization condition:

$$\int_0^\infty f_{c'} dc' = \int_0^\infty \frac{4\beta^3}{\sqrt{\pi}} c'^2 e^{-\beta^2 c'^2} dc' = \frac{4}{\sqrt{\pi}} \int_0^\infty \widehat{c}^2 \exp\left(-\widehat{c}^2\right) d\widehat{c} = 1 \tag{5.78}$$

having made the change of variable $\widehat{c} = \beta c'$.

The probability $P(\widehat{c}_1, \widehat{c}_2)$ of having a sample molecule with velocity in the range $(\widehat{c}_1, \widehat{c}_2)$ is given by

$$P(\widehat{c}_1, \widehat{c}_2) = \frac{4}{\sqrt{\pi}} \int_{\widehat{c}_1/\beta}^{\widehat{c}_2/\beta} \widehat{c}^2 \exp\left(-\widehat{c}^2\right) d\widehat{c}. \tag{5.79}$$

This theoretical probability distribution, computed in the range $\widehat{c}$ (0, 3), subdivided in 30 intervals, is shown in Fig. 5.5 together with the distribution obtained sampling function 5.82 with the exception-rejection method (see 1.8 in Chap. 1). The maximum of the distribution corresponds to the most probable velocity,

$$c'_m = \frac{1}{\beta} = \sqrt{2RT} = \sqrt{\frac{2kT}{m}}. \tag{5.80}$$

The average molecular speed is given by

$$\overline{c}' = \int_0^\infty c' f_{c'}\, dc' = \frac{4\beta^3}{\sqrt{\pi}} \int_0^\infty c'^3 e^{-\beta^2 c'^2}\, dc' = \frac{2}{\sqrt{\pi}\beta} = \frac{2}{\sqrt{\pi}} c'_m = \sqrt{\frac{8RT}{\pi}}, \tag{5.81}$$

having used $\int_0^\infty c'^3 e^{-\beta^2 c'^2}\, dc' = 1/(2\beta^4)$.

The root mean square thermal velocity is defined by

$$c'_s = \sqrt{\overline{c'^2}} = \sqrt{\int_0^\infty c'^2 f_{c'}\, dc'} = \sqrt{\frac{4\beta^3}{\sqrt{\pi}} \int_0^\infty c'^4 e^{-\beta^2 c'^2}\, dc'} = \sqrt{3RT} = \sqrt{\frac{3}{2}\frac{1}{\beta}}$$
$$= \sqrt{\frac{3}{2}} c'_m = \sqrt{\frac{3\pi}{8}} \overline{c}'. \tag{5.82}$$

In 5.82, we used $\int_0^\infty c'^4 e^{-\beta^2 c'^2}\, dc' = \frac{3\sqrt{\pi}}{8\beta^5}$.

The mean relative velocity among molecules is

$$\overline{c}_r = \sqrt{2}\overline{c}', \tag{5.83}$$

5.4.2 New Approach of the THEPHOR-3 Code Version

We introduced a scale factor SF, which in the computational praxis has been varied in the range 10–100, and used it in the code to scale the following quantities:

(i) Velocities of gas molecules and, consequently, their momenta are scaled (divided) by SF.

(ii) Temperatures of all material surfaces and the initial temperature distribution in the aerosol particle are scaled by SF^2. This is consistent with the relationship between the translational kinetic temperature of the molecules and their velocities.

(iii) The most probable thermal velocity of a gas molecule, obtained by sampling a Maxwellian distribution at the scaled temperature, is automatically scaled.

(iv) When gas molecules impinge upon the aerosol particle their momenta are scaled back (multiplied by SF) to ensure that the proper momenta are imparted to the particle.

(v) The temperature distribution of the aerosol particle is recalculated every time step using the scaled temperature field.

The most important calculation that is not affected by the scaling is the calculation of the collision frequency between molecules.

A key point in the numerical treatment is that *every sample* molecule in the simulation is representative of a given number (FN) of real molecules. Depending on the computing capabilities available to the numerical analysts, FN may be as large as 10^{18} or as low as 10^6.

Taking into account item *iv* above, we complement the definition of "sample molecule as a molecule that is representative of FN real molecules and carries a momentum SF times larger than in unscaled numerical simulations."

Based on these considerations, we distinguish between *pristine* variables or physical quantities, which are not scaled in the numerical simulation, and *scaled* variables.

Pristine variables are:

- All geometrical quantities defining the domain of calculation and the size of the particle.
- Mass of the molecules.
- Mass and density of the aerosol particle.
- Mean free path of the molecules and Knudsen number of the gas.
- The number FN of real molecules represented by a sample molecule.

Scaled variables are:

- Temperatures of all material surfaces and the particle.
- Velocities of the molecules.
- The most probable thermal velocity of a gas molecule obtained by sampling the Maxwellian velocity distribution.

Implicitly become also scaled the momenta of the molecules, which however are scaled back in case of impact against the aerosol particle.

5.4.3 Code Verification with Equilibrium Properties

A necessary, though insufficient, condition for the correctness of a molecular dynamic code is the capability to reproduce as closely as possible the theoretical values of some important physical quantities. Among these are: (i) the most probable thermal velocity c'_m; (ii) the average molecular speed $\overline{c}'$; (iii) the root mean square thermal velocity c'_s; and (iv) the mean relative velocity among molecules $\overline{c}_r$.

In the geometrical conditions of the reference calculation explained in the next section, we used scaled variables to produce Maxwellian velocity distributions of the air molecules. From these Maxwellian distributions, we computed the above-mentioned physical quantities in the fictitious scaled dimensions. Upon scaling back to physical units, we obtained the values shown in Table 5.3, where they are compared with the theoretical values. In this case, the number of sample molecules, 10^6, was much larger than in the numerical applications explained later.

The largest difference between theoretical and computed values is that of the mean relative velocity between molecules $\overline{c}_r$. This depends on the fact that, out of the $10^{12}/2$ possible couples of the sample molecules, we considered only 10^7 randomly chosen couples for averaging.

Table 5.3 Computed and theoretical values of most probable thermal velocity c'_m, the average molecular speed $\bar{c}'$, of the root mean square thermal velocity c'_s, and of the mean relative velocity among molecules $\bar{c}_r$

	Computed	Theoretical	Equation
Most probable thermal velocity c'_m [m/s]	426.37	426.80	(Eq. 5.80)
Average molecular speed $\bar{c}'$[m/s]	481.05	481.59	(Eq. 5.81)
Root mean square thermal velocity c'_s[m/s]	522.19	522.72	(Eq. 5.82)
Mean relative velocity among molecules $\bar{c}_r$[m/s]	679.42	681.07	(Eq. 5.83)

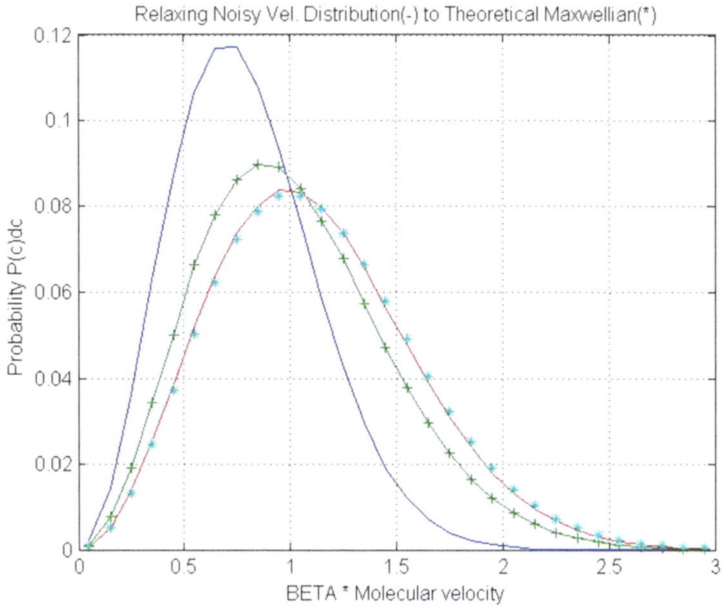

Fig. 5.6 Relaxation of a noisy velocity distribution of molecules (uppermost at left), computed at time $t_0 = 0$, to a theoretical Maxwellian velocity distribution (*). Intermediate plot (+) refers to time $t_1 = 0.1s$; third computed plot (almost superimposed to the Maxwellian distribution) refers to time $t_3 = 0.3s$

A second check of the approach's correctness concerned verification of the relaxation of a nonequilibrium velocity distribution of molecular velocities to a Maxwellian distribution when the gas undergoes molecular collisions in a domain with appropriate temperature boundary conditions. Again, we simulated the geometrical conditions of the reference calculation with a domain bounded by upper and lower material plates, 7000 μm apart, at the temperatures of 335.05 K and 295.55 K, respectively. In this domain, we introduced a nonequilibrium velocity distribution sampling a Maxwellian and superimposing noise. The initial velocity distribution of the molecules is shown by the uppermost left plot of Fig. 5.6. Starting from this distribution, it is expected that the intermolecular collisions let the distribution relax

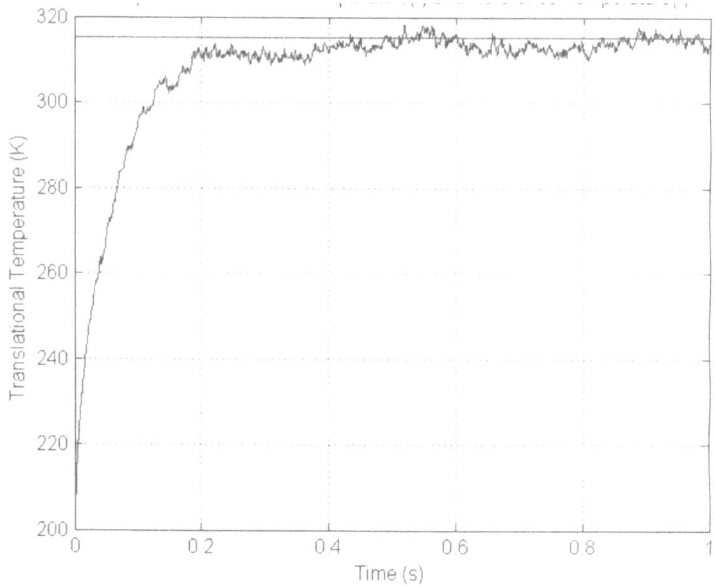

Fig. 5.7 Translational temperature of molecules versus time, computed with an initial nonequilibrium velocity distribution of molecules, which relaxes to a Maxwellian distribution

to the Maxwellian distribution corresponding to the mean temperature in the domain (315.30 K). As shown in Fig. 5.6, after a time of 0.3 s, the computed temperature distribution is almost superimposed with the theoretical Maxwellian distribution.

The initial noisy velocity distribution of the molecules with the maximum at $\beta c'_m \approx 0.8$ (instead of the value $\beta c'_m = 1$ at equilibrium), corresponds to a nonequilibrium translational temperature of 193.67 K. Fig. 5.7 shows that, as the situation evolves to an equilibrium Maxwellian distribution, because of the intermolecular collisions, the translational temperature rises to the equilibrium value of 315.30 K, imposed by the boundary conditions at the upper and lower material plates.

The distributions shown in Figs. 5.6 and 5.7 have been computed both for a domain filled with 21,000 sample molecules (on average 15 molecules per cell) and with 10^6 sample molecules (on average, 714 molecules per cell). Results plotted in Fig. 5.6 are almost indistinguishable between the two cases, while the rise of the translational temperature to the equilibrium value (shown in Fig. 5.7) displays a larger scatter in the case of a smaller number of sample molecules.

5.4.4 Reference Numerical Simulation

Numerical simulations of thermophoretic forces and the relative displacements were repeated for the reference case made in 2005 to compare new results with those

Table 5.4 Physical characteristics of the aerosol particle and data about computational domain, boundary conditions, and numerical simulation

Physical properties of the aerosol particle
Radius [μm] 0.5
Density [kg/m^3] 890.0
Specific heat [J/kg K] 2140.0
Thermal conductivity [W/m K] 0.15
Physical conditions of air in the definition domain
Mean free path λ [m] 4.86×10^{-8}
Knudsen number Kn 2.5×10^{-4}
Temperature ($t = 0$) [K] 315.30
Density ρ [kg/m^3] 1.2868
Dynamic viscosity [kg/(m s)] 1.450×10^{-5}
Kinematic viscosity [m^2/s] 1.127×10^{-5}
Boundary conditions and temperature gradient
Temperature of top surface [K] 335.05
Temperature of bottom surface [K] 295.55
ΔT [K] 39.5
$\nabla T = \Delta T/(zmax - zmin)$ [K/m] 5.642×10^3
$\nabla T/T$ [m^{-1}] 17.89
Domain (x, z), discretization and calculation conditions
$Xmin - xmax$ [μm] 3490–3510
$Zmin - zmax$ [μm] 0–7000
Δx [μm] 10
Δz [μm] 10
Total Nr. Of cells 1400
Nr. Of simulated molecules 21,000
Mean number of molecules per cell 15
Time step [s] 10^{-8} (without scale factor); 10^{-6} (with scale factor = 100)

obtained previously, thus inferring the improvements of the numerical simulations made with the new code version.

In the reference case, an aerosol particle of 0.5 μm radius is initially set with zero velocity in the center of the domain bounded by an upper and lower material plate and by lateral geometrical surfaces. In the simulations, the vertical two-dimensional cross-section (x, z) of this domain was modeled. Table 5.4 summarizes the physical characteristics of the particle and data about the computational domain and boundary conditions.

The reference calculation was repeated at first, for a problem time of only 0.1 s, with the same code version (THEPHOR-2) used in the year 2005, but with a modern personal computer with peak performance of 13 Gcycles, to ascertain the improvement solely due to better hardware. The improvement was large indeed, reducing the computational time from more than 20 days (Calculation 1) to less than 2 days (Calculation 2). Then the calculation was performed for a problem time of 10 s using the new scaled program version (THEPHOR-3), thus ascertaining the improvement in computational speed achieved with the code modifications described.

Table 5.5 summarizes the comparison of computational speed thus achieved.

Table 5.5 Comparison of the performances of code version THEPHOR-3 to version THEPHOR-2

	CALCULATION (Ref. case 2005)	CODE VERSION	PROBLEM TIME (S)	TIME STEP (S)	CPU TIME (days)	SCALE SF	GAIN (*)
1.		THEPHOR-2	0–0.1	10^{-8}	20.80	–	–
2.		THEPHOR-3	0–0.1	10^{-8}	1.71	1	–
3.	(**)	THEPHOR-3	0–10	10^{-6}	1.19	100	143.7

(*)Gain in CPU time for Calculation 2
(**)Calculation 4 is the continuation of Calculation 3 up to 10 s problem time. The gain reported (143.7 = 171 / 1.19) refers to the CPU time of Calculation 2 extrapolated to 10 s problem time

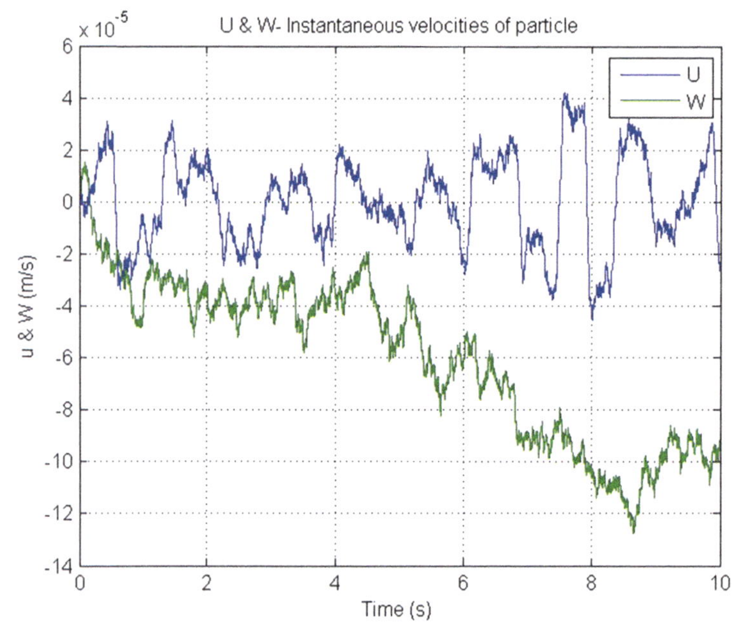

Fig. 5.8 Horizontal (U) and vertical (W) velocity components of aerosol particle versus time (reference calculation with an average of 15 sample molecules per cell)

The improvement in computational time between Calculation 1 and Calculation 2 is due only to better hardware. Using a scale factor of 100, in Calculation 3, the improvement in computational time is of the order of magnitude of the scale factor itself.

Upon application of the Waldmann formula (Waldmann, 1959) for thermophoretic forces, one finds that the thermophoretic speed $V_{th} = 10.3 \times 10^{-5}$m/s.

Results obtained with Calculation 3 of Table 5.5 are shown in the following two figures. Figure 5.8 shows the computed horizontal (U) and vertical (W) instantaneous velocity components of the aerosol particle versus time, sampled at intervals of 1000 time steps. The vertical velocity component seems to stabilize at

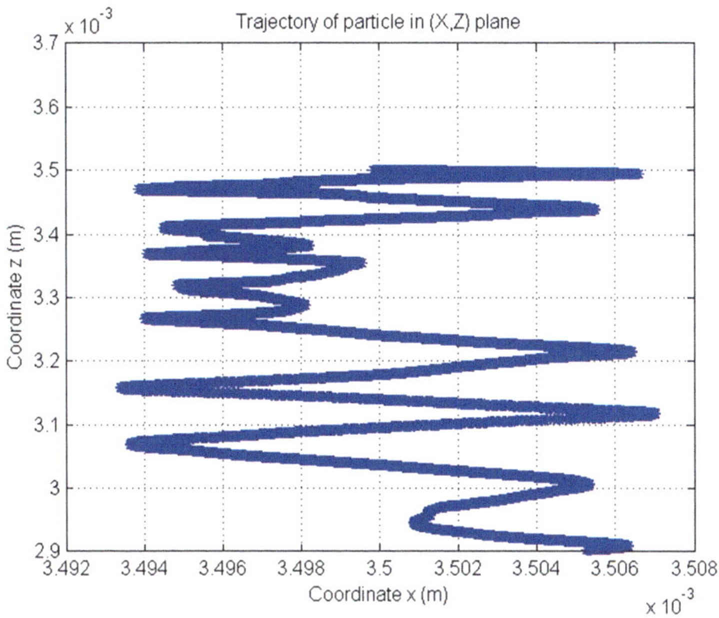

Fig. 5.9 Displacement of aerosol particles in the (x, z) plane

a level that agrees well with the value provided by the Waldmann formula for the thermophoretic speed. Figure 5.9 shows the trajectory of the aerosol particle in the (x,z) plane. The aerosol particle oscillates to and fro in the horizontal x direction while it moves almost steadily backward in the z direction. The time parameter t evolves from 0 to 10 s along the plot from the starting point P0(3.5×10^{-3}, 3.5×10^{-3}) to the endpoint P10(3.505×10^{-3}, 2.9×10^{-3}). The mean thermophoretic velocity in this time interval (0–10 s) is $\overline{V}_{th} = 6.05 \times 10^{-5}$m/s.

5.4.5 *State of the Code and Further Developments*

The current version of the code, though advanced in computational efficiency compared to the previous version, still presents two main flaws, which are explained and discussed in this section.

The first flaw consists of the fact that the simulation is two-dimensional—the thermophoretic displacement of the simulated particle occurs in a 2D domain, the (x, z) plane. Molecular velocity components, following collisions or boundary reflections, are computed or sampled in 3D (otherwise Maxwellian equilibrium distributions would never be achieved), but the displacements occur only in the (x, z) plane.

The third dimension, the y direction, is apparently irrelevant for the calculation of the thermophoretic displacement in the z direction, but in the reality of the simulation, it has an impact that cannot be neglected. The total number of real molecules in the computational domain depends on their volume and hence on the depth of the simulated domain in the y direction. The same is true for the number (FN) of real molecules represented by one sample molecule. In the numerical calculations referred to in the previous section, the only mesh simulated in the y direction was given the same dimension (10 μm) of the cells in the x and z directions, which implies a definite value of the total volume, hence of the FN parameter.

Moreover, the probability of molecular collisions with the aerosol particle depends on the movement of the sample molecules in the 3D space, and it is not a two-dimensional problem. However, because our trajectories are planar, we estimated the impact of molecular collisions upon the particle with a probabilistic approach, evaluating the probability of molecules hitting the particle in the 3D mesh in which the particle is located. The probabilistic approach is based on the comparison between the cell volume and the volume spaced by sample molecules within the time step, given by $V_s = \sigma_T c_m \Delta t$, where σ_T is the geometrical cross-section and c_m is the mean velocity of the sample molecules in the cell where the particle is located.

The second flaw exists in the narrow domain simulated in the x direction, consisting, in the calculations reported above, of only two meshes of 10 μm width. In case the particle approached the bounding (y, z) surfaces within one particle diameter, the particle was assumed to rebound elastically, by reversing the x-velocity component. This happened in some simulations but not in the one documented in the previous section.

The simulation of a narrow domain, including a particle, has also an implication for the calculation of the translational temperature in the computational domain. The presence of the particle, with multiple reflections of molecules from its surface, is a perturbation of the equilibrium state of the gas, which is reflected by an anomaly in the calculation of the translational kinetic temperature. The narrower the domain, the larger the anomaly is. This can be shown computationally by progressively increasing the size of the domain and recording the corresponding time behavior of the translational temperature. Figure 5.10 summarizes the calculations we made to illustrate this issue. To save some computational time, the size of the domain in the z direction has been reduced, according to the reference calculation documented in the previous section, from 700 to 500 meshes of 10 μm, by retaining the same temperature gradient. The temperature boundary conditions of the upper and lower plate are in this case 329.41 K and 301.19 K, respectively. The reference temperature (mean temperature in the domain) remains the same as before (315.30 K).

The size of the domain in the x direction has been increased progressively in subsequent steps from 2 meshes up to 200 meshes. The number of sample molecules has been increased likewise to maintain an average number of 15 sample molecules per cell. From Fig. 5.10, it is evident that increasing the size of the computational domain reduces progressively the anomaly in the calculation of the translational temperature and at the same time reduces the statistical scattering of the computed values. In the case of only two meshes in the x direction, the anomaly is reduced from

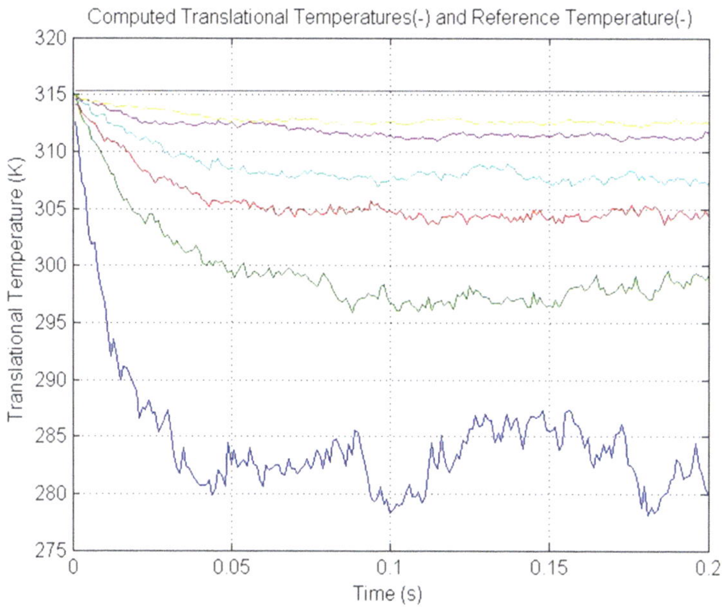

Fig. 5.10 Computed translational temperatures in the domain are dependent on its size. From the bottom upwards plots refer to the following number of meshes IMAX, KMAX (in the x and z directions) and the number NMOL of sample molecules: First, lowermost plot: IMAX = 2, KMAX = 500, NMOL = 15,000. Second plot: IMAX = 10, KMAX = 500, NMOL = 75,000. Third plot: IMAX = 20, KMAX = 500, NMOL = 150,000. Fourth plot: IMAX = 40, KMAX = 500, NMOL = 300,000. Fifth plot: IMAX = 100, KMAX = 500, NMOL = 750,000. Sixth, uppermost plot: IMAX = 200, KMAX = 500, NMOL = 1,500,000

a large anomaly of about 35 K to about 3 K in the case of 200 meshes in the x direction. The computational time for the cases plotted goes from 53 min for the first case to about 1 day for the last one.

The anomaly in the computed translational temperatures may depend in part on the sheer presence of the aerosol particle that reflects the air molecules according to the laws of diffuse (not specular) reflection and perturbs the theoretical Maxwellian velocity distribution. This is a topic that should be further investigated.

It has not been investigated whether a narrow domain and the consequent anomaly in the calculation of the translational kinetic temperature has or has not had a remarkable impact on the calculation of thermophoretic forces. A correct answer to this issue implies a large computational effort and goes beyond the purpose of this work. It seems, however, to be a good rule to simulate a computational domain large enough to minimize the impact of boundary conditions.

Both flaws we have discussed are going to be eliminated in our subsequent work, which aims to perform full 3-D simulations. The use of massively parallel computers will be unavoidable, but even then the performance of every single processor can be enhanced with the use of the scaled variables as explained in this section.

5.5 Negative Thermophoretic Force

Let us consider again a temperature gradient in the bulk of a gas, induced by a temperature difference between a lower horizontal plate at uniform and constant temperature T_1 and an upper horizontal plate at a temperature $T_2 > T_1$, at a distance L measured along a vertical reference axis z. Let λ be the mean free path of the molecules in the gas. We consider in the gas a spherical particle of radius r_p and particle Knudsen number $K_n = \lambda / r_p \ll 1$, hence a large particle.

The thermal conductivity K_p of the particle is of the order of magnitude of the thermal conductivity K_f of the gas, hence the ratio $\Lambda = K_p/K_f \approx 1$. The surface temperature of the particle is subjected to the same temperature gradient as the bulk of the gas in the immediate proximity of the particle. In this case, the proximity of the surface of the particle develops a thermal creep flow directed in an upward direction, and the particle is subjected to a thermophoretic force oriented downwards, opposite to the thermal gradient vector. This is the case of the well-known positive thermophoretic force first described by Waldmann (1959).

If, conversely, the thermal conductivity of the particle is very large compared to that of the gas, $\Lambda = K_p/K_f \gg 1$, the particle, including its surface, has a rather uniform temperature distribution. Because of the thermal gradient in the bulk of the gas, however, a *thermal stress slip flow* develops, as first described by Sone (1972), and induces a thermophoretic force directed in an upward direction, hence in the same direction as the thermal gradient vector. The force, which has been predicted on the base of theoretical considerations and confirmed by numerical results, but never proved experimentally, is referred to as the *negative thermophoretic force*.

5.6 A Step Towards a Parallel THEPHOR Code

5.6.1 General Approach to the Problem of Parallelization

We return to the topic of experimental investigations of thermophoretic forces acting upon aerosol particles in the atmosphere carried out in cooperation between the Institute of Atmospheric Sciences and Climate (ISAC) of the National Research Council (CNR) of Italy, the Department of Physics of the University of Ferrara, Italy, and the Microgravity Research Centre of the Université Libre de Bruxelles, Belgium.

This experimental task, aiming at understanding the behavior and distribution of aerosol particles in the atmosphere, is complemented by the development of a computer program simulating the thermophoretic forces with the Direct Simulation Monte Carlo (DSMC) method of molecular dynamics. The purpose of the computational effort is a better understanding of the experimental data and analysis of thermophoretic forces in regimes, like intermediate Knudsen numbers, where experimental data are not available. In this section, we explain the need for and advantages of the parallel version of the program, along with details of the application of the DSMC method to the simulation of thermophoretic forces and the calculation of the temperature distribution in the simulated aerosol particle, which is essential to compute the energy exchanged between particle and colliding air molecules. Preliminary computational data provide information about the suitability of the computational approach followed. The order of magnitude of the computed thermophoretic velocities agrees fairly with experimental data.

The programming language for the current version of the code THEPHOR is FORTRAN 77. This programming language has some disadvantages from the viewpoint of memory allocation, but since the original code THEMIS (Muramatsu, 1995) was coded in FORTRAN 77 in Japan, for easy data transfer and portability, the same programming language was used. The size of the code varies with the number of molecules and the total number of cells in the computational domain.

The main problem of parallelization consists of the distribution of tasks among several processors to save the running time of the code. Before parallelizing any code, it is necessary to recognize the parts of the code that require parallelization. The decision can be made based on time taken for computations or memory storage. After this, the process of parallelization precisely involves the exchange of information amongst the processors. This is achieved with the help of Message Passing Interface (MPI). In MPI procedures, each processor is assigned a number called the *task number,* and the processor that controls the transfer of information in the whole process is called the *root processor*. The root processor is usually given the task number 0. The code is run on the root processor, which transfers the required information to other processors by calling MPI_BCAST, which passes the same information from the main processor to other ones. For one-to-one communication, MPI_SEND and MPI_RECIEVE can be called. The parallelized version of THEPHOR also calls MPI_GATHER and MPI_REDUCE. As the name suggests, MPI_GATHER gathers information from all processors and saves it at one processor specified in the call. MPI_REDUCE is used to reduce some fields from all processors to a single field at one processor and under some specified operation, mentioned in the call. When all jobs concerned with communications among the processors are completed, a call to MPI_FINALIZE is made which cleans up all MPI states. In case of urgent termination of code, MPI_ABORT is called to abort all tasks in a process. During the process of parallelization, the amount of data sent and received should be synchronized. Differences in the number or type of data sent and received may lead to strange behavior in the code.

5.6.2 Computational Platform at CINECA

A preliminary version of the parallelized code was tested on the CINECA's IBM SP5 supercomputer in June–July 2006. The name IBM SP5 was a short name for the IBM SP 1600 series with Power5+ processors. This leading (at that time) supercomputer class was built as a *cluster* architecture—constituted by an ensemble of connected computing units, called *nodes*. Each SMP (Smaller Symmetric Multiprocessing) node owned 8 powerful p5-575 CPUs; each CPU was an IBM Power5+ 64-bit processor at 1.9 GHz with 2 GB of local RAM, amounting to 16 GB of RAM per node. Each CPU also accessed 1.9 MB of L2 and 36 MB of L3 dedicated caches. The nodes were interconnected by a high-performance switch (HPS). The cluster architecture provided a highly scalable, high-performance computing (HPC) solution for UNIX and Linux environments, able to manage up to 128 computing nodes in its largest configuration (which could also accommodate the 16 CPUs per node version). The one hosted at CINECA, at the time the parallel THEPHOR code was tested, was configured with 8-core 64 nodes, for a total of 512 Power5+ CPUs working together, with >1.2 TB of total RAM.

CINECA chose this kind of supercomputer to meet the demands of a broad range of compute-intensive and memory bandwidth-intensive applications: in fact, the 8-core node design allowed not only fast processing but also rapid and continuous access to large amounts of data, ensured by its >25 Gbps peak memory bandwidth per processor. System performances were enhanced by the DDR2 memory DIMMs (newly available at that time), with a maximum memory capacity of 256 GB per node and a peak memory data transfer speed of over 200 Gbps. Each node was also equipped with IBM's Error Checking and Correction (ECC) and Chipkill technologies: both were designed to detect and automatically correct memory errors (including the caches) to prevent system crashes due to memory chip failure. The system also integrated 10 TB of fast-access distributed disk space with 800 MB/s of measured bandwidth, managed by the GPFS filesystem as a short-term, temporary storing area.

Overall, the peak performance of the CINECA's IBM SP5 supercomputer was 3.89 Tflops.

Aside from the hardware, the supercomputer was run by the UNIX-like AIX 5.2 operating system. Along with the IBM and Intel compilers for both FORTRAN and C/C++ programming languages, the system was equipped with several high-performance cluster management software tools. These consisted of: the Parallel Environment (PE), a high-function development and execution environment for parallel message-passing applications; the LoadLeveler, a dynamic job scheduling and workload balancing software supporting thousands of jobs within the cluster; the GPFS, a high-performance, shared-disk file system providing fast data access to all nodes in the cluster; and the High Availability Cluster Multiprocessing (HACMPTM), a failure-recovery tool providing continuous access to data and applications.

On the scientific applications side, among many other specialized software libraries, the system provided the ESSL (Engineering and Scientific Subroutine Library) and Parallel ESSL mathematical libraries, to enhance the performances of both serial and parallel applications. Moreover, the standard BLAS (Basic Linear Algebra Subroutines), LAPACK (Linear Algebra PACKage), PBLAS (Parallel BLAS), and ScaLAPACK (Scalable LAPACK) libraries were readily available, together with the support for MPI and OpenMP.

Hence, the CINECA's IBM SP5 supercomputer was a great platform for HPC applications such as oceanographic studies, meteorology, computational fluid dynamics, energy research, data mining, and other bandwidth-intensive work that requires transferring, accessing, and rapidly analyzing large quantities of data.

Additional features for the visualization of computed results are packages of the MATLAB software, providing the simplest ways to plot graphs. This software allows you to read data from other files and plot them, with the additional advantages of colored graphs, graphs with dots, crosses, and several other forms for easy readability. Time-dependent visualization, available in the program, helps in identifying logical errors made in the code. If the graph does not progress in the form expected, it gives information about inherent problems. This feature proved to be a big advantage in debugging the code.

5.6.3 Preliminary Results

This section gives a short explanation of the preliminary results of a few simulations.

In all cases, a spherical aerosol particle of 0.5 μm radius is initially set with zero velocity ($u = w = 0$) in the center of the domain with $z_c = 3500$ μm. The physical properties of the aerosol particle and data of the computational domain are summarized in Table 5.6.

Table 5.6 Physical properties of the aerosol particle and computational domain

Physical properties of particle:
Density [kg/m^3]: 890.0 Specific heat [J/kgK]: 2140.0 Thermal conductivity [W/mK]: 0.15
Definition domain and time step:
Mean free path [m]: 4.86×10^{-8} Knudsen number [Kn]: 2.5×10^{-4} Air density [kg/m^3]: 1.2 Dynamic viscosity [μ] [kg/(ms)]: 1.45×10^{-5} Kinematic viscosity [ν] [m2/s]: 1.13×10^{-5} $\Delta T/(z_{\max} - z_{\min})$ [K/m]: 5.642×10^3 $\Delta T/T$ [1/m]:= 17.89 Time step [s]: 10^{-8}
Domain:
Size: $x_{\max} \times z_{\max} = 1000 \times 7000$ μm, Discretization: $\Delta x = \Delta z = 10$ μm Number of cells: 70,000 Nr. of simulated molecules: 700,000 Mean number of molecules per cell: 10
Boundary conditions:
Temperature of top surface [K]: 335.05 Temperature of bottom surface [K]: 295.55 Air temperature ($t = 0$) [K]: 315.30

Data in the above Table 5.6 refer to the calculation performed on the ATHOS4, which was the name of a computer installed in the Physics Department of the University of Ferrara with a peak performance of 5 GHz.

A comparison of computational times was made with the MARK personal computer of one of the authors with a peak performance of 1.3 GHz.

Normalizing to 1 the computer time needed for one calculation with the MARK computer, the calculation times of the other runs were as follows: (i) 1 / 2.5 = 0.4, with the ATHOS4 computer; (ii) 1 / (2.5 × 7) = 0.057, with 8 CPUs of the parallel computer at the CINECA center. Thus, roughly speaking, the code parallelization allowed an almost optimal speed-up on a single node (8 CPUs) of the CINECA's IBM SP5, with a gain of factor 7 in the computational performance of the parallel computer compared with the sequential calculation. The parallel code was also tested on more processors, ranging from 10 to 16 (two nodes), but the performances did not improve. The reason was that increasing the number of processors above a given threshold, without contemporarily enlarging the amount of data to be elaborated by the single processor, caused the inter-CPU communication time to dominate the computational time. Moreover, in the particular case of cluster architectures such as that of the IBM SP5, internode communications are inherently slower than intra-node communications (even when managed by a high-performance switch, as was the case with CINECA's supercomputer). Hence, to observe satisfactory speed-ups even for more than 8 CPUs, a necessary (but not sufficient) condition is to enlarge both the number of subdomains in the domain decomposition and the number of molecules to be simulated in each subdomain, so that the per-CPU computational burden dominates the communications time. However, this was not the only thing to take care of—internode communications called for additional code optimizations, which have not been implemented yet.

In the course of about 3 months, a postgraduate student, Aparna Goyal (Goyal et al., 2006), under the supervision of Prof. Zanghirati, with the cooperation of the co-authors, managed to improve the performances of the parallel calculation. It turned out that: (i) The simulation of the computational domain should be sufficiently large (theoretically infinite) in the x direction to prevent boundary-effect disturbances, and (ii) the number of sample molecules should be as large as possible to minimize statistical scattering.

After the termination of her work with Prof. Zanghirati, Aparna Goyal moved to her home country of India, where she remained for about 7 years, before joining the Applied Micro Electronics (AME) Company in Eindhoven, The Netherlands, where she is still working with management responsibilities. Mrs. Goyal has been informed about our purpose in publishing this work. We would like to thank her once more for her contribution and her commitment to the work.

5.6.4 Open Problems and Future Code Developments

5.6.4.1 Three-Dimensional Calculations

Rigorous Monte Carlo calculations require a three-dimensional treatment of the problem; otherwise, approximations are made in the velocity distribution of the simulated molecules, which does not correctly reproduce the Maxwellian distribution at the given temperature. In a two-dimensional approximation, in which one velocity component is systematically set to 0, the resulting velocity distribution of the molecules deviates from the rigorous Maxwellian distribution. Consequently, the definition of the kinetic temperature is also approximated.

This modeling limitation is effective in the subroutines of the code which treat the collisions of the sample molecules with the material walls. In the specific case of the simulation of the thermophoretic force acting upon an aerosol particle, material surfaces are the upper and lower plate, kept at different temperatures, and the surface of the aerosol changes temperature due to the kinetic energy transferred from the impinging molecules and subtracted by the reflected molecules. The subroutines used in the code to treat the collisions with the material walls and with the aerosol particles are SPECDI and DIFFUS, respectively.

In the code, all fields have been dimensioned to allow a three-dimensional treatment of the problem, but in the preliminary calculations made thus far, the two-dimensional approximation has been made to save computer time. With the parallel version of the code, three-dimensional computations, and therefore more rigorous simulations of thermophoretic forces, are possible.

5.6.4.2 Improvement of Numerical Solution for Temperature Distribution in the Aerosol Particle

The temperature distribution in the aerosol particle is computed in spherical coordinates, by assuming a spherical particle and partitioning it in a given number of radial, latitudinal, and azimuthal meshes. The number of these meshes is typically 10 for the radial, and 36 for the latitudinal and azimuthal directions, respectively. In the case of a two-dimensional calculation, the number of azimuthal meshes can be reduced to only 3, the minimum that formally allows a discretization of the heat diffusion equation. This heat diffusion equation, written for every computational cell, is formally integrated into the cell volume and transformed into a linear algebraic equation for the unknown temperature defined in the center node of the cell. The system of algebraic equations is solved with the iterative Successive Over Relaxation (SOR) method. No attempt was made to optimize the relaxation parameter to improve the computational speed. The calculation of the temperature distribution in the aerosol particle is made in the subroutine SPTEMP.

Due to the very small time step necessary for the computation of the molecular collisions with the aerosol particle, the number of iterations needed to get a converged solution within the time step is generally small and therefore the computational time spent by the program within this subroutine is limited to only a few percents. In case, however, the code reaches a high degree of efficiency in the parallelization, the time spent in the subroutine SPTEMP becomes not negligible concerning the total time and therefore an improvement of the solution in this subroutine becomes a priority.

An alternative to the SOR method is provided by direct techniques, such as the Cholesky matrix decomposition, or by more advanced iterative techniques like the Alternating Direction Implicit (ADI) method. The latter might be more suitable for parallelization, by assigning to different processors the tasks of integrating in the three coordinate directions.

5.6.4.3 Simulation of Swarms of Aerosol Particles and Relative Interactions

In some atmospheric conditions, the concentration of aerosol particles is very high ranging from 10^3 to 10^5 per cubic centimeter. In the presence of temperature gradients in the atmosphere, migration of the particles, imposed by thermophoretic forces, implies interparticle collisions with the formation of ever-larger aggregates of fractal characteristics. Given a possible integration of the code into a more general program describing a variety of microphysical phenomena in the atmosphere, the treatment of aerosol particles in a realistic concentration, therefore, of swarms of particles, might become a topic of higher priority. Each particle would be treated by using the same subroutines as would be applied to a single particle. Interparticle collisions would however require code improvements to simulate the aggregation efficiency, dependending on the stream characteristics.

5.6.4.4 Memory Allocation to Individual Processors

Memory has not been allocated efficiently during the process of parallelization of the THEPHOR code. The main objective of parallelizing the code has been to reduce the run time. For this purpose, more emphasis is laid on transferring information read by the main processor to other processors that have memory allocated for them. Each processor allocates a large amount of memory containing the same information. Also, the code is written in FORTRAN 77, which lacks the facility of dynamic memory allocation. Therefore, in some parts of the code, complete fields are stored in each processor, although the calculation would require only a part of them. It is expected that in the future, with further improvements in the code, these problems of optimizing memory allocation will be solved.

The parallel version of the THEPHOR code has been documented in an internal unpublished report from the Department of Mathematics and Computer Science of the University of Ferrara. It can be delivered from Professor Gaetano Zanghirati to interested readers or researchers upon written request, provided an agreement about the development and use of the code can be agreed upon

5.7 Synopsis of Computational Codes of the THEMIS Family

The computer program THEPHOR is the last development of a series of codes originating from a code version named THEMIS developed at the beginning of the 1990s at the Japan Nuclear Cycle (JNC) research center of O-Arai, Ibaraki-ken, Japan. In this section, we summarize the most salient characteristics of the available variants of the codes originating from the prototypical code THEMIS. We also give a summary of the subsequent code versions and information about available documentation in the form of publications.

(i) **Original Code THEMIS**

Numerical simulations of fluid dynamics concerning the safety of nuclear reactors were conducted with the DSMC method by a research group of the JNC (Japan Nuclear Cycle Development Institute) at the O-Arai research center in Ibaraki-ken, Japan. These investigations were documented in the computational code THEMIS, which was made available to the author during the academic year 1996–1997, which the author spent as a guest professor at the Tokyo Institute of Technology (Tokodai) in the frame of a cooperation between KfK and the O-Arai research center.

The THEMIS program presents an analysis of thermal striping phenomena characterized by stationary random temperature fluctuations in the region immediately above the core exit of a Liquid Metal-cooled Fast Reactor (LMFR) due to the interaction of cold sodium flowing out of a control rod assembly and hot sodium flowing out of adjacent fuel assemblies. Components located in the core outlet region must be protected against random processes that induce high-cycle fatigue. Because experimental analysis of these thermal striping phenomena is both difficult and expensive to perform, the computer code THEMIS was developed and tested to assess its applicability for the safety analysis of fast nuclear reactors (Muramatsu, 1995).

(ii) **Code THEMIS-Annular**

A two-dimensional version of the code THEMIS was adapted to model an annular domain representing a small bundle of a fast reactor. The inner surface of the annulus is representative of the central hot pins of the bundle, while the outer surface is representative of the outer hexagonal canning (Bottoni, 1998).

(iii) **Code THEMIS/3D.1**

The computer code THEMIS, originally developed for two-dimensional problems, was then extended to the third dimension and adapted to numerical simulations of evaporation and condensation problems for applications to safety analysis of atrium-cooled fast breeder reactors. This code version is based on a three-dimensional calculation of sodium evaporation and condensation in a 60-degree sector of a 7-pin bundle for a fast nuclear reactor.

Traditional approaches used in safety analysis codes to simulate sodium vapor condensation and vaporization rely upon the application of macroscopic semiempirical correlations for heat transfer and vapor condensation or evaporation rates. As an alternative to this macroscopic approach, the code THEMIS/3D.1 is based on the application of the basic laws of the kinetic theory for the determination of the evaporation and condensation rates of vapor in a reactor bundle. This microscopic approach is based upon a Monte Carlo simulation of the molecular trajectories, and collision rates between vapor molecules and the molecules with the surfaces of the claddings of the pins of a reactor bundle (Bottoni, 1999, 2000, 2004, 2007).

(iv) **Code TH3SL**

This code is a three-dimensional version of the code THEMIS, dedicated to the mixing of air, water vapor, and hydrogen in a control volume representative of the conditions within the containment of a pressurized water reactor.

The program consists of the simulation of three species, water vapor, air, and hydrogen, and was written to simulate the deflagration of hydrogen in a mixture with air and water vapor in case of a hypothetical accident in a pressurized water reactor. For this reason, all fields allocated in the program are three-fold fields, treating all at the same times for all species. The code can be run both in 2D and 3D options.

The code, developed by the first author at KfK, was then delivered to INET (Institute of Nuclear Energy Technology) of the Tsinghua University in Beijing (Gan et al., 2002; Chen et al., 2002).

(v) **Code THEMIS/3D.2**

This code originated from version THEMIS/3D.1. In a subsequent development of the program, thermophoretic forces modeled fuel particles (uranium oxide) fragmented in case of a hypothetical nuclear incident occurring in a gas mixture with sodium vapor and fixion products.

Three-dimensional calculations of the thermophoretic forces and computations of the velocity of debris of nuclear fuel sputtering from pellets of a fuel bar of a fast nuclear reactor under hypothetical accident conditions in the reactor core were added to the previous code versions. It simulates a 60-degree sector of a 7-pin bundle (Bottoni & Bottoni, 2005; Bottoni, 2007b).

(vi) **Code THEMIS/3D-B7**

The code is based on a three-dimensional calculation of thermophoretic force and velocity of debris of nuclear fuel sputtering from pellets of a fuel bar of a fast nuclear reactor under hypothetical accident conditions in the reactor core. The structure of the code is similar to that of THEMIS/3D.2 but a full 7-pin bundle is simulated (Bottoni, 2007).

(vii) **Code THEPHOR**

This code is based on a bidimensional calculation of thermophoretic force acting upon an aerosol particle of spherical shape, suspended in a gas with an assigned temperature gradient, in microgravity conditions, simulating the upper atmosphere's physical conditions. The code is fully operational for sequential calculations. A parallel version of the code has been arranged and runs on eight processors with a gain in computational time of a factor of 7 (Bottoni, 2006; Bottoni et al., 2005, 2018).

References

Bird, G. A. (1994). *Molecular gas dynamics and the direct simulation of gas flow*. Clarendon Press.

Bottoni, M. (1998). Monte Carlo calculation of sodium vapour condensation and rewetting simulation in equivalent LMFBR bundle under hypothetical accident conditions (In Japanese). In *1998 Spring Meeting of the Japanese Atomic Power Society (JAPS)*, Osaka.

Bottoni, M. (1999). Monte Carlo calculation of sodium vapour condensation and simulation of rewetting in LMFBR bundle under hypothetical accident conditions (In Japanese). In *1999 Spring Meeting of the Japanese Atomic Power Society (JAPS)*, Hiroshima, .

Bottoni, M. (2000). Molecular approach to sodium vapour condensation, rewetting and vaporisation in LMFBR bundle under hypothetical accident conditions. In *Proceedings of the Eighth International Colloquium on Numerical Analysis and Computer Science with Applications*, Plovdiv.

Bottoni, M. (2004). Molecular approach to sodium vapour condensation, rewetting and vaporization in LMFBR bundle under hypothetical accident conditions. *Journal of Nuclear Science and Technology, 41*(5).

Bottoni, M. (2006). Thirty years of endevour on numerical simulation of molecular dynamics. In *Invited lecture delivered at the 18th Meeting of the "International Institute for Advanced Studies (IIAS) in Systems Research and Cybernetics"*, Baden-Baden, Germany.

Bottoni, M. (2007). Three dimensional Monte Carlo simulation method for LMFBR fuel bundle heat transfer under LOCA conditions (in Chinese). *Atomic Energy Science and Technology Journal, 41*(2).

Bottoni, M. (2007b). Molecular dynamic simulation of thermal creep in 7-pin LMFBR bundle under hypothetical accident conditions. In *15th International Conference On Nuclear Engineering, (ICONE15)*, Nagoya, Japan.

Bottoni, M., & Bottoni, C. (2005). Direct Monte Carlo simulation of fuel particles relocation in nuclear reactor bundle during hypothetic core disruptive accident. In *13th International Conference On Nuclear Engineering (ICONE13)*, Beijing.

Bottoni, M., Porcu, F., Prodi, F., Santachiara, G., & Travaini, S. (2005). Bi-dimensional simulation of thermophoretic forces on aerosol particle with the direct simulation Monte Carlo (DSMC) method. *Computational Fluid Dynamics Journal, 14*, 255.

Bottoni, M., Mantovani, S., & Cavicchi, M. (2018). Thermophoretic calculations with pristine and scaled variables. *Academia Journal of Scientific Research, 6*(12).

Chen, L., Yao, Z., & Bottoni, M. (2002). Three-dimensional simulation of steam condensation in water reactor containment with the Monte Carlo computer program TH3SL. In *Spring Meeting of the Japanese Atomic Power Society (JAPS)*, .

Choi, S. J., Rader, D. J., & Geller, A. S. (1996). Massively parallel simulations of Brownian dynamics transport in low pressure parallel plate reactors. *Journal of Vacuum Science and Technology A, 14*(2), 660 665.

De Laplace, P. S. (1951). *A phylosophical essay on probabilities*. Dover.

Gan, X., Bottoni, M., & Zhou, Z. (2002). TH3SL: Simulation of gas distribution inside a water reactor containment (in Chinese). *Atomic Energy Science and Technology, 36*(1), 70–75.

Goyal, A., Zanghirati, G., & Bottoni, M. (2006). *Parallel version of the computer program THEPHOR*. (Internal documentation of the Physics Department of the Ferrara University).

Kaburaki, H., & Yokokawa, M. (1993). Numerical simulation of near continuum gas flow by the direct simulation Monte Carlo method. In *Proceedings of the Joint International Conferenceon Mathematical Methods and Supercomputing in Nuclear Applications, M&C + SNA '93*, April 19–23, 1993, Karlsruhe, Germany, Edited by Forschungszentrum Karlsruhe (FZK).

Kaburaki, H., Yamamoto, H., & Yokokawa, M., Arisawa, T. (1990). A new collision cell system for the direct simulation Monte Carlo method. In *Proceedings of First International Conference on Supercomputing in Nuclear Applications (SNA '90)*, Mito, March 12–16, 1990, Edited by Japan Atomic Energy Research Institute (JAERI).

Muramatsu, T. (1995). A Direct Simulation Monte Carlo code THEMIS for the evaluation of nonstationary heat transfer phenomena in LMFR thermohydraulics. In *8-th IAHR Working Group Meeting on Advanced Nuclear Reactors Thermal Hydraulics, Nuclear Research Institute Rezplc*, Czech Republic, June 6–8.

Nambu, K. (1980). Direct simulation scheme derived from the Boltzmann equation, (I): Monocomponent gases. *Journal of the Physical Society of Japan, 49*(5), 2042–2049.

Rapaport, D. C. (2004). *The art of molecular dynamics simulation* (2nd ed.). Cambridge University Press.

Sone, Y. (1972). Flow induced by thermal stress in rarefied gas. *Physics of Fluids, 15*, 1418.

Vincenti, W. G., & Kruger, C. H. (1965). *Introduction to physical gas dynamics*. Krieger Publishing Company.

Waldmann, L. (1959). Über die Kraft eines inhomogenen Gases auf kleine suspendierte Kugeln. *Zeitschrift für Naturforschung, 14a*, 589.

Watanabe, T., Kaburaki, H., & Yokokawa, M. (1994). Simulation of two-dimensional Rayleigh-Benard system using the direct simulation Monte Carlo method. *Physical Review E, 49*(5).

Chapter 6
Thermal Confinement

Contents

The original version of the chapter has been revised. A correction to this chapter can be found at https://doi.org/10.1007/978-3-031-84567-3_8

Text by M. Bottoni and S. Mantovani

M. Bottoni et al., *Molecular Dynamics - Theory and Applications*, Mechanical Engineering Series, https://doi.org/10.1007/978-3-031-84567-3_6

This chapter was originally intended to document an international cooperation between the Microgravity Centre of the Université Libre of Bruxelles in the section headed by Dr. Andrei Vedernikov and the MEEO (Meteorological Earth Environment Observation) Company, situated in Ferrara, Italy, aimed at developing a device for trapping cosmic particles floating in microgravity condition in the outer Earth atmosphere. Dr. Vedernikov had also working relationships with Institute of Atmospheric Sciences and Climate (ISAC) of the Italian National Research Council (CNR) in Bologna, Italy, and with the Department of Physics of the Ferrara University.

For reasons that were never well understood, the cooperation between the Université Libre di Bruxelles and the counterparts in Italy was de facto terminated without a preliminary investigation of the problem originally proposed.

The team formed at the MEEO, supported by Dr. Bottoni as consultant, did not drop the undertaken and continued the investigation of the problem, till a tentative solution could be proposed and discussed for its feasibility.

The chapter is divided thematically into four major parts.

(i) The first part recalls the theoretical backgrounds needed to understand the following applications.
(ii) The second part documents the structure of the two-dimensional program developed to investigate the problem of capturing the cosmic particles.
(iii) The third part is dedicated to the verification of the code by computing a series of test cases consisting of two groups of problems. The first group consists of classical examples, like the Bénard problem, simulation of thermal creep, and thermophoretic problems in rarefied gas conditions. These classical problems, well known from a variety of publications, intend to give the beginners who wish to enter the domain of molecular simulations, a useful background of knowledge. The other group of these cases simulate with the method of molecular dynamic three test cases (Gallis et al., 2002) well known and documented in the literature.
(iv) The fourth part documents the effort to optimize the performances of trapping cosmic particles by analysing the computational performances of a variety of test conditions, mainly based on modifying size and location of the heating elements of the ICAPS device under investigation.

The analysis performed and documented in this chapter did not lead to a conclusive judgement about the performance of the ICAPS experiments but may be considered as a feasibility report about the computational performances of the program developed for the purpose.

Thus far our effort remains the only attempt to provide an answer to the problem of trapping cosmic particles. No other documentation, published or unpublished or in development, is known to the authors.

6.1 The ICAPS Experiments

6.1.1 Introduction

The European Space Agency (ESA) in an "Invitation to Tender" dated August 15, 2006, with an addendum dated August 21, 2006 (ESA, 2006), illustrated a

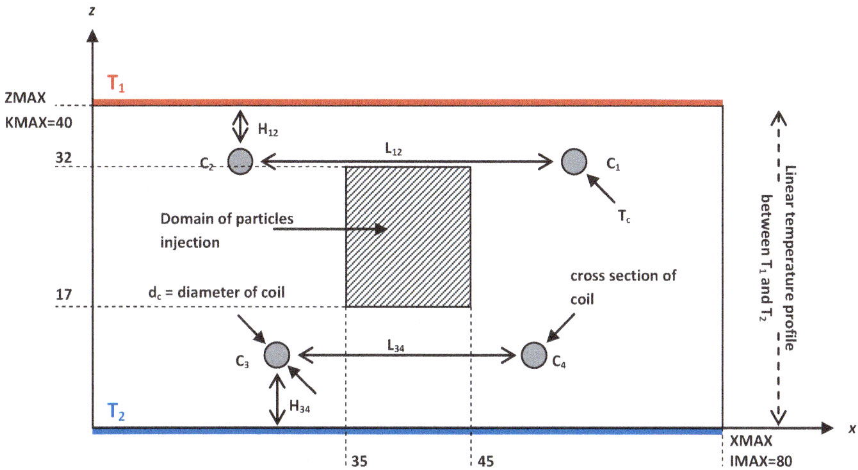

Fig. 6.1 Computational domain for the ICAPS experiments

feasibility study of the Interaction in Cosmic and Atmospheric Particle System (ICAPS) experiment. The geometrical configuration of this experiment, with tentative physical dimensions of the components, is illustrated in the Fig. 6.1.

The experiment consists essentially of thermal elements, called "Gas Flow Driver" contained in a cylindrical box, the centre of which should be a trapping area for thermally confined particles floating in a rarefied gas within the chamber. The "gas flow drivers" consist of toruses (hereafter referred to as "coils") placed in the cylindrical case of the experiment.

In the cross section of a 2-D representation the coils reduce to couples of circles. Figure 6.1 shows a sketch of the 2D computational domain with the circular cross sections of the coils and the area of injection of the particles. Temperatures of upper and lower plates (T_1, T_2), coil diameter (d_c) and positions (H_{12}, L_{12}, H_{34}, L_{34}), temperature of coil surfaces (T_c), size of the domain (XMAX, ZMAX), and partitioning of the domain in cells (IMAX, KMAX) are free parameters to be investigated to optimize the thermal confinement of the particles.

Based upon the physical principles of thermal creep, or flow induced by thermal gradients along material surfaces, and of thermophoretic forces induced from thermal gradients across the bulk of the gas, particles carried from the gas should be concentrated in and collected from the trapping area.

As explained in the Note from the ESA, the physical dimensions of the experiment and of its components, the positioning of the thermal elements within the chamber, the temperature of the thermal elements are free parameters that should be experimentally and numerically investigated in order to enhance the efficiency of the experiment in collecting particles in the trapping area. To enable numerical investigations of the proposed device a two-dimensional computer program called

THERCONF-2D (THERmal CONFinement in 2D) has been build up as a first step towards a full three-dimensional representation of the experiment.

This code version has been developed from a previous code version called THEPHOR (Bottoni et al., 2005) which describes the displacement due to thermophoresis of a tiny particle in a gas domain subjected to a temperature gradient. The gas behaviour and its interaction with material surfaces are modelled with the statistical methods of molecular dynamics, based on the "Direct Simulation Monte Carlo" (DSMC) (Bird, 1994). The main reasons for the choice of the statistical approach, as an alternative to the traditional methods based on numerical solutions of the Navier-Stokes equations, are explained in the following section.

Besides simulating the geometrical configuration of the thermal confinement experiment proposed by the ESA, the THERCONF-2D computer program retains all capabilities of the previous version THEPHOR, and also includes options to simulate particular physical phenomena like the insurgence of thermal instability in the Bénard problem and the aforementioned thermal creep flow induced by temperature gradients along heated material surfaces. The present code version is therefore rather versatile, being able to simulate a variety of physical situations. In order to assess the capability of the code and especially for checking its correctness after every major development or inclusion of a new model, a set of standard test cases has been included in the code. The program has been organized as to enable computing of all test cases, 8 at present, in a single run, to minimize the time and manpower to be invested in the accomplishment of the task. A short description of the test cases is given in Sect. 6.10.1.

6.1.2 Molecular Dynamic Approach

Fluid-dynamic calculations are traditionally based on the numerical solution of the Navier-Stokes equations in terms of primitive variables (velocity components of the fluid) and pressure. In most cases these solutions are obtained by integrating the equations of mass and momentum conservation upon cell volumes and then deriving from the volume averaged equations an algebraic system of finite difference equations defined on the grid mesh, which partitions the definition domain of the problem. In case the definition domain is bounded by curved surfaces which cannot be represented rigorously in a convenient (f. i. spherical or cylindrical) coordinate system, these bounding surfaces must be approximated by stepwise surfaces, which introduce numerical error and increase the so-called numerical diffusion.

In case of heated fluids an energy conservation equation is added to the mass and momentum balance equations. In turbulent flow regimes a turbulence model, usually based upon at least two transport equations, for the turbulent kinetic energy and its dissipation, is added to the previous system. This traditional approach, which might be called "macroscopic approach", is based upon the assumption of a continuum medium and fails in case of rarefied gases in which the scale of the phenomena to be described, like eddies in the flow, becomes comparable in size or smaller than the mean free path of the molecules. In terms of the system Knudsen number Kn_L characterizing the flow, the limit of applicability of the macroscopic approach is usually assumed to be $Kn_L = \lambda/L = 0.1$. Flow regimes characterized by a system Knudsen number just above this threshold, in the range $0.1 < Kn_L < 10$, are referred to as "intermediate regime flows", while larger Knudsen number characterize very rarefied flows, almost collisionless flows or "free-molecule regimes". In case of heated flows, the traditional macroscopic approach relies upon the application of fluid-structure heat transfer coefficients which are generally known with sufficient approximation under usual temperature and pressure conditions, but with only crude approximation in the intermediate and rarefied flow regimes.

One might therefore conclude that the traditional approach is affected by three inherent limitations, regarding: (i) the rigorous representation of the geometry; (ii) the use of semi-empirical heat transfer and momentum transfer (turbulence model) correlations; (iii) domain of applicability restricted to continuum medium, which rules out applications to rarefied gases.

As an alternative to the macroscopic approach we propose a "microscopic approach" based upon the application of the Direct Simulation Monte Carlo (DSMC) methodology in the frame of the Molecular Dynamic (MD) method (Bird, 1994) which relies upon the strict application of the basic laws of the kinetic theory (Vincenti & Kruger, 1965). The main advantages of the microscopic method, in comparison with the traditional one, can be summarized as follows. The geometry can be represented rigorously, in whatever coordinate system, as long as the equations defining the surfaces bounding the definition domain are known. The definition domain is partitioned in cells, like in the traditional approach, but the cells are not used to derive difference equations. Therefore irregular cells do not infirm the accuracy of numerical results. Cells are defined for the only purpose of determining the collision rates among the molecules within the cells. Cumbersome representation of geometry is therefore avoided in the microscopic approach. The DSMC method does not imply the use of heat transfer coefficients or of turbulence models. Heat and momentum transfer are derived directly from the application of the kinetic laws. The domain of application, in terms of flow regimes, is unbounded, including very rarefied gases.

The microscopic or molecular approach is based upon the Monte Carlo simulation of molecular trajectories, collision rates among the molecules, collisions of the molecules with ideal (symmetry) or material surfaces. Ideal surfaces are treated as specularly reflecting surfaces, while material surfaces are treated as diffusively reflecting surfaces. Diffusive reflections imply that molecules reflected from material surfaces are assigned velocities obtained by sampling a Maxwellian velocity distribution corresponding to the temperature of the surface in the impact point.

6.1.3 Simulation of Thermophoretic Forces in Atmospheric Conditions

Distribution and scavenging of aerosol particles in the atmosphere of the Earth play an important role for cloud formation, providing condensation nuclei for droplets and ice crystals, and for the accretion of hydrometeors. Knowledge of the mechanisms governing the migration and accretion of aerosol particles in aggregates of fractal structure is therefore important both from the fundamental viewpoint and for the development and verification of computer programs simulating particles diffusion. Among the transport mechanisms of aerosol particles in the atmosphere, thermophoretic forces originating from temperature gradients are, together with diffusiophoretic forces, among the most important ones. Theoretical analysis of thermophoretic forces has been made in case of atmospheres characterized by small or intermediate Knudsen numbers, and led to the analytical expression of thermophoretic velocity provided by the so-called Waldmann formula (Waldmann, 1959).

In case of large Knudsen numbers, therefore in the higher elevations of the atmosphere, theoretical analysis of thermophoretic forces has not led to satisfactory analytical formulations. To comprehend and characterize thermophoretic forces in these domains, and especially in the so-called transition region of intermediate Knudsen numbers, it is therefore necessary to complement the analytical work with experimental investigations and with computational work.

Experimental investigations of thermophoretic forces under microgravity conditions can be conducted either in drop-tower facilities or in parabolic flights. Both kinds of these experiments were conducted under cooperation between the Institute of Atmospheric Sciences and Climate (ISAC) of the National Research Council (CNR) of Italy and the Microgravity Research Centre of the Université Libre de Bruxelles, Belgium. Preliminary results were published (Prodi et al., 2004).

As an alternative to the experimental work, numerical solutions of transport equations for the gas flow around a particle would provide insight into the problem of describing and quantifying thermophoretic forces. Earlier numerical attempts were mainly based upon solution of simplified forms of the Boltzmann equation, including the BGK approximation (Bhatnagar et al., 1954). Among these earlier attempts we just recall the numerical solutions by Yamamoto and Ishihara (1988) and Ohwada and Sone (1992).

The experimental investigations are paralleled by a computational effort aiming at simulating thermophoretic forces and migration of aerosol particles in zero-gravity conditions. This effort led to the development of the aforementioned THEPHOR program which has been the basis for the development of the THERCONF code documented in this chapter.

6.2 Theoretical Background

The most important basic concepts, already introduced in Chap. 1, are recalled and summarized in this section for the sake of continuity of the presentation.

6.2.1 *Preliminaries and Basic Concepts*

We consider a fluid with number density (number of molecules per unit volume) n. If δ is the mean average spacing between the molecules of the fluid, then it holds

$$\delta^3 = \frac{1}{n} \tag{6.1}$$

hence

$$\delta = n^{-1/3} \tag{6.2}$$

The mass of a molecule is

$$m = \frac{M}{N_\circ} \tag{6.3}$$

where M is the molecular weight and $N_o \approx 6.02252 \times 10^{26}$ is the Avogadro number. It holds

$$n = \rho \frac{N_\circ}{A}, \tag{6.4}$$

where A is the atomic number and the density of the fluid is

$$\rho = nm. \tag{6.5}$$

Letting d be the diameter of the molecules, the geometrical collision cross section is

$$\sigma_T = \pi d^2. \tag{6.6}$$

Let us define: $\vec{c}$ velocity of a molecule, $\vec{c}_\circ$ mean stream velocity, $\vec{c}' = \vec{c} - \vec{c}_\circ$ velocity of a molecule relative to the mean stream velocity, $\vec{c}_r$ relative velocity between one chosen molecule ("test" or "sample" molecule) and the field molecules.

The mean collision rate per molecule, or collision frequency, ν, is obtained multiplying the mean volume swept by a molecule in the unit time ($\sigma_T c_r$) with the number density, n:

$$\nu = n\sigma_T c_r = \frac{1}{t_c}, \tag{6.7}$$

where t_c is the mean collision time, or mean time interval between collisions.

The total number of collisions per unit time, N_c, is obtained by multiplying the collision frequency, ν, with the number density, n, and dividing by 2 (to avoid to count twice the same couple of molecules):

$$N_c = \frac{1}{2}n\nu = \frac{1}{2}n^2\sigma_T c_r. \tag{6.8}$$

The mean free path of a molecule is obtained by dividing its velocity relative to the mean stream by the collision frequency:

$$\lambda = \frac{c'}{\nu} = \frac{c'}{n\sigma_T c_r}. \tag{6.9}$$

If Eq. (6.6) holds, one obtains

$$\lambda = \frac{c'}{n\pi d^2 \overline{c}_r}. \tag{6.10}$$

In case of a gas in thermodynamic equilibrium the mean relative velocity is given by

$$\overline{c}_r = \sqrt{2}\overline{c}', \tag{6.11}$$

hence, from Eq. (6.9)

$$\lambda = \frac{1}{\sqrt{2}n\sigma_T} = \frac{m}{\sqrt{2}\rho\sigma_T}. \tag{6.12}$$

The pressure tensor is defined by

$$\dddot{P} = nm\,\overrightarrow{c}'\overrightarrow{c}' = \rho\,\overrightarrow{c}'\overrightarrow{c}', \tag{6.13}$$

or, referring to its components, by

$$P_{ij} = \rho c_j' c_j' = nmc_i' c_j'. \tag{6.14}$$

The components of the vector $\vec{c}\,'$ will be denoted u', v', w'. The scalar pressure is given by

$$p = \frac{1}{3}\rho\left(u'^2 + v'^2 + w'^2\right) = \frac{1}{3}\rho c'^2 = \frac{1}{3}nmc'^2. \tag{6.15}$$

The viscous stress tensor $\ddot{\tau}$ is defined by the equation

$$\dddot{P} = -\ddot{\tau} + p\,\dddot{\delta}, \tag{6.16}$$

where $\dddot{\delta}$ is the unit tensor. Hence

$$\tau_{ij} = -P_{ij} + p\delta_{ij}. \tag{6.17}$$

The average kinetic energy is

$$E = \frac{1}{2}mc'^2. \tag{6.18}$$

The translational kinetic energy, or specific energy, is defined by

$$e_{tr} = \frac{1}{2}c'^2. \tag{6.19}$$

Combining Eqs. (6.15) and (6.19) one derives

$$p = \frac{2}{3}\rho e_{tr}. \tag{6.20}$$

The translational kinetic temperature T_{tr} is defined for a gas by

$$\frac{3}{2}RT_{tr} = e_{tr} = \frac{1}{2}c'^2, \tag{6.21}$$

where R is the gas constant. Using the Boltzmann constant

$$k = mR \tag{6.22}$$

Eq. (6.21) becomes

$$
\begin{aligned}
\frac{3}{2}kT_{tr} &= me_{tr} = \frac{1}{2}mc'^2 = \frac{1}{2}m\left(u'^2 + v'^2 + w'^2\right) = \\
&= \frac{1}{2}m\left((u-u_0)^2 + (v-v_0)^2 + (w-w_0)^2\right) = \\
&= \frac{1}{2}m\left[u^2 + v^2 + w^2 - \left(u_0^2 + v_0^2 + w_0^2\right)\right] = \\
&= \frac{1}{2}m\left(c^2 - c_0^2\right)
\end{aligned}
\tag{6.23}
$$

The translational kinetic temperature can be computed, using Eq. (6.23) by

$$
\begin{aligned}
T_{tr} &= \frac{m}{3k}\langle c'^2\rangle = \frac{m}{3k}\left(\langle c^2\rangle - c_0^2\right) = \frac{m}{3k}\left(\frac{1}{N}\sum_{i=1}^{N} c_i^2 - \frac{1}{N}\sum_{i=1}^{N} c_0^2\right) \\
&= \frac{m}{3k}\left(\frac{1}{N}\sum_{i=1}^{N} c_i^2 - c_0^2\right),
\end{aligned}
\tag{6.24}
$$

where the symbol $\langle\rangle$ denotes mean value and

$$
c_0 = \frac{1}{N}\sum_{i=1}^{N} c_i. \tag{6.25}
$$

In the applications we consider a number N of sample molecules per unit volume, each one representative of a number F_N of real molecules. Thus it holds$n = NF_N$.

6.2.2 *Implementation of the Larsen-Borgnakke Method in the Code*

Some major topics, concerning the rotational temperature of diatomic molecules, the Larsen- Borgnakke phenomenological method (Larsen & Borgnakke, 1974) and the CLL (Cercignani-Lampis-Lord) model (Cercignani & Lampis, 1974; Lord, 1991) were treated in Chap. 1 and therfore we skip to the implementation of these concepts in the code.

The preliminary version of the code THEPHOR was used only for simulation of the thermophoretic force acting on one particle. The code THEPHOR was strictly two-dimensional in the sense that motions of the sample molecules occur in the coordinate plane $y = 0$ and the velocity components of the molecules in the y-direction are not even defined.

As long as the only defined temperature is the translational temperature and the inner energy of the molecules in neglected, hence the rotational temperature is not defined, the two-dimensional approximation is acceptable. The translational temperature is associated to the kinetic energy of each one of the three coordinate directions by means of $E_{tr} = (\zeta_{tr}/2)RT_{tr}$, where ζ_{tr} is the number of the translational degrees of freedom. This relation can be equally well applied for a two-dimensional or a three-dimensional flow.

In case the rotational molecular energies are accounted for and the rotational temperature is defined, the two-dimensional flow representation is useless, because the rotational energy and the rotational temperature cannot be associated to the translational degrees of freedom. Therefore, a three-dimensional definition of molecular velocities is necessary and intermolecular collisions need to be calculated in three-dimensional case, even if the y-component of the velocity is not used for the calculation of the displacements of molecules in the y coordinate direction.

For these reasons, in the implementation of the calculation of inner rotational energies and rotational temperature in the THERCONF program, the first important development of the code consisted in defining all through the program the y-component of molecular velocities and in introducing three-dimensional collisions.

The subroutine used for elastic intermolecular collisions has been renamed COLMOD_1 and a new subroutine called COLMOD_2 has been written for the simulation of the inelastic collisions, based upon the Larsen-Borgnakke method. The calculation of the post collision translational and rotational energies for a couple of molecules is made in the Subroutine INELRS. The post collision molecular velocities have been computed with the Variable Hard Sphere (VHS) model. An implementation of the Variable Soft Sphere (VSS) model is envisaged. The calculation of the translational, rotational, and overall temperatures is made in Subroutine TEMPER.

Test case A8 (see Sect. 6.10.1) is a check of the calculation of the rotational and overall temperatures, in addition to the translational temperature.

6.3 Calculation of Effective Cross Section of Air Molecules

For this section, see Sect. 1.3.3 of Chap. 1.

6.4 Interaction Between Gas Molecules and Particles

6.4.1 Kinetic Energy Incident upon and Reflected from Particles

We compute the kinetic energy transferred from a gas molecule to a particle upon impact, the kinetic energy reflected, associated to a molecule emerging from the particle, and their difference, which is the net energy transferred from the air molecule to the particle.

The kinetic energy of a sample molecule incident with velocity c_i upon the particle is

$$e_{kin,i} = \frac{1}{2} m c_i^2 \tag{6.26}$$

Let F_N be the number of real molecules represented by a sample air molecule. The real kinetic energy associated to a sample molecule is

$$E_{kin,i} = F_N e_{kin,i} = \frac{1}{2} m F_N c_i^2. \tag{6.27}$$

Similarly, the kinetic energy associated to a molecule reflected from the particle (subscript r) with velocity c_r is

$$E_{kin,r} = F_N e_{kin,r} = \frac{1}{2} m F_N c_r^2. \tag{6.28}$$

In the code the components u_r, w_r of the velocity c_r are computed sampling a Maxwellian velocity distribution corresponding to the temperature of the impact point.

6.4.2 Momentum Exchange

Let $\vec{C}_p$ and M_p denote velocity and mass of a particle and indices 1 and 2 refer to the velocities of an air molecule and of the particle before and after a collision. The momentum conservation implies that

$$m \vec{c}_1 + M_p \vec{C}_{p1} = m \vec{c}_2 + M_p \vec{C}_{p2}. \tag{6.29}$$

From Eq. (6.29) the post-collision velocity components of the particle are derived as

$$W_{p2} = W_{p1} + \frac{m}{M_p}(w_1 - w_2), \tag{6.30}$$

$$U_{p2} = U_{p1} + \frac{m}{M_p}(u_1 - u_2). \qquad ((6.31)$$

6.5 Wall Reflection

Section 6.5 is treated in Sect. 1.5 of Chap. 1.

6.6 Collisionless Free Motion and Collision Model

In the numerical simulation, the principle of uncoupling is applied by splitting the calculation in the collisionless motion of the molecules followed by the simulation of collision events among the molecules.

6.6.1 Collisionless Free Motion

Following the collisionless motion of the molecules, their position coordinates are updated by

$$\begin{aligned} x^{n+1} &= x^n + u\Delta t \\ z^{n+1} &= z^n + w\Delta t. \end{aligned} \tag{6.32}$$

6.6.2 Collision Frequency

Let us consider a cell of the computational domain with volume V_c containing a number $N = nV_c/F_N$ of sample molecules, where F_N is the number of real molecules represented by every sample molecule. The collision number between all molecule pairs in the cell is (Bird, 1994)

$$C_N = \frac{N^2}{2} F_N \frac{\sigma_T c_r \Delta t}{V_c}, \tag{6.33}$$

where σ_T is the collision cross section and c_r is the relative velocity of the molecules. The maximum theoretical collision number $C_{N,\,\max}$ is given by the same formula (6.33) taking the maximum theoretical value $(\sigma_T c_r)_{\max}$ of the product between the collision cross section and the relative velocity of the molecules. In the numerical computations the maximum number $C_{N,\,\max}$ of collisions is considered and, among this theoretical number of collisions events, only those are accepted which satisfy the condition

$$R_1 \le \frac{C_N}{C_{N,\max}} = \frac{\sigma_T c_r}{(\sigma_T c_r)_{\max}} \tag{6.34}$$

where R_1 is a random deviate.

6.6.3 Collision Dynamics (Elastic Collisions)

In this section we consider the case of elastic intermolecular collisions.

Within a computational cell collision pairs are chosen randomly among the sample molecules. Let indices 1 and 2 refer to the two molecules of a random chosen couple and let $\vec{c}_r = \vec{c}_1 - \vec{c}_2$ be the relative velocity before collision. For molecules of equal mass the centre of mass velocity is $\vec{c}_m = \left(\vec{c}_1 + \vec{c}_2\right)/2$. Let us denote with a star (*) as superscript the post-collision velocities. For the collision dynamics we use the isotropic "hard sphere" collision model in which the post-collision relative velocity has the same magnitude as the pre-collision relative velocity: $\left|\vec{c}_r^{\,*}\right| = \left|\vec{c}_r\right|$. In an isotropic scattering model the components of the post-collision relative velocity can be computed by multiplying a unit vector generated with the Marsaglia technique (Marsaglia, 1972), with the module of the relative velocity:

$$\begin{aligned} u_r^* &= S c_r \cos\beta \\ w_r^* &= C_s c_r, \end{aligned} \tag{6.35}$$

using two uniform deviates R_1 and R_2 and computing

$$\begin{aligned} \beta &= 2\pi R_1 \\ C_s &= 1 - 2R_2 \\ S &= \sqrt{1 - C_s^2}. \end{aligned} \tag{6.36}$$

Because the components u_m and w_m of the centre of mass velocity are known from the pre-collision velocity components, the components of the post-collision velocities of the two molecules are given by

$$\begin{aligned} u_1^* &= u_m + \frac{u_r^*}{2} \\ u_2^* &= u_m + \frac{u_r^*}{2} \end{aligned} \tag{6.37}$$

and

$$w_1^* = w_m + \frac{w_r^*}{2}$$
$$w_2^* = w_m + \frac{w_r^*}{2} \tag{6.38}$$

6.7 Layout of Particles Simulation

6.7.1 Injection and Motion of Particles

The number of simulated particle is arbitrary, from one to an upper limit set only by space requirements and computational time. The latter increases more than linearly with the number of simulated particles not because of the computational time needed to move and manage the particles, but because of the increasing number of collisions between molecules and particles. The code provides for two options regarding the injection of the particles into the computational domain: The injection can occur at time $t = 0$ (option $IT\_INJECT = 0$ from INPUT File) or at a specified time step $IT = IT\_INJECT$, larger than zero. The second option is preferable because it allows for the preliminary calculation of the temperature distribution in the computational domain, unperturbed by the presence of the particles, which corresponds to the experimental situation where the gas is in thermal equilibrium with the bounding surfaces before particles injection.

Particles are injected in a domain defined by ($I1PAR \leq I \leq I2PAR$; $K1PAR \leq K \leq K2PAR$), where I and K are mesh indices for the x- and z-directions, respectively. Coordinates of particles are assigned at random in the given domain, radiuses R_NPAR of particles are assigned at random in the domain ($RMIN\_NPAR \leq R\_NPAR \leq RMAX\_NPAR$), where the boundaries of the range are INPUT parameters. Initial velocities of the particles can be assigned either zero or using the Waldmann formula for the thermophoretic velocity

$$\vec{V}_{th} = \frac{3\nu}{4T} \frac{\nabla T}{1 + \frac{\pi}{8}}. \tag{6.39}$$

where ν is the kinematic viscosity of the gas and ∇T is the temperature gradient.

If a particle leaves the computational domain, it is considered lost for the calculation and treated as specified below. The general management of the particles follows the following rules.

(i) **At injection**. Every particle NP, besides being assigned initial coordinates X_NPAR, velocity C_NPAR and size R_NPAR, hence (given the density) mass $AMASS\_NPAR$, is assigned two parameters:

$$IND\_NPAR(NP) = 1$$

$$FNU\_NPAR(NP) = 1.0$$

The parameter *IND _ NPAR*, as long as it is equal to 1, signals that the particle is still retained in the active particle's calculation. Otherwise it is by-passed for the reasons explained below. The parameter *FNU _ NPAR* gives the "weight" of a particle, or the number of particles it actually represents in the calculation (see below for further explanation).

(ii) **Upon inter-particle collisions**. The code provides for two assumptions concerning the collision and coalescence of particles. Coalescence of particles is assumed either when the distance of the centres of the particles is less than a given threshold (typically 0.5 μm plus the sum of the radii) or when two particles are in the same computational mesh. The second option is a crude one, but still acceptable for all practical purposes, if the mesh size is small (about 10 μm). Coalescence of particles is simulated according to the following rules.

Let assume that two particles with indices $NP1$ and $NP2$, with ($NP1 < NP2$), collide. The particle with lower index is retained with unchanged *IND _ NPAR* and assumes a "weight" *FNU _ NPAR*($NP1$) incremented by 1 [more precisely, incremented by the weight of particle $NP2$ which might have already increased above 1, following previous collisions]. In the subsequent calculations of the momentum of inertia of the particles with respect to their centre of mass, particle $NP1$ participates with its increased weight. Particle $NP2$ receives index equal to $-NP1$ and henceforth the coordinates of particle $NP1$. The mass of particle $NP2$ is set to zero, so that it does not contribute any more to the calculation of the inertial momentum of all particles with respect to their centre of mass. Therefore after collision and coalescence the parameters of the two particles become as follows:

$$IND\_NPAR(NP1) = 1$$

$$FNU\_NPAR(NP1) = FNU\_NPAR(NP1)|_{old-value} + FNU\_NPAR(NP2)$$

$$AMASS\_NPAR(NP1) = unchanged$$

$$IND\_NPAR(NP2) = -NP1$$

$$FNU\_NPAR(NP2) = 0$$

$$AMASS\_NPAR(NP2) = 0$$

(iii) **Upon collision of a particle with a coil**. If a particle NP collides with a coil, its index *IND _ NPAR*(NP), its mass *AMASS _ NPAR*(NP), and its weight *FNU _ NPAR*(NP) are set to zero. Its coordinates are fixed to the last values before collision. Hence:

$$IND\_NPAR(NP) = 0$$

$$FNU\_NPAR(NP) = 0$$

$$AMASS\_NPAR(NP) = 0$$

(iv) **Upon exiting the computational domain.** If a particle *NP* exits the computational domain of particles its index *IND _ NPAR(NP)* is set to -99,999, its mass *AMASS _ NPAR(NP)* and its weight FNU_NPAR(np) are set to zero. Its coordinates are fixed to the last values before exiting. Hence:

$$IND\_NPAR(NP) = -99,999$$

$$FNU\_NPAR(NP) = 0$$

$$AMASS\_NPAR(NP) = 0$$

These computational rules allow to apply to the particles the same "equivalence principle" applied to the sample molecules. A sample molecule is equivalent to 'Fn' real molecules (Typically *Fn* is in the range $10^{12} \div 10^{15}$). Likewise, every particle which has retained index $IND\_NPAR = 1$ is equivalent to a number of particles given by *FNU _ NPAR*. Following multiple inter-particle collisions the number of particles decreases, but the sum of all indices *FNU _ NPAR* remains equal to the total number of particles initially injected into the domain. There is however, besides the order of magnitude of the equivalence, an important difference between molecules and particles. The equivalence number for molecules is a fixed number set at the beginning of the calculation and depending on the ratio between the number of real molecules and of sample molecules. The equivalence number of particles changes during the calculation because it depends on the computed coalescences.

The computer program monitors with a given frequency the sum of particles still considered in the calculation (with indices $IND\_NPAR = 1$), coalesced (with indices $IND\_NPAR = -NN$, where *NN* is the index of the particle with which coalescence occurred), stuck on coils (with indices $IND\_NPAR = 0$), lost from the computational domain of particles (with indices $IND\_NPAR = -99,999$). In a calculation simulating good experimental conditions, where the temperature gradients allow for efficient thermal confinement of the particles, the number of particles stuck on coils or lost from the computational domain should be a small fraction of the total number of particles. Theoretically, at the end of a "perfect" calculation, all particles should be collected by particle number 1 in the position of the minimum of the temperature distribution. We will see in the following that this perfect result can be achieved in the frame of a calculations based on the classical Waldmann-treatment of the thermophoretic forces, if the distribution of temperature, computed with the molecular dynamic approach, is smoothed by means of some numerical technique which gets rid of small local minima. This test, based on the classical approach, has been made to prove the correctness of all algorithms programmed for the management of the motion, collision, and coalescence of the particles.

Following particles injection (Subroutine INJECT_NPAR) motion of particles is simulated in the subroutine MOVE_NPAR, where the following calculations are made.

(i) Increment of the particles velocity, based on conservation of the quantity of motion in the collisions molecule-particle. The updated velocity of one particle which underwent collision is given by

$$\vec{C}_p^{\,n+1} = \vec{C}_p^{\,n} + \frac{\sum_s \vec{c}_s m}{M_p}, \tag{6.40}$$

where subscript p refers to the particle, subscript s denotes the number of collisions occurred between molecules ant the particle considered, m and $M_p$are the masses of molecules and particle and superscript n refers to the dime discretization.

(ii) Coalescence of particles is computed either with the criterion of the minimum distance or for particles in the same mesh.

(iii) The momentum of inertia of all particles, $AMOM\ \_\ NPAR$, is computed (Subroutine XBARM) and the ratio $RATIO\ \_\ AMOM$ between this value and the initial momentum at the time of injection is computed. This ratio is one of the key parameters to judge the efficiency of the thermal confinement of particles.

(iv) Because of the displacement of particles, which may pass from one mesh to another one, the geometry of the computational domain is recomputed (Subroutine GEOM_NPAR).

Particles of small dimension, with very small thermal capacity, have a rather uniform temperature distribution. They are assigned the temperature of the mesh in which they are located. In case of the simulation of a large particles, where the temperature distribution cannot be considered uniform, the code provides for the calculation of this distribution taking into account the kinetic energy exchanged with the colliding molecules.

6.7.2 Temperature Distribution in Large-Size Particles.

For Sect. 6.7.2 see Sect. 5.2.6 of Chap. 5.

6.8 Convergence Criterion

For Sect. 6.8 see Sect. 5.2.7 of Chap. 5.

6.9 Structure of the Computational Code THERCONF

The code, written in FORTRAN 77 of 100 subroutines for a total of about 14,000 statements, out of which about one-third are comments or descriptions of the main tasks performed within the subroutines. The main subroutine of the code, called THERCONF, which acts as the driving program, is shown in a somewhat simplified form in the block diagram of Fig. 6.2. This block diagram includes several modules, from Block A till Block I, which are explained in the following.

Block A refers to the initialization of the code through the input data which are read from file INPUT for the main (standard) calculation and from files INPUT_TEST_AN ($N = 1, 2,\ldots, 8$) for the test cases (see Sect. 6.10.1).

Block B refers to the restart file, which is read if the option IREST = 1 is chosen, which implies that the calculation being started is a continuation of a previously run. For the standard calculation the file OUTRES is read and than saved in OUTRES_OLD. For the test cases the files OUTRES_TEST_AN ($N = 1, 2,\ldots,8$) are read and are not saved.

Block C is the main initialization of the code, before the time loop starts. It includes definition of physical constants (Subroutine CONSTANTS), calculation of physical properties of the fluid (Subroutine PHYS_PROP), injection of particles (Subroutine INIT_NPAR) in case it was chosen to inject them at time zero, definition of the standard geometry (Subroutine GEOMZE), modification of the geometry in case of presence of particles (Subroutine GEOMZE_NPAR), and modification of geometry in case coils are simulated (Subroutine GEOM_COIL). Initialization of the fluid in the computational domain is made with Subroutine INITF, which assigns random positions to the molecules and velocity components obtained by sampling a Maxwellian distribution at the initial reference temperature. Maximum, minimum, and mean values of the velocity components of the molecules and translational temperature (Subroutine CHKMET) are computed for initial checks. All fields which at time zero do not store relevant information are zeroed (Subroutine ZEROFL).

The time-dependent calculation starts with the time loop at pointer P. Counter IT ($1 \leq IT \leq ITMAX$) refers to the sequence of time steps. The first important block of the time loop, **Block D** computes the displacement of the molecules. A loop with counter M ($1 \leq M \leq NMOL$) runs over all molecules, treating them sequentially (unless a parallel calculation, not documented in this section, is made). Following the displacement of a molecule, Subroutine SCREEN (not shown in the block diagram) checks that the molecule is still within the computational domain. In this case (flag INDEX = 0) the subsequent molecule is treated. Otherwise flag INDEX is set different from zero and conveys the following information:

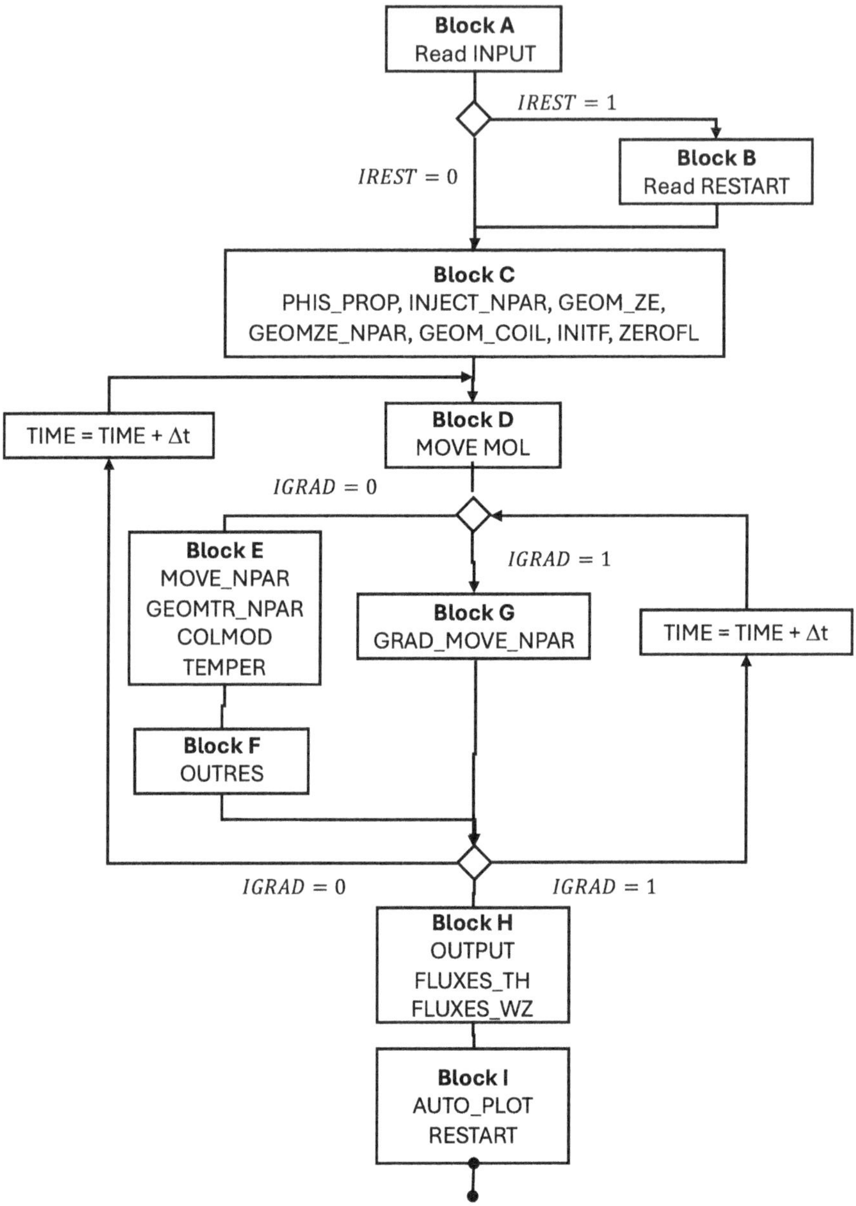

Fig. 6.2 Block diagram of the THERCONF code

INDEX = 1, 2,…,6, molecule exited crossing one of the bounding surfaces of the domain;
INDEX = 7, molecule collided with the particle with minimum index NP = 1;

INDEX = 11,...,16, molecule collided with one of the simulated coil cross sections;
INDEX ≥ 1000, molecule collided with one of the particles with index $NP \geq 2$.

In case of collision of the molecule with one of the bounding walls (INDEX = 1, 2,...,6) specular or diffuse reflection from the walls is computed (Subroutines SPECDI). In case of collision of the molecule with a particle (INDEX = 7 or INDEX ≥ 1000) diffusion of the molecule from the surface of the particle is computed (Subroutine DIFFUS_NPAR). In case of collision of the molecule with a coil cross section (INDEX = 11,...,16) diffusion of the molecule from the surface of the coil is computed (Subroutine DIFFUS_COIL). If INDEX ≥ 1000 diffusion of the molecule from the parrticle is computed.

Block E is the pendant of block D, for the particles. Injection of particles (INIT_NPAR) is made in case it was chosen to initialize particles at $IT \geq 2$. After injection, particles move under the impact of the momenta transferred to them from colliding molecules (Subroutine MOVE_NPAR). Particles are coalesced (Subroutine COALESCE_NPAR) in case there is more than one particle per computational cell or (optionally) in case the minimum distance between particles is below a given threshold. The inter-molelcular collision model is then applied. It is coded in Subroutine COLMOD_1 for monoatomic molecules and in Subroutine COLMOD_2 for biatomic molecules. At pointer C14 computational data are sampled with a chosen frequency for allowing final presentation of statistically averaged results. Temperature of particles is computed (Subroutine TEMP_NPAR). In case of small particles it is assumed that their temperature is the mean gas temperature in the cell. In case of large particles their temperature can be computed taking into account the heat exchanged with the surrounding gas. Finally the geometry of the computational domain is updated taking into account the presence of particles (Subroutine GEOM_NPAR).

In **Block F** intermediate restart files are written for safety with a frequency IFRERE, chosen from INPUT. Optionally automatic plot generation (Subroutine AUTO_PLOTS) can be made every million time steps.

Block G is used as alternative to the main path of the calculations to simulate the displacement of the particles under the impact of forces arising from temperature gradients in the gas domain. In this case the classical Waldmann formula for thermophoretic forces is applied. After injection of the particles (Subroutine INIT_NPAR), the temperature field in the computational domain is smoothed by filtering out small temperature oscillations due in part to the fact that the calculation was not long enough to reach a stable state and in part due to the shadowing effect of the particles. Smoothing of the temperature field aims at realizing the ideal situation in which the Waldmann formula is applicable. Main Subroutines for smoothing the temperatures are GRAD_SMOOTHX_TEMPER and GRAD_SMOOTHZ_TEMPER (not shown in the block diagram) which make use of a Fourier transform to eliminate high frequency oscillations. The components of the temperature gradients are computed in Subroutine GRAD_TEMPER. Particles are then moved under the impact of the forces arising from these temperature gradients (Subroutine GRAD_MOVE_NPAR). Particles are coalesced (Subroutine COALESCE_NPAR) in case there is more than one particle per computational cell.

The end of the time loop is reached in **Block H** either for IT = ITMAX or in case flag ISTOP was turned to 1 in the INPUT file, which is re-read every 200 time steps. After exiting from the time loop the code makes final averaging of sampled data (Subroutine SAMPLING_AV), prints information about final state of the particles (Subroutine OUTLOOK_NPAR), computes thermal fluxes from the walls (Subroutine FLUXES_TH) and molecular fluxes in the direction normal to the z-axis (Subroutine FLUXES_WZ).

Eventually, in **Block I**, final restart files are written, OUTRES for the standard calculation and OUTRES_TEST_AN ($N = 1, 2, \ldots, 8$) for the test cases. A summary of warning flags is written (Subroutine WARNINGS) and optionally a final set of plots is generated in subroutine AUTO_PLOT for the standard calculation and in subroutine GRAD_AUTO_PLOT for the code option in which thermal gradients were used for displacement of the particles.

6.10 Preliminary Results

A series of test cases, TEST_A1 through TEST_A8, has been prepared with input data selected in a way allowing to test all major branches of the code, for instance both specular and diffusive reflections, thermal creep, buoyancy effects in the Bénard problem, thermophoretic effects with one or more particles, with or without the simulation of the coils, etc. These test cases are presented in Sect. 6.10.1.

A quantitative verification of the code has been attempted with the calculation of heated fluxes in the test conditions documented in Gallis et al. (2002), hereafter referred to as Gallis-Tests. These tests are presented and discussed in next Sect. 6.10.1.

6.10.1 Test Cases A1–A8

As explained above, a series of eight test cases has been selected to test the major models of the code. These test cases are rerun quickly, in a few hours, for about 30,000 time steps, to check the code after every major code development. The program has been organized to compute all test cases in a single computational run.

The main INPUT parameters, which refer to the models of the wall reflections, and to the presence or absence of particles and coils, are summarized in Table 6.1:

In the following we will discuss briefly the relevance of these tests. All tests were run for 50,000 time steps, with exception of test A8 which was run for 10,000 time steps.

TEST_A1. The Bénard problem between two parallel plates with a fluid heated from below is a classical test for computer programs based both on the numerical solution of the Navier-Stokes equations and on the molecular dynamic simulations. We made only a qualitative assessment of the results obtained for normalized

Table 6.1 Overview of test cases A1–A8

	IPARTICLE				
	|	ICIRCLE			
	|	|	ISIDEWALL		
	| | |	| | |	| | |	APPLICATION	
T2 < T1	0	0	0	Square cavity, Benard problem	TEST_A1
T2 > T1	0	0	1	Thermal creep	TEST_A2
T2 > T1	1	0	0	Thermophoresis between plates	TEST_A3
T2 > T1	1	1	1	Thermophoresis with circles C1–C4 and thermal creep contribution. Diffusive reflection from side walls	TEST_A4
T2 > T1	1	1	0	Thermophoresis with circles C1-C4 without th. creep contribution. Specular reflection from side walls	TEST_A5
T2 = T1	1	1	1	Thermophoresis with only circles C1-C4 without thermal creep contribution. Diffusive reflection from side walls	TEST_A6
T2 = T1	0	0	3	All surfaces specularly reflecting	TEST_A7
T2 = T1	0	0	3	Calculation of molecular rotational energy and temperature	TEST_A8

Key	Value	Description
T1		Temperature of lower surface
T2		Temperature of upper surface
IPARTICLE	0	No particles are simulated
	1	Particles are simulated
ICIRCLE	0	No coils are simulated
	1	Coils are simulated
ISIDEWALL	0	Upper and lower surfaces are diffusively reflecting. Side surfaces are specular
	1	All surfaces are diffusively reflecting
	3	All surfaces are specular

temperature distribution, molecular densities, flow pattern, velocity distribution. The typical Bénard cells are clearly formed as can be seen from the plots. The plot displaying the vertical velocity component shows the convective cells. The plot displaying the horizontal velocity component shows the typical anti-symmetric pattern (see Figs. 6.3, 6.4, 6.5, 6.6, 6.7, and 6.8).

Further work: Assess quantitatively size and number of convective cells forming in dependence on the aspect ratio, computed velocities, heat fluxes, frequency of oscillations of the convective cells, etc.

TEST_A2. This test is intended to verify the capability of the code to simulate flow induced by thermal creep. For this purpose the temperature of the side walls is set to change linearly between the temperatures of the lower and of the upper surfaces. The resulting velocity plot (see Fig. 6.9) shows clearly ascending flows at the side walls, along the temperature gradient, and descending flow at the centre of the domain. The situation is clearly illustrated by the plot of the vertical velocity component and that of the anti-symmetric distribution of the horizontal velocity component.

Further work: Assess quantitatively the computed velocities. See for instance the reference (Papadopoulos & Rosner, 1995) (Figs. 6.10, 6.11, 6.12, 6.13 and 6.14).

TEST_A3. This test is intended to simulate the thermophoretic forces acting upon a single particle and the resulting velocity and displacement of the particle.

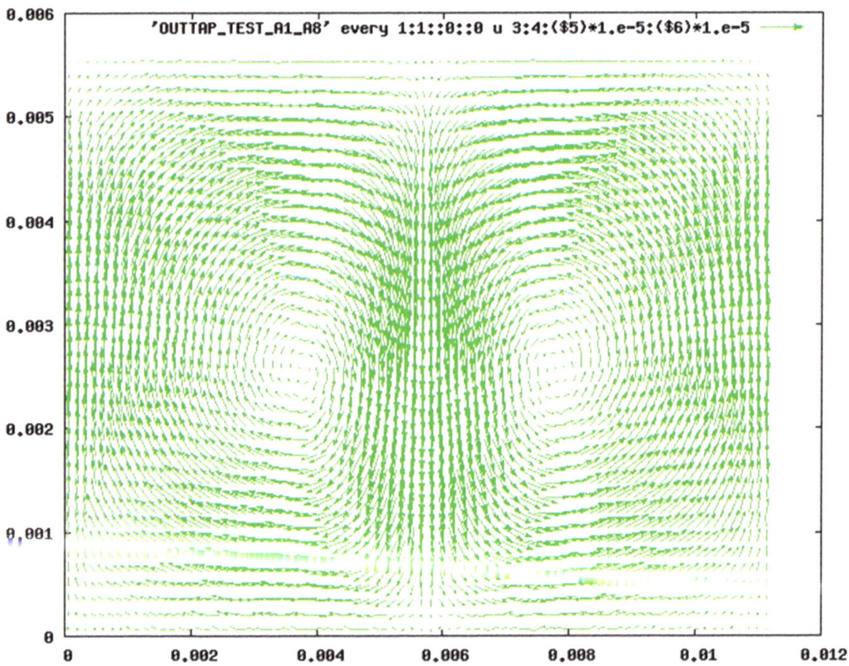

Fig. 6.3 TEST_A1: Velocity plot for Bénard problem

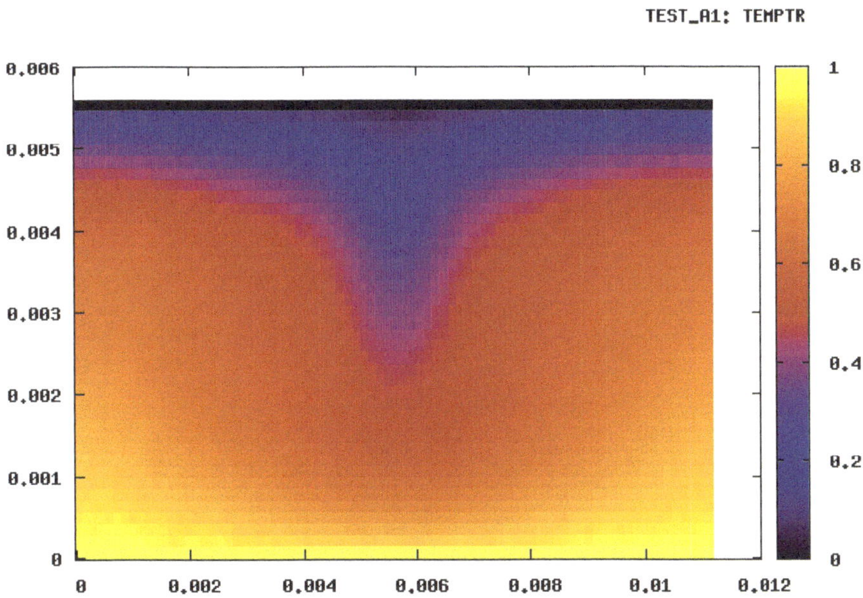

Fig. 6.4 TEST_A1: Normalized temperature distribution for Bénard problem

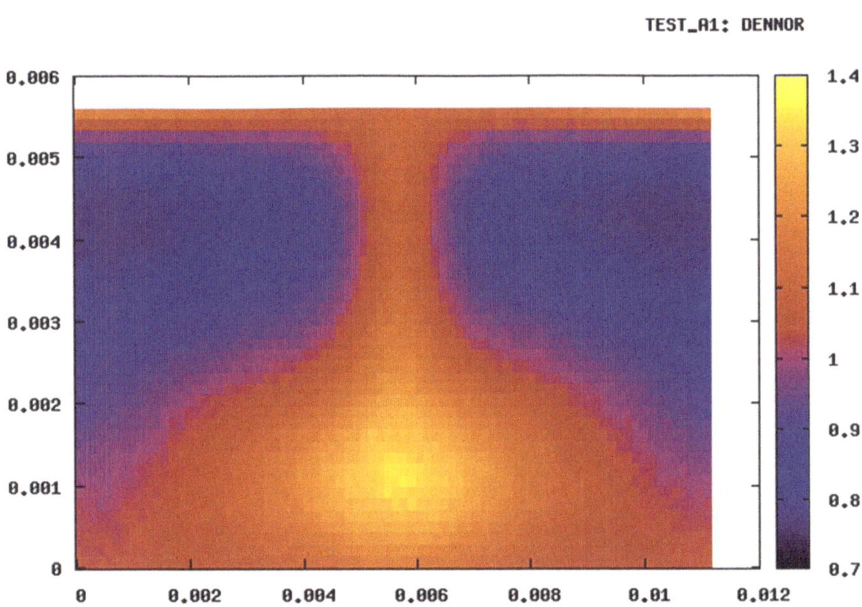

Fig. 6.5 TEST_A1: Normalized molecular density for Bénard problem

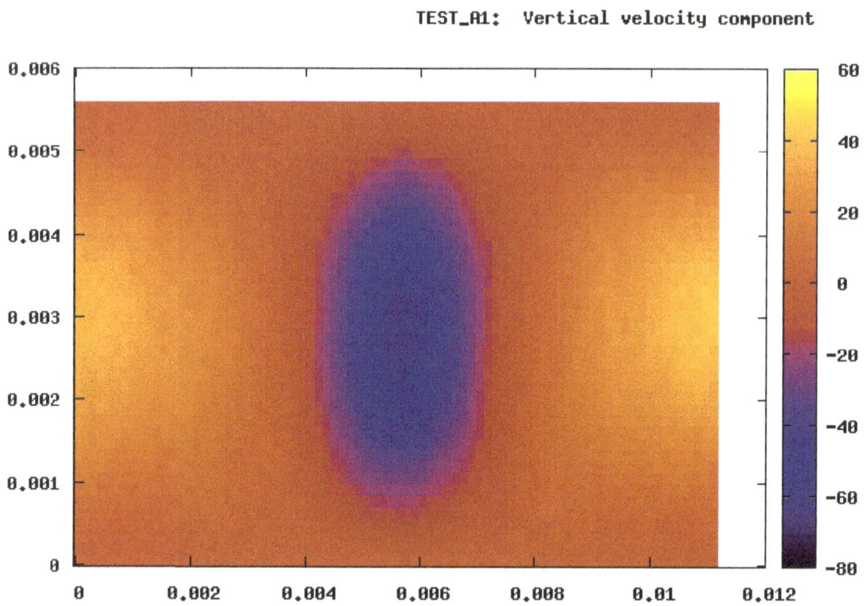

Fig. 6.6 TEST_A1: Distribution of vertical velocity component for Bénard problem

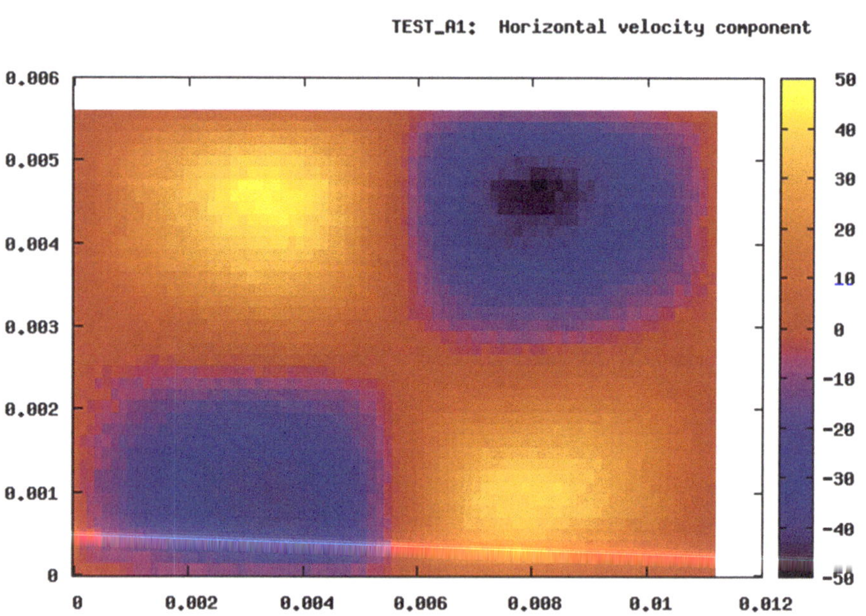

Fig. 6.7 TEST_A1: Distribution of horizontal velocity component for Bénard problem

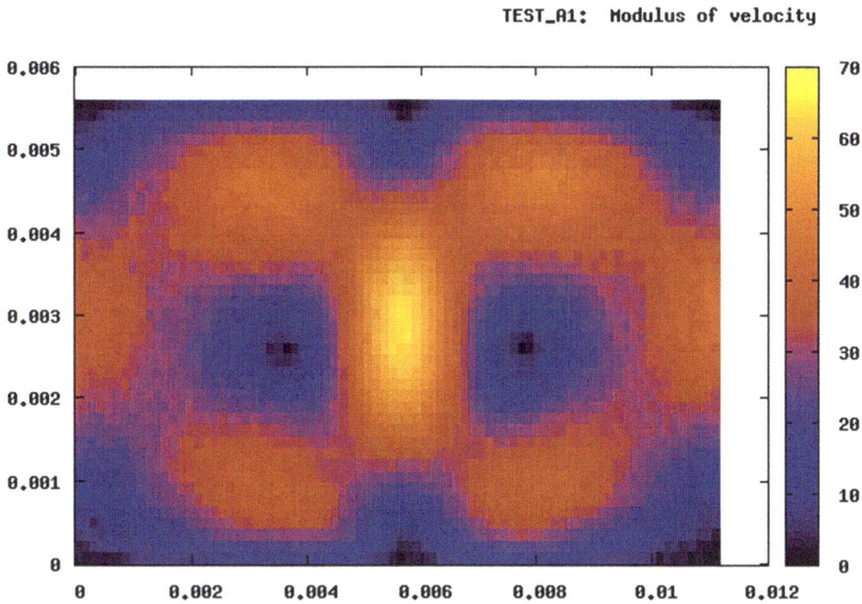

Fig. 6.8 TEST_A1: Distribution of modulus of molecular velocities for Bénard problem

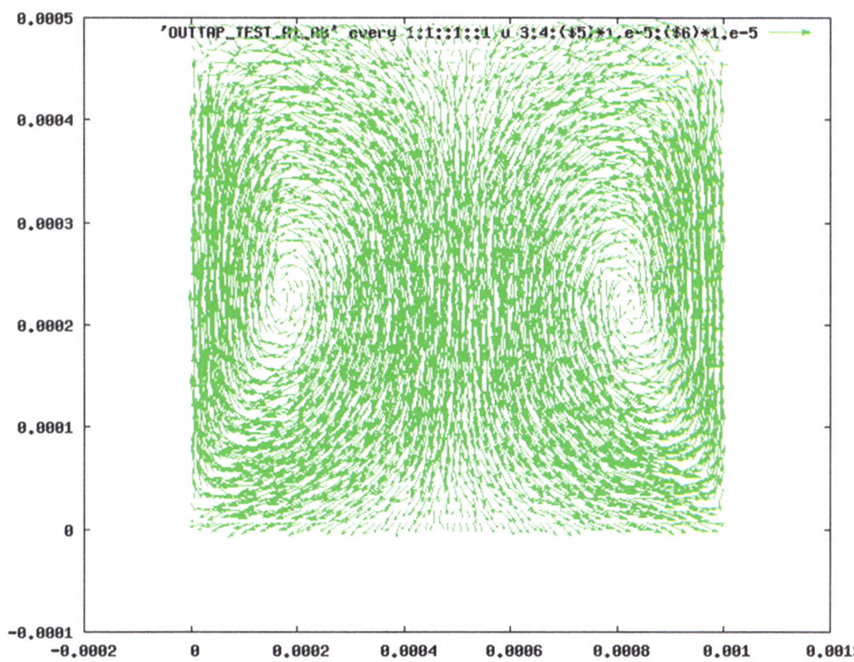

Fig. 6.9 TEST_A2: Velocity plot for thermal creep induced by thermal gradients along the vertical walls

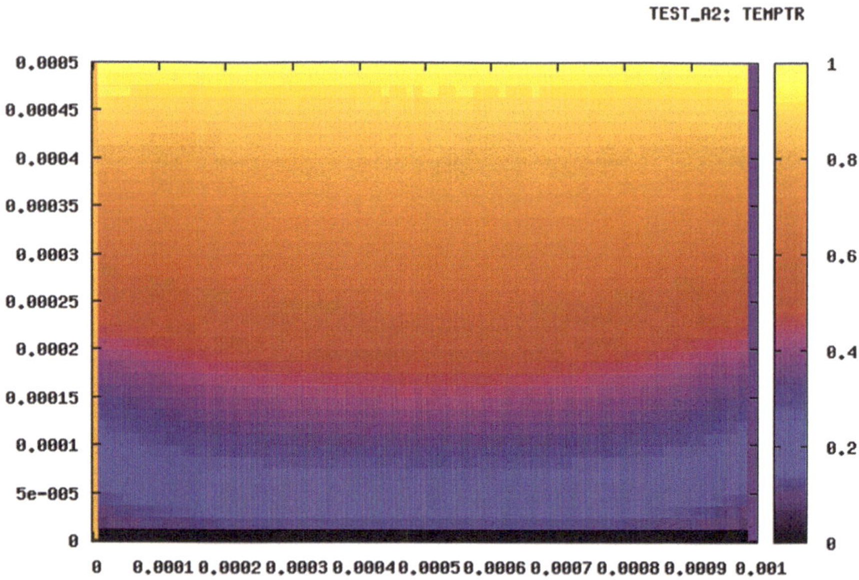

Fig. 6.10 TEST_A2: Normalized temperature distribution for thermal creep problem

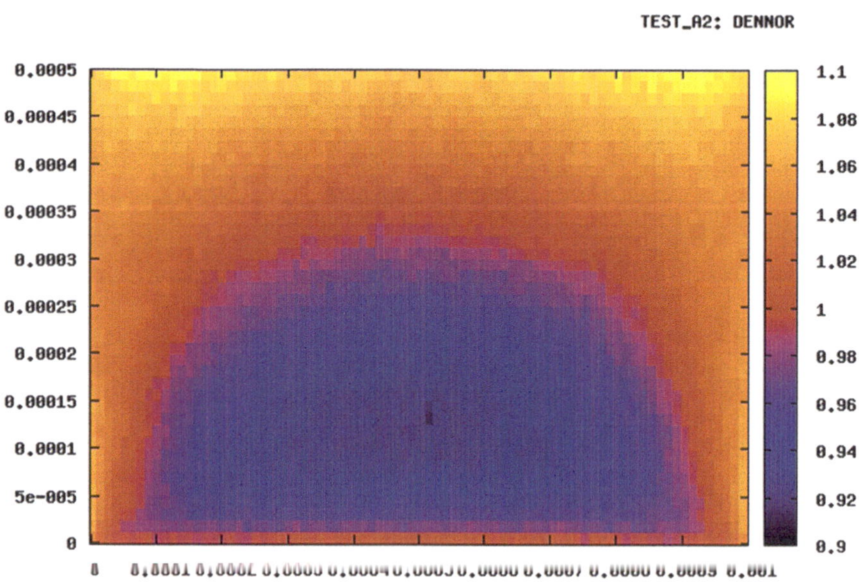

Fig. 6.11 TEST_A2: Normalized distribution of molecular density for thermal creep problem

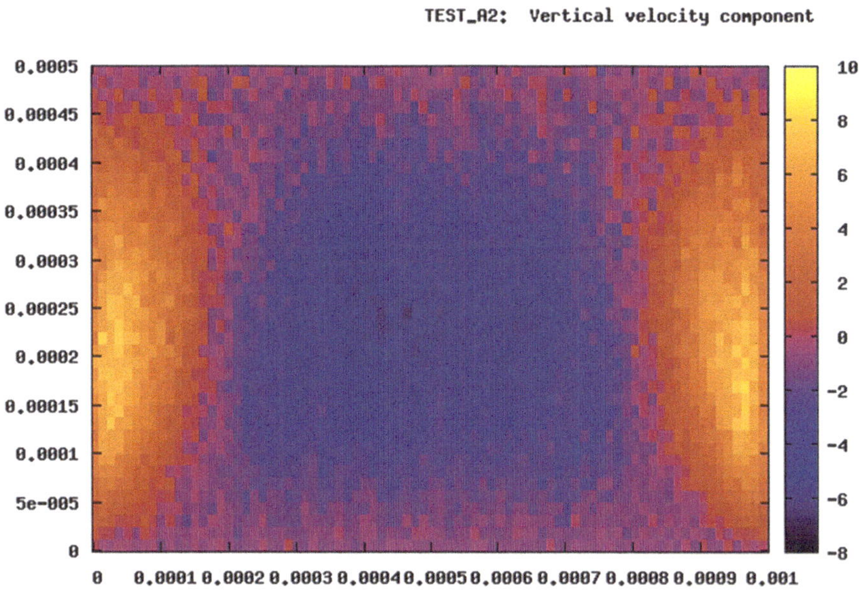

Fig. 6.12 TEST_A2: Vertical velocity component for thermal creep problem

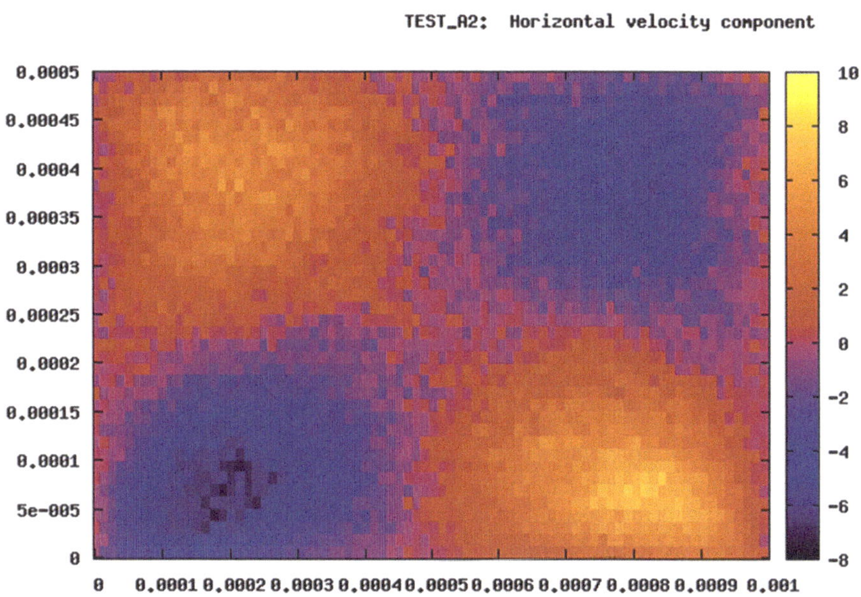

Fig. 6.13 TEST_A2: Horizontal velocity component for thermal creep problem

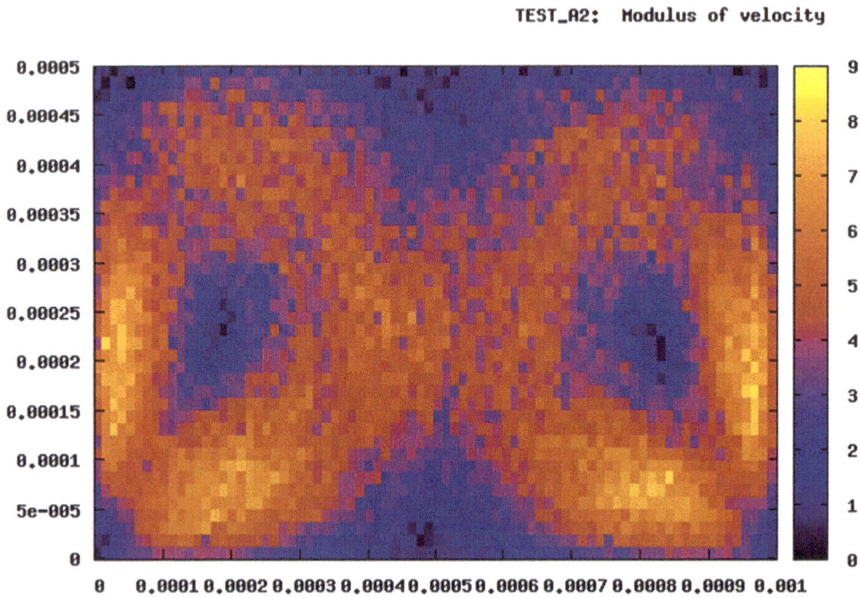

Fig. 6.14 TEST_A2: Distribution of modulus of molecular velocities for thermal creep problem

This test, qualitatively correct for the short time simulated, should be carried on for a longer computational time (Fig. 6.15).

TEST_A4. This test is intended to simulate the thermophoretic forces acting upon a cloud of particles, in presence of the heated coils, and the resulting velocity and displacement of the particles. In the short time simulated the displacement of the particles is minimum, though correctly oriented against the temperature gradient. This test case may be the most important one for the simulation of the ESA ICAPS experiment, and for this reason it has been chosen for the longer calculation discussed in Sect. 0. An example of scattering of molecules from the surface of one sample particle is given in Fig. 6.16.

TEST_A5. This test is intended to assess the importance of the thermal creep contribution for the flow pattern and therefore for the displacement of the particles. Thus far, it seems that the impact of thermal creep upon the stream of the molecules is negligible in case of the presence of the heated coils.

TEST_A6. Setting equal temperatures for upper and lower surfaces (see Fig. 6.17), this test is intended to assess the impact of the size and temperature of the coils upon the displacement of the particles. The test has been simulated earlier up to problem times of about 0.4 s for a cloud of 100 particles. Trajectories of the particles were rather random, suggesting that the thermophoretic forces exerted by the simulated coils alone was not sufficient to thermally confine particles.

Further work: Analyse parametrically the size and temperatures of the coils necessary to thermally confine particles in the centre of the domain.

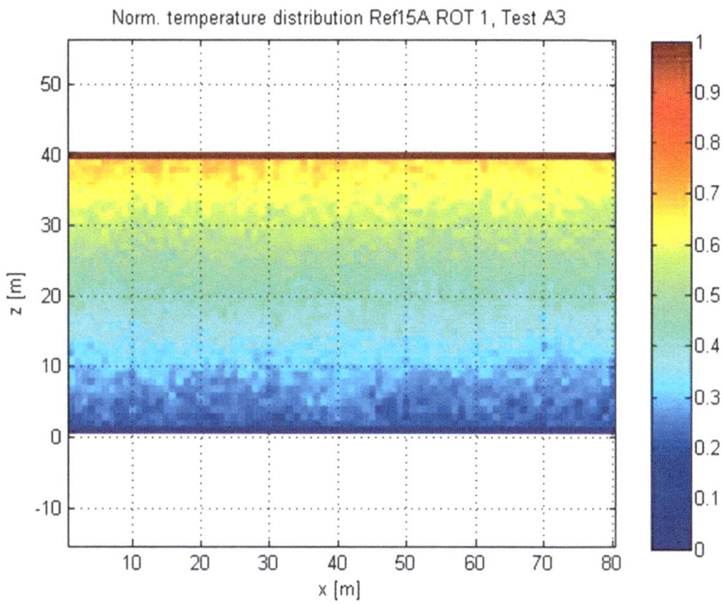

Fig. 6.15 TEST_A3: Normalized temperature distribution in simulation of thermophoresis between parallel plates

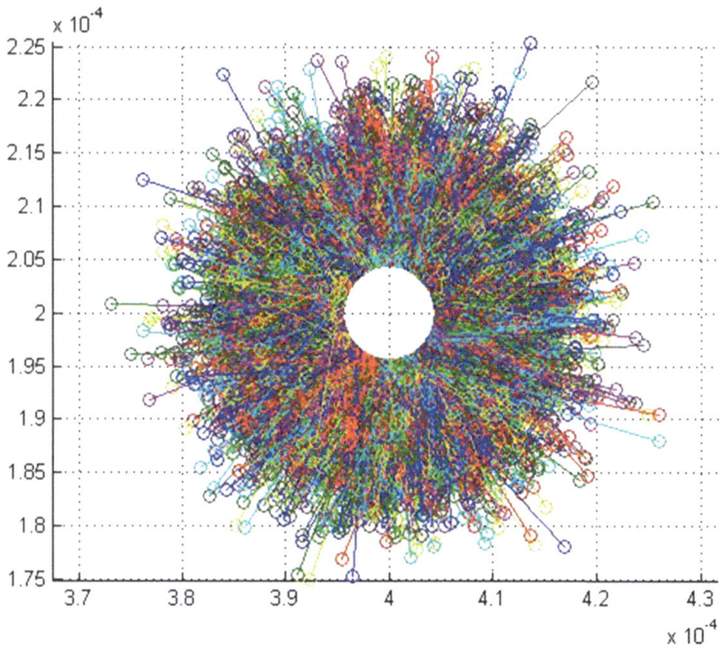

Fig. 6.16 TEST_A4: Scattering of molecules from the surface of one sample particle

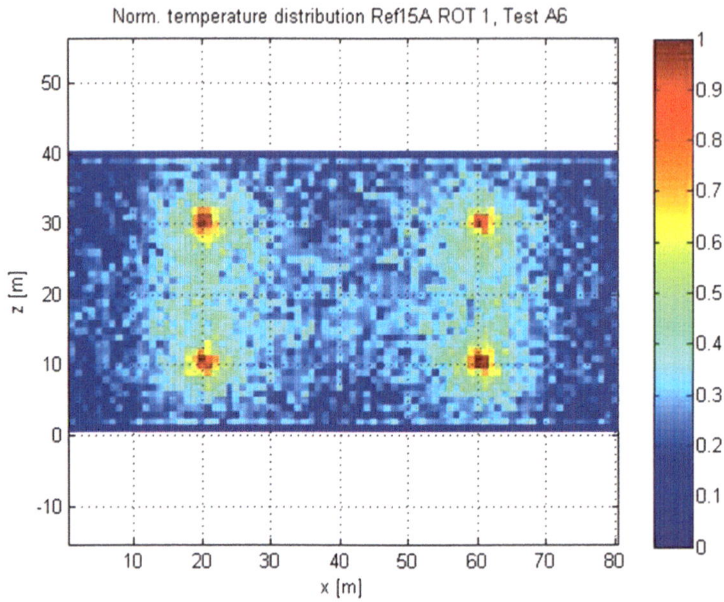

Fig. 6.17 TEST_A6: Temperature distribution in case of thermophoresis induced by coils, with uniform temperature of bounding walls

TEST_A7 (Test with specular reflections). This test reproduces Example 1.5.2.1 from Chap. 1.

TEST_A8. (Check of rotational temperature calculation). This test is intended to check the calculation of the internal rotational energy of diatomic molecules, hence of the rotational temperature of the gas. The computational domain, without particles or coils, is assumed to be bounded by specularly reflecting ideal surfaces and is filled with nitrogen. The nitrogen molecules are assumed to have an initial translational temperature of 500 K and an initial rotational temperature of 0 K (ideally, the molecules have no rotational energy at time point $t = 0$). As the calculation of intermolecular collisions starts, the initial translational energy is transferred to the internal rotational modes and an equilibrium between translational and rotational energies must be reached at the temperature of 300 K.

The plot of Fig. 6.18 shows that after 2×10^{-4} s the equilibrium of temperatures is reached with great precision. After 50,000 time steps with $\Delta t = 2 \times 10^{-8}$ s, hence at time $t = 10^{-3}$ s, the mean values in the domain of the translational, rotational, and overall temperatures (having used 20 sample molecules per computational cell) are:

T_translational:	300.23 K
T_rotational:	300.16 K
T_overall:	300.20 K

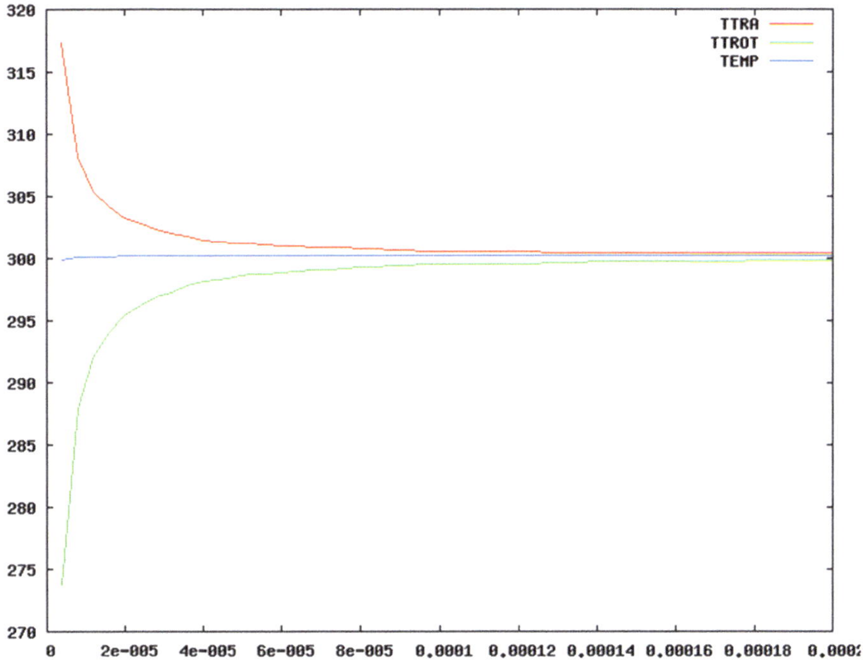

Fig. 6.18 TEST_A8: Translational, rotational and overall temperatures versus time

The maximum error of 0.23 K is for the translational temperature and, referred to the temperature span of 500 K, is of 0.00046 or 0.046%.

Relevant data concerning tests A1–A8 are summarized in Table 6.2.

6.10.2 *Energy Balance*

The code THERCONF provides for an overall energy balance. The power transferred to the computational domain through the hot surfaces (the surfaces kept at higher temperature) must be balanced by the power lost from the system through the cold surfaces. Surfaces reflecting specularly do no transfer energy to or from the system. Surfaces reflecting diffusively do transfer energy.

In case the inner coils are not simulated only four bounding surface (in 2D) can transfer energy. In case the coils are simulated, the energy delivered by the coils is added to that transferred through the bounding surfaces.

Simulated particles, if any, do not affect the energy balance as long as their temperature is assumed to be equal to the temperature of the gas, an assumption which is acceptable in case of particle of small size. For these small particles we do not compute the temperature distribution in the particles.

Table 6.2 Summary of data for tests A1–A8

TEST	IP	IC	IS	IMAX	KMAX	# CELLS	NMOL	XMAX (m)	ZMAX (m)	Δx (μm)	Δz (μm)	Δt (s)
A1	0	0	0	80	40	3200	64,000	0.0112	0.0056	140	140	1,00E−07
A2	0	0	1	80	40	3200	64,000	0.001	0.005	12,5	12,5	1,00E−08
A3	1	0	0	80	40	3200	64,000	0.0008	0.0004	10	10	1,00E−08
A4	1	1	1	80	40	3200	64,000	0.0008	0.0004	10	10	1,00E−08
A5	1	1	0	80	40	3200	64,000	0.0008	0.0004	10	10	1,00E−08
A6	1	1	1	80	40	3200	64,000	0.0008	0.0004	10	10	2,00E−08
A7	0	0	3	80	40	3200	64,000	0.02	0.01	250	250	5,00E−08
A8	0	0	3	80	40	3200	64,000	0.0008	0.0004	10	10	2,00E−08

TEST	Ttop (K)	Tbot (K)	P (N/m^2)	ρ (kg/m^3)	FND	λ (μm)	KnL	Rp (μm)	Rc (μm)/Tc (K)
A1	80	300	100	1,85E−03	3,86E+22	33.83	0.012	-	–
A2	300	80	100	1,85E−03	3,86E+22	33.83	0.135	-	–
A3	335	295	100	1,12E−03	2,33E+22	56.1	0.28	8	–
A4	335	295	100	1,12E−03	2,33E+22	56.1	0.28	5	1,25/350
A5	335	295	100	1,12E−03	2,33E+22	56.1	0.28	5	1.25/350
A6	273	273	100	1,28E−03	2,68E+22	48.6	0.24	0.25	2.5/350
A7	273	273	133,3	1,71E−03	3,58E+22	36.5	0.0036	–	–
A8	500	500	100	1,28E−03	1,46E+22	89	0.22	–	–

Key: IP = IPARTICLE; IC = ICIRCLE; IS = ISIDEWALL

At the beginning of a calculation, energy can be stored in the gas or delivered from the gas to the material surfaces. When a quasi steady state is approached the kinetic energy stored in the gas approaches a limiting value and the energy exchange gas-surfaces tends to zero. For this reason the energy stored in the gas can be neglected in the energy balance, as long as one does not simulate strong transient situations.

The energy balance, as well as the sampling of molecular velocities, can be skipped from the initial phase of the calculation choosing the input parameter *ISTD* large enough (typically a few thousands). Only after *ISTD* time steps the energy balance, and the sampling of physical quantities, start.

We checked the energy balance in the code on the basis of a variant of the TEST A4, the only difference being the location of the lower circles, closer to each other than the upper ones. The coordinates of the lower circles were: C3[0.000340, 0.000120], C4[0.000460, 0.000120].

The following documentation of the energy balance is divided in two parts referring to monoatomic and to diatomic molecules, respectively. However, the tests for the monoatomic case have been made using air molecules, treated as elastic spheres.

6.10.2.1 Balance of Translational Energy for Monoatomic Molecules

6.10.2.1.1 Energy Through Material Surfaces (Plates and Coils)

Calculation E1. Without simulation of the inner circles.

Figure 6.19 shows the transient behaviour of the power (called "Fluxes" in the figure) transferred to (positive) or from (negative) the system during 5 ms of problem time. Flux1 (0.3733 W, at the end of the calculation) and flux2 (−0.3940 W) refer to the upper and lower surfaces, flux3 (0.0104 W) and flux5 (0.0097 W) refer to the side surfaces, left and right, respectively. Flux3 and flux5 should be identical for symmetry reasons. The computed total and relative error are −0.00062 W and 0.00158, respectively. The relative error, of the order of magnitude of one per thousand, is satisfactorily.

Calculation E2. With simulation of the inner circles.

The normalized temperature distribution is shown in Fig. 6.20. The transient power through the side surfaces is shown in Fig. 6.21. The power released from the circles is shown in Fig. 6.22. Because of the temperature distribution induced by the coils, the side surfaces subtract a small amount of power (−0.04 W) from the system. The upper surface transfers power to the system, but far less than in the case of absence of coils. Surface 1 (lower plate) subtracts from the system about 0.6 W. The total and relative errors, at the end of the calculation, are in this case, 0.0036 W and 0.00344 [−], respectively. The relative error is about the double of that in the previous calculation without coils.

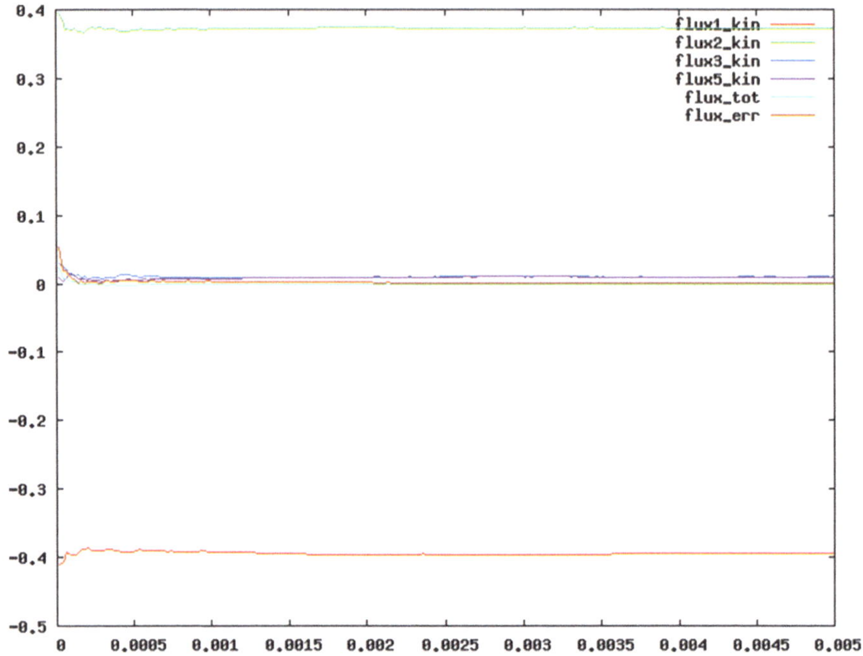

Fig. 6.19 Calculation E1—Power balance in case of simulation without inner coils (20 sample molecules per mesh)

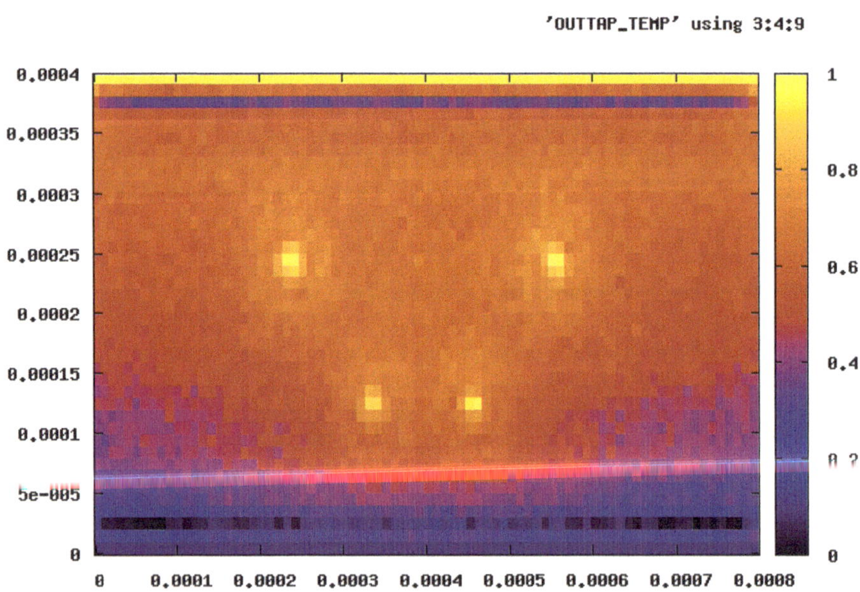

Fig. 6.20 Calculation E2—Normalized temperature distribution in the variant of Test A4, with closer lower coils

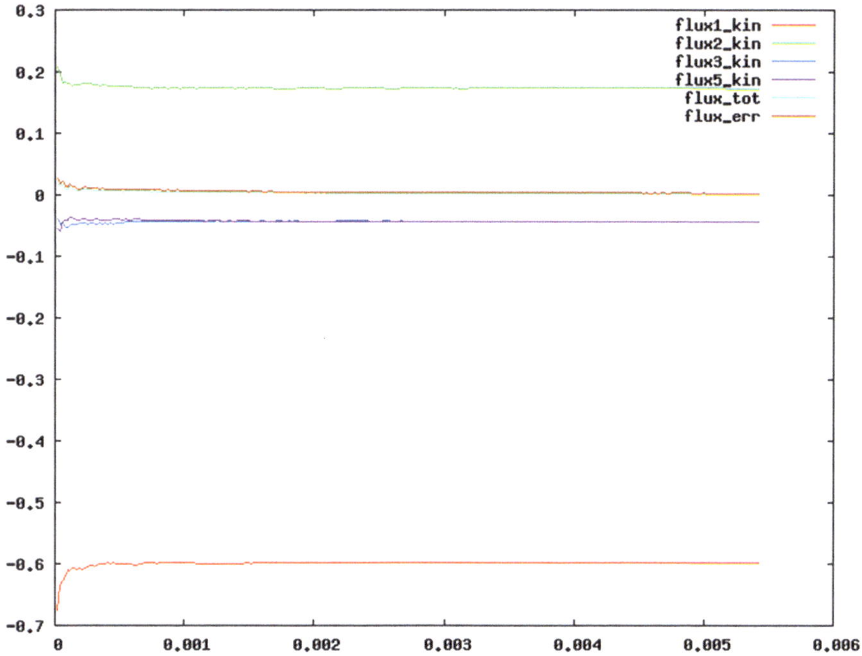

Fig. 6.21 Calculation E2—Power through the bounding surfaces in case of simulation wit inner coils (20 sample molecules per mesh)

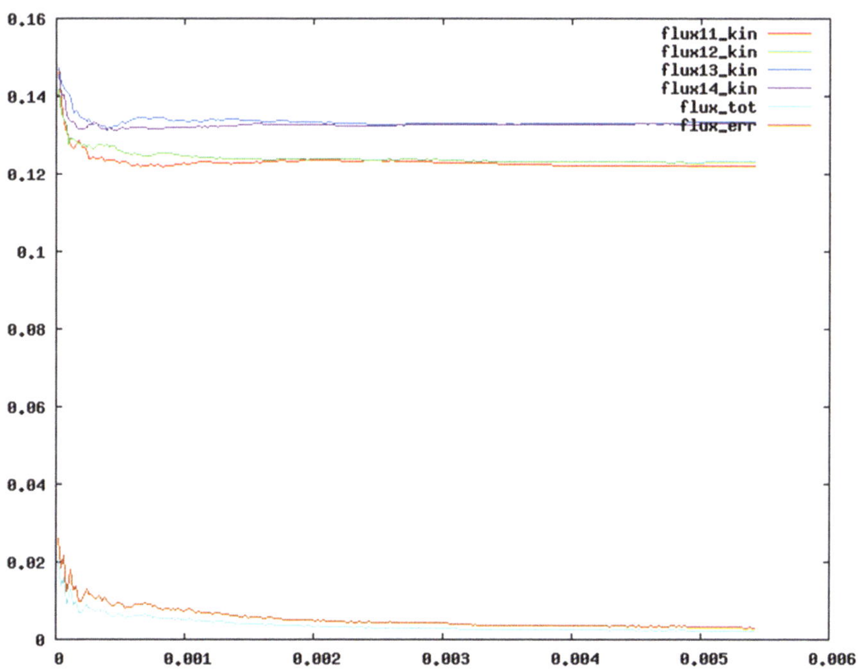

Fig. 6.22 Calculation E2—Power released by the coils (20 sample molecules per mesh)

Calculations E3. Miscellaneous Calculations.

To assess the impact of the number of sample molecules per mesh on the accuracy of the results, we performed the following calculations, obtaining the relative errors (Cfr. Fig. 6.23) given Table 6.3.

As remarked above, the best result, in terms of relative error, is obtained without coils.

Comparing the calculations with coils, and time step of 10×10^{-8} s, one deduces that passing from 10 to 20 sample molecules per mesh the results are substantially

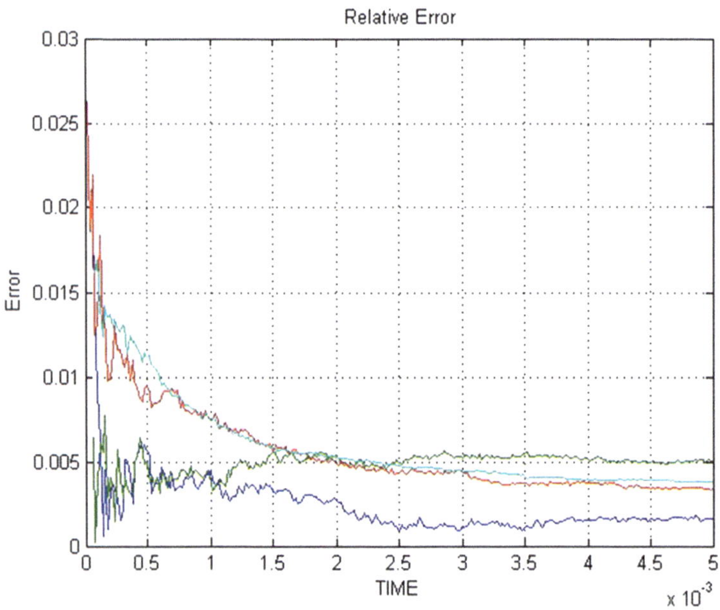

Fig. 6.23 Transient errors for the calculations E3.1 (lowermost plot at the right, 20 molecules/mesh, without coils), E3.2 (uppermost, 10 molecules/mesh), E3.3a (third from top, 20 molecules/mesh), E3.4 (second from top, 100 molecules /mesh)

Table 6.3 Comparison of calculations with different number of sample molecules and different time steps

Calculation	With/without coils	Molecules per mesh	Relative error	Δt (s)	N_DTR	CPU time (s)
E3.1 (*)	Without	20	1.58×10^{-3}	10×10^{-8}	5	15,194
E3.2	With	10	5.05×10^{-3}	10×10^{-8}	5	9,313
E3.3a	With	20	3.44×10^{-3}	10×10^{-8}	5	18,627
E3.3b	With	20	1.96×10^{-3}	2×10^{-8}	1	50,456
E3.4	With	100	3.79×10^{-3}	10×10^{-8}	5	98,372

CPU time refers to the simulation of 0.005 s problem time

(*) Calculation E3.1 is identical with Calculation E1

improved, but going from 20 to 100 sample molecules per mesh results do not further improve. The difference between calculations E3.3a and E3.3b consists in the use of different time steps. In calculation E3.3.a the macro time step of 10×10^{-8} s is divided in a maximum of 5 micro-time steps N_DTR for the fastest molecules, while in calculation E3.3b the time step of 2×10^{-8} s is used for all molecules, slow and fast. In the latter calculation the relative error is reduced by almost a factor of two, however, at the expense of a computational time almost three times larger. Figure 6.23 shows the time distributions of the errors for calculation E3.1 through E3.4.

6.10.2.1.2 Energy Balance for the Gas

The kinetic energies of the gas molecules are initialized sampling the Maxwellian velocity distribution at the temperature of the points occupied by the molecules. This temperature is obtained with a linear interpolation between the temperatures of the upper and lower heated plates. In the course of a calculation the temperature of the fluid, hence its kinetic energy, adjusts to the new temperature distribution, imposed for instance by the presence of the heated coils. When a steady state situation is approached the kinetic energy of the gas approaches a limiting value, hence the difference between the kinetic energy at time $t_0 = 0$ and the kinetic energy at the current problem time, divided by the problem time, must vanish. We verified that in the calculations this result is obtained.

Figure 6.24, which refers to the calculation E3.3a of the above table, shows the oscillations of the kinetic energy of the fluid computed in time intervals of $200\Delta t$, compared to the total kinetic energy. Although the peaks of these oscillations may occasionally reach about 2% of the total kinetic energy, their mean value is rather close to zero. Accordingly, as shown in Fig. 6.25, the difference between the total kinetic energy of the fluid at time $t_0 = 0$ and the kinetic energy at the current problem time, divided by the problem time (i. e. the mean power delivered to the gas) tends to zero. The final computed values oscillate around $\pm 1.4 \times 10^{-3}$. In calculation E3.3b (see Fig. 6.26), with the smaller time step of 2×10^{-8} s, the final computed values oscillate around $\pm 1.5 \times 10^{-5}$.

6.10.2.2 Balance of Translational and Rotational Energies for Diatomic Molecules

Calculation E5 is comparable with calculation E3.3b but differs from it for the simulation of the bi-atomic air molecules (*NATOMS* = 2) which entails the calculation of the rotational inner energy of the molecules and of the rotational contribution to the heat transfer at the walls. In a steady state situation the kinetic and rotational temperatures should become equal. Figure 6.27 shows the mean values in the computational domain of the translational energy (*TTRAM*), rotational energy (*TROTM*) and overall energy (*TEMPM*) versus time. The discrepancy between the

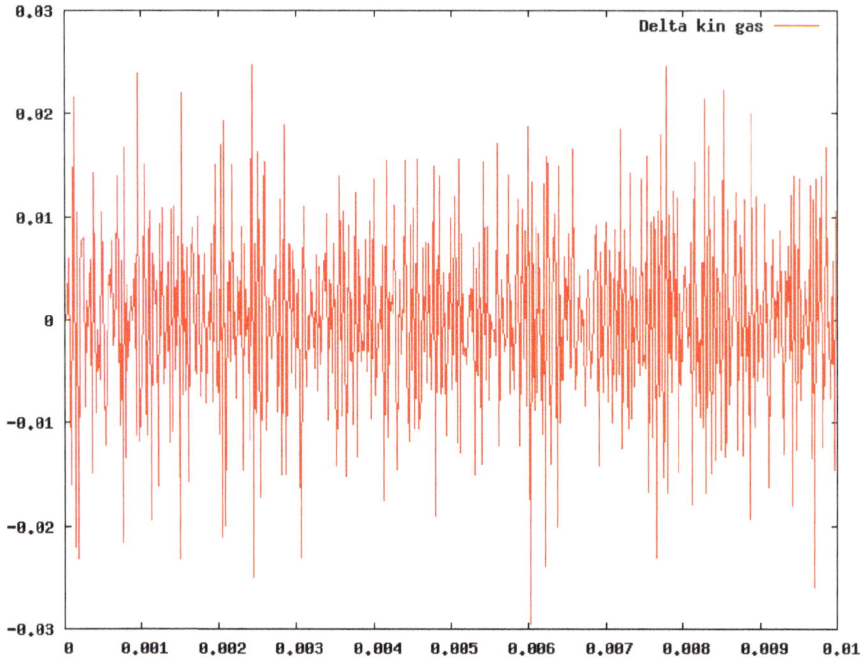

Fig. 6.24 Calculation E3.3a. Oscillations versus time of the kinetic energy of the fluid, computed in time intervals of $200\Delta t$. The ordinates show the percentage of the excursion of the kinetic energy, compared to the total kinetic energy of the fluid

computed kinetic and rotational temperatures becomes, after 4 ms of problem time, less than a tenth of a degree.

Figure 6.28 shows the kinetic power transferred from the bounding surfaces to the gas. Figure 6.30 shows the kinetic power transferred from the circles to the gas. The amount of the rotational power from the bounding surfaces and from the circles is shown in Figs. 6.29 and 6.31, respectively. This last figure shows that the rotational contribution to the power transferred from the circles to the gas is about half of the kinetic contribution, hence very large. The consequence is that the uppermost gas region is very hot and the uppermost surface (surface Nr. 2) at $324K$ does not contribute to heat the system but subtracts energy, though in very small amount. The same holds for the lateral surfaces Nr. 3 and 5. The only bounding surface which subtracts a considerable amount of energy from the system is the lowermost plate (surface Nr. 1) with about −0.51 W.

The overall energy balance is shown in Fig. 6.32, which displays the absolute and relative errors. The relative error oscillates around the values $\pm 1.0 \times 10^{-7}$ and is therefore better than in the corresponding calculation E3.3b made with monoatomic molecules. Figure 6.33 shows the difference between the total kinetic energy of the fluid at time $t_0 = 0$ and that at the current problem time, divided by the problem time (i. e. mean power released to the gas). At the end of the calculation this value oscillates around $\pm 1.0 \times 10^{-5}$.

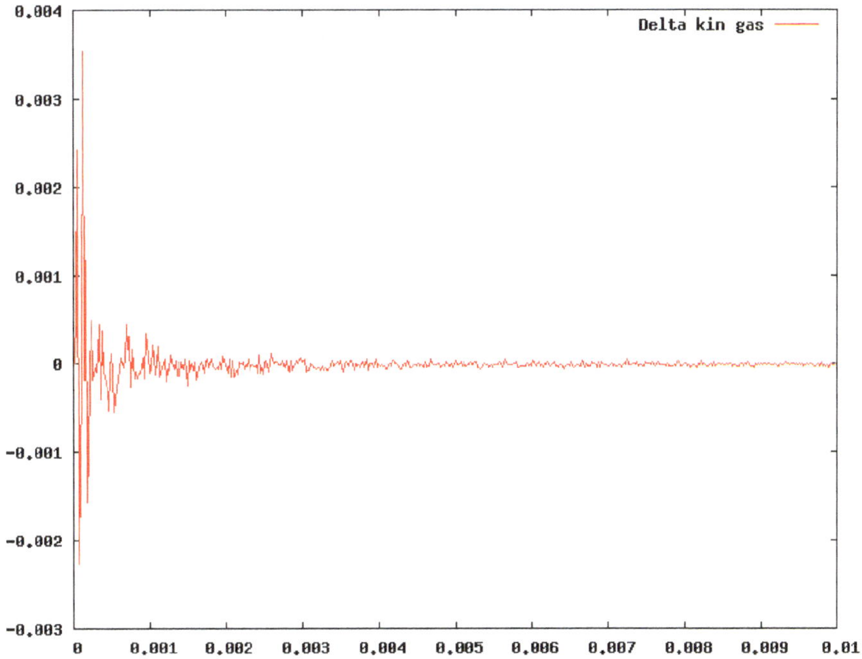

Fig. 6.25 Calculation E3.3a. Difference between the total kinetic energy of the fluid at time $t_0 = 0$ and at the current problem time, divided by the problem time (i. e. mean power released to the gas)

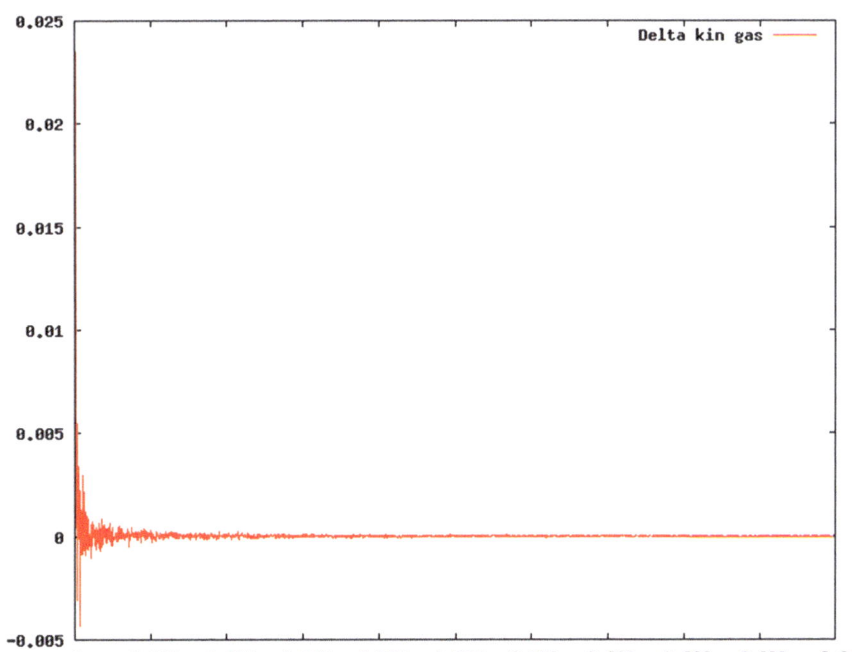

Fig. 6.26 Calculation E3.3b. Difference between the total kinetic energy of the fluid at time $t_0 = 0$ and at the current problem time, divided by the problem time (i. e. mean power released to the gas)

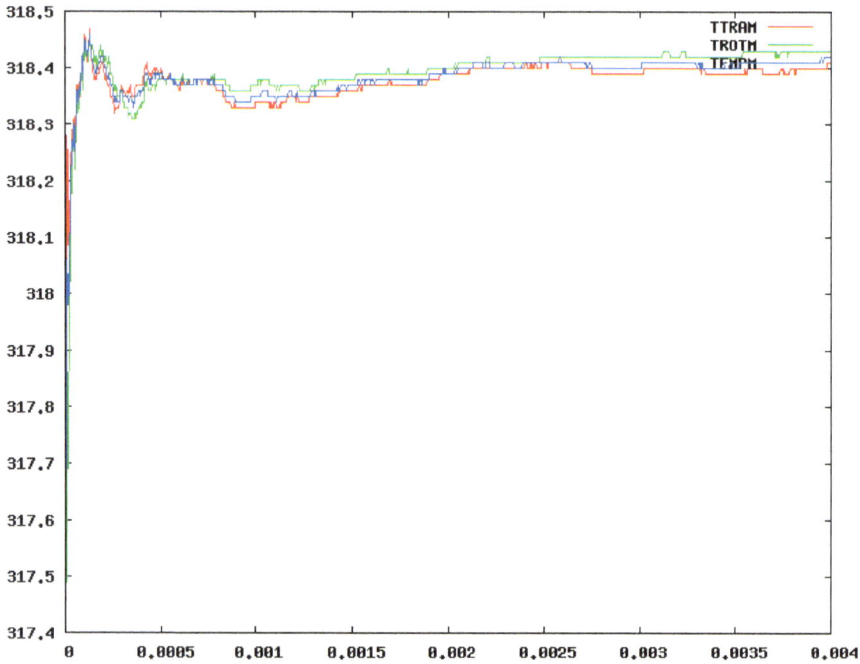

Fig. 6.27 Calculation E5. Mean values in the computational domain of the translational energy (TTRAM), rotational energy (TROTM), and overall energy (TEMPM) versus time. Units in Kelvin

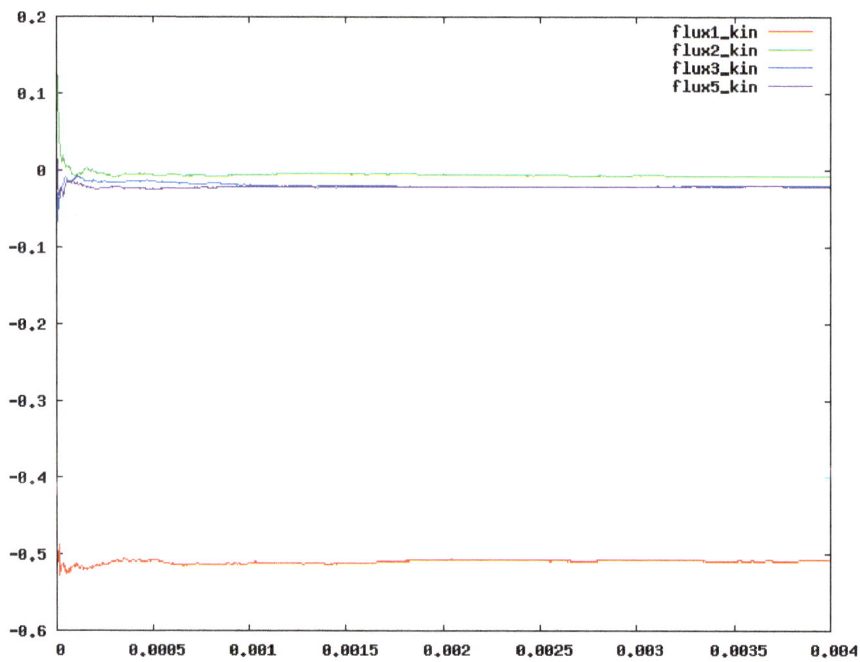

Fig. 6.28 Calculation E5. Kinetic power transferred to the system from the bounding surfaces (1 = top surface, 2 = bottom surface, 3 = left surface, 5 = right surface) versus time. Units in watts

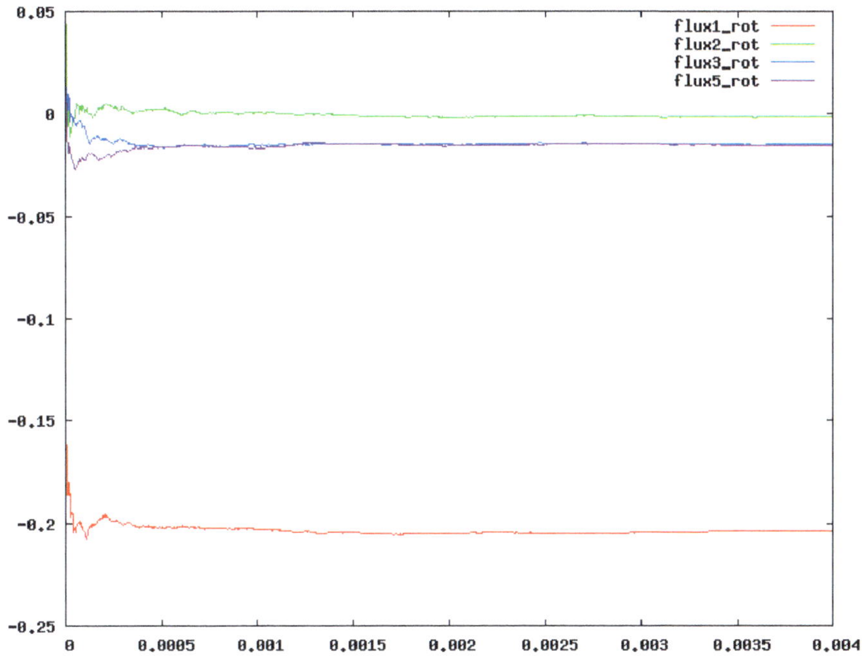

Fig. 6.29 Calculation E5. Rotational power transferred to the system from the bounding surfaces (1 = top surface, 2 = bottom surface, 3 = left surface, 5 = right surface) versus time. Units in watts

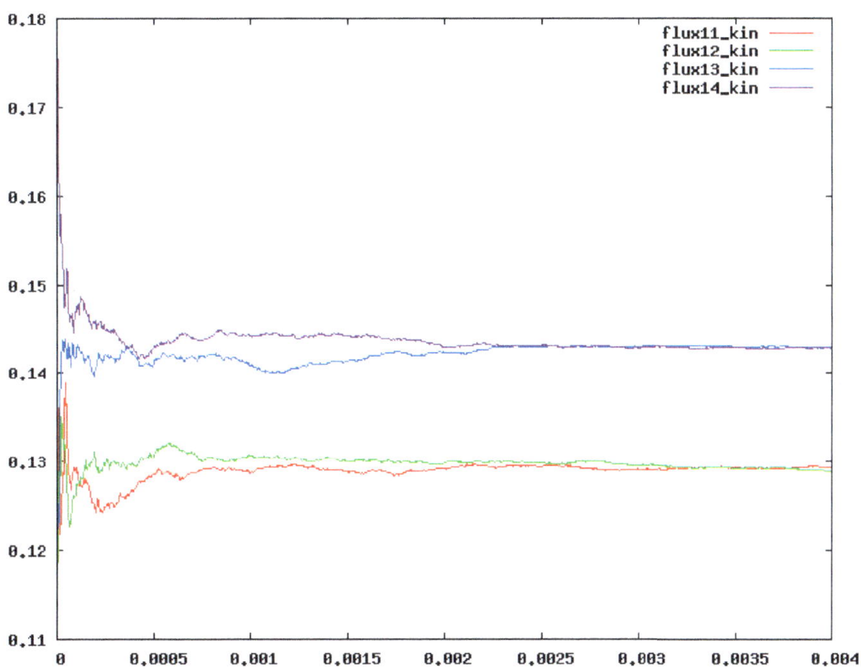

Fig. 6.30 Calculation E5. Kinetic power transferred to the system from the surfaces of the circles (11 and 12 = uppermost circles, 13 and 14 lowermost circles) versus time. Units in watts

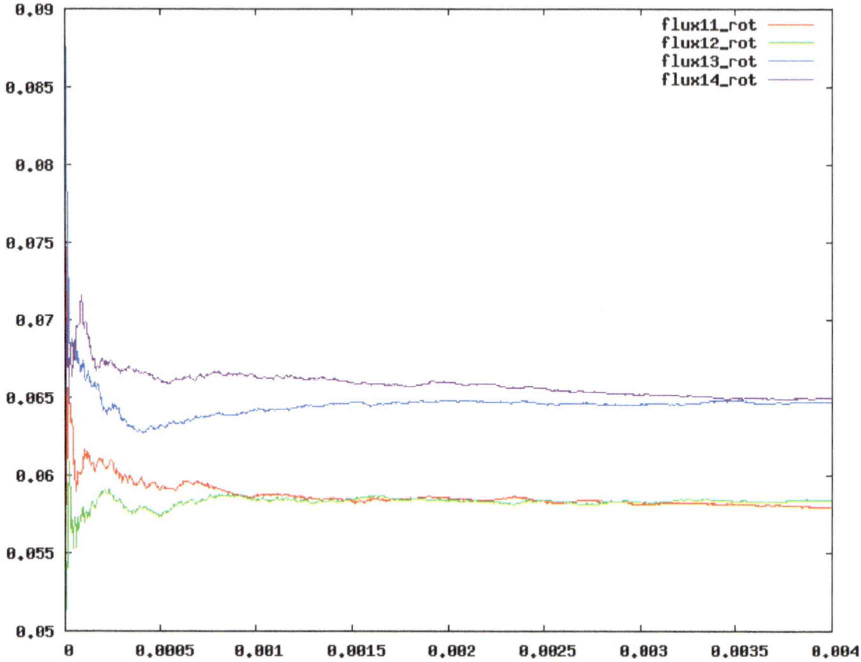

Fig. 6.31 Calculation E5. Rotational power transferred to the system from the surfaces of the circles (11 and 12 = uppermost circles, 13 and 14 lowermost circles) versus time. Units in watts

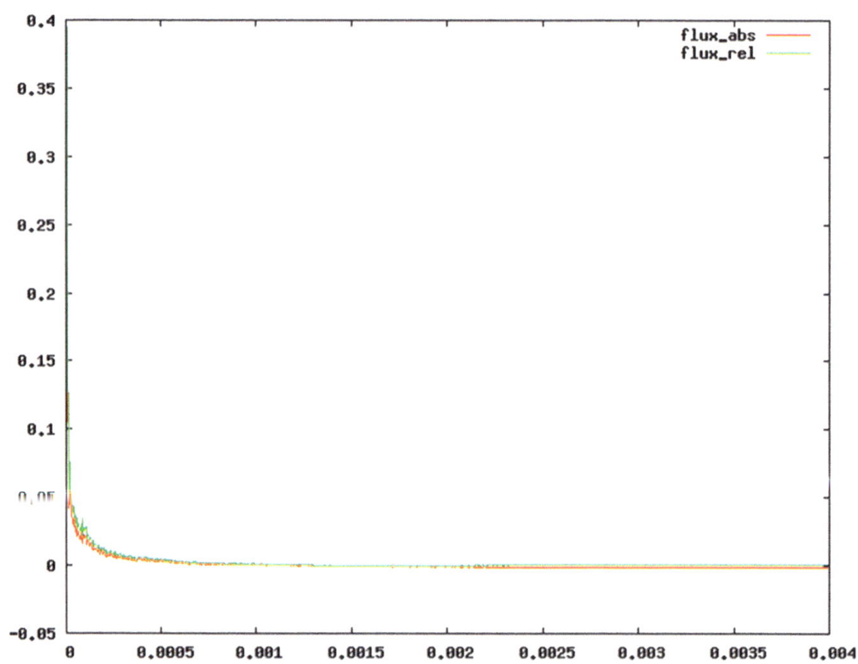

Fig. 6.32 Calculation E5. Absolute and relative errors in the global power balance versus time

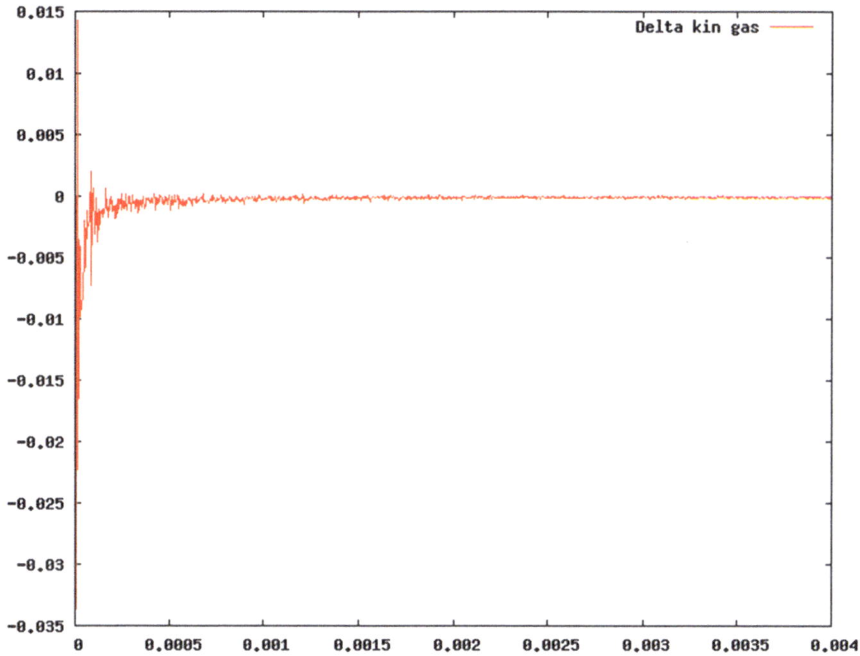

Fig. 6.33 Calculation E5. Difference between the total kinetic energy of the fluid at time $t_0 = 0$ and at the current problem time, divided by the problem time (i. e. mean power released to the gas)

6.10.3 Calculation of Gallis Test Cases

6.10.3.1 Simulation Data and Calculation Conditions

The code has been used to simulate three of the test cases reported by Gallis et al. (2002). In all cases the fluid was nitrogen with a diatomic molecule of mass $m = 4.65 \times 10^{-26}$ kg. Three test cases were simulated, differing from each other only in the gas pressure which was of 1000 mT, 10 mT, and 0.1 mT (or 133.3 Pa, 1.333 Pa, and 0.0133 Pa), respectively. Simulations were made in a rectangular domain of 500 × 10,000 μm, divided in 50,000 square cells of 10 μm late, using $N_{mol} = 1, 250, 000$ sample molecules, therefore with 25 molecules per cell, in average.

The upper and lower bounding surfaces of the computational domain were material surfaces kept at the temperatures of 283 K and 263 K, respectively, while the lateral surfaces were symmetry surfaces. Calculations were made with time steps varying between 1×10^{-8} and 5×10^{-8} s, the lower figure being applied to fast molecules and the larger time step to slower molecules.

The following test cases have been computed:

Table 6.4 Gallis calculations G1, G2, G3. Summary of main parameters for the three test cases

Domain:	$0 \div XMAX = 0 \div 500$ μm
	$0 \div YMAX = 0 \div 1$ m
	$0 \div ZMAX = 0 \div 10{,}000$ μm $= 0 \div 10$ mm
Number of cells:	50 in x-direction, $\Delta x = \Delta z = 10$ μm
	1 in y-direction, $\Delta y = 1$ m
	1000 in z direction
Number of simulated molecules ($NMOL$):	1,250,000 (25 molecules per cell, in average)
Temp. of upper plate	283 K
Temp. of lower plate	263 K
Temp. of fluid (at $t = 0$ s)	273 K
Time step:	$1 \times 10{-}^{8}$ s

Calculation	G1	G2	G3
Mean free path λ (m):	3.72×10^{-5}	3.72×10^{-3}	3.72×10^{-1}
Knudsen Number ($Kn = \lambda/Z_{\max}$)	3.72×10^{-3}	3.72×10^{-1}	$3.72 \times 10^{+1}$
Density number n (mol/m^3)	3.58×10^{22}	3.58×10^{20}	3.58×10^{18}
Molecules in domain ($_{Nd}$)	1.79×10^{17}	1.79×10^{15}	1.79×10^{13}
Fn ($=Nd/NMOL$)	1.43×10^{11}	1.43×10^{9}	1.43×10^{7}
Density ρ (kg/m^3)	1.66×10^{-3}	1.66×10^{-5}	1.66×10^{-7}
Mass in domain (kg)	8.32×10^{-9}	8.32×10^{-11}	8.32×10^{-13}

1. GALLIS_1 (G1) with pressure of 1000 mT (133.3 Pa).
2. GALLIS_2 (G2) with pressure of 10 mT (1.333 Pa).
3. GALLIS_3 (G3) with pressure of 0.1 mT (0.0133 Pa).

Data used for the simulations are given in Table 6.4.

The following data hold for the three simulations:

Calculations were made by taking into account the rotational energy of the molecules around two axes normal to the geometrical axis joining the centres of the two constituent atoms. The theoretical background upon which the calculation of the inner rotational energy of the molecules is based is the Larsen-Borgnakke phenomenological method.

The results show clearly that the total available energy of the molecules is shared between the translational and the rotational modes. The kinetic and rotational temperatures of the molecules are given by

$$T_r = \frac{2E_i}{\zeta R} = \frac{2ME_i}{\zeta k}, \tag{6.41}$$

$$T_{tr} = \frac{2E_{tr}}{3R}, \tag{6.42}$$

where E_i, E_{tr} are the inner (rotational) and translational energies, k is the Boltzmann constant, m the mass of a molecule, R the gas constant, and the degree of freedom ζ is 2 for a diatomic molecule.

In a situation of thermal equilibrium in which the translational and rotational temperatures have equalized, the overall temperature assumes the same value

$$T_{\sigma v} = \frac{3T_{tr} + \zeta T_r}{3 + \zeta}. \tag{6.43}$$

In this situation of thermal equilibrium Eqs. (6.41) and (6.42) give

$$E_i = \frac{2}{3} E_{tr}. \tag{6.44}$$

Moreover it holds for the total energy of the molecules

$$E_{tot} = E_i + E_{tr} = E_{tr}\left(1 + \frac{2}{3}\right) \tag{6.45}$$

The theoretical result, according to which, in thermal equilibrium state, the inner rotational energy is 2/3 of the kinetic energy, is reproduced very well by the numerical results, as shown below.

6.10.3.2 Results of Gallis-1 Simulation

Figure 6.34 shows the computed transient behaviour of the translational, rotational, and overall temperatures up to 5 ms of problem time. The plots do not start at time $t = 0$, because data are stored after some initial transient intended to let the situation to stabilize somewhat. At time $t = 5$ ms the expected value (mean value of the temperatures of the upper and lower plates) of 273.0 K has been reached with a 0.5 K accuracy by all temperatures.

Figure 6.35 shows the computed kinetic and rotational energy of all the molecules in the system. These results are obtained summing up the energies of the sample molecules and multiplying the results by Fn (number of real molecules represented by one sample molecule). The sum of the rotational and translational energies is the total energy of the gas. The plot shows how the total available energy in the gas is partitioned between the translational and rotational modes. The last computed values for the rotational, translational, and total energy are 1.012×10^{-3}, 0.674×10^{-3}, and 1.686×10^{-3} J, respectively. These values satisfy well the above Eqs. (6.44) and (6.45).

The results of Fig. 6.35 can be verified in the following way. The most probable speed of the nitrogen molecules at the temperature of 273 K is $c_m = 402.63$ m/s. The average thermal speed is $\overline{c}' = 1.12838 \times c_m = 454.32\,\text{m/s}$. The corresponding

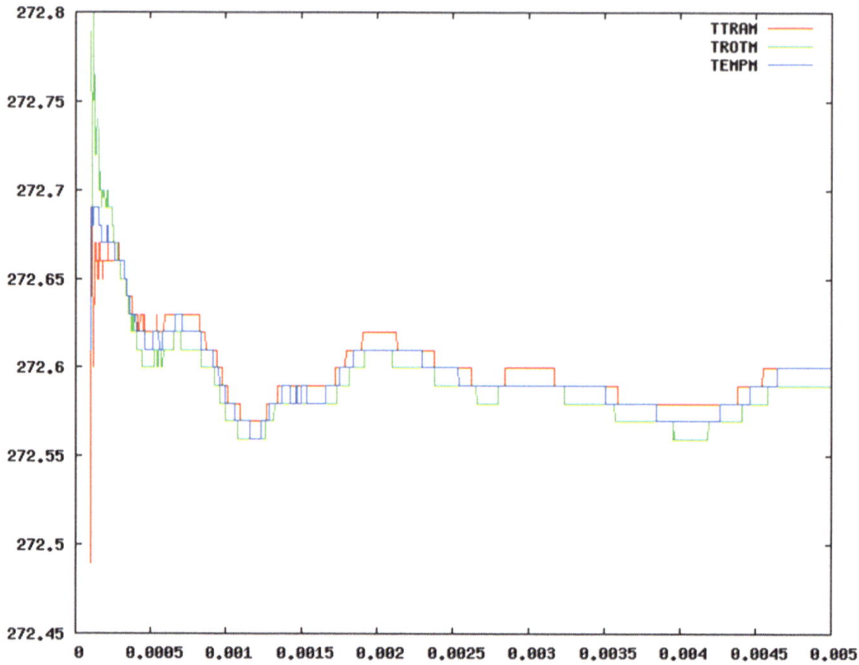

Fig. 6.34 Calculation G1—Computed transient mean temperatures in domain. Uppermost plot: Rotational temperature; Middle plot: Overall temperature; Lower plot: Translational temperature

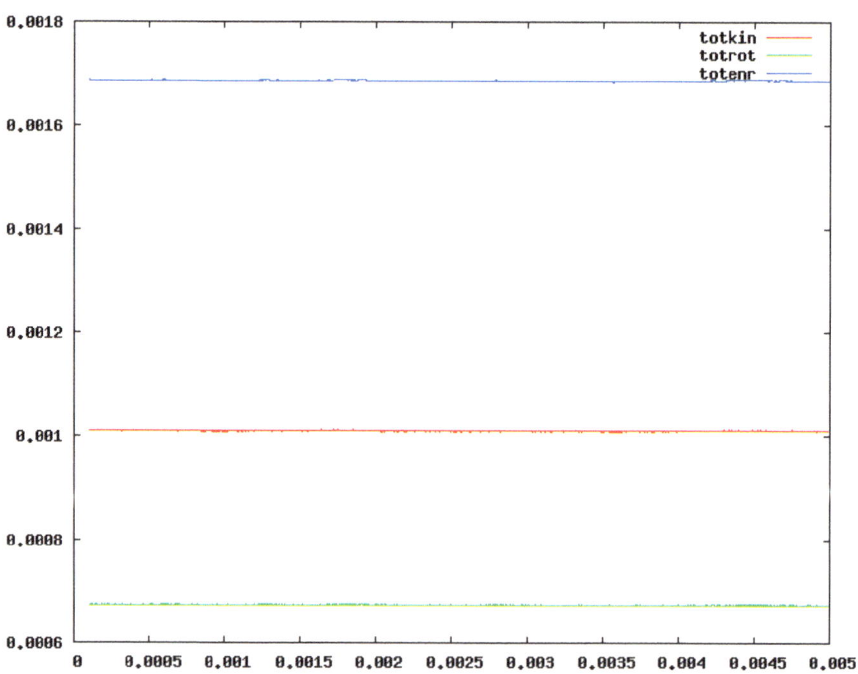

Fig. 6.35 Calculation G1—Computed transient energies in domain. Uppermost plot: Total energy; Middle plot: Kinetic energy; Lower plot: Rotational energy

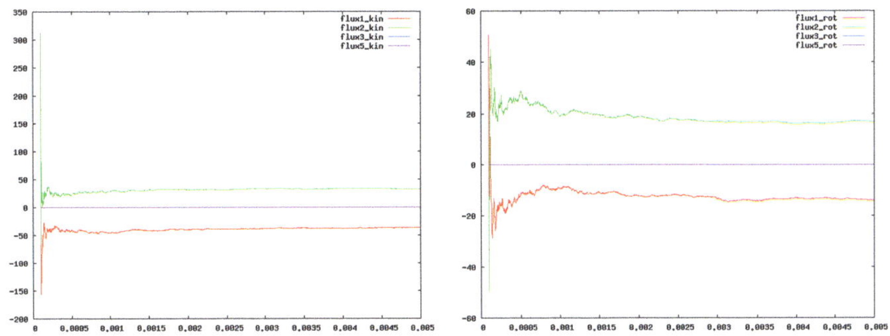

Fig. 6.36 Calculation G1. Computed transient energy fluxes at upper and lower plate. Left plot: Translational energy fluxes; Right plot: Rotational energy fluxes

kinetic energy is $e_{kin} = (1/2)m\overline{c}'^2 = (1/2)4.6510^{-26}(454.32)^2 = 4.798 \times 10^{-21}$J. The computed value of the mean kinetic energy of the molecules is $PK_{mean} = 5.692 \times 10^{-21}$J. This value is larger than e_{kin} because the molecules in the range of temperatures between 273 and 283 K contribute to the average kinetic energy, via the quadratic factor, proportionally more than those in the range 263–273 K. Multiplying PK_{mean} by the number Nd of the molecules in the domain ($Nd = NMOL * Fn = 1.25 \times 10^6 \times 1.43 \times 10^{11} = 1.79 \times 10^{17}$) one obtains for the total kinetic energy $E_{kin} = PK_{mean} \cdot Nd = 5.692 \times 10^{-21} \cdot 1.79 \times 10^{17} = 1.019 \times 10^{-3}$, which is close to the plotted value of 1.012×10^{-3}.

Figure 6.36 shows the computed transient energy fluxes at the lower plate (Surface Nr. 1) and upper plate (Surface Nr. 2). Lateral surfaces (Nr. 3 and 5) are symmetry surfaces and do not contribute to the fluxes. The kinetic energy flux is due to the difference in kinetic energy of the reflected and incident molecules. The rotational energy flux is due to the difference in rotational energy of the reflected and incident molecules. The reflection of the rotational energy of the molecules from the wall has been computed with the Cercignani-Lampis_Lord (CLL) model. The last computed values for the kinetic, rotational, and total energy fluxes at the upper plate are 16.96, 32.87, and 49.83 (W/m^2), respectively. Accordingly, one deduces that about 34% of the energy is transferred via the rotational energy change of the molecules, and about 66% via the kinetic energy change of the molecules.

Figure 6.37 shows the mean collision frequency of the molecules in the cells of the domain, indexed sequentially from the bottom to the top of the domain. Because of the rarefaction of the gas the computed collision frequencies are more than a factor of ten smaller than the collision frequency at atmospheric pressure. For a comparison, the latter can be evaluated from

$$\nu = n\sigma_T c_r, \tag{6.46}$$

where $n = 2.688 \times 10^{25}$ is the molecular density at atmospheric pressure and 273 K, $\sigma_T = \pi d^2$ ($d = 4.11 \times 10^{-10}$ molecular diameter) is the collision cross section, and c_r

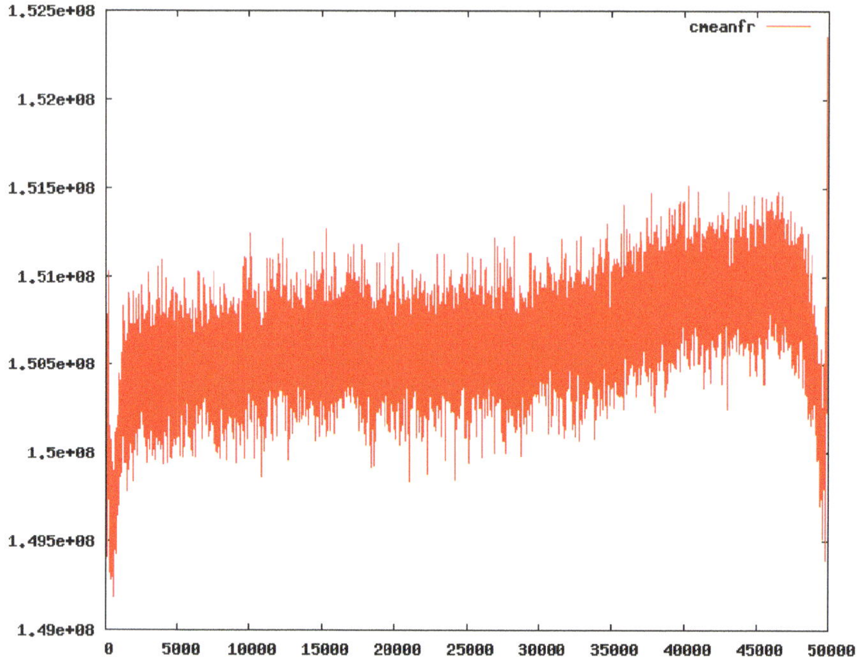

Fig. 6.37 Calculation G1. Mean collision frequency of one molecule in the cells of the domain. Mean collision frequency of one molecule in the cells of the domain. The cells are numbered sequentially row by row from the bottom to the top of the domain. The values of every fourth cell are plotted

is the mean value of the relative velocity between colliding molecule pairs. The latter is given by

$$c_r = \sqrt{2}\overline{c}'. \tag{6.47}$$

The average thermal speed has already been evaluated above as $\overline{c}' = 1.12838 \times c'_m = 454.32$ m/s. Hence $c_r = \sqrt{2}\overline{c}' = \sqrt{2} \times 454.32 = 642,50$ m/s. With these values one obtains the average collision frequency of one molecule $\nu = n\sigma_T c_r = 9.16 \times 10^9 [\mathrm{s}^{-1}]$.

Figure 6.38 shows the minimum, mean (over the full calculation) and maximum numbers of accepted collisions occurring in the cells per time step. For a comparison, the same figures for the case of atmospheric pressure at 273 K were computed as: about 20, 80 and 180 collisions per time step.

Figure 6.39 shows the percentage of accepted (lower plot) and rejected intermolecular collisions, versus time, over the full calculation. The number of accepted collisions is only about 22 % of the total.

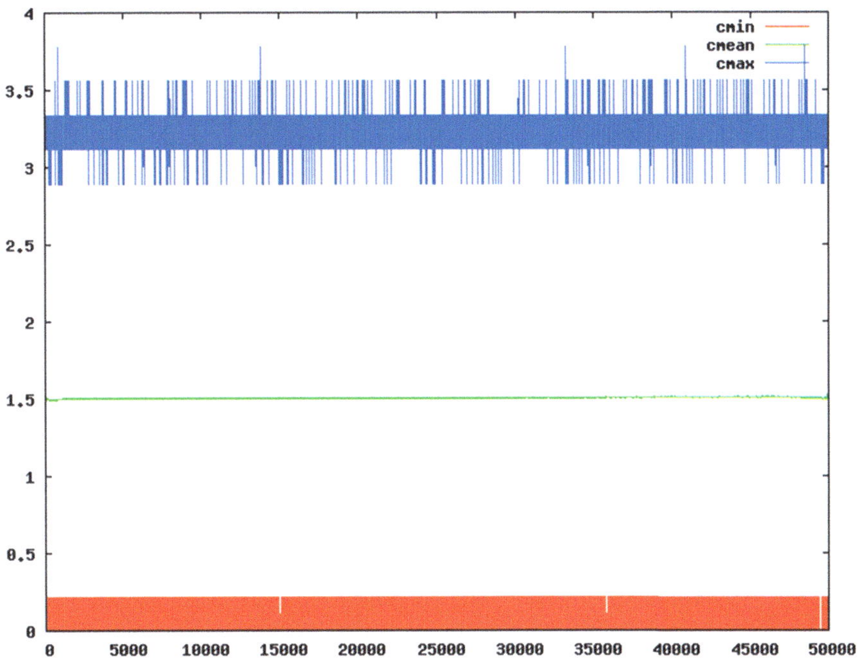

Fig. 6.38 Calculation G1. Minimum, mean and maximum numbers of collisions in the cells of the domain. Minimum, mean and maximum numbers of collisions in the cells of the domain. The cells are numbered sequentially row by row from the bottom to the top of the domain. The values of every fourth cell are plotted

6.10.3.3 Results of Gallis-2 Simulation

The figures of the simulations made for the G2 and G3 test cases are reported in the following without further comments, unless some particular remark is necessary.

Errore. L'origine riferimento non è stata trovata. Figure 6.40 shows the translational, rotational and overall temperatures, averaged over the full computational domain. These values are about 1.5 K higher than expected, although the computational fluxes from the plates, as seen below, are in good agreement with the experimental values (Figs. 6.41 and 6.42).

Errore. L'origine riferimento non è stata trovata. Figure 6.43 shows the mean collision frequency of the molecules in the cells of the domain. Due to the rarefaction of the gas the collision frequencies are reduced by a factor of 100 with respect to those for the case G1 (Figs. 6.44 and 6.45).

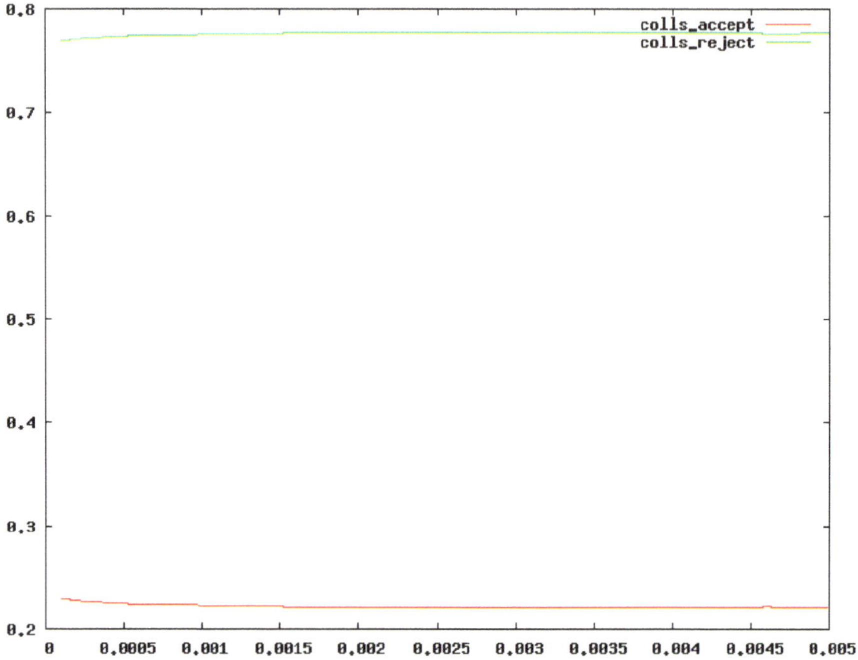

Fig. 6.39 Calculation G1. Percentage of accepted (lower plot) and rejected inter-molecular collisions, versus time

6.10.3.4 Results of Gallis-3 Simulation

In case of the G3 test case the mean values of both the kinetic and the rotational temperatures, hence the mean value of the overall temperature, stabilize at a value larger than the expected 273 K. The discrepancy is about 1.8 K (See Fig. 6.46). Besides, the slope of the temperature distribution across the domain remains much larger than reported in Gallis et al. (2002). Consequently the heat flux Q_TAV, based on the gradient of the temperature distribution, is a factor of 10 larger than expected, while the value $Q\_TAV = Q\_KIN + Q\_ROT$ almost matches the expected value. This effect has been obtained in all calculations performed thus far, even those (not documented here) with finer meshes and larger number of molecules. There is not explanation thus far for this discrepancy. The problem will be further investigated (Figs. 6.47 and 6.48).

Figure 6.49 shows the mean collision frequency of the molecules in the cells of the domain. There is again a reduction of a factor 100 with respect to the G2 case. The oscillations from mesh to mesh are in this later case much larger than in the previous cases because in the very rarefied gas the collision events are rare and a much longer calculation would be needed to obtain smooth averaged values.

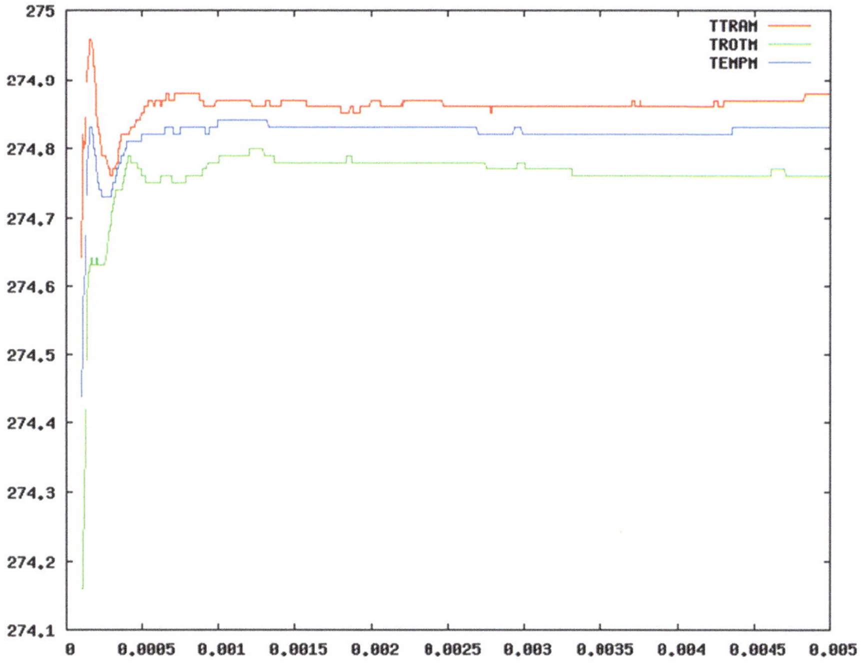

Fig. 6.40 Calculation G2—Computed transient mean temperatures in domain. Uppermost plot: Translational temperature; Middle plot: Overall temperature; Lower plot: Rotatio2nal temperature

The results of Errore. L'origine riferimento non è stata trovata. Figure 6.50 show that the flow in this test case is almost collisionless (Fig. 6.51).

6.10.3.5 Overall Results

Figure 6.52 shows the temperature profiles obtained for the three test cases. Both the plots of G3 (lowermost at left) and of G2 (middle at left) are shifted towards the higher temperatures.

A calculation has been made for the test G3 using only one mesh in the x direction and 400 meshes in the z-direction with 2,000,000 sample molecules, hence 2500 molecules per cell. Practically the same results were obtained as in the case with 25 molecules per cell. This proves that the discrepancy is not due to insufficient number of simulated molecules.

The heat fluxes were calculated with two different approaches:

Evaluation made with reference to the gradient of the temperature profiles but eliminating the meshes close to the walls, where the temperature jump is displayed. Results are referred to as *Q_TAV* (Temperature based on AVerage gradient) in the table below.

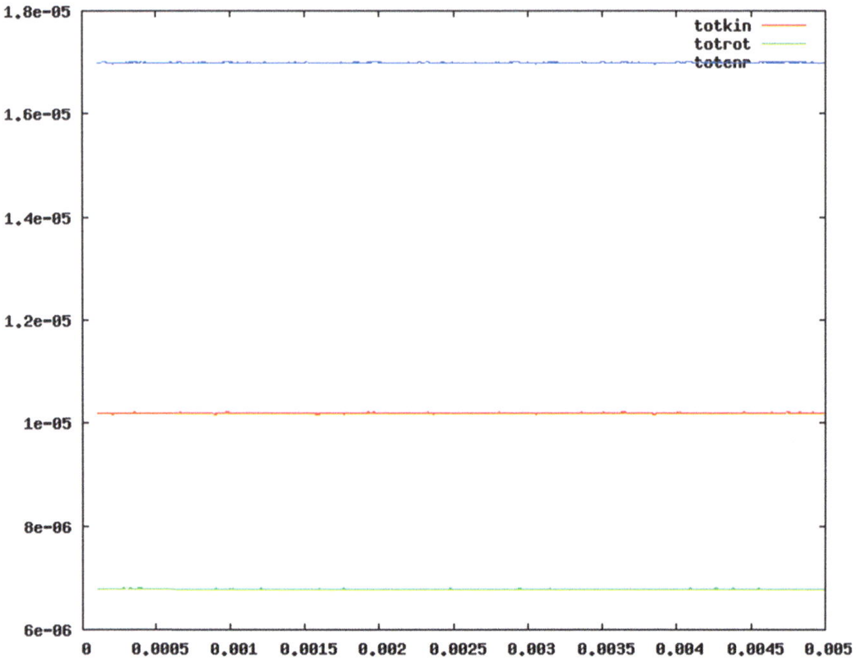

Fig. 6.41 Calculation G2. Computed transient energies in domain. Uppermost plot: Total energy; Middle plot: Kinetic energy; Lower plot: Rotational energy

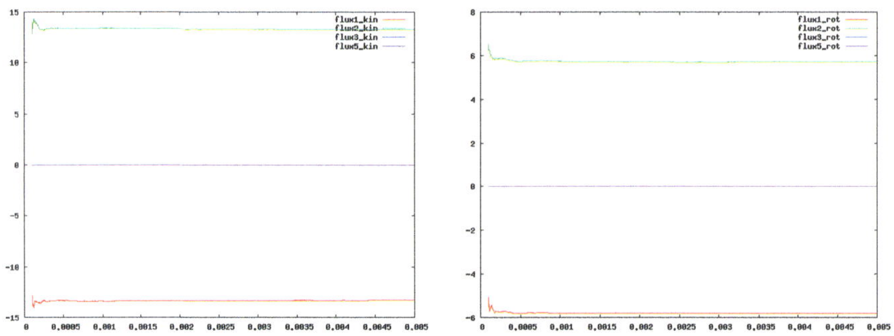

Fig. 6.42 Calculation G2. Computed transient energy fluxes at upper and lower plate. Left plot: Translational energy fluxes; Right plot: Rotational energy fluxes

Evaluation based on the difference between the kinetic and rotational energies of the impinging and reflected molecules at the upper and lower plates (two values *Q_KIN* and *Q_ROT* of opposite sign: the negative value refers to the lower plate, the positive value to the upper plate).

Results about the heat fluxes were as shown in Table 6.5.

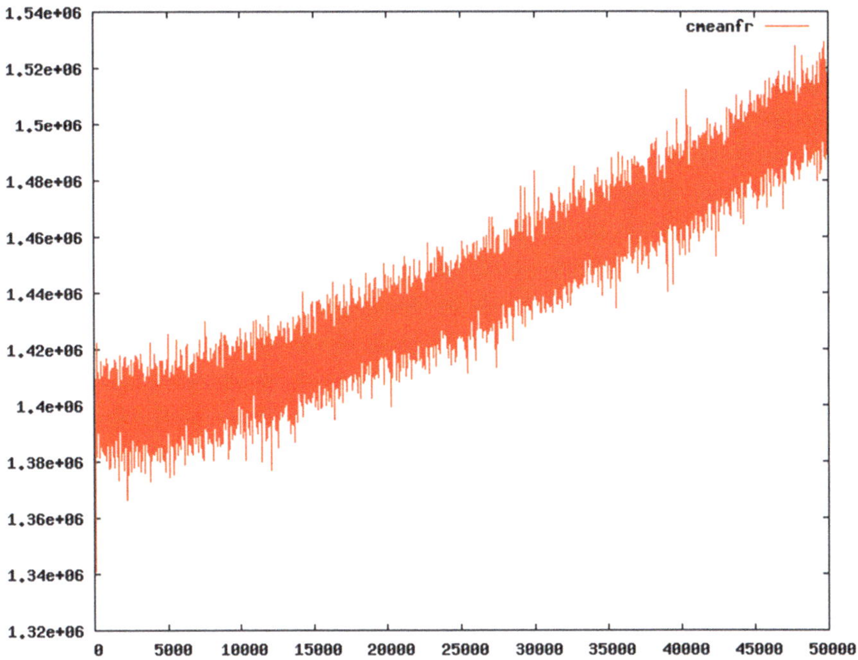

Fig. 6.43 Calculation G2. Mean collision frequency of one molecule in the cells of the domain. The cells are numbered sequentially row by row from the bottom to the top of the domain. The values of every fourth cell are plotted

The computed rotational [*PR*] and kinetic [*PK*] energies and their ratios are given for the three test cases in Table 6.6. The ratios of computed rotational and kinetic energies are very close to the theoretical value of 2/3.

On the basis of the results discussed above, the following conclusion can be drawn: The results for the heat fluxes obtained with the molecular model are satisfactorily. In case of the more diluted gas, cases G2 and G3, the temperature distribution remains steeper. The results provided by the molecular simulations are better than the macroscopic results obtained on the basis of the mean temperature distributions.

6.10.4 Calculations of Particle Trajectories

6.10.4.1 Preliminary Calculations

(i) **Calculation C10**
Preliminary results of particle trajectories were obtained with a two-dimensional computation with the data shown in Table 6.7 (Calculation C10 with 10 particles simulated.

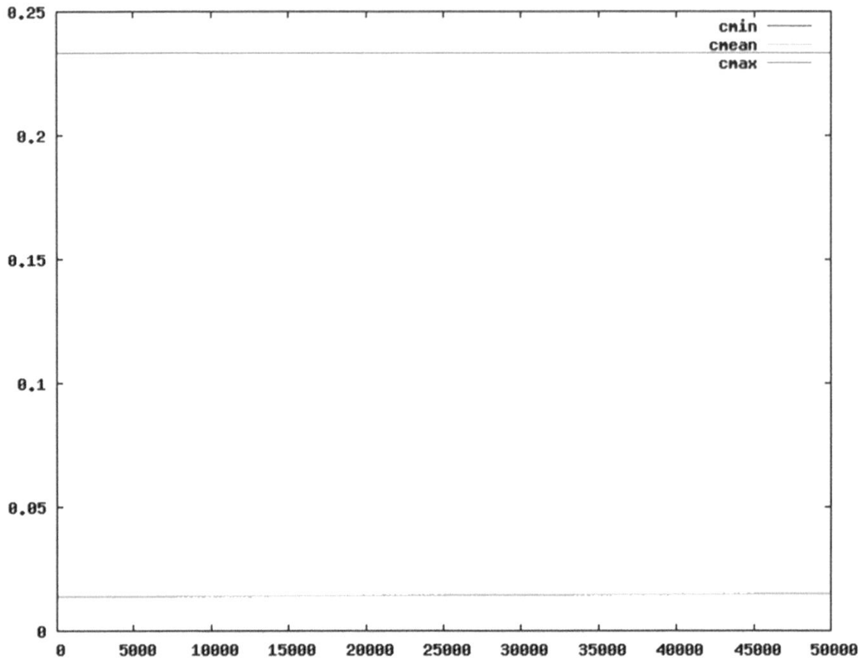

Fig. 6.44 Calculation G2. Minimum (zero values), mean and maximum numbers of collisions in the cells of the domain. The cells are numbered sequentially row by row from the bottom to the top of the domain. The values of every fourth cell are plotted

Figure 6.53 shows the computed normalized temperature distribution obtained with the input data given above.

Figure 6.54 shows the trajectories of the particles computed for 5 s of problem time. Some trajectories circulate around the centre of the domain, while other ones tend to escape. The particle at the right lower corner in this figure is particle Nr. 9. This particle seems to be attracted towards the lower right circle. Its trajectory is shown in detail in Fig. 6.55. It is not clear whether this result can be interpreted as a case of negative thermophoresis.

(ii) **Calculation C10_22e1_500**

A question arises whether a stronger temperature difference between plates and circles may help in better collimating the particles. To try to answer this question, calculation C10b has been made with the same data as for the above calculation (Table 6.7), with exception of the temperatures of plates and circles which were as follows: Upper plate: 324 K, Lower plate: 312 K, Circles (all): 500 K. Besides, 2 rings with different diameters were simulated.

Figure 6.56 shows the computed normalized temperature distribution obtained with these temperature at the boundaries.

Figure 6.57 shows the computed trajectories of the particles for the first 3 s of problem time. All trajectories move upwards. They slightly collimate towards the

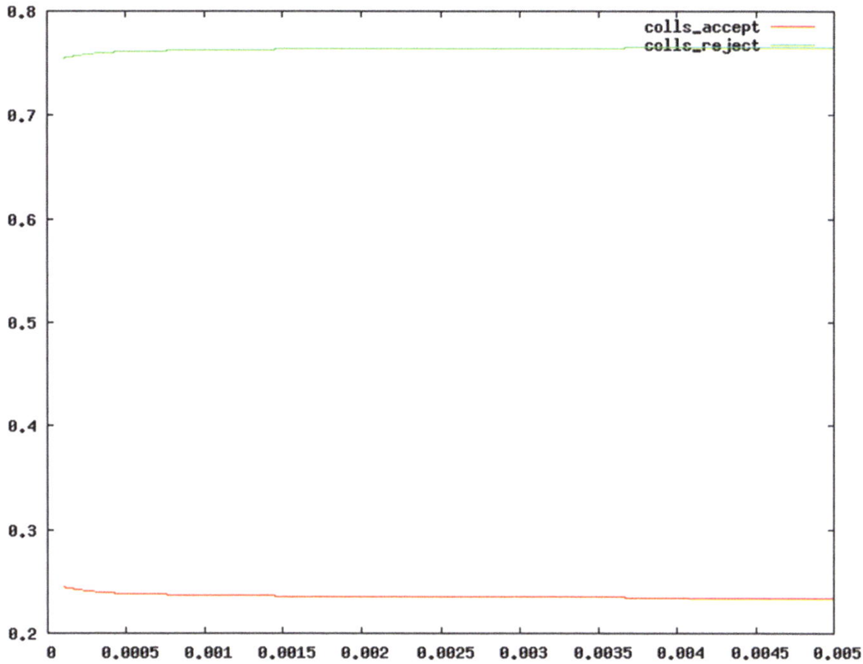

Fig. 6.45 Calculation G2. Percentage of accepted (lower plot) and rejected inter-molecular collisions, versus time

central plane ($x = 400$ μm) but the upward motion is not stopped by the temperature gradient in vertical direction because in the upper part of the domain a large plateau is formed where the temperatures are rather uniform and therefore the temperature gradients, hence the forces acting upon the particles, rather faint.

It is informative, for assessing the tendency of the particles to collimate, to consider the ratio of inertial momentum of rotation of particles with respect to their centre of mass at current time 't' to the moment at $t = 0$. This ratio (in the following referred to simply as "momentum ratio") is shown versus time in Errore. L'origine riferimento non è stata trovata (Fig. 6.58). The plot decreases up to a minimum of about 0.3, but than starts oscillating hinting that a further concentration of particles does not occur.

These results and many other ones obtained with similar conditions hint at the difficulty of finding the right conditions (temperatures of circles and plates, size and location of circles) for collimating the particles. But these results also suggest that the option of heating the coils at higher temperature than the upper and lower plates may not be the best one. The alternative option of having the coils at temperature lower than that of the plates is investigated in the next section.

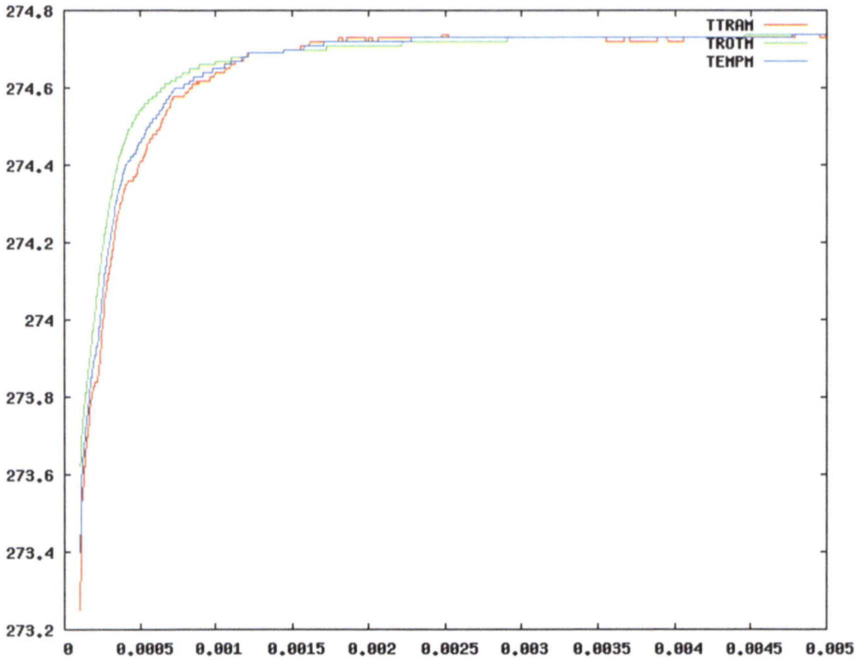

Fig. 6.46 Calculation G3. Computed transient mean temperatures (translational, rotational and overall) in domain, versus time

6.10.4.2 Search for Optimum Temperature Gradients

We explored the idea of simulating coils at temperature lower than that of the bounding plates, in the attempt to obtain a temperature distribution in the gas with a unique minimum a the centre of the domain. The temperature trough would imply strong temperature gradients, hence forces acting upon the particles, capable of bringing them to collimate to the point of minimum temperature. This idea was pursued defining a series of four test cases, referred to as ASME-TEST_01 through ASME-TEST_04, because made in the frame of the preparation of results to be presented at the ASME conference (Bottoni & Mantovani, 2008). We run these test cases at first without simulation of particles to assess the temperature distribution arising from the molecular dynamic calculations unperturbed by the presence of particles.

Data for the **ASME-TEST_01** were as reported in Table 6.8.

Data for **ASME-TEST_02** were as those of ASME- TEST_01 but for the location of the upper coils:

C1 (575, 350), C2 (225, 350).

Data for **ASME-TEST_03** were as those of ASME- TEST_01 but for the temperatures of the surfaces, which were as follows: T(plates) = 335 K, T(coils)

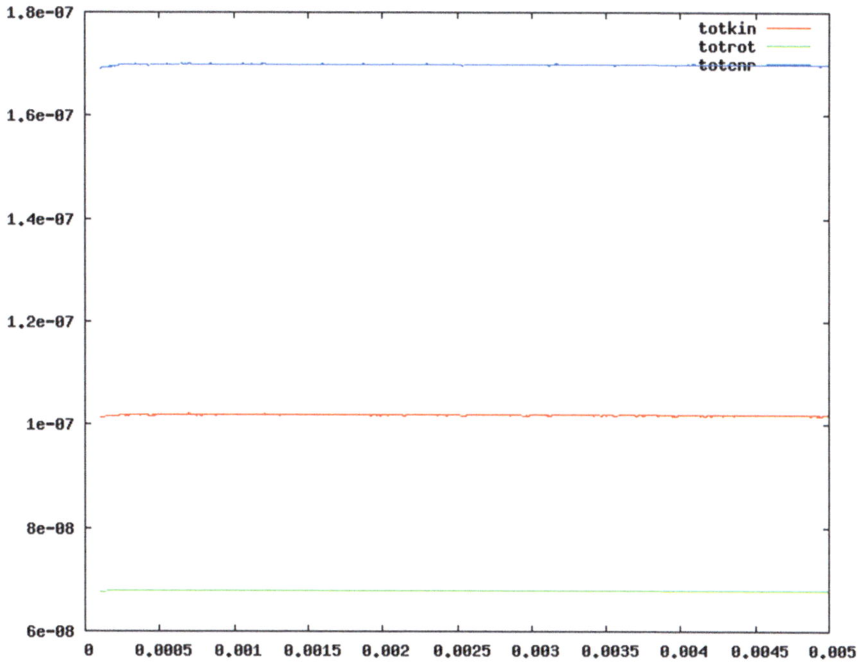

Fig. 6.47 Calculation G3. Computed transient energies in domain. Uppermost plot: Total energy; Middle plot: Kinetic energy; Lower plot: Rotational energy

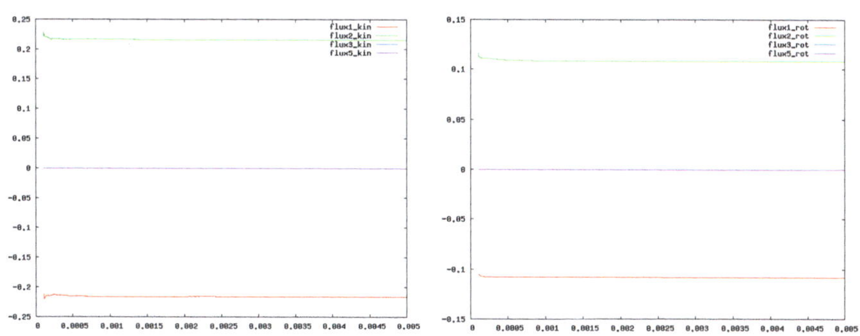

Fig. 6.48 Calculation G3. Computed transient energy fluxes at upper and lower plate. Left plot: Translational energy fluxes; Right plot: Rotational energy fluxes

= 275 K. The overall temperature difference was therefore 10 K larger than in ASME-Test_01.

Data for **ASME-TEST_04** were as those of ASME- TEST_02 but for the temperatures of the surfaces, which were as follows: T(plates) = 335 K, T(coils) = 275 K. The overall temperature difference was therefore 10 K larger than in ASME-TEST_02.

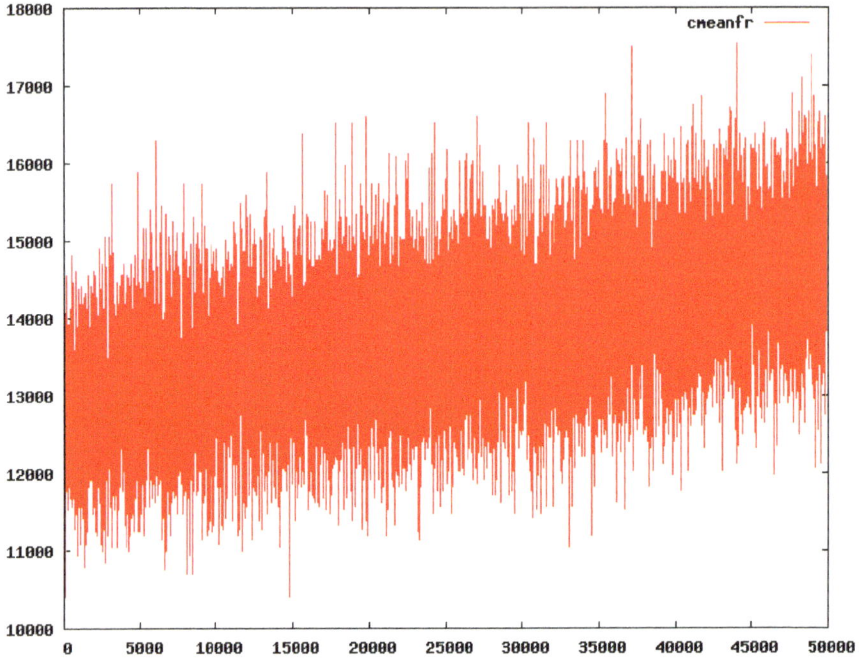

Fig. 6.49 Calculation G3. Mean collision frequency of one molecule in the cells of the domain. The cells are numbered sequentially row by row from the bottom to the top of the domain. The values of every fourth cell are plotted

Computing for 10^7 time steps of length $\Delta t = 10^{-7}$s, hence for a problem time of 1.0 s, we obtained for the four test cases the temperature distributions shown in Errore. L'origine riferimento non è stata trovata. Figure 6.59 for a mid-plane through the domain in vertical direction and for a mid-plane in horizontal direction. The temperature troughs between maximum and minimum values of the gas temperature are normalized so that 10 K correspond to a sink of one unity in the ordinates of the plots. The deepest temperature troughs, hence the most favourable conditions for the collimation of the temperature under the forces arising from the temperature gradients, are obtained for the test cases 3 and 1. The less favourable conditions are for tests 4 and 2. The most important feature of these distributions, from the viewpoint of the efficiency of the thermal confinement, is the absence of significant local minima of the temperature distributions. The presence of particles, if injected at the beginning, does perturb the temperature distribution inducing two or more distinct minima. This effect is shown in Errore. L'origine riferimento non è stata trovata. Figure 6.60, obtained with a calculation where the particles were injected at time $t = 0$.

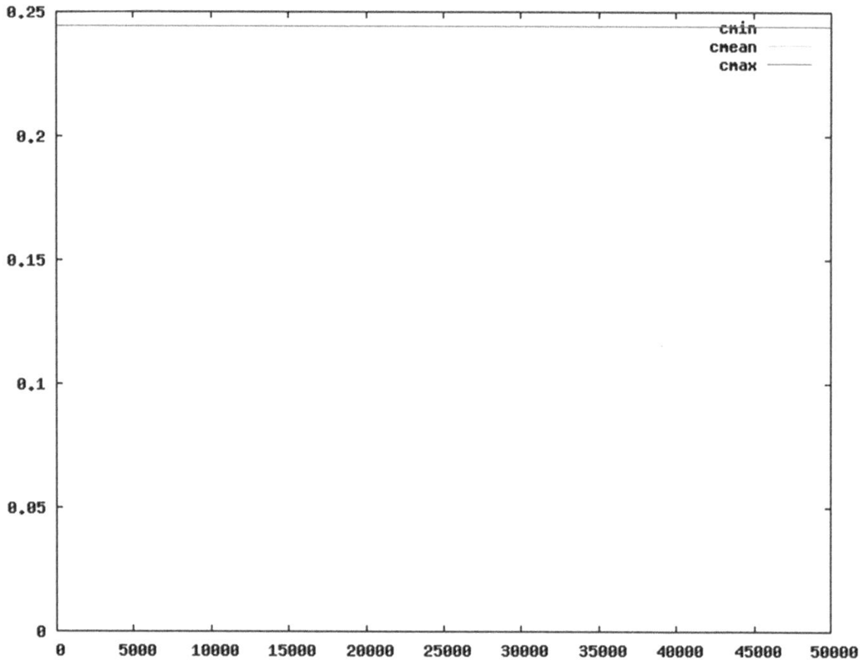

Fig. 6.50 Calculation G3. Minimum (zero values), mean (about 0.13×10^{-3}) and maximum numbers of collisions in the cells of the domain. The cells are numbered sequentially row by row from the bottom to the top of the domain. The values of every fourth cell are plotted

6.10.4.3 Comparative Test Case Based on the Waldmann Formula for Thermophoretic Forces

A separate code version, referred to as TH_GRAD_REF3, has been written to investigate the results of a simulation based on thermal confinement of particles due to forces arising from temperature gradients evaluated without taking into account the perturbations due to the presence of particles. The simulation has been made as follows, using the input data of the ASME-TEST_04 reported earlier.

A preliminary calculation, referred to as TEST_GRAD3, is made with the statistical molecular dynamics method up to a time problem of 0.05 s, simulating 500,000 time steps of $\Delta t = 10^{-7}$s, in absence of particles. This simulation allows for computing the temperature distribution in the fluid shown in normalized scales in Fig. 6.61. The temperature distributions along a vertical and a horizontal plane through the mid-point of the domain are shown by the red plots in Fig. 6.62. These plots are not as smooth as those of Fig. 6.59, because the problem time was limited to 0.05 instead of 1.0 s and because the calculation was made with a relatively small number of 20 sample molecule for computational cell. To simulate the ideal temperature distribution which would be achieved by a very long calculation with a large number of sample molecules, at least one hundred per cell, we

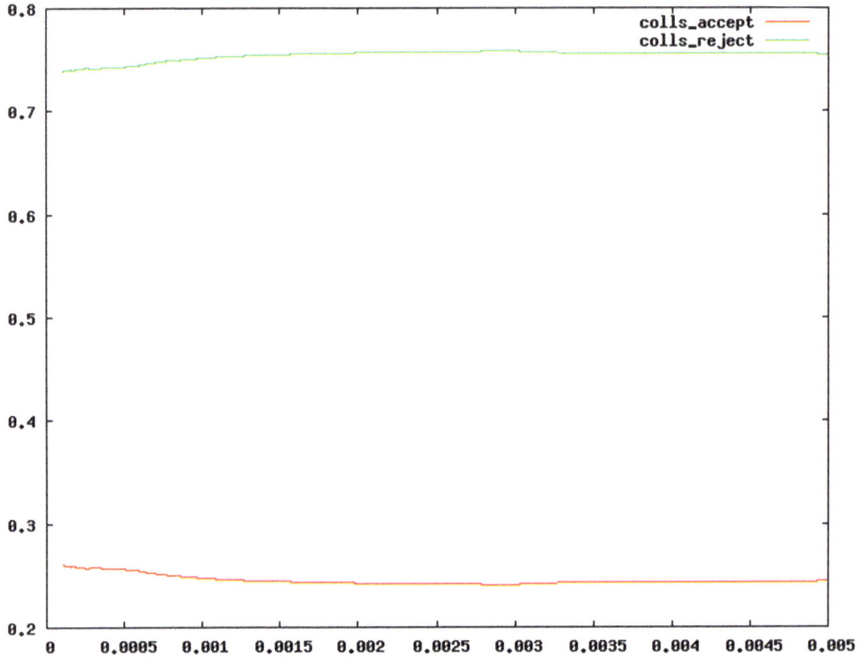

Fig. 6.51 Calculation G3. Percentage of accepted (lower plot) and rejected inter-molecular collisions, versus time

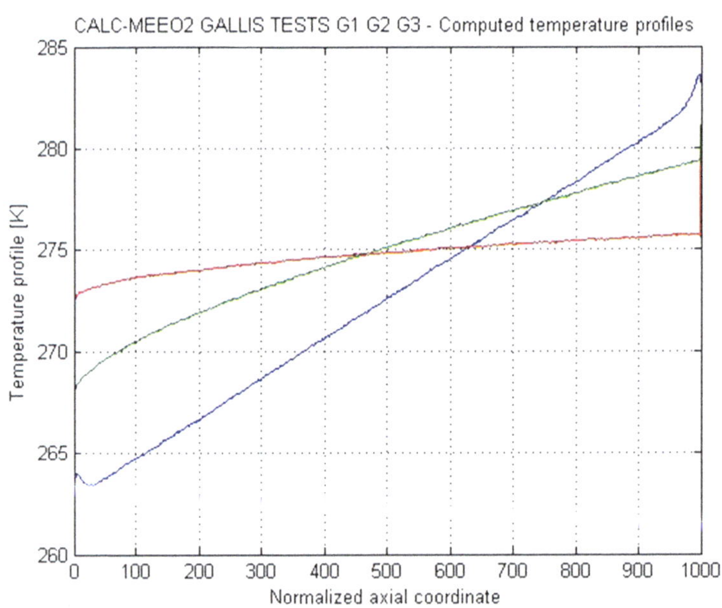

Fig. 6.52 Computed temperature profiles for the test cases G1 (uppermost at left boundary), G2, G3 versus number of cells in axial direction

Table 6.5 Gallis calculations G1, G2, G3. Summary of computed heat fluxes ad comparison with expected values (Q_Gallis)

	Q_TAV (W/m^2)	Q_KIN (W/m^2)	Q_ROT (W/m^2)	Q_TOT (W/m^2)	Q_Gallis
G1	47.2.13	+32.87/−36.32	+16.95/−13.73	+49.83/−50.04	48.6
G2	26.05	+13.30/−13.26	+5.73/−5.78	+19.03/−19.02	18.4
G3	6.75	+0.216/−0.216	+0.108/−0.108	+0.324/−0.324	0.333

Table 6.6 Computed rotational energy (PR) and kinetic energy [PK] in the computational domain, for the three test cases

	PR (J)	*PK* (J)	*PR/PK*
G1	3.761×10^{-21}	5.646×10^{-21}	0.6661
G2	3.791×10^{-21}	5.693×10^{-21}	0.6659
G3	3.794×10^{-21}	5.642×10^{-21}	0.6665

Table 6.7 Relevant data for the calculation C10

Domain	
Dimension of domain:	*XMAX* = 800 μm, *ZMAX* = 400 μm
Scale factor (ratio between real and simulated size of domain):	200
Number of cells:	80 × 40
Number of sample molecules:	64,000 (20 molecules per cell)
Gas (air) pressure:	100 N/m^2
Type of gas simulation:	Monoatomic
Particles	
Number of particles:	10
Injection range:	Between meshes 37–44 in *x*-direction and between meshes 17–24 in *z* direction
Size of particles:	0.3 μm
Plates and coils	
Temperature of plates:	Top = 324 K, bottom = 300 K
Position of circles from the upper-right one (C1) counter-clockwise:	C1 (560, 240) (Coordinates (*x*, *z*) in μm) C2 (240, 240) C3 (340, 120) C4 (460, 120)
Temperature of upper circles:	350 K
Temperature of lower circles:	450 K

applied to the computed temperature distribution a smoothing technique aiming at getting rid of small local oscillations. The smoothing technique is based on the application of the subroutine SMOOFT described in Press et al. (1986). The smoothed temperature distributions are shown in the same Fig. 6.62 by the green plots.

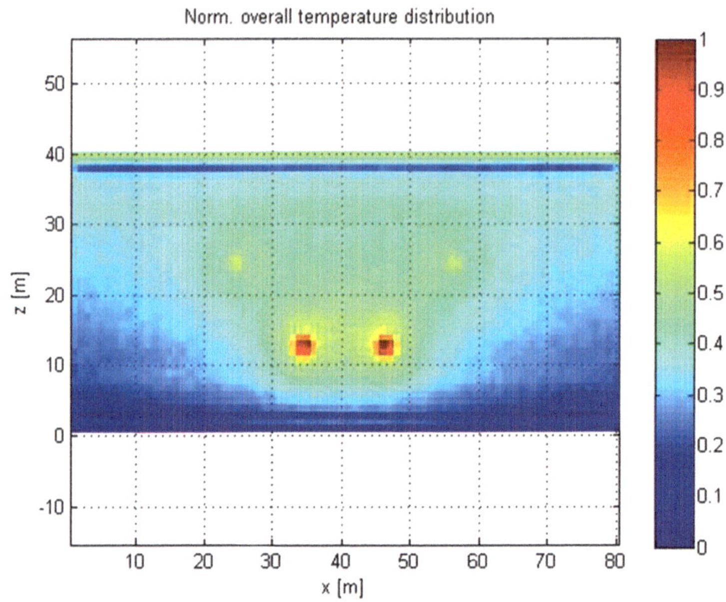

Fig. 6.53 Calculation C10. Normalized temperature distribution in the domain, induced by the upper and lower plates and by the four simulated rings (shown in cross section)

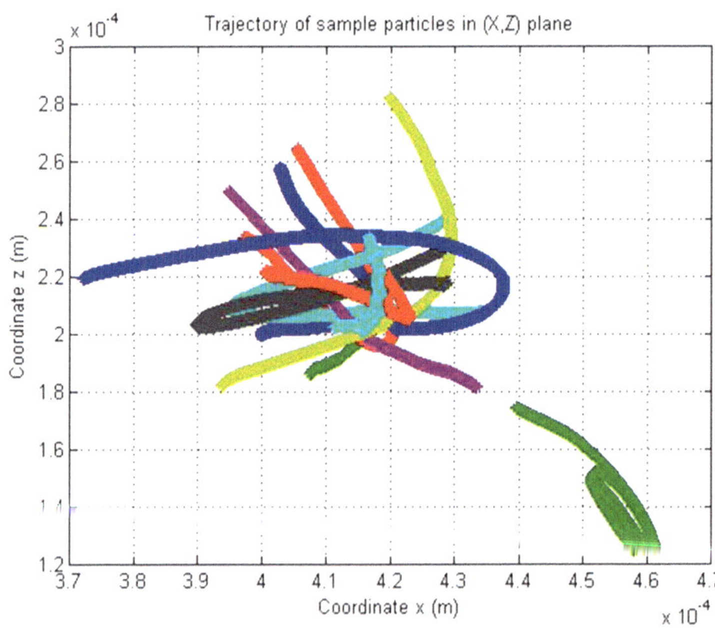

Fig. 6.54 Calculation C10. Trajectories of the particles during 4.5 s of problem time

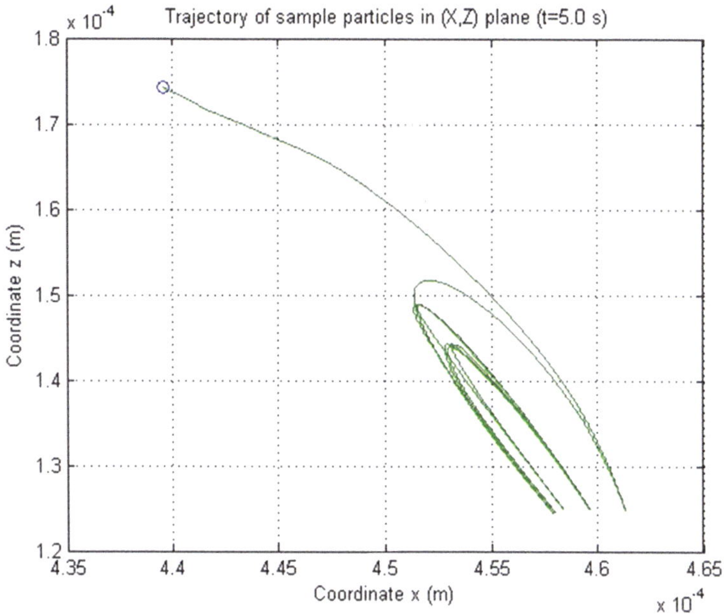

Fig. 6.55 Calculation C10. Trajectory of particle Nr. 9

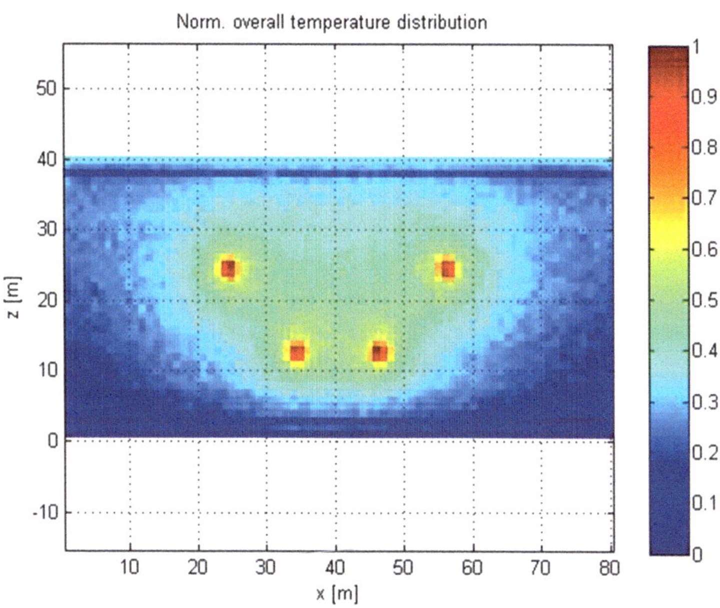

Fig. 6.56 Calculation C10b. Normalized temperature distribution in the domain, induced by the upper and lower plates and by the four simulated rings (shown in cross section)

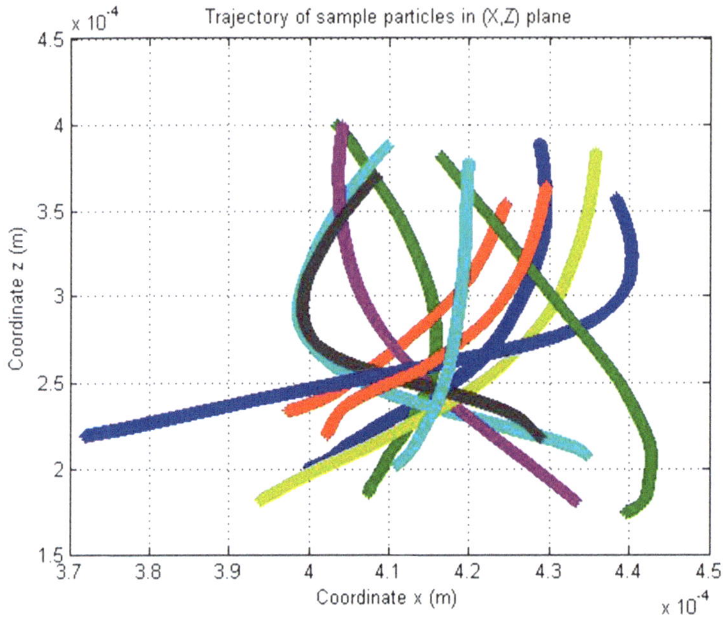

Fig. 6.57 Calculation C10b. Upwards trajectories of the particles during 3.0 s of problem time

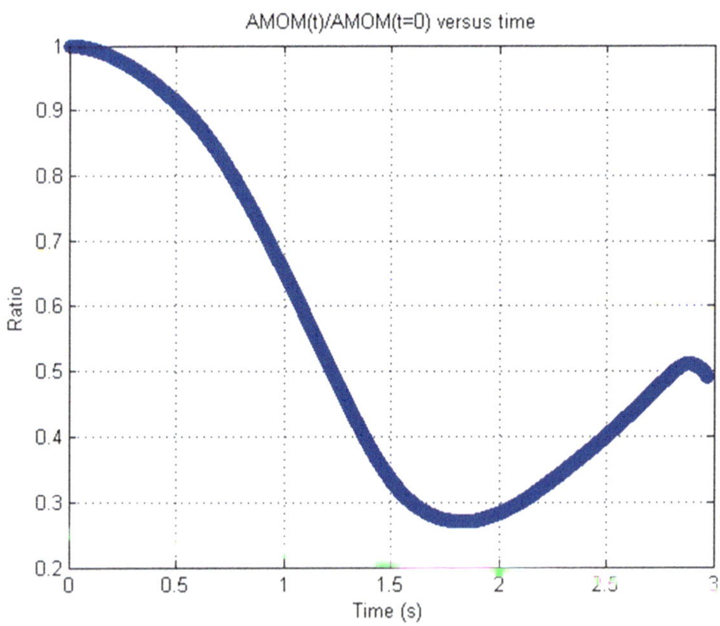

Fig. 6.58 Calculation C10b. Ratio of inertial momentum of rotation of particles with respect to their centre of mass at time t, to the momentum at time $t = 0$

Table 6.8 Relevant data for the calculation ASME-Test-01

Domain	
Dimension of domain:	*XMAX* = 800 μm, *ZMAX* = 400 μm
Scale factor (ratio between real and simulated size of domain):	200
Number of cells:	80 × 40
Number of sample molecules:	64,000 (20 molecules per cell)
Gas (air) pressure:	100 N/m^2
Time step	
Δt (s)	10^{-7}
Number of time steps	ITMAX = 10^7(Calculation run till 1 s problem time)
Plates and coils	
Temperature of all plates:	330 K
Position of coils from the upper-right one (C1) counter-clockwise:	C1 (480, 280) (Coordinates (x, z) in μm) C2 (320, 280) C3 (320, 120) C4 (480, 120)
Temperature of all coils:	280 K

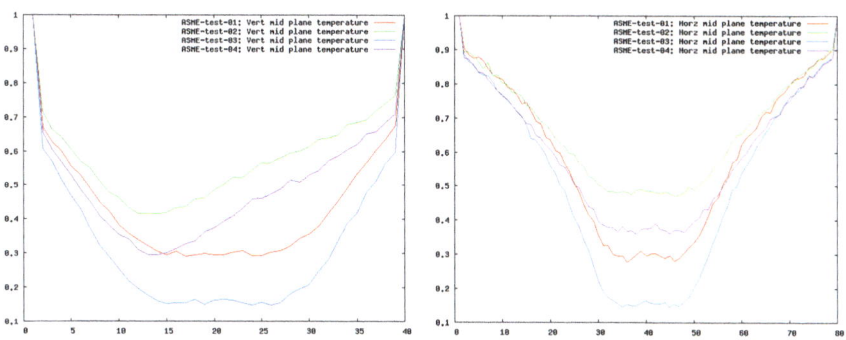

Fig. 6.59 Calculations ASME-TEST_01 through ASME-TEST_04. Computed normalized temperature distributions in a vertical (left) and in a horizontal mid-plane (right) through the domain

The smoothed temperature distribution in the full domain is used to compute the thermal gradients in the x- and z-direction, which are shown in Fig. 6.63. This figure must be interpreted keeping in mind that the scaled colour range (0–0.5) applies to negative gradients, while the range (0.5–1) is for positive gradients (hence, a vector component directed along the positive axis in either direction).

Forces acting on a particle injected in the field with the given temperature distribution are computed using the Waldmann formula

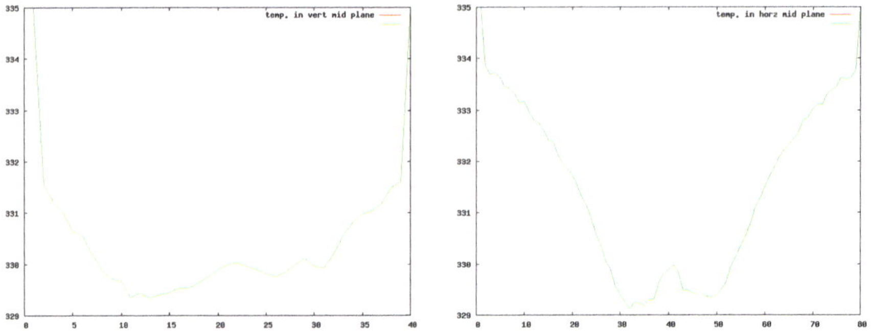

Fig. 6.60 Comparative calculation for ASME- TEST_03 with presence of particles. Computed normalized temperature distributions in a vertical (left) and in a horizontal mid-plane (right) through the domain

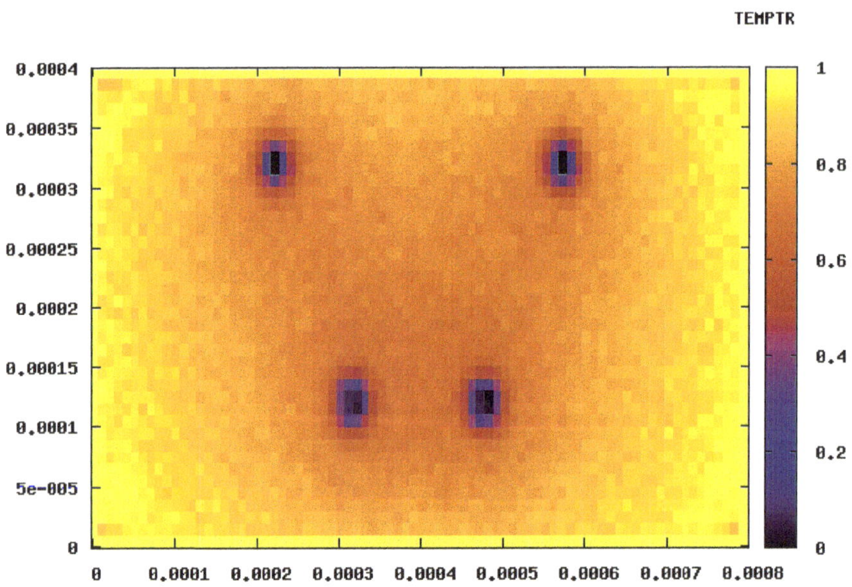

Fig. 6.61 Calculation TEST_GRAD3. Temperature distribution obtained at = 0.05 s in absence of particles

$$\vec{F} = -\frac{16\sqrt{\pi}}{15}\beta r_p^2 K \frac{dT}{d\overline{x}}, \tag{6.48}$$

where $\beta = 1/c'_m$ is the inverse of the most probable thermal speed, K is thermal conductivity of the gas and $d\overline{x} = (dx, dz)$.

Velocities of the particles moved by this thermophoretic force have been computed according to the formula (6.39).

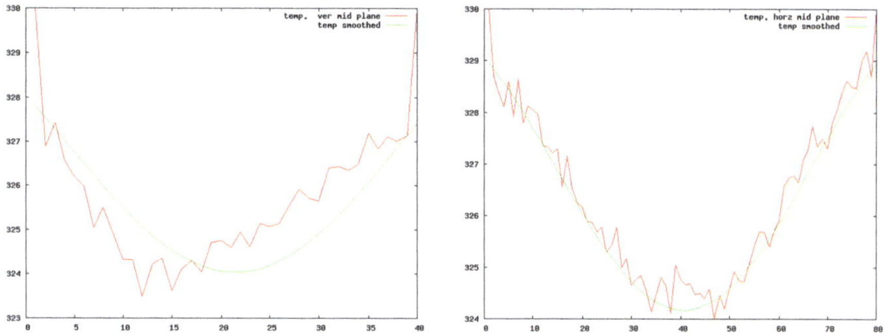

Fig. 6.62 Calculation TEST_GRAD3. Computed and smoothed temperature profiles in a vertical plane (left) and in a horizontal plane (right) through the mid-point of the domain

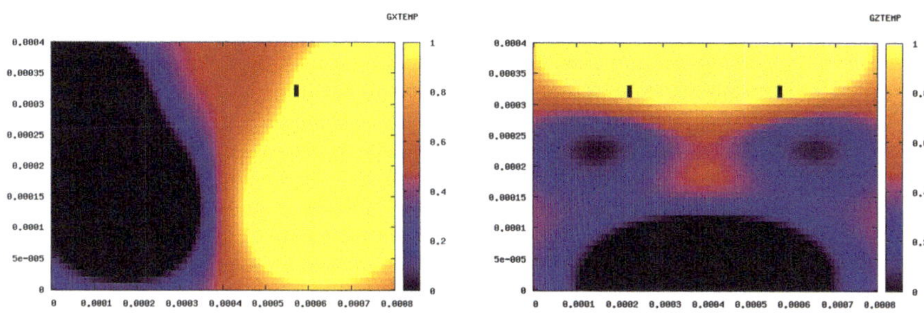

Fig. 6.63 Calculation TEST_GRAD3. Temperature gradients in the x-direction (left) and in the z-direction (right) based on smoothed temperature distribution. Key: Colour scale in the interval (0–0.5) refers to negative gradients, in the interval (0.5–1) refers to positive gradients

The results of the application of formula (6.48) to the temperature gradients obtained from the smoothed temperature distribution are shown in Errore. L'origine riferimento non è stata trovata (Fig. 6.64). It is evident that the forces acting on a hypothetical particle injected into the field would transport it towards the middle region of the domain where gradients and forces vanish in a rather well marked plateau.

The simulations we performed proceeded from this point by injecting into the domain 100 sample particles of 0.3 μm radius and letting them move and coalesce. The assumption is made that the presence of the particles does not perturb the computed temperature distribution and, according to this assumption, we in fact by-passed for all the rest of the calculation the molecular dynamic simulation of the molecules, hence the computation of updated temperature distributions. In the frozen temperature field particles are moved and concentrated by the thermophoretic forces towards the region of the minimum temperature. Subsequent coalescence of couple of particles reduce their final number to only one. Fig. 6.65, left, shows the distribution of particles just after injection in the domain when only a few

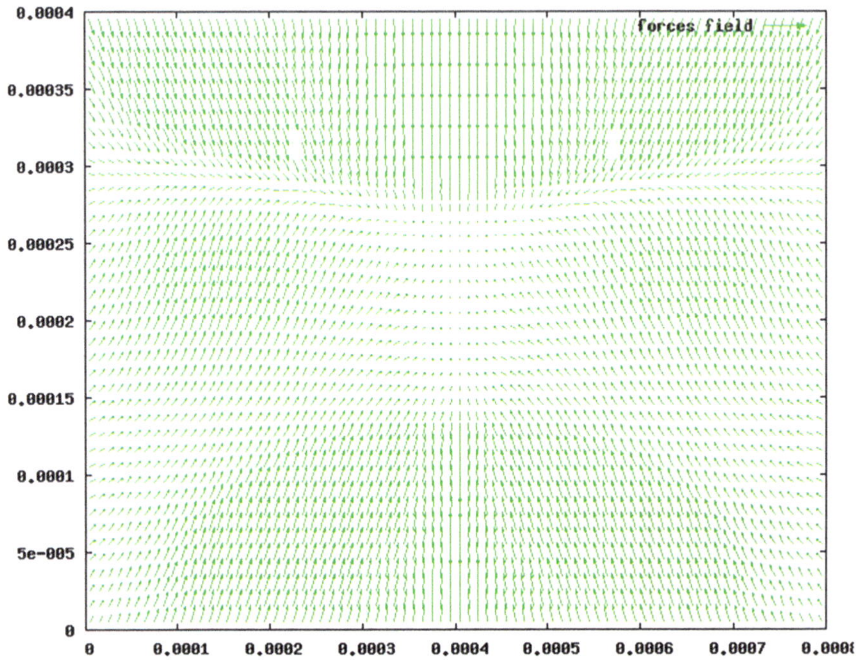

Fig. 6.64 Calculation TEST_GRAD3. Thermophoretic forses arising from the temperature gradients shown in Fig. 6.63

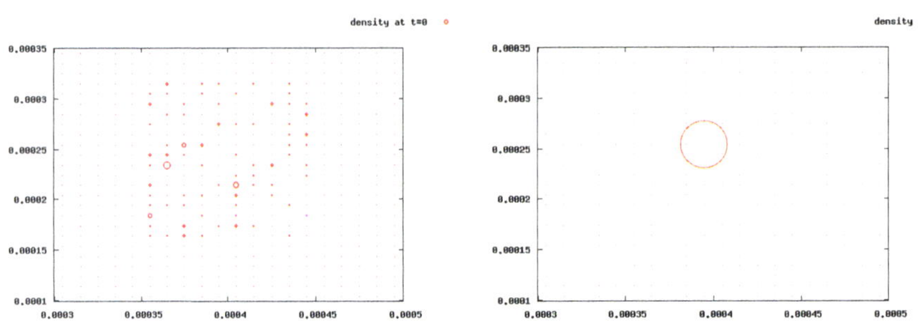

Fig. 6.65 Calculation TEST_GRAD3. Distribution of particles density just after injection (left) and just before the end of the calculation at $t \approx 8$ s (right)

coalescence were made. The size of the circles is proportional to the number of the particles represented. The right part of this figure shows the last particle left at the end of the simulation, representative of all 100 injected particles.

Information about the rapidity and efficiency of the coalescence of the particles is given by Errore. L'origine riferimento non è stata trovata. Figure 6.66, which shows the ratio of the inertial momentum of rotation of particles with respect to their centre

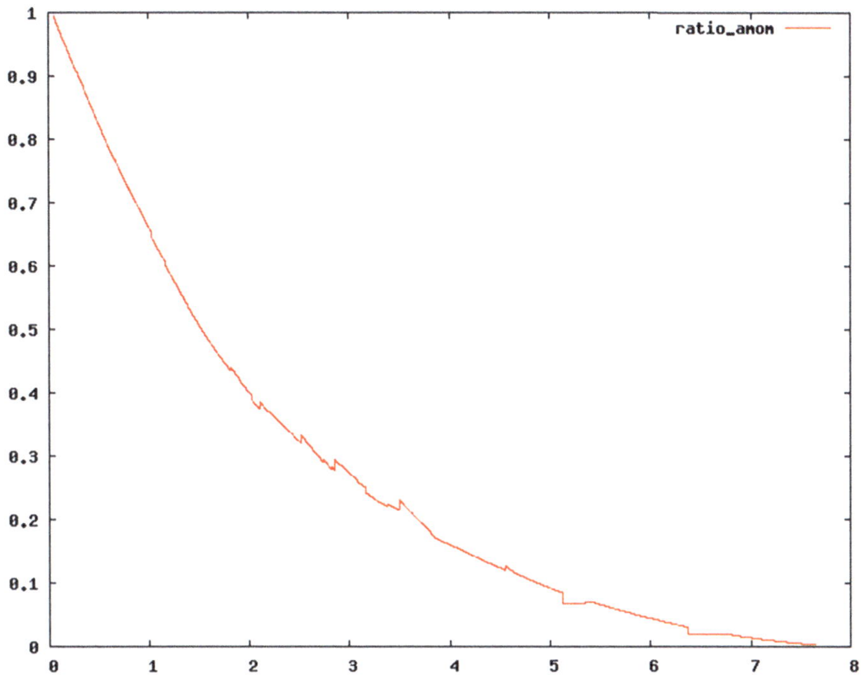

Fig. 6.66 Calculation TEST_GRAD3. Ratio of inertial momentum of rotation of particles with respect to their centre of mass at time t, to the momentum at time $t = 0$

of mass at time t to the momentum at time $t = 0$. With complete coalescence of all particles into only one at time $t \approx 8\ \text{s}$, the ratio of the momenta is reduced to zero.

The main purposes of this comparative calculation based on the application of the classical formulas by Waldmann for the thermophoretic forces and velocities are the following:

(i) Prove that in the idealized situation of a smooth temperature distribution, unperturbed by the presence of the particles, the algorithms coded in the computer program yield the correct results of concentrating all particles in the point of the minimum of the temperature distribution.
(ii) Provide a reference ideal situation which should be considered as the limiting case of performance of a more realistic simulation based on the application of the molecular dynamic calculation yielding a temperature field perturbed by the presence of the particles.

Results obtained with the application of the realist case of molecular dynamics combined with the simulation of a relatively large number of particles are discussed in the next section.

Table 6.9 Relevant data for the calculation ASME-TEST_03 with 2 coils

Domain	
Dimension of domain:	*XMAX* = 800 μm, *ZMAX* = 400 μm
Scale factor (ratio between real and simulated size of domain):	200
Number of cells:	80 × 40
Number of sample molecules:	64,000 (20 molecules per cell)
Gas (air) pressure:	100 N/m^2
Type of gas simulation:	Monoatomic
TIME STEP	
Δt (s)	10^{-7}
Number of time steps	ITMAX =16 × 10^7 (Calculation run till 16 s problem time)
Particles	
Number of particles:	100
Injection range:	Between meshes 37–44 in x-direction and between meshes 17–24 in z-direction
Size of particles:	0.3 μm
Particles injection:	At time t = 0.03
Inertial range simulation:	INORM = 25,000 IBLOCK = 100,000
Plates and coils	
Temperature of plates:	Top = 335 K, bottom = 335 K
Position of circles from the upper-right one (C1) counter-clockwise:	C1 (480, 280) (Coordinates (x, z) in μm) C2 (320, 280) C3 (340, 120) C4 (480, 120)
Temperature of upper circles:	275 K
Temperature of lower circles:	275 K

6.10.4.4 Results from Calculation ASME-TEST_03

6.10.4.4.1 Results from Calculation ASME-TEST_03 with Two Coils

For the reasons explained above the experimental conditions realized in the ASME-TEST_03 seem thus far to be the most promising for obtaining an efficient thermal confinement of particles. We therefore simulated this test for about 16 s problem time, with injection into the domain of 100 particles at time t = 0.003 s. For saving computational time the calculation was run with the monoatomic gas option. The main data for this reference calculation are shown in Table 6.9.

The computed normalized temperature distribution in the domain, is shown in Errore. L'origine riferimento non è stata trovata (Fig. 6.67). The kinetic power transferred to the system from the bounding surfaces, at higher temperature than the coils, are shown versus time in Fig. 6.68. The kinetic power subtracted from the system from the coils is shown in Fig. 6.69.

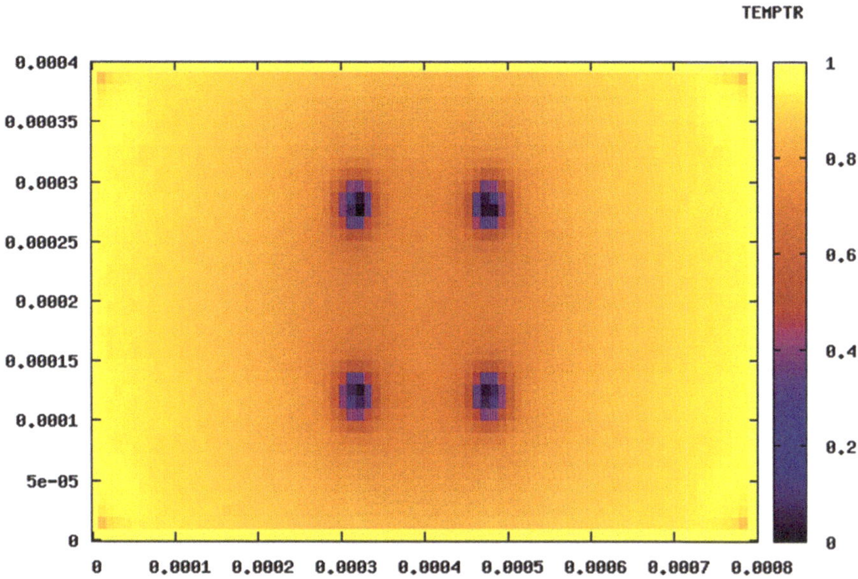

Fig. 6.67 Calculation ASME-TEST_03. Normalized temperature distribution obtained at = 16 s in presence of particles

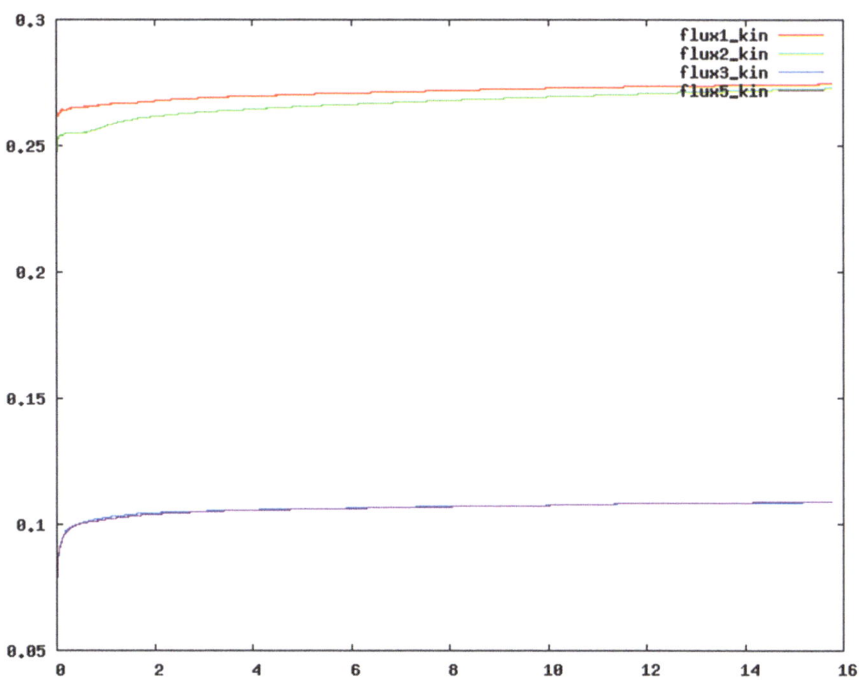

Fig. 6.68 Calculation ASME-TEST_03. Kinetic power transferred to the system from the bounding surfaces (1 = top surface, 2 = bottom surface, 3 = left surface, 5 = right surface) versus time (s). Units in watts

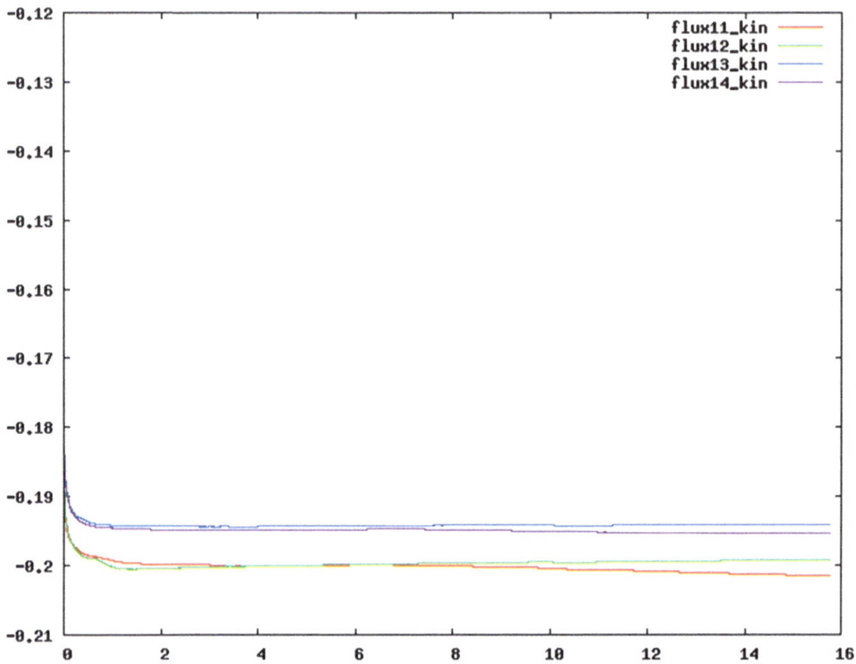

Fig. 6.69 Calculation ASME-TEST_03. Kinetic power transferred to the system from the surfaces of the circles (11 and 12 = uppermost circles, 13 and 14 lowermost circles) versus time (s). Units in watts

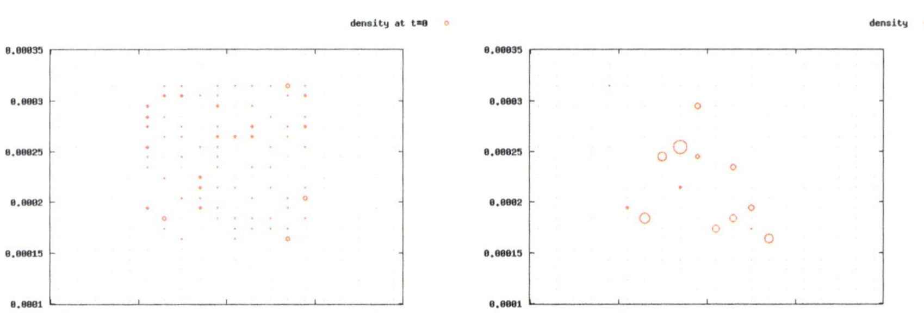

Fig. 6.70 Calculation ASME-TEST_03. Distribution of particles density just after injection (left) and at $t \approx 16$ s (right)

The distribution of particles density just after injection and the distribution at time $t \approx 16$ s are shown in Fig. 6.70. In the left plot of this figure are represented 77 particles, left after the first coalescences which occur immediately after particles injection into the domain. In the right plot are shown 14 particles left at the end of calculation at $t \approx 16$ s. In the course of calculation 11 particles were stuck to the

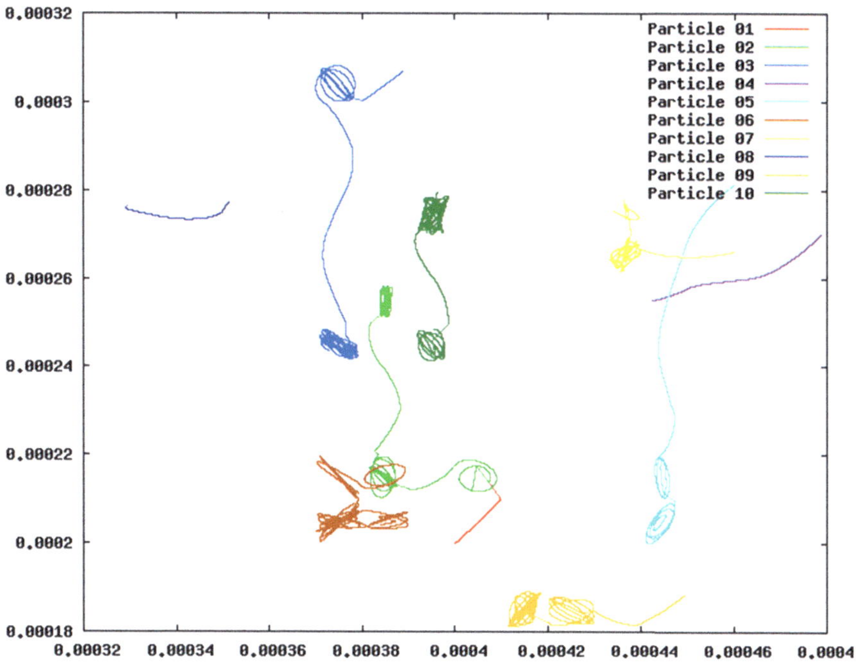

Fig. 6.71 Calculation ASME-TEST_03. Trajectories of a sample of 10 particles (with the lowermost indices 1–10) (Coordinates in meters)

coils, attracted by the strong temperature gradients in the immediate proximity of the cold coil surfaces. These particles are considered as lost from the viewpoint of the thermal confinement. There were 75 coalescences reducing the remaining 89 particles to 14.

The trajectories of the ten particles with the lowermost indices (1–10) are shown in Fig. 6.71 in a sub-domain close to the geometrical centre of the full computational domain. In this figure time is a parameter monotonically increasing along the trajectories. Most of the particles, starting from the original position at time of injection, first move and reach a relative minimum in the computed temperature distribution, in which position they start rotating and are finally trapped. These circumstances prevent the remaining particles to approach each other and therefore prevent further coalescences.

The ratio of inertial momentum of rotation of all particles with respect to their centre of mass at time t, to the momentum at time $t = 0$ is shown in Fig. 6.72. This picture can be easily interpreted considering that the rotational momentum of the whole bunch of particles increases when one particle, attracted by the temperature well around the coils, moves in outwards direction, gets stuck to a coil and is considered as lost from the viewpoint of the thermal confinement. Accordingly, these particles are neglected from further consideration. These "particle-loss" events are 11 in all the calculation and correspond to the local small peaks in the plot of

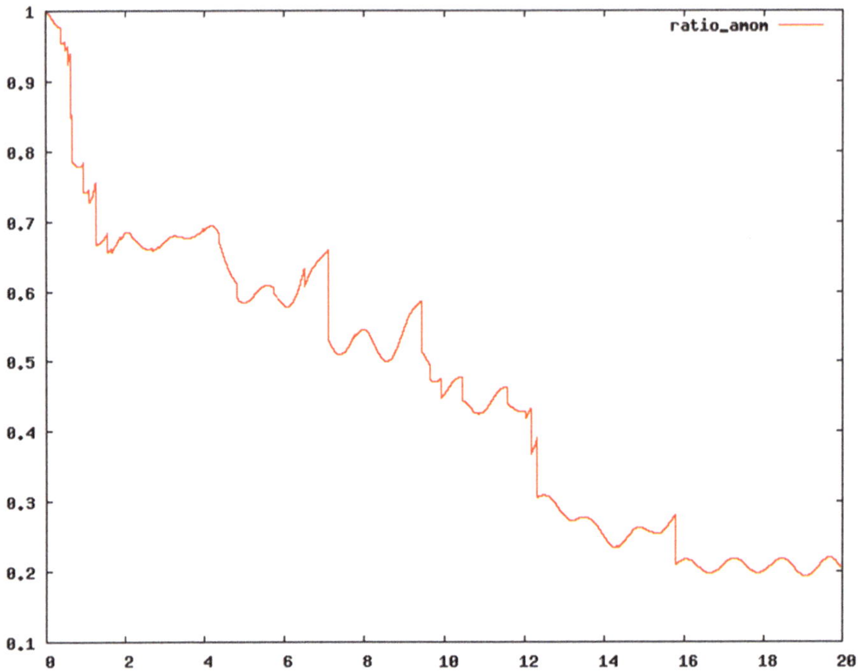

Fig. 6.72 Calculation ASME-TEST_03. Ratio of inertial momentum of rotation of particles with respect to their centre of mass at time t, to the momentum at time $t = 0$

Fig. 6.72. After about 16 s into the transient, the remaining particles always rotate in the niches of minima of the temperature distribution in which they fell and correspondingly the whole rotational momentum keeps oscillating, but does not reduce any more substantially. Thus the ratio of the current rotational momentum to its initial value does not decrease below a value of about 0.20.

These computational results can therefore be evaluated by stating that:

(i) The efficiency of the thermal confinement, in the chosen configuration of the coils and with the given temperatures of the boundary surfaces, is of about 90% (11 particles, out of 100, lost from the thermal confinement).
(ii) In terms of packing of the whole bunch of particles, measured by the ratio of the rotational momenta, the given configuration allows to reach an efficiency of about 80%.

The comprehension and justification of these limits of the efficiency of thermal confinement were made possible by the careful examination of the computed results in the simulated experimental conditions. This computational experience thus formed the necessary basis for directing further choices of initial and boundary conditions, in subsequent numerical experiments, towards the goal of improving the thermal confinement by coping with its inherent limitations.

Table 6.10 Relevant data for the calculation ASME-TEST_03 with 3 coils

Domain	
Dimension of domain:	XMAX – 800 μm, ZMAX = 400 μm
Scale factor [real size]/[simulated size]:	200
Number of cells:	80 × 40
Number of sample molecules:	320,000 (100 molecules per cell)
Gas (air) pressure:	100 N/m^2
Mean free path:	59.6 μm
TIME STEP	
Δt:	10^{-5} s
Particles	
Number of particles:	100
Injection range:	Meshes 35–45 in x-direction
	Meshes 17–28 in z direction
Size of particles:	1 μm
Particles injection:	At time $t = 0.3$ s
Plates and coils	
Temperature of all plates:	335 K
Position (Coordinates (x, z) in μm) and temperatures (K) of circles:	C1 (600,300) 240 C2 (200,300) 270 C3 (200,100) 270 C4 (600,100) 240 C5 (650,200) 210 C6 (150,200) 330
Radius of coils (μm)	0.3

6.10.4.4.2 Results from Calculation ASME-TEST_03 with Three Coils

The code was than implemented to treat up to 3 coils with 6 intersections, as small circles, with the plane of the two-dimensional simulation. Besides, different temperatures could be assigned to the single circles which means, in the practical realization of the experiment, that the coils can be split in two semi-coils kept at different temperatures. The use of three coils allows for a smoother temperature distribution with more uniform temperature gradients and thus better performances in the thermal confinement. Referring to the same ASME-TEST_03, the new calculation conditions are as reported in Table 6.10.

A digression is made at this point on the physical implications of modelling a scaled experiment. In our calculations, physical quantities, like length, velocity, and temperature, were scaled according to the appropriate scale factor. The only physical quantities which were not scaled are the following: (i) Inter-molecular collision frequency; (ii) Size and mass of the particles; (iii) Momentum exchange between molecules and particles. This last item implies that the momentum exchange is computed using real, unscaled, velocities of the molecules, which thus implies the

Table 6.11 Root mean square molecular velocity SUMC2, theoretical value SC2THE and mean relative velocity of molecules GMEA in scaled and unscaled sample calculations

	Unscaled	Scaled (Scale factor 100)
SUMC2 = $\sqrt{\overline{c'^2}}$ (m/s)	538.25	5.38
SC2THE = $\sqrt{3p/\rho}$ (m/s)	534.74	5.34
GMEA = $\overline{g}$ (m/s)	804.30	8.04

calculation of the correct thermophoretic velocity of the particles. All results displayed in the plots, made automatically at the end of a calculation, are presented in real, unscaled, physical dimensions. Because temperature scales as the square of molecular velocities, calculations were made at a very low scaled temperature. The question therefore arises as to the legitimacy of this modelling approach which consists in assuming that a fluid of molecules behaves at very low temperatures as it would at the real temperatures. The question of the legitimacy can only be answered considering whether or not the characteristics of a Maxwellian distribution at the scaled temperature remain or not the same as those of the Maxwellian distribution at the real temperature.

For this purpose we made a sample calculation in a domain of $160 \times 80 = 12{,}800$ meshes of 10 μm size, with 128,000 molecules, hence only 10 molecules per mesh. We computed, both in the real dimensions and in a scaled domain with a scale factor 100, the following quantities: (i) The root mean square value SUMC2 = $\sqrt{\overline{c'^2}}$ of molecular velocity, and compared to its theoretical value SC2THE = $\sqrt{3p/\rho}$. (ii) The mean value GMEA of the relative velocity of the molecules.

Results are given in Table 6.11.

Because the physical parameters of Table 6.11 compared well in the scaled and unscaled test calculations we concluded that the main characteristics of the Maxwellian distribution of molecular velocities are not infirmed by the application of the scale factor to the molecular velocities. Therefore also the computed temperature distributions remain faithful. Used scaled-down computational domains one reduces the total number of cells and therefore the computing time.

We return now to the simulation of the ASME-TEST_03. Figure 6.73 shows a prospective view of the computed normalized temperature distribution in the two-dimensional domain with the conditions reported in Table 6.10. Besides the deep temperature wells due to the cold circles, the computed temperature distribution shows a well defined sink in the middle of the domain.

In the course of a calculation, the code THERCONF-2D writes the file 27 (OUTPAR_FILM) which stores at subsequent time points the coordinates of the particles. This file can be used, in a post-processing program package, to plot the trajectories of the particles and to display in a transient film their motion and coalescences. From the film produced for the experiment ASME-TEST_03 over 4.47 s of problem time we present in Fig. 6.74 four plots which show the situation of the particles at different time points: (i) At $t = 0.3$ s, just after the injection of the

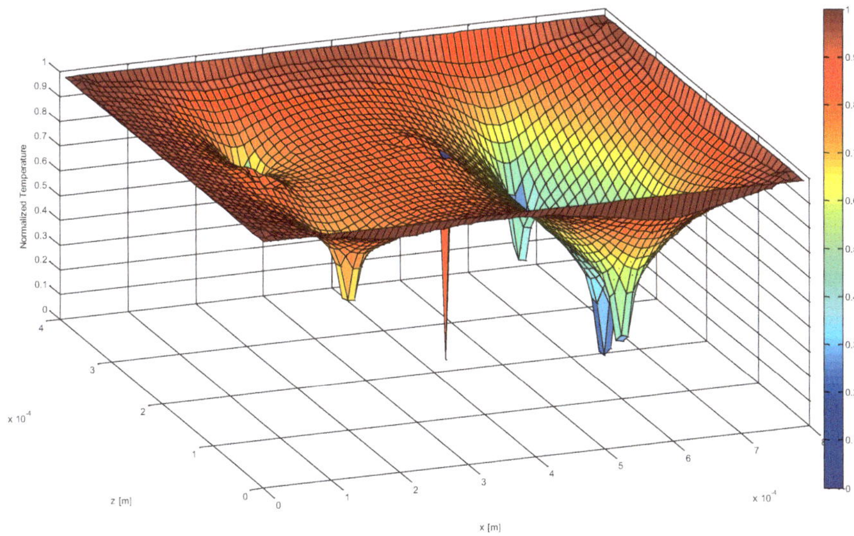

Fig. 6.73 Temperature distribution in the two-dimensional domain for the experiment ASME-TEST_03 with 3 coils

particles into the domain. Because of the tight loading of the particles in a small domain, several coalescences are already simulated just after the particles injection. (ii) At $t = 0.6$ s. (iii) At $t = 1.3$ s. (iv) At $t = 4.47$ s, at the end of the calculation. In these plots the size of the circles representing the particles is proportional to the number of particles coalesced into one. From the last plot one can see that just two particles were left. The largest one is representative of 90 initial particles, the smaller one of 10. This smaller particle is actually trapped in the local temperature minimum corresponding to the mesh 20 in the plots of the horizontal temperature distribution. The efficiency of the calculation, in terms of the percentage of the particles thermally confined and coalesced into a single one, is in this calculation of 90%. Generally speaking, in the large number of test calculations made with several parameter variations, the efficiency of thermal confinement ranged between 70% and 90%. It is evident that better performances can only be reached if more appropriate temperature distributions can be achieved by varying positions and temperature of the simulated coils. This consideration implicitly states the kind of work to be performed successively in a future three-dimensional code version.

The THERCONF code has been fully documented in an internal unpublished report of the MEEO (Meteorological Earth Environment Observation) of Ferrara. It can be delivered from the co-authors Dr. Simone Mantovani and Dr. Maurizio Bottoni to interested readers or researchers, upon written request, provided an agreement about development and use of the code can be agreed upon.

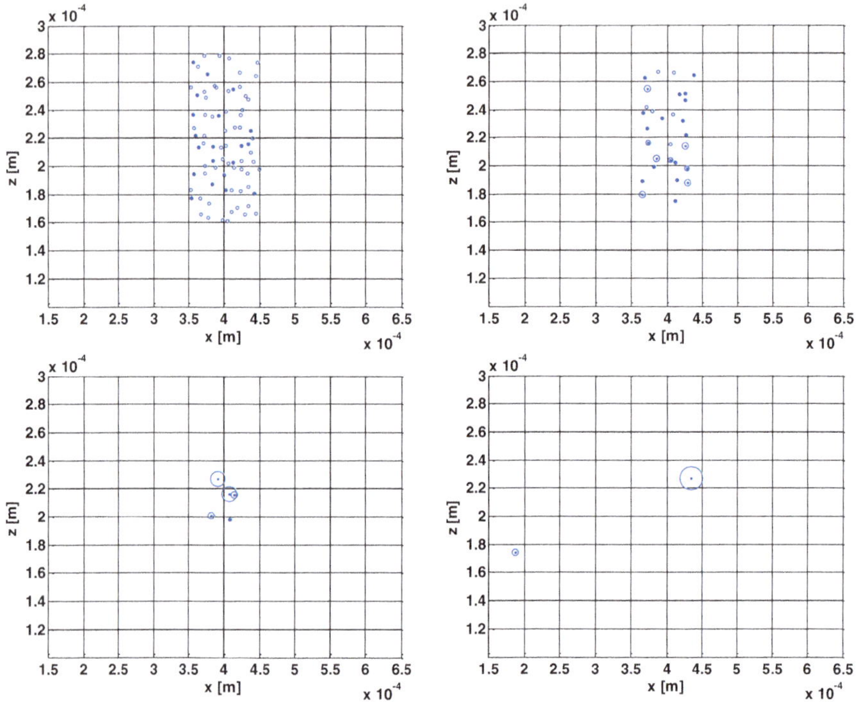

Fig. 6.74 Distributions of the particles at different time points: (i) At $t = 0.3$ s (upper left), just after the injection of the particles into the domain. (ii) At $t = 0.6$s (upper right). (iii) At $t = 1.3$ s (lower left). (iv) At $t = 4.47$ s (lower right), at the end of the calculation

References

Bhatnagar, P. L., Gross, E. P., & Krook, M. (1954). A model for collision processes in gases. I. Amplitude processes in charged and neutral one-component systems. *Physical Review, 94*(3), 511–525.

Bird, G. A. (1994). *Molecular gas dynamics and the direct simulation of gas flow*. Clarendon Press.

Bottoni M., & Mantovani, S. (2008). *Molecular dynamic simulations of thermal confinement of cosmic dust particles*. Accepted at the ASME 2008 Conference, Boston, Massachusetts, USA, October 31–November 6, 2008.

Bottoni, M., Porcu, F., Prodi, F., Santachiara, G., & Travaini, S. (2005). Bi-dimensional simulation of thermophoretic forces on aerosol particle with the Direct Simulation Monte Carlo (DSMC) method. *Computational Fluid Dynamics Journal, 14*(3).

Cercignani, C., & Lampis, M. (1974). Rarefied gas dynamics. In Karamcheti (Ed.), New York: Academic Press.

ESA (2006). *Invitation to Tender: AO/1-5045/06/NL/VK, feasibility study phase B0 of the Interaction in Cosmic and Atmospheric Particle System (ICAPS) Experiment, 15th August, 2006.* Addendum of 21st August 2006.

Gallis, M. A., Rader, D. J., & Torczynski, J. R. (2002). Thermophoresis in rarefied gas flows. *Aerosol Science and Technology, 36*, 1099–1117.

Larsen, P. S., & Borgnakke, C. (1974). Rarefied gas dynamics. In M. Becker & M. Fiebig (Eds.), *Paper A7*. DFVLR Press.

Lord, R. G. (1991). (1991). Rarefied gas dynamics. In A. E. Beylich (Ed.), *Proceedings of the seventeenth international symposium on rarefied gas dynamics*. VCH Verlagsgesellschaft mbH.

Marsaglia, G. (1972). Choosing a point from the surface of a sphere. *Annals of Mathematical Statistics, 43*, 645–646.

Ohwada, T., & Sone, Y. (1992). Analysis of thermal stress slip flow and negative thermophoresis using the Boltzmann equation for hard-sphere molecules. *Mechanics of Fluids Journal B, 11*, 389–414.

Papadopoulos, D. H., & Rosner, D. E. (1995). Enclosure gas flow driven by non-isothermal walls. *Physics of Fluids, 7*(11), 2535–2537.

Press, W. H., Flannery, B. P., Teukolsky, S. A., & Vetterling, W. T. (1986). *Numerical recipes*. Cambridge University Press.

Prodi, F., Santachiara, G., Travaini, S., Vedernikov, A., Dubois, F., & Legros, J. C. (2004). Measurements of phoretic velocities of aerosol particles in microgravity conditions. In *Proceedings of the 14th International Conference on Clouds and Precipitations* (Vol. I).

Vincenti, W. G., & Kruger, C. H. (1965). *Introduction to physical gas dynamics*. Krieger Publishing Company.

Waldmann, L. (1959). Ueber die Kraft eines inhomogenen Gases auf kleine suspendierte Kugeln. *Zeitschrift fuer Naturforschung, 14*(7), 589–599.

Yamamoto, K., & Ishihara, Y. (1988). Thermophoresis of a spherical particle in a rarefied gas of a transition regime. *Physic of Fluids, 31*, 3618–3624.

Chapter 7
Look into the Future

Contents

In this section, we recall some applications of the molecular dynamics method that are suitable for further developments, in part already undertaken in the past years, in part not fully exploited but in any case promising for the advancement of this discipline.

7.1 Reaction Rate Theory

Numerical analysis of collisionless molecular dynamics is based upon prediction of the trajectories of molecules or particles. One sample molecule or particle is treated as representative of a state of the system or of a large number of molecules or particles of the system. A generalization of the concept of trajectory, as locus of subsequent points representative of the system in phase space, is applied in the Transition State Theory (TST), or Reaction-Rate Theory RRT), which has been reviewed in the article by Haenggi et al. (1990), starting from the seminal publication by Kramers (1940). The typical situation described in the frame of the TST is governed by a potential U(x) with two meta-stable states, A and C, separated by an energy barrier B. The bottoms of the potential wells characterizing the states A and C are separated from the barrier B by energy barriers E_{b+} and E_{b-}, respectively. Reaction rates for the forward (A $\rightarrow$ C) and backward (C $\rightarrow$ A) transitions are

Text by M. Bottoni

M. Bottoni et al., *Molecular Dynamics - Theory and Applications*, Mechanical Engineering Series, https://doi.org/10.1007/978-3-031-84567-3_7

described by reaction rate coefficients k^+ and k^-, respectively. We refer to x as a generalized coordinate characterizing the dynamic state of a system and to $x(t)$ as a random walker describing a trajectory in phase space. Under suitable assumptions, the Transition State Theory leads to an equation, describing the state of the system, known as generalized Langevin equation for non -linear Brownian motion:

$$\frac{dx}{dt} = v, \tag{7.1a}$$

$$\frac{d^2x}{dt^2} = -\frac{1}{m}\frac{\partial U}{\partial x} - \gamma v + \frac{1}{m}\xi(t), \tag{7.1b}$$

where m is the mass of the particle, $\gamma v = \gamma(\partial x/\partial t)$ is a friction force, and γ a velocity relaxation rate. The function $\xi(t)$ represents a white Gaussian noise with zero mean value and correlation function given by

$$\langle \xi(t)\xi(s) \rangle = 2k_B T\gamma\delta(t - s), \tag{7.2}$$

where k_B is the Boltzmann constant, T the absolute temperature, and brackets $\langle\rangle$ denote time mean value.

Equations (7.1a) and (7.1b) must be solved either analytically or numerically. Analytical methods rely upon a variety of approaches, which are summarized in Haenggi et al. (1990). Among these, the techniques referred to as "flux-over population method" and "reactive flux method" were widely used in the past decades. Simulations based upon the reactive flux method were made by Berne (1985), among others.

Numerical methods are based on two different techniques:

(i) Eigenvalue method for the master equation

The stochastic dynamics resulting from Eqs. (7.1a, 7.1b) and (7.2) for the reaction coordinate x and the velocity $v = \dot{x}$ are Markovian. The time evolution of their probability density is governed by the Klein-Kramers equation (Risken, 1989), also referred to as "master equation". In case of the single-event probability $p(x,t)$ the master equation can be written in the compact form

$$\frac{dp(x,t)}{dt} = \Gamma p(x,t), \tag{7.3}$$

where Γ is a time-independent operator with eigenvalues and eigenfunctions denoted by λ_i and $\Phi_i(x)$, which can be computed numerically. Space and time evolution of the probability $p(x,t)$ is then given by

$$p(x,t) = \sum\nolimits_i \Phi_i(x)e^{-\lambda_i t}. \tag{7.4}$$

(ii) Method of stochastic trajectories

Let a stochastic equation of motion be represented, as generalization of Eq. (7.1a), by

$$x^{\cdot} = A(x) + \xi, \tag{7.5}$$

where $A(x)$ represents a drift vector and ξ a random noise source.

Procedures for solving numerically stochastic differential equations of this type with the molecular dynamics methods have been developed. See, for instance (Evans, 1983; Nose, 1984)

7.2 Advanced Software Packages

In recent years H. Kaburaki and his co-workers (Shimizu et al., 2000, 2002) developed software for molecular dynamic simulation methods based on first principles to carry out large-scale fracture simulations which enabled the analysis of a wide range of fracture mechanisms, from brittle to ductile fractures. Besides, a numerical simulation environment has been constructed to share computational tools and experimental data among researchers located at distant sites. The interweaving connections realized for this purpose have been tested for efficiency on the Virtual Private Network (VPN).

Among the most advanced software packages, the "Parallel Molecular Dynamics Stencil" (PMDS) (Shimizu et al., 2002) allows for simulation of short-range molecular dynamic of solids. PMDS is designed to separate parts of the tasks, such as interprocessor communications, for parallel algorithm and to conceal them within the algorithm, so that parallel programming can be done essentially in the same way as serial programming. As an example, in an array of atoms, a "particle decomposition algorithm" is chosen at runtime, so that atoms are indexed, partitioned in subsets, and allotted to the processors. Bookkeeping of all atoms in the given set is eased by assigning to each processor not only a given subset of atoms but also sets of "image atoms" of all other processors. This spatial decomposition algorithm was used to model system of copper atoms, ranging from a few thousand up to one billion, by simulating their interactions, based, for instance, on the Lennard-Jones potential.

Advanced software for molecular dynamics simulations are being developed by the Prof. Raczynski and his co-workers at the Universidad Panamericana of Mexico City, where a wide-purpose numerical simulation system named PASION was developed and tested (Raczynski, 2002).

7.3 The Maxwell Demon Revisited

The question of whether the second principle of thermodynamics holds or not at microscopic level has been, since the time of Maxwell (1871), one of the most frequently debated problem of science. The hypothetical Maxwell demon, operating at microscopic level, immersed in a gas contained in a reservoir initially at uniform temperature, could let fast molecules pass a gate between two chambers of the reservoir but shutting the gate to prevent slow molecules to pass. The overall effect would be separating cold and hot molecules thus violating the second principle of thermodynamics. We do not enter into the details of the discussions based on the argument that the demon must fetch information about the speed of the approaching molecules and in so doing produces entropy which offsets the entropy decrease originated by the separation of the molecules. It is more interesting to observe that at present, thanks to the capabilities offered by the methods of molecular dynamics, the Maxwell demon, not constrained any more to be a challenging thought experiment, becomes a feasible numerical experiment. As remarked in Van den Broeck et al. (2005) the Maxwell demon, or rather a modern representation of it, *"…should be amenable to extremely precise molecular dynamic simulations…"*. The modern version of the demon, which could be realized, thanks to the progress of nanotechnology, consists in a porous sheet immersed in a gas between two parallel plates. The sheets lets molecules with a kinetic energy above a given threshold pass through, while molecules below the threshold are specularly reflected. In the latter case, momentum is shared between the molecules and the sheet which thus sets into movement. As computation progresses, the temperature of the gas at either parts of the sheet become different, thus violating the second principle of thermodynamics. The above-given reference, Van den Broeck et al. (2005), reports details and preliminary computational results.

7.4 Molecular Dynamics and Biological Applications

In a 1959 lecture entitled "*There's plenty of room at the bottom*" (Feynman, 1960), devoted to an audience of biologists, Nobel laureate Prof. Richard Feynman expressed the following view about the advent of more powerful microscopes and their application in biological research:

> *It is very easy to answer many of these fundamental biological questions: you just look at the thing! You will see the order of bases in the chain; you will see the structure of the microsome. Unfortunately, the present microscope sees at a level which is just a bit too crude. Make the microscope one hundred times more powerful, and many problems of biology would be made very much easier.*

As years passed, the improvement of a hundred times has not yet been achieved, nonetheless the progress in molecular biology has been greatly enhanced by two advances in scientific methodology. The first is the advent of the Atomic Force Microscope (AFM), which allows for measurements on single molecules. The second is the introduction of the methods of molecular dynamic to probe the statistical properties of molecules. Work in this field is progressing (Private communication by Prof. S. M. Lindsay), for instance, at the Department of Physics and Astronomy, Arizona State University, Tempe.

7.5 Conclusions

Molecular dynamics does allow insights into the microscopic world comparable to those first afforded by the advent of the electron microscope. Strength and domain of applicability of molecular dynamic techniques increase together with the increasing speed of modern computer and computer clusters, allowing parallel computations. It may be argued that even at the moment in which, in the years to come, technological advances and computer performances will level off, molecular dynamics will still progress, at least for some time, thanks to the development of new mathematical algorithms. As remarked in the beautiful book by Rapaport (2004), which enlightens the *"Art of Molecular Dynamics Simulations,"* visualization of computational data, in the form of two- or three-dimensional pictures, animation, etc., will increase in importance with the pace of the mathematical algorithms. Time-dependent problems will become increasingly important. At the advent of the computers, calculations were limited to the simulation of processes, like wave disturbances, characterized by problem times proportional to the length scale of the system under investigation. Diffusive processes characterised by problem time proportional to the square of the length scale could be investigated in subsequent upgrades of the computational techniques. The advent of molecular dynamic simulation, besides allowing us to enter into the microscopic world, will soon allow treatment of non-Newtonian fluids, as for instance polymer molecules, and of other processes, which require large computational times for their treatment.

References

Berne, B. J. (1985). Molecular dynamics and hybrid Monte Carlo in systems with multiple time scales and long range forces: Reference system propagator algorithm. In J. U. Brackbill & B. I. Cohen (Eds.), *Multiple time scales* (pp. 419–425). Academic Press.

Evans, D. J. (1983). Computer experiment for nonlinear thermodynamics of Couette flow. *The Journal of Chemical Physics, 78*, 3297–3302.

Feynman, R. (1960). There's plenty of room at the bottom. *Engineering and Science, XXIII(5)*.

Haenggi, P., Talkner, P., & Borkovec, M. (1990). Reaction-rate theory: fifty years after Kramers. *Reviews of Modern Physics, 62*(2), 251–341.

Kramers, H. A. (1940). Brownian motion in a field of force and the diffusion model of chemical reactions. *Physica (Utrecht), 7*, 284.

Maxwell, J. C. (1871). *Theory of heat*. Longmans Green.

Nose, S. (1984). A unified formulation of the constant temperature molecular-dynamics methods. *The Journal of Chemical Physics, 81*, 511–519.

Raczynski, S. (2002). The PASION simulation system. *Simulation News Europe, 35*(36), 24–28.

Rapaport, D. C. (2004). *The art of molecular dynamics simulation* (2nd ed.). Cambridge University Press.

Risken, H. (1989). *The Fokker Plank equation* (2nd ed.). Springer Verlag.

Shimizu, F., Kimizuka, H., Kaburaki, H., Li, J., & Yip, S. (2000). Parallel molecular dynamic simulation on elastic properties of solid argon. In *Proceedings of the 4th International Conference SNA2000*.

Shimizu F, Kimizuka H, Kaburaki H, Arakawa C. (2002). Development of parallel molecular dynamics stencil (In Japanese). *Transactions of the Japan Society for Computational Engineering and Science*, #20020015.

Van den Broeck, C., Meurs, P., & Kawai, R. (2005). From Maxwell demon to Brownian motor. *New Journal of Physics, 7*(1), 10. (pp. 1–24).

Correction to: Molecular Dynamics - Theory and Applications

Correction to:
M. Bottoni et al., *Molecular Dynamics - Theory and Applications*, Mechanical Engineering Series, https://doi.org/10.1007/978-3-031-84567-3

The book was inadvertently published with a few errors in Sections 5.3.2, 6.1.1, 6.10.1, and 6.10.2.1.1. The term "access" should read "assess" in these sections. Also, the figure captions of Figures 6.3 to 6.18 have been published without the experiment labels. The errors in the sections and the figure captions have now been corrected in the chapter.

The updated version of this chapters can be found at
https://doi.org/10.1007/978-3-031-84567-3_5
https://doi.org/10.1007/978-3-031-84567-3_6

M. Bottoni et al., *Molecular Dynamics - Theory and Applications*, Mechanical Engineering Series, https://doi.org/10.1007/978-3-031-84567-3_8

Appendix A: List of Computer Programs

The computer programs given in following lists (Appendixes A.1–A.4) in FORTRAN V were written only for didactic purposes. It has not been tried to optimize the coding style, but to preserve the clarity of the logic.

The programs are self-sufficient and do not need any link (to the expense of some duplicated subroutines).

The subroutine CPU_TIME is a system subroutine. Whenever used on a different platform, it must be replaced with a subroutine from the other system.

Computations have been made by the author with a personal computer with peak performance of 13 Gcycles. All examples have been rerun at the MEEO (Meteorological and Environmental Earth Observation) with the following platform:

CPU: Intel(R) Xeon(R) Silver 4210R CPU @ 2.40GHz

Compiler: GNU Fortran 13.2.0.

All results matched well with previous calculations.

The following computer programs are listed in italics, while general information and results of the calculations are given in cursive characters.

Appendix A.1

```
      PROGRAM DIFFLA

c     Diffusion limited aggregation. Let us consider a
c     2D lattice with a seed inside it. A random walker is started from
c     the edge of the lattice. If the random walker gets to one of the
c     nearest neighbours of the seed it stiks to it. Then a new random
c     walker is released. One analyses the evolution of the aggregate.
c
```


M. Bottoni et al., *Molecular Dynamics - Theory and Applications*, Mechanical Engineering Series, https://doi.org/10.1007/978-3-031-84567-3

```
c   ----------------------------------------------------------------
c   LIST of SUBROUTINES and Funtions

c   SCREEN  Checks if a random walker is inside the domain.
c   SEED    Sets a seed
c   STARTP  Sets a start point
c   STP     Set new stick points
c   RF (Function)
c   WRITEDENS Writes density files
c   ================================================================
c   Definition of main variables
c   IMAX=JMAX= Number of lattice point in domain.
c   IW, JW  = Denote the position of a particle during the random walk.
c   NACCR   = Number of lattice point in the aggregate.
c   LATTICEW  = Defines the grid upon which the random walks occur.
c         Nodes of lattice with seed or aggregates are set = 3.
c         Sticking nodes are set = 2.
c         Nodes which show the path of the walker are set = 1.
c         After one walk is completed, nodes with 1 are turned to 0.
c   PACCR   = NACCR/(IMAX * JMAX) = Fraction of lattice nodes in the
c         aggregation.
c   ================================================================
c
c
c   File 10 = OUTPL0       Stores PACCR versus number of walks.
c   File 11 = OUTPL_DENS    Stores as 1 the density of nodes in the
c                 aggregate. Other nodes have density 0.
c   File 60  = OUTPUT_DIFFLA
c   File 72 = OUTPL_TRJ    Stores particle trajectory
c   File 111 - 114 = OUTPL_DENS1 - OUTPL_DENS4  Store Density Files
c
      implicit real*8 (a-h, o-z)
      parameter (imax=100, jmax=100)
      parameter (nfft=16384)   ! 2**14  Chose if imax=100

      dimension lattice(imax,jmax)

      open(10,file='OUTPL0',      status='unknown')
      open(11,file='OUTPL_DENS',  status='unknown')
      open(60,file='OUTPUT_DIFFLA',status='unknown')
      open(72, file='OUTPL_TRJ', status='unknown')

c   Additional files for prinning intermediate results:
       open(111,file='OUTPL_DENS1',  status='unknown')
       open(112,file='OUTPL_DENS2',  status='unknown')
```

```
    open(113,file='OUTPL_DENS3',  status='unknown')
    open(114,file='OUTPL_DENS4',  status='unknown')

   iname1 = 111    ! Used for printing density files
   iname2 = 112
   iname3 = 113
   iname4 = 114
c
c-----------------------------------------------------------------
   nw1= 9000 ! Pointer for print when iwalk=iwalk1
   nw2=18000 ! Pointer for print when iwalk=iwalk2
   nw3=27000 ! Pointer for print when iwalk=iwalk3
   nw4=36000 ! Pointer for print when iwalk=iwalk4
c-----------------------------------------------------------------

   nwalk_max= 38100

   call CPU_TIME(tcorr1)
   arf=RF(-3)         ! Initialise RF
   isave_trj=1        ! Flag for writing trajectory
   iwrite= 0         ! Alternative for printing = 1

   xmax=1.d0
   ymax=1.d0
   dx=xmax/float(imax)
   dy=ymax/float(jmax)

c  ---------------------------------------------------------------
   call SEED(imax,jmax,iseed0,jseed0,lattice) ! Set new stick point
c  ---------------------------------------------------------------

   noutr=0  ! Counter of lost walkers
   naccr=0  ! Counter of accretion points
   nwalk=0  ! Counter of walks
c
50  continue               ! *** Start Main Branch AAA ***
   nwalk=nwalk+1
c
c  Start a random walker (particle) from one side chosen at random
c  and from a position on the side also chosen at random. Esclude the
c  corners and the aggregates as starting points.
c  iw, jw denote the node where the random walker is.
c
```

```
c   ------ ir=3 ----P(imax,jmax)
c   |        |
c   |        |
c  ir=2      ir=0
c   |        |
c   |        |
c  P(0,0)-- ir=1 ----P(imax,0)
c

c  ----------------------------------------------------------------
   call STARTP(imax,jmax,itrials,iwrite,iw,jw,lattice)
c  ----------------------------------------------------------------
   if(itrials .eq. 1) go to 900

   lattice(iw,jw)=1
c                        ir=3
c                         |
c                         |
c   Start random walk      ir=2 <-- P(iw,jw) --> ir=0
c                         |
c                         |
c                        ir=1
100  continue           ! Move particle

   ir=rf(0)*4

   if(ir.eq.0) iw=iw + 1
   if(ir.eq.2) iw=iw - 1
   if(ir.eq.3) jw=jw + 1
   if(ir.eq.1) jw=jw - 1

c  ------------------------------ ------------------------------
   if(nwalk .eq. 20 000) write(72,1012) iw,jw
1012  format(2i6)
c  -------------------------------------------------------------

c   -------------------------- ------------------------------
   call screen(iw,jw,imax,jmax,icode)
c   -------------------------------------------------------

c   Random walk terminates because particle went out of domain.
   if(icode.ne.0) noutr = noutr + 1
   if(icode.ne.0) go to 650
```

```
   if(lattice(iw,jw) .lt. 2) then     ! Continue the walk
     lattice(iw,jw) = 1
     go to 100
   endif

    if(lattice(iw,jw).eq.2) then ! Start Branch BBB
      lattice(iw,jw) = 3      ! Set for formation of aggregate

c   Random walk terminates because particle sticked to aggregation.
     naccr = naccr + 1

c   -------------------------------------------------------------
   call STP(imax,jmax,iw,jw,lattice) ! Set new stick points
c   -------------------------------------------------------------

    paccr=float(naccr)/float(imax*jmax)

c   ---------------------------------------------------------------
      go to 650
    endif                   ! End Branch BBB
c
650  continue
c
c
-----------------------------------------------------------------
c   Zero lattice if lattice=1 to prepare for next walk.

   do 860 i=1,imax
       do 860 j=1,jmax

   if(lattice(i,j) .eq.1) lattice(i,j)=0
860  continue
c
   if(mod(nwalk,100) .eq.1) write(10,2010) nwalk, paccr
2010 format( i8, e12.5)

c-----------------------------------------------------------------
   nw=nwalk ! Print Density Files

    if(nw.eq.nw1 .or. nw.eq.nw2 .or. nw.eq.nw3 .or. nw.eq.nw4) then
      if(nw.eq.nw1) call WRITEDENS(imax,jmax,lattice,dx,dy,iname1)
      if(nw.eq.nw2) call WRITEDENS(imax,jmax,lattice,dx,dy,iname2)
      if(nw.eq.nw3) call WRITEDENS(imax,jmax,lattice,dx,dy,iname3)
      if(nw.eq.nw4) call WRITEDENS(imax,jmax,lattice,dx,dy,iname4)
```

```
   endif
c-------------------------------------------------------------

   if(nwalk.gt.nwalk_max) go to 900 ! *** End Branch AAA ***
   go to 50 ! if(paccr.lt.0.2d0) go to 50
c  ======================== End of loop on walks ==========
c
900 continue
C  ---------------------------------------------
c  Final printing. Set to 0 all nodes outside the aggregation.
   do 910 i=1,imax
      do 910 j=1,jmax
         if(lattice(i,j) .ne.3) lattice(i,j)=0
910 continue
c
c  ------------   Write File OUTPL_DENS  ---------------

     do 950 i=1,imax
   do 950 j=1,jmax
             xi= (i-1)*dx + 0.5d0 * dx
    yj= (j-1)*dy + 0.5d0 * dy
    if(lattice(i,j) .eq. 3) dens=1.d0
    if(lattice(i,j) .lt. 3) dens=0.d0
    write(11,1011) i,j,xi,yj,dens
1011  format(2i6, 3f12.5)
950 continue

c  -------------------------------------------------------
   write(60,*) 'nwalk,naccr,noutr =', nwalk,naccr,noutr
   write(* ,*) 'nwalk,naccr,noutr =', nwalk,naccr,noutr
   write(60,*)
   write(* ,*)

      call CPU_TIME(tcorr2)
   dtcorr=tcorr2 - tcorr1
   write(*,*) 'DIFFLA - dtcorr', dtcorr
   write(60,*) 'DIFFLA - dtcorr', dtcorr
   write(60,*)
      write(* ,*)

      close(10)
   close(11)
      close(60)
      close(72)
```

```
      close(111)
   close(112)
   close(113)
      close(114)

   write(60,*) 'DIFFLA Calculation Done'
   write(* ,*) 'DIFFLA Calculation Done'

   stop
   end

   subroutine screen(iw,jw,imax,jmax,icode)
   implicit real*8 (a-h,o-z)
c   icode = 0 particle at position P(iw,jw) is inside the lattice or
c        on the boundary.
c   icode = 1 particle at position P(iw,jw) is outside of the domain.
   icode =0
   if(iw.lt.1 .or. iw.gt.imax .or. jw.lt.1.or.jw.gt.jmax) icode=1
   return
   end

   Subroutine SEED(imax,jmax,iseed0,jseed0,lattice) ! Set new stick
point

   implicit real*8 (a-h, o-z)
   dimension Lattice(imax,jmax)

   do 10 i=1,imax
   do 10 j=1,jmax
    lattice(i,j)=0
10  continue

15  continue  ! The seed is assumed to be inside the lattice.
   iseed0=1+rf(0)*imax
   jseed0=1+rf(0)*imax
   if(iseed0.lt.2 .or. iseed0.gt.imax-2) go to 15
   if(jseed0.lt.2 .or. jseed0.gt.imax-2) go to 15
c
   lattice(iseed0,  jseed0)  = 3    ! Set position of walker
   lattice(iseed0 +1, jseed0 +1) = 2    ! Set stick points
   lattice(iseed0 -1, jseed0 -1) = 2
   lattice(iseed0,  jseed0 +1) = 2
   lattice(iseed0,  jseed0 -1) = 2
   lattice(iseed0 +1, jseed0)  = 2
   lattice(iseed0 -1, jseed0)  = 2
```

```
      lattice(iseed0 +1, jseed0 -1) = 2
      lattice(iseed0 -1, jseed0 +1) = 2
c
        write(*,*)  ' DIFFLA - Position of seed ', iseed0, jseed0
        write(60,*) ' DIFFLA - Position of seed ', iseed0, jseed0
        write(60,*)
        write(* ,*)
c
      return
      end

      Subroutine STARTP(imax,jmax,itrials,iwrite,iw,jw,lattice)

      implicit real*8 (a-h, o-z)
      dimension Lattice(imax,jmax)

      ntrials = 0  ! Counter for finding new starting point
      ntrials_max = 1000
      itrials = 0
60    continue
      ntrials=ntrials + 1
c     Terminate calculation when ntrials more than 1000. This happens
c     when practically all bounding nodes are labelled 2 or 3.

      if(ntrials .gt. ntrials_max) then
        write(60,*) ' Stop because ntrials ', ntrials
          write( *,*) ' Stop because ntrials ', ntrials
        write(60,*)
        write(* ,*)
        itrials = 1
        go to 100
      endif

      ir=rf(0)*4
c
      if(ir.eq.0) then
      jw=1 + rf(0)*jmax
      iw=imax
      endif
c
      if(ir.eq.1) then
      jw=1
      iw=1+ rf(0)*imax
      endif
```

```
c
   if(ir.eq.2) then
   jw=1 + rf(0)*jmax
   iw=1
   endif
c
   if(ir.eq.3) then
   jw=jmax
   iw=1 + rf(0)*imax
   endif
c
c   Exclude starting points in the corners of the domain.
   if(iw.eq.1 .and. jw.eq.1 .or. iw.eq.1 . and. jw.eq.jmax
  1 .or.iw.eq.imax.and.jw.eq.1.or.iw.eq.imax.and.jw.eq.jmax)then
    go to 60
   endif
c
c   Exclude starting points which are already accretion points or
c   stick points
   if(lattice(iw,jw).eq.2 .or. lattice(iw,jw).eq.3) go to 60
c

100 continue
   return
   end

   Subroutine STP(imax,jmax,iw,jw,lattice)  ! Set new stick points

   implicit real*8 (a-h, o-z)
   dimension Lattice(imax,jmax)

   call screen(iw+1,jw,imax,jmax,icode1)
   if(icode1.eq.0 .and. lattice(iw+1,jw).lt.3) lattice(iw+1,jw)=2

   call screen(iw-1,jw,imax,jmax,icode1)
   if(icode1.eq.0 .and. lattice(iw-1,jw).lt.3) lattice(iw-1,jw)=2

   call screen(iw,jw+1,imax,jmax,icode1)
   if(icodc1.eq.0 .and. lattice(iw,jw+1).lt.3) lattice(iw,jw+1)=2

   call screen(iw,jw-1,imax,jmax,icode1)
   if(icode1.eq.0 .and. lattice(iw,jw-1).lt.3) lattice(iw,jw-1)=2

   call screen(iw+1,jw+1,imax,jmax,icode1)
  if(icode1.eq.0.and.lattice(iw+1,jw+1).lt.3) lattice(iw+1,jw+1)=2
```

```
      call screen(iw-1,jw-1,imax,jmax,icode1)
     if(icode1.eq.0.and.lattice(iw-1,jw-1).lt.3) lattice(iw-1,jw-1)=2

      call screen(iw+1,jw-1,imax,jmax,icode1)
     if(icode1.eq.0.and.lattice(iw+1,jw-1).lt.3) lattice(iw+1,jw-1)=2

      call screen(iw-1,jw+1,imax,jmax,icode1)
     if(icode1.eq.0.and.lattice(iw-1,jw+1).lt.3) lattice(iw-1,jw+1)=2

      return
      end

      function rf(IDUM)
*--generates a uniformly distributed random fraction between 0 and 1
*----IDUM will generally be 0, but negative values may be used to
*------re-initialize the seed
c     Cfr. function RUN3 of Numerical Recipes, p. 199.
      implicit real*8 (a-h,o-z)
      SAVE MA,INEXT,INEXTP
      PARAMETER (MBIG=1000000000,MSEED=161803398,MZ=0,FAC=1.E-9)
      DIMENSION MA(55)
      DATA IFF/0/
      IF (IDUM.LT.0.OR.IFF.EQ.0) THEN
       IFF=1
       MJ=MSEED-IABS(IDUM)
       MJ=MOD(MJ,MBIG)
       MA(55)=MJ
       MK=1
       DO 50 I=1,54
        II=MOD(21*I,55)
        MA(II)=MK
        MK=MJ-MK
        IF (MK.LT.MZ) MK=MK+MBIG
        MJ=MA(II)
50      CONTINUE
       DO 100 K=1,4
        DO 60 I=1,55
         MA(I)=MA(I)-MA(1+MOD(I+30,55))
         IF (MA(I).LT.MZ) MA(I)=MA(I)+MBIG
60       CONTINUE
100     CONTINUE
       INEXT=0
       INEXTP=31
      END IF
```

```
200  INEXT=INEXT+1
   IF (INEXT.EQ.56) INEXT=1
   INEXTP=INEXTP+1
   IF (INEXTP.EQ.56) INEXTP=1
   MJ=MA(INEXT)-MA(INEXTP)
   IF (MJ.LT.MZ) MJ=MJ+MBIG
   MA(INEXT)=MJ
   RF=MJ*FAC
   IF (RF.GT.1.E-8.AND.RF.LT.0.99999999) RETURN
   GO TO 200
   END

   Subroutine WRITEDENS(imax,jmax,lattice,dx,dy,iname)
   implicit real*8 (a-h, o-z)
   dimension Lattice(imax,jmax)
    do 931 i=1,imax
    do 931 j=1,jmax
      xi= (i-1)*dx + 0.5d0 * dx
      yj= (j-1)*dy + 0.5d0 * dy
      if(lattice(i,j) .eq. 3) dens=1.d0
      if(lattice(i,j) .lt. 3) dens=0.d0
      write(iname,1011) i,j,xi,yj,dens
931  continue

1011 format(2i6, 3f12.5)
    return
    end
```

Information on OUTPUT files:

OUTPL0 Stores PACCR versus number of walks.

OUTPL_DENS Stores as 1 the final density of nodes in the aggregate.
OUTPL_DENS1 - OUTPL_DENS4 Store Density Files
at the 4 choesen numders of walks.
OUTPL_TRJ Stores a choosen particle trajectory.

From OUTPUT_DIFFLA:

DIFFLA - Position of seed 50 30
nwalk,naccr,noutr = 38101 1217 36884
DIFFLA - dtcorr 0.98280599999999996

Appendix A.2

```
      PROGRAM DLA_1

c     Diffusion limited aggregation. Let us consider a
c     2D lattice with a seed inside it. A random walker is started from
c     the edge of the lattice. If the random walker gets to one of the
c     nearest neighbours of the seed it stiks to it. Then a new random
c     walker is released. One analyses the evolution of the aggregate.
c     ------------------------------------------------------------------
c     Short description of subroutines.

c     DLA_1   Diffusion Limited Accretion. Like DIFFLA but particles
c             going out of domain are reflected specularly into domain.
c             Walkers are originated at the border of domain, but not
c             in the corners.
c     SCREEN1 Checks if a random walker is inside domain.
c     SETLAT2
c     WRITEDENS
c     RF (Function)
c     ------------------------------------------------------------------
c     Definition of main variables
c     IMAX=JMAX= Number of lattice point in domain.
c     IW, JW  = Denote the position of a particle during the random walk.
c     NACCR   = Number of lattice point in the aggregate.
c     LATTIC  = Defines the grid upon which the random walks occur.
c                 Nodes of lattic with seed or aggregates are set = 3.
c                 Sticking nodes are set = 2.
c                 Nodes along the pattern of a walker are set to 1.
c     PACCR   = NACCR/(IMAX * JMAX) = Fraction of lattice nodes in the
c                aggregation.
c     ------------------------------------------------------------------
c
      implicit real*8 (a-h, o-z)
      parameter (imax=100, jmax=100)
      parameter (nfft=16384)   ! 2**14  Chose if imax=100
      dimension lattice(imax,imax)

c      parameter (nfft= 65536)    ! 2**16  Chose if imax=200
c      parameter (nfft=131072)    ! 2**17  Chose if imax=300
c      parameter (nfft=262144)    ! 2**18  Chose if imax=400 or 500
c      parameter (nfft=1048576)   ! 2**20  Chose if imax=1000
c
c     File 10 = OUTPL0      Stores PACCR versus number of walks.
```

```
c   File 11 = OUTPL_DENS   Stores as 1 the density of nodes in the
c                aggregate. Other nodes have density 0.
c   File 60 = OUTPUT_DLA_1
c   File 111 - 114      OUTPL_DENS1 - OUTP_DENS4

    open(10,file='OUTPL0',        status='unknown')
    open(11,file='OUTPL_DENS',    status='unknown')
    open(60,file='OUTPUT_DLA_1', status='unknown')

    open(111,file='OUTPL_DENS1',  status='unknown')
    open(112,file='OUTPL_DENS2',  status='unknown')
    open(113,file='OUTPL_DENS3',  status='unknown')
    open(114,file='OUTPL_DENS4',  status='unknown')

    iname1 = 111    ! Used for printing density files
    iname2 = 112
    iname3 = 113
    iname4 = 114
c
c-------------------------------------------------------------
    nw1=250 ! Default=0, Pointer for print when iwalk=iwalk1
    nw2=500 ! Default=0, Pointer for print when iwalk=iwalk2
    nw3=750 ! Default=0, Pointer for print when iwalk=iwalk3
    nw4=1000 ! Default=0, Pointer for print when iwalk=iwalk4
c-------------------------------------------------------------

    nwalk_max=38000 ! imax*jmax/10

    write(60,*) 'DLA_1 -  IMAX=', imax
    write( *,*) 'DLA_1 -  IMAX=', imax

    icount3=0         ! Counter of particles in all aggregations

    call CPU_TIME(CPU_TIME1)

    arf=RF(-mean)          ! Re-initialise RF

    xmax=1.d0
    ymax=1.d0
    dx=xmax/float(imax)
    dy=ymax/float(jmax)

    do 10 i=1,imax
    do 10 j=1,jmax
```

```
      lattice(i,j)=0
10    continue

c     iseed_set=0 Seed position is determined at random
c     iseed_set=1 Seed is positioned.

      iseed_set=1

      if(iseed_set.eq.0) then
5       continue
        iseed0=1+rf(0)*imax
        jseed0=1+rf(0)*imax
        if(iseed0.lt.2 .or. iseed0.gt.imax-1) go to 5
        if(jseed0.lt.2 .or. jseed0.gt.imax-1) go to 5
        lattice(iseed0,jseed0) = 3
      else
        iseed0=1
        jseed0=1
        lattice(iseed0,jseed0) = 3
      endif

      if(iseed0.eq.1 .and. jseed0.eq.1) then
        lattice(iseed0 +1,  jseed0 +1) = 2
        lattice(iseed0,     jseed0 +1) = 2
        lattice(iseed0 +1,  jseed0)    = 2
      else       ! Initial seed middle in domain
        lattice(iseed0 +1, jseed0 +1) = 2
        lattice(iseed0 -1, jseed0 -1) = 2
        lattice(iseed0,    jseed0 +1) = 2
        lattice(iseed0,    jseed0 -1) = 2

          lattice(iseed0 +1, jseed0)    = 2
        lattice(iseed0 -1,  jseed0)    = 2
        lattice(iseed0 +1,  jseed0 -1) = 2
        lattice(iseed0 -1,  jseed0 +1) = 2
      endif
c
      write(*,*) 'DLA_1 - Position of seed ', iseed0, jseed0
      write(60,*)'DLA_1 - Position of seed ', iseed0, jseed0

      paccr_max=0.2d0

      nwalk=0
      naccr=0
```

```
      no_accr=0
50    continue

      nwalk=nwalk+1

c     Start a random walker (particle) from one side chosen at random
c     and from a position on the side also chosen at random. Esclude the
c     corners as starting points.
c     iw, jw denote the node were the random walker is.
c
c     ------ ir=3 ----P(imax,jmax)
c     |               |
c     |               |
c    ir=2           ir=0
c     |               |
c     |               |
c   P(0,0)-- ir=1 ----P(imax,0)
c

      ntrials=0
60    continue
      ntrials=ntrials + 1

      if(ntrials.gt.1000) then
      write( *,*) 'DLA_1 - Calc. stop, nwalk, ntrials', nwalk,ntrials
      write(60,*) 'DLA_1 - Calc. stop, nwalk, ntrials', nwalk,ntrials
      go to 900
      endif

      ir=rf(0)*4
      if(ir.eq.0) then
      jw=1 + rf(0)*jmax
      iw=imax
      endif
c
      if(ir.eq.1) then
      jw=1
      iw=1+ rf(0)*imax
      endif
c
      if(ir.eq.2) then
      jw=1 + rf(0)*jmax
      iw=1
      endif
c
```

```
      if(ir.eq.3) then
      jw=jmax
      iw=1 + rf(0)*imax
      endif
c
c     Exclude starting points in the corners of the domain.
      if(iw.eq.1 . and. jw.eq.1 .or. iw.eq.1 . and. jw.eq.jmax
     1 .or. iw.eq.imax . and. jw.eq.1 .or. iw.eq.imax . and. jw.eq.jmax)
     2  go to 60

c     Exclude starting points which are already accretion points or stick
c     points.
      if(lattice(iw,jw).eq.2 .or. lattice(iw,jw).eq.3) go to 60

c     New walker must be distant from aggegate
      idist_min=100000

      i1=iw - imax/10
      i2=iw + imax/10
      j1=jw - jmax/10
      j2=jw + jmax/10

      i1=max(i1,1)
      i1=min(i1,imax)
      i2=max(i2,1)
      i2=min(i2,imax)

      j1=max(j1,1)
      j1=min(j1,imax)
      j2=max(j2,1)
      j2=min(j2,imax)

      do 80 i=i1,i2
      do 80 j=j1,j2
      if(lattice(i,j) .eq. 3) then

      a=float(i-iw)
      b=float(j-jw)
      dist=dsqrt( a**2 + b**2)
      idist=dist
      if(idist.lt.idist_min) idist_min=idist
      endif
 80   continue
      if(idist_min.lt.imax/10) go to 60          ! NB
```

```
   lattice(iw,jw)=1 ! Marks the path of the particle
c
c                                        ir=3
c                                          |
c                                          |
c   Start random walk     ir=2 <-- P(iw,jw) --> ir=0
c                                          |
c                                          |
c                                        ir=1

100  continue   ! Move particle
   ir=rf(0)*4

   if(ir.eq.0) iw=iw + 1
   if(ir.eq.2) iw=iw - 1
   if(ir.eq.3) jw=jw + 1
   if(ir.eq.1) jw=jw - 1

   call screen1(iw,jw,imax,jmax,icode)

c   -------------------- Reflection from walls --------------
   if(icode.eq.-1) jw=1
   if(icode.eq.-2) iw=1
   if(icode.eq.-3) jw=jmax
   if(icode.eq.-4) iw=imax
c   ----------------------------------------------------------

   if(lattice(iw,jw).eq.0) then
   lattice(iw,jw)=1
   go to 100
   endif

   if(lattice(iw,jw).eq.1) go to 100

   if(lattice(iw,jw).eq.3) then ! Point is already in the aggregate
   no_accr=no_accr + 1
   go to 50
   endif

   if(lattice(iw,jw).eq.2) THEN
    lattice(iw,jw) = 3
    naccr=naccr+1
    paccr=float(naccr)/float(imax*jmax)
C  ----------------------------------------------------------------
     CALL SETLAT2(iw,jw,imax,jmax,lattice)
```

```
C ------------------------------------------------------------
   ENDIF

c   Avoid sticking point wind indices smaller than iseed0,jseed0

   if(iseed0.gt.1) then
   do 200 i=1,iseed0-1
   do 200 j=1,jmax
   lattice(i,j)=0
200  continue
   endif

   if(jseed0.gt.1) then
   do 210 i=1,imax
   do 210 j=1,jseed0-1
   lattice(i,j)=0
210  continue
   endif

c  -----------------------------------------------------------
c   Zero LATTICE(i,j) if =1, to prepare for next walk
   do 860 i=1,imax
   do 860 j=1,jmax
   if(lattice(i,j).eq.1) lattice(i,j)=0
860  continue

   if(mod(nwalk,50) .eq.1) write(10,2010) nwalk, paccr
2010  format( i8, e12.5)

c------------------------------------------------------------
   nw=nwalk  ! Print Density Files

    if(nw.eq.nw1 .or. nw.eq.nw2 .or. nw.eq.nw3 .or. nw.eq.nw4) then
      if(nw.eq.nw1) call WRITEDENS(imax,jmax,lattice,dx,dy,iname1)
      if(nw.eq.nw2) call WRITEDENS(imax,jmax,lattice,dx,dy,iname2)
      if(nw.eq.nw3) call WRITEDENS(imax,jmax,lattice,dx,dy,iname3)
      if(nw.eq.nw4) call WRITEDENS(imax,jmax,lattice,dx,dy,iname4)
    endif
c-----------------------------------------

   if(nwalk.gt.nwalk_max .or. paccr.gt.paccr_max) go to 900
   go to 50

c  ====================== End of loop on walks ==========
c
```

```
900  continue
   write(* ,*) 'DLA_1 - nwalk,naccr,no_accr ', nwalk,naccr,no_accr
   write(60,*) 'DLA_1 - nwalk,naccr,no_accr ', nwalk,naccr,no_accr

c  ---------------   Write File OUTPL_DENS  ----------------
   do 930 i=1,imax
   do 930 j=1,jmax
   xi= (i-1)*dx + 0.5d0 * dx
   yj= (j-1)*dy + 0.5d0 * dy

   if(lattice(i,j) .eq. 3) then
   dens=1.d0
   icount3=icount3 + 1
   write(20,1020) mean, nwalk, i,j        ! File 20 = STORE_IWJW
1020 format(4i10)
   endif

   if(lattice(i,j) .lt. 3) dens=0.d0
   write(11,1011) i,j,xi,yj,dens
1011 format(2i6, 3f12.5)
930  continue

c  ---------------------------------------------------------------

   call CPU_TIME(CPU_TIME2)
   delta_time=cpu_time2 - cpu_time1
   write( *,*) 'DLA_1 - delta_time ', delta_time
   write(60,*) 'DLA_1 - delta_time ', delta_time

   close(10)
   close(11)
   close(60)

   close(111)
   close(112)
   close(113)
   close(114)
c  ----------------------------------------------------------------
   stop
   end

   subroutine screen1(iw,jw,imax,jmax,icode)
   implicit real*8 (a-h,o-z)
```

```
c           ICODE = -3
c
c              ------ ir=3 ----P(imax,jmax)
c              |                |
c              |                |
c ICODE = -2 ir=2             ir=0   ICODE = -4
c              |                |
c              |                |
c           P(0,0)-- ir=1 ----P(imax,0)
c
c           ICODE = -1
c

c  icode = 0 particle at position P(iw,jw) is inside the lattice or
c       on the boundary.
c  icode = -1 particle at position P(iw,jw) is below j=1
c  icode = -2 particle at position P(iw,jw) is left of i=1
c  icode = -3 particle at position P(iw,jw) is above j=jmax
c  icode = -4 particle at position P(iw,jw) is right of i=imax

      icode =0
      if(jw.lt.1)     icode=-1
      if(iw.lt.1)     icode=-2
      if(jw.gt.jmax) icode=-3
      if(iw.gt.imax) icode=-4

      return
      end

      Subroutine SETLAT2(iw,jw,imax,jmax,lattice)
       implicit real*8 (a-h, o-z)
       dimension Lattice(imax,jmax)

c           |---------------|
c           | 8     3     1 | Numbering of neigh. nodes
c           |               |
c           | 6  (iw,jw)  5 |
c           |               |
c           | 2     4     7 |
c           |---------------|
       i1=iw+1                ! Set new sticking nodes
       j1=jw+1
       if(lattice(i1,j1).ne.3) lattice(i1,j1)=2
```

```
      i2=iw-1
      j2=jw-1
      if(i2.ge.3.and.j2.ge.3.and.lattice(i2,j2).ne.3)
     1                        lattice(i2,j2)=2
c
      i3=iw
      j3=jw+1
      if(lattice(i3,j3).ne.3) lattice(i3,j3)=2

      i4=iw
      j4=jw-1
      if(j4.ge.3 .and. lattice(i4,j4).ne.3) lattice(i4,j4)=2

      i5=iw+1
      j5=jw
      if(lattice(i5,j5).ne.3) lattice(i5,j5)=2

      i6=iw-1
      j6=jw
      if(i6.ge.3 .and. lattice(i6,j6).ne.3) lattice(i6,j6)=2

      i7=iw+1
      j7=jw-1
      if(j7.ge.3 .and. lattice(i7,j7).ne.3) lattice(i7,j7)=2

      i8=iw-1
      j8=jw+1
      if(i8.ge.3 .and. lattice(i8,j8).ne.3) lattice(i8,j8)=2

      return
      end

    function rf(IDUM)
*--generates a uniformly distributed random fraction between 0 and 1
*----IDUM will generally be 0, but negative values may be used to
*------re-initialize the seed
c   Cfr. function RUN3 of Numerical Recipes, p. 199.
    implicit real*8 (a-h,o-z)
    SAVE MA, INEXT, INEXTP
    PARAMETER (MBIG=1000000000,MSEED=161803398,MZ=0,FAC=1.E-9)
    DIMENSION MA(55)
    DATA IFF/0/
    IF (IDUM.LT.0.OR.IFF.EQ.0) THEN
     IFF=1
     MJ=MSEED-IABS(IDUM)
```

```
      MJ=MOD(MJ,MBIG)
      MA(55)=MJ
      MK=1
      DO 50 I=1,54
       II=MOD(21*I,55)
       MA(II)=MK
       MK=MJ-MK
       IF (MK.LT.MZ) MK=MK+MBIG
       MJ=MA(II)
50    CONTINUE
      DO 100 K=1,4
       DO 60 I=1,55
        MA(I)=MA(I)-MA(1+MOD(I+30,55))
        IF (MA(I).LT.MZ) MA(I)=MA(I)+MBIG
60      CONTINUE
100   CONTINUE
      INEXT=0
      INEXTP=31
     END IF
200  INEXT=INEXT+1
     IF (INEXT.EQ.56) INEXT=1
     INEXTP=INEXTP+1
     IF (INEXTP.EQ.56) INEXTP=1
     MJ=MA(INEXT)-MA(INEXTP)
     IF (MJ.LT.MZ) MJ=MJ+MBIG
     MA(INEXT)=MJ
     RF=MJ*FAC
     IF (RF.GT.1.E-8.AND.RF.LT.0.99999999) RETURN
     GO TO 200
     END

     Subroutine WRITEDENS(imax,jmax,lattice,dx,dy,iname)
     implicit real*8 (a-h, o-z)
     dimension Lattice(imax,jmax)
      do 931 i=1,imax
      do 931 j=1,jmax
        xi= (i-1)*dx + 0.5d0 * dx
        yj= (j-1)*dy + 0.5d0 * dy
        if(lattice(i,j) .eq. 3) dens=1.d0
        if(lattice(i,j) .lt. 3) dens=0.d0
        write(iname,1011) i,j,xi,yj,dens
931   continue
```

```
1011 format(2i6, 3f12.5)
     return
     end
```

Information on OUTPUT files:

OUTPL0 Stores PACCR versus number of walks.

OUTPL_DENS Stores as 1 the final density of nodes in the aggregate.
OUTPL_DENS1 - OUTPL_DENS4 Store Density Files
at the 4 choesen numders of walks.

From OUTPUT_DLA_1

```
DLA_1 - IMAX=        100
DLA_1 - Position of seed        1        1
DLA_1 - Calc. stop, nwalk, ntrials     1165      1001
DLA_1 - nwalk,naccr,no_accr      1165      1157        7
DLA_1 - delta_time   0.87360499999999996
```

Appendix A.3

```
   PROGRAM DLA_2          ! Version 11.06.2024

c  Diffusion limited aggregation. Let us consider a
c  2D lattice with a seed inside it. A random walker is started from
c  the edge of the lattice. If the random walker gets to one of the
c  nearest neighbours of the seed it stiks to it. Then a new random
c  walker is released. One analyses the evolution of the aggregate.
c
c  ------------------------------------------------------------------
c  Short description of Subroutines and Functions.

c  DLA_2   Diffusion Limited Accretion. Like DLA_1 but walkers are
c         originates within the domain in position chosen at random,
c         but not belonging to the aggregation nor to stick points
c  SCREEN1 Checks if a random walker is inside domain. Used by DLA_2.
c  SETLAT2 Sets newn sticking nodes
c  WRITEDENS Writes density files

c  Function RF Computes random deviates
c  ------------------------------------------------------------------
```

```
c
c   Definition of main variables

c    IMAX=JMAX= Number of lattice point in domain.
c   IW, JW      = Denote the position of a particle during the random walk.
c    NACCR       = Number of lattice point in the aggregate.
c    LATTICE     = Defines the grid upon which the random walks occur.
c           Nodes of lattice with seed or aggregates are set = 3.
c           Sticking nodes are set = 2.
c           Nodes along the pattern of a walker are set to 1.
c    PACCR       = NACCR/(IMAX * JMAX) = Fraction of lattice nodes in the
c           aggregation.
c   -----------------------------------------------------------------
    implicit real*8 (a-h, o-z)
    parameter (imax=100, jmax=100)
    dimension lattice(imax,imax)
c
c    File 10 = OUTPL0           Stores PACCR versus number of walks.
c    File 11 = OUTPL_DENS       Stores as 1 the density of nodes in the
c                 aggregate. Other nodes have density 0.
c    File 60 = OUTPUT file
c    Files 111 - 114 = Write density files at given steps

    open(10,file='OUTPL0',     status='unknown')
    open(11,file='OUTPL_DENS', status='unknown')
    open(60,file='OUTPUT_DLA_2', status='unknown')

    open(111,file='OUTPL_DENS1',   status='unknown')
    open(112,file='OUTPL_DENS2',   status='unknown')
    open(113,file='OUTPL_DENS3',   status='unknown')
    open(114,file='OUTPL_DENS4',   status='unknown')

    iname1 = 111     ! Used for printing density files
    iname2 = 112
    iname3 = 113
    iname4 = 114
c-----------------------------------------------------------------
    nw1=250 ! Default=0, Pointer for print when iwalk=iwalk1
    nw2=500 ! Default=0, Pointer for print when iwalk=iwalk2
    nw3=750 ! Default=0, Pointer for print when iwalk=iwalk3
    nw4=1000 ! Default=0, Pointer for print when iwalk=iwalk4
c-----------------------------------------------------------------

    nwalk_max= 1140  ! imax*jmax/10
    write(60,*) 'nwalk_max', nwalk_max
```

```
   write(* ,*) 'nwalk_max', nwalk_max
   write(60,*)
   write(* ,*)
222  continue
   call CPU_TIME(time1)

   icount3=0      ! Counter of particles in aggregations

   do 950 mean=1,1

   arf=RF(-mean)          ! Re-initialise RF

   xmax=1.d0
   ymax=1.d0
   dx=xmax/float(imax)
   dy=ymax/float(jmax)

   do 10 i=1,imax
   do 10 j=1,jmax
   lattice(i,j)=0
10  continue

c   iseed_set=0 Seed position is determined random
c   iseed_set=1 Seed is positioned.
   iseed_set=1

   if(iseed_set.eq.0) then
5   continue
   iseed0=1+rf(0)*imax
   jseed0=1+rf(0)*imax
   if(iseed0.lt.2 .or. iseed0.gt.imax-1) go to 5
   if(jseed0.lt.2 .or. jseed0.gt.imax-1) go to 5
   lattice(iseed0,jseed0) = 3

   else

   iseed0=1
   jseed0=1
   lattice(iseed0,jseed0) = 3
   endif

   if(iseed0.eq.1 .and. jseed0.eq.1) then
   lattice(iseed0 +1, jseed0 +1) = 2
   lattice(iseed0,  jseed0 +1) = 2
   lattice(iseed0 +1, jseed0)   = 2
```

```
      else       ! Initial seed middle in domain

      lattice(iseed0 +1, jseed0 +1) = 2
      lattice(iseed0 -1, jseed0 -1) = 2
      lattice(iseed0,   jseed0 +1)  = 2
      lattice(iseed0,   jseed0 -1)  = 2

      lattice(iseed0 +1, jseed0)    = 2
      lattice(iseed0 -1, jseed0)    = 2
      lattice(iseed0 +1, jseed0 -1) = 2
      lattice(iseed0 -1, jseed0 +1) = 2
      endif
c
      write(* ,*) ' DLA_2 - Position of seed ', iseed0, jseed0
      write(60,*) ' DLA_2 - Position of seed ', iseed0, jseed0
      write(60,*)
      write(* ,*)

      paccr_max=0.2d0

      if(imax.le.500) ifreq=100    ! Frequency for time control
      if(imax.gt.500) ifreq= 50
      nwalk=0
      naccr=0
      no_accr=0
50    continue ! ---------------------- START LOOP ON NWALK ------

      nwalk=nwalk+1

      if(mod(nwalk,ifreq).eq.0) then
       call CPU_TIME(timen)
       dtime=timen - time1
       write( *,*) 'DLA_2 - nwalk, dtime ', nwalk,dtime
       write(60,*) 'DLA_2 - nwalk, dtime ', nwalk,dtime
      endif

      ntrials=0
60    continue
      ntrials=ntrials + 1

      if(ntrials.gt.1000) then
      write( *,*) 'DLA_2 - Calc. stop, nwalk, ntrials', nwalk,ntrials
      write(60,*) 'DLA_2 - Calc. stop, nwalk, ntrials', nwalk,ntrials
      write(60,*)
      write(* ,*)
```

```
      go to 900
      endif

      iw=1 + rf(0)*imax
      jw=1 + rf(0)*jmax

c
c      Exclude starting points in the corners of the domain.
      if(iw.eq.1 . and. jw.eq.1 .or. iw.eq.1 . and. jw.eq.jmax
     1 .or. iw.eq.imax . and. jw.eq.1 .or. iw.eq.imax . and. jw.eq.jmax)
     2  go to 60

c      Exclude starting points which are already accretion points or stick
c      points.
      if(lattice(iw,jw).eq.2 .or. lattice(iw,jw).eq.3) go to 60

c      New walker must be distant from aggegate
      idist_min=100000

      i1=iw - imax/10
      i2=iw + imax/10
      j1=jw - jmax/10
      j2=jw + jmax/10

      i1=max(i1,1)
      i1=min(i1,imax)
      i2=max(i2,1)
      i2=min(i2,imax)

      j1=max(j1,1)
      j1=min(j1,imax)
      j2=max(j2,1)
      j2=min(j2,imax)

      do 80 i=i1,i2
      do 80 j=j1,j2
      if(lattice(i,j) .eq. 3) then

      a=float(i-iw)
      b=float(j-jw)
      dist=dsqrt( a**2 + b**2)
      idist=dist
      if(idist.lt.idist_min) idist_min=idist
      endif
```

```
80  continue
   if(idist_min.lt.imax/10) go to 60

   lattice(iw,jw)=1  ! Marks the path of the particle
c
c                                       ir=3
c                                         |
c                                         |
c   Start random walk      ir=2 <-- P(iw,jw) --> ir=0
c                                         |
c                                         |
c                                       ir=1
100  continue   ! Move particle
   ir=rf(0)*4

   istep=1

   if(ir.eq.0) iw=iw + istep
   if(ir.eq.2) iw=iw - istep
   if(ir.eq.3) jw=jw + istep
   if(ir.eq.1) jw=jw - istep

   call screen1(iw,jw,imax,jmax,icode)

c   ------------------- Reflection from walls -------------
   if(icode.eq.-1) jw=1
   if(icode.eq.-2) iw=1
   if(icode.eq.-3) jw=jmax
   if(icode.eq.-4) iw=imax
c   ---------------------------------------------------------

  if(lattice(iw,jw).eq.0) then
   lattice(iw,jw)=1
   go to 100
   endif

   if(lattice(iw,jw).eq.1) go to 100

   if(lattice(iw,jw).eq.3) then  ! Point is already in the aggregate
   no_accr=no_accr + 1
   go to 50
   endif
c
c
   if(lattice(iw,jw).eq.2) THEN
```

```
      lattice(iw,jw) = 3
      naccr=naccr+1
      paccr=float(naccr)/float(imax*jmax)
c   -----------------------------------------------------------
      call SETLAT2(iw,jw,imax,jmax,lattice)
c   -----------------------------------------------------------
    ENDIF
c
c   ------------------------------------------------------------
c   Zero LATTICE(i,j) if =1, to prepare for next walk
   do 860 i=1,imax
   do 860 j=1,jmax
   if(lattice(i,j).eq.1) lattice(i,j)=0
860  continue

   if(mod(nwalk,50) .eq.1) write(10,2010) nwalk, paccr
2010 format( i8, e12.5)

c--------------------------------------------------------------
   nw=nwalk  ! Print Density Files

    if(nw.eq.nw1 .or. nw.eq.nw2 .or. nw.eq.nw3 .or. nw.eq.nw4) then
      if(nw.eq.nw1) call WRITEDENS(imax,jmax,lattice,dx,dy,iname1)
      if(nw.eq.nw2) call WRITEDENS(imax,jmax,lattice,dx,dy,iname2)
      if(nw.eq.nw3) call WRITEDENS(imax,jmax,lattice,dx,dy,iname3)
      if(nw.eq.nw4) call WRITEDENS(imax,jmax,lattice,dx,dy,iname4)
    endif
c-----------------------------------------

   if(nwalk.gt.nwalk_max .or. paccr.gt.paccr_max) go to 900

   go to 50

c   ======================== End of loop on walks ==========
c
900  continue
   write( *,*) 'DLA_2 - nwalk,naccr,no_accr ', nwalk,naccr,no_accr
   write(60,*) 'DLA_2 - nwalk,naccr,no_accr ', nwalk,naccr,no_accr
   write(60,*)
   write(* ,*)
c   ---------------   Write File OUTPL_DENS  ----------------
   do 930 i=1,imax
   do 930 j=1,jmax
   xi= (i-1)*dx + 0.5d0 * dx
   yj= (j-1)*dy + 0.5d0 * dy
```

```
   if(lattice(i,j) .lt. 3) dens=0.d0
   if(lattice(i,j) .eq. 3) dens=1.d0
   write(11,1011) i,j,xi,yj,dens         ! File 11 = OUTPL_DENS
1011 format(2i6, 3f12.5)
930  continue

c   -------------------------------------------------------

   call CPU_TIME(CPU_TIME2)
   delta_time=cpu_time2 - cpu_time1
   write( *,*) 'DLA_2 - delta_time ', delta_time
   write(60,*) 'DLA_2 - delta_time ', delta_time
   write(60,*)
   write(* ,*)

950  continue

   close(10)
   close(11)
   close(60)

   close(111)
   close(112)
   close(113)
   close(114)

   stop
   end

   subroutine screen1(iw,jw,imax,jmax,icode)
   implicit real*8 (a-h,o-z)

c          ICODE = -3
c
c             ------ ir=3 ----P(imax,jmax)
c             |                |
c             |                |
c ICODE = -2 ir=2            ir=0   ICODE = -4
c             |                |
c             |                |
c       P(0,0)-- ir=1 ----P(imax,0)
c
c          ICODE = -1
c
```

```
c   icode = 0 particle at position P(iw,jw) is inside the lattice or
c         on the boundary.
c   icode = -1 particle at position P(iw,jw) is below j=1
c   icode = -2 particle at position P(iw,jw) is left of i=1
c   icode = -3 particle at position P(iw,jw) is above j=jmax
c   icode = -4 particle at position P(iw,jw) is right of i=imax

      icode =0
      if(jw.lt.1)     icode=-1
      if(iw.lt.1)     icode=-2
      if(jw.gt.jmax) icode=-3
      if(iw.gt.imax) icode=-4

      return
      end

      Subroutine SETLAT2(iw,jw,imax,jmax,lattice)
      implicit real*8 (a-h, o-z)
      dimension Lattice(imax,jmax)

c         |---------------|
c         | 8     3     1 | Numbering of neigh. nodes
c         |               |
c         | 6  (iw,jw)  5 |
c         |       |
c         | 2     4     7 |
c         |---------------|

       i1=iw+1                  ! Set new sticking nodes
       j1=jw+1
       if(lattice(i1,j1).ne.3) lattice(i1,j1)=2

       i2=iw-1
       j2=jw-1
       if(i2.ge.3.and.j2.ge.3.and.lattice(i2,j2).ne.3)
     1                        lattice(i2,j2)=2
c
       i3=iw
       j3=jw+1
       if(lattice(i3,j3).ne.3) lattice(i3,j3)=2

       i4=iw
       j4=jw-1
       if(j4.ge.3 .and. lattice(i4,j4).ne.3) lattice(i4,j4)=2
```

```
      i5=iw+1
      j5=jw
      if(lattice(i5,j5).ne.3) lattice(i5,j5)=2

      i6=iw-1
      j6=jw
      if(i6.ge.3 .and. lattice(i6,j6).ne.3) lattice(i6,j6)=2

      i7=iw+1
      j7=jw-1
      if(j7.ge.3 .and. lattice(i7,j7).ne.3) lattice(i7,j7)=2

      i8=iw-1
      j8=jw+1
      if(i8.ge.3 .and. lattice(i8,j8).ne.3) lattice(i8,j8)=2

      return
      end

      function rf(IDUM)
*--generates a uniformly distributed random fraction between 0 and 1
*----IDUM will generally be 0, but negative values may be used to
*------re-initialize the seed
c     Cfr. function RUN3 of Numerical Recipes, p. 199.
      implicit real*8 (a-h,o-z)
      SAVE MA,INEXT,INEXTP
      PARAMETER (MBIG=1000000000,MSEED=161803398,MZ=0,FAC=1.E-9)
      DIMENSION MA(55)
      DATA IFF/0/
      IF (IDUM.LT.0.OR.IFF.EQ.0) THEN
       IFF=1
       MJ=MSEED-IABS(IDUM)
       MJ=MOD(MJ,MBIG)
       MA(55)=MJ
       MK=1
       DO 50 I=1,54
        II=MOD(21*I,55)
        MA(II)=MK
        MK=MJ-MK
        IF (MK.LT.MZ) MK=MK+MBIG
        MJ=MA(II)
50      CONTINUE
       DO 100 K=1,4
        DO 60 I=1,55
         MA(I)=MA(I)-MA(1+MOD(I+30,55))
```

```
      IF (MA(I).LT.MZ) MA(I)=MA(I)+MBIG
60    CONTINUE
100   CONTINUE
     INEXT=0
     INEXTP=31
   END IF
200  INEXT=INEXT+1
   IF (INEXT.EQ.56) INEXT=1
   INEXTP=INEXTP+1
   IF (INEXTP.EQ.56) INEXTP=1
   MJ=MA(INEXT)-MA(INEXTP)
   IF (MJ.LT.MZ) MJ=MJ+MBIG
   MA(INEXT)=MJ
   RF=MJ*FAC
   IF (RF.GT.1.E-8.AND.RF.LT.0.99999999) RETURN
   GO TO 200
   END

   Subroutine WRITEDENS(imax,jmax,lattice,dx,dy,iname)
   implicit real*8 (a-h, o-z)
   dimension Lattice(imax,jmax)
    do 931 i=1,imax
    do 931 j=1,jmax
      xi= (i-1)*dx + 0.5d0 * dx
      yj= (j-1)*dy + 0.5d0 * dy
      if(lattice(i,j) .eq. 3) dens=1.d0
      if(lattice(i,j) .lt. 3) dens=0.d0
      write(iname,1011) i,j,xi,yj,dens
931   continue

1011 format(2i6, 3f12.5)
    return
   end
```

Information on OUTPUT files:

OUTPL0 *Stores PACCR versus number of walks.*

OUTPL_DENS *Stores as 1 the final density of nodes in the aggregate.*
OUTPL_DENS1 - OUTPL_DENS4 Store Density Files
at the 4 choesen numders of walks.

From OUTPT_DL_2

```
nwalk_max     1140

DLA_2 - Position of seed       1       1

DLA_2 - nwalk, dtime       100 0.14040100000000000
DLA_2 - nwalk, dtime       200 0.18720099999999998
DLA_2 - nwalk, dtime       300 0.32760200000000000
DLA_2 - nwalk, dtime       400 0.34320200000000001
DLA_2 - nwalk, dtime       500 0.35880199999999995
DLA_2 - nwalk, dtime       600 0.48360299999999995
DLA_2 - nwalk, dtime       700 0.48360299999999995
DLA_2 - nwalk, dtime       800 0.60840400000000006
DLA_2 - nwalk, dtime       900 0.60840400000000006
DLA_2 - nwalk, dtime      1000 0.60840400000000006
DLA_2 - nwalk, dtime      1100 0.73320399999999997
DLA_2 - Calc. stop, nwalk, ntrials     1128     1001

DLA_2 - nwalk,naccr,no_accr     1128     1126      1

DLA_2 - delta_time   0.87360499999999996
```

Appendix A.4

```
      PROGRAM CORR1D

c     List of Subroutines and Functions:

c     CORREL       ! Computes auto-correlation
c     FOUR1
c     REALFT
c     TWOFFT
c
      implicit real*8 (a-h, o-z)
      Dimension data1(64), data2(64), ans(128)

      open(60, file='OUTPUT_CORR1D' , status='unknown')

      [illegible]

c     Retain the chance to compute correlation and autocorrelation
      do 13 i = 1,n
        data1(i) = 0.d0
```

```
      data2(i) = 0.d0
13    continue

      data1(1) = 1
      data1(2) = 1
      data1(3) = 1
      data1(4) = 1
      data1(5) = 1

      data1(9)  = 1
      data1(10) = 1

      do 20 i=1,n
20    data2(i) = data1(i)

      do 22 i=1,n
       if(data1(i) .ne. 0.d0) then
          write(60,1005) i, data1(i)
       write(* ,1005) i, data1(i)
      endif
1005   format(' i, data1, data2 ', i5,e12.5)
22    continue

          call correl(data1, data2, n, ans)

c    Write only correlations at increasing positive lags which are
c    in ANS(1) (zero lag) on up to ANS(n)
      do 40 i=1,n/2
      im1 = i - 1
      if(dabs(ans(i)) .lt. 1.d-10) ans(i)=0.d0

       if(ans(i) .gt. 0) write(60,1040) im1, ans(i)
       if(ans(i) .gt. 0) write( *,1040) im1, ans(i)

1040   format( i8, e12.5)
40    continue

      stop
      end

      subroutine correl(data1,data2,n,ans)
      implicit real*8(a-h,o-z)
c
      dimension data1(n), data2(n)
```

```
c   Length of FFT is the maximum anticipated FFT size.
    complex*16 FFT(n), ans(n)

c   Check that n is a power of two
    do 10 i=1,20
      ipower=2**i
      if(n.eq.ipower) go to 15
10  continue
      write(60,1000)
1000    format(' CORREL - N is not a power of two or too small, STOP')
    STOP
15  continue
c
c   FFT both data1,data2 vectors at once
    call twofft(data1,data2,fft,ans,n)

c   Compute FFT of their correlation
    ano2=dfloat(n)/2.d0
   do 11 i=1,n/2+1
   ans(i)=fft(i)*dconjg(ans(i))/ano2
11  continue
c
c   Pack first and last into one lement
    ans(1)=dcmplx(dreal(ans(1)),dreal(ans(n/2+1)))

c   Apply inverse transform to find correlation
    call realft(ans,n/2,-1)
c
    return
   end

    subroutine four1(dataa,nn,isign)
    implicit real*8(a-h,o-z)
c
c   if isign= 1 replaces dataa by its discrete fourier transform
c   if       =-1 replaces data by nn times its inverse discrete
c          fourier transform
c
c   dataa is a complex array of length nn or a real array of length
c      2*nn
c   nn must be an integer power of 2
c
    dimension dataa(*)
c
c   check that nn is a power of two
```

```
   do 5 i=1,20
   npar=2**i
   if(npar.eq.nn) go to 3
5   continue
   write(6,1003)
1003 format(' FOUR1 - , stop because nn is not a power of two')
3   continue
c
   twopi=8.*atan(1.d0)
   n=2*nn
   j=1
   do 11 i=1,n,2
    if(j.gt.i)then
     tempr=dataa(j)
     tempi=dataa(j+1)
     dataa(j)=dataa(i)
     dataa(j+1)=dataa(i+1)
     dataa(i)=tempr
     dataa(i+1)=tempi
    endif
    m=n/2
1     if ((m.ge.2).and.(j.gt.m)) then
     j=j-m
     m=m/2
    go to 1
    endif
    j=j+m
11   continue
   mmax=2
2   if (n.gt.mmax) then
    istep=2*mmax
    theta=twopi/(isign*mmax)
    wpr=-2.d0*dsin(0.5 *theta)**2
    wpi=dsin(theta)
    wr=1.d0
    wi=0.d0
    do 13 m=1,mmax,2
     do 12 i=m,n,istep
      j=i+mmax
      tempr=wr *dataa(j)  -wi *dataa(j+1)
      tempi=wr *dataa(j+1)+wi *dataa(j)
      dataa(j)=dataa(i)-tempr
      dataa(j+1)=dataa(i+1)-tempi
      dataa(i)=dataa(i)+tempr
      dataa(i+1)=dataa(i+1)+tempi
```

```
12     continue
     wtemp=wr
     wr=wr*wpr-wi*wpi+wr
     wi=wi*wpr+wtemp*wpi+wi
13   continue
    mmax=istep
   go to 2
   endif
   return
   end

   subroutine realft(dataa,n,isign)
   implicit real*8(a-h,o-z)
c
   dimension dataa(*)
   pi=4.d0*atan(1.d0)
   theta=pi/dfloat(n)
  c1=0.5d0
  if(isign.eq.1) then
   c2=-0.5d0
   call four1(dataa,n,+1)
  else
   c2=0.5
   theta=-theta
  endif
   wpr=-2.d0*dsin(0.5d0*theta)**2
  wpi=dsin(theta)
  wr=1.d0+wpr
  wi=wpi
  n2p3=2*n+3
  do 11 i=2,n/2+1
  i1=2*i-1
  i2=i1+1
  i3=n2p3-i2
  i4=i3+1
  wrs=wr
  wis=wi
  h1r=c1*(dataa(i1)+dataa(i3))
  h1i= c1*(dataa(i2)-dataa(i4))
  h2r=-c2*(dataa(i2)+dataa(i4))
  h2i= c2*(dataa(i1)-dataa(i3))
  dataa(i1)=h1r+wrs*h2r-wis*h2i
  dataa(i2)=h1i+wrs*h2i+wis*h2r
  dataa(i3)=h1r-wrs*h2r+wis*h2i
  dataa(i4)=-h1i+wrs*h2i+wis*h2r
```

```
  wtemp=wr
  wr=wr*wpr-wi*wpi+wr
  wi=wi*wpr+wtemp*wpi+wi
11  continue
c
   if(isign.eq.1) then
    h1r=dataa(1)
    dataa(1)=h1r+dataa(2)
    dataa(2)=h1r-dataa(2)
  else
     h1r=dataa(1)
    dataa(1)=c1*(h1r+dataa(2))
    dataa(2)=c1*(h1r-dataa(2))
    call four1(dataa,n,-1)
  endif
c
  return
  end

   subroutine twofft(data1,data2,fft1,fft2,n)
   implicit real*8(a-h,o-z)
c
   dimension data1(n), data2(n)
  complex*16 fft1(n), fft2(n),h1,h2,c1,c2
  c1= dcmplx(0.5d0,  0.d0)
  c2= dcmplx(0.0d0, -0.5d0)
c
   do 11 j=1,n
  fft1(j)=dcmplx(data1(j),data2(j))
11  continue
c
   call four1(fft1,n,1)
  fft2(1)=dcmplx(dimag(fft1(1)),0.d0)
c  fft1(1)=dcmplx(dreal(fft1(1)),0.d0)
   fft1(1)=dcmplx(dreal(fft1(1)),0.d0)

  n2=n+2
  do 12 j=2,n/2+1
  h1=c1*(fft1(j)+ dconjg(fft1(n2-j)))
  h2=c2*(fft1(j)- dconjg(fft1(n2-j)))
  fft1(j)=h1
  fft1(n2-j)= dconjg(h1)
  fft2(j)=h2
  fft2(n2-j)= dconjg(h2)
12  continue
```

```
c
   return
  end
```

Information on OUTPUT files:

From OUTPUT_CORR1D

```
i, data1, data2    1 0.10000E+01
i, data1, data2    2 0.10000E+01
i, data1, data2    3 0.10000E+01
i, data1, data2    4 0.10000E+01
i, data1, data2    5 0.10000E+01
i, data1, data2    9 0.10000E+01
i, data1, data2   10 0.10000E+01

      0 0.70000E+01
      1 0.50000E+01
      2 0.30000E+01
      3 0.20000E+01
      4 0.20000E+01
      5 0.20000E+01
      6 0.20000E+01
      7 0.20000E+01
      8 0.20000E+01
      9 0.10000E+01
```

Appendix A.5

```
      PROGRAM DLA_2_POSPRO

c     Density auto-correlation is made from pre-existing OUTPL_DENS
c     and file OUT_CORR is written.
c     File OUTPL_DENS can be generated with the program DLA_2
c     of Appendix 3.3 using a lattice of dimensions (512,512).

c     -------------------------------------------------------------
c     Short description of subroutines:

c     CORREL  Used for computing autocorrelation of two sets of data.
c     FOUR1   Makes Fourier transform or anti-transform of one array of
data.
c     MINMAX  Finds minimum and maximum values from an array af data.
```

```
c   REALFT  Makes Fourier anti-transform of an array of complrx numbers.
c   TWOFFT  Makes Fourier transform of two real arrays of data.
c   -----------------------------------------------------------------
c
    implicit real*8 (a-h, o-z)

    parameter (imax=512, jmax=512)
    dimension data1(512), data2(512), ans(1024), aux(512)
    dimension correxp(512),lattice(512,512)

    n = 512
    nfft= 512
    open(11,file='OUTPL_DENS',  status='old')        ! Reads OUTPL_DENS
    open(50,file='OUTPUT_CORR', status='unknown')  ! Writes
OUTPUT_CORR
    open(60,file='OUTPUT_PSPRO', status='unknown')  ! Writes
OUTPUT_PSPRO

    write(* ,*) ' imax,jmax', imax,jmax
    write(60,*) ' imax,jmax', imax,jmax

    call CPU_TIME(time_corr0)

    do 3 nn=1,nfft
     aux(nn) =0.d0
3   continue

    do 4 ii=1,imax
    do 4 jj=1,jmax
     read(11,1011) i,j,xi,yj,dens        ! Reads OUTPL_DENS
     lattice(i,j)=0
     if(dens.ne.0.d0) lattice(i,j) = 1   ! Matrix LATTICE is re-used.
1011   format(2i6, 3f12.5)
4   continue

c   *******************************************************
c   Fill arrays DATA1, with rows from matrix OUTPL_DENS and
c   process each row in turn.

    ncount=0
    do 500 j=1,512

     do 50 i=1,imax
      data1(i)=lattice(i,j)
```

```
     data2(i)=lattice(i,j)
50   continue

    call minmax(imax, data1, amin,amax)
      if(amax.eq.0.d0) then
      write(* ,*) ' Skipped column j =', j
      write(60,*) ' Skipped column j =', j
     go to 200
    endif

    ncount=ncount+1
c  --------------------------------------------------------------
    call CORREL(DATA1,DATA2,n,ANS)
c  --------------------------------------------------------------
    call minmax(imax,ans,smin,smax)

    do 85 nn=1,512
      if(smin .lt. 1.0) then
        ANS (nn) = 0.d0
        smin   = 0.d0
      endif
85   continue

    do 90 nn=1,nfft
     aux(nn)=aux(nn)+ans(nn)  ! The values AUX(nn) are incremented
90   continue
200  continue
500  continue

    coeffa=0.31d0
   do 19 nn=1,imax/2  ! nfft

    aux(nn)=aux(nn)/dfloat(ncount) ! The values AUX(nn) are averaged

    if(dabs(aux(nn)).gt.1.d-10 .and.nn.le.nfft/2) then

c   Compare computed autocorrelation with expected value
c   CORREXP= <RHO(r + s) * RHO(r)> = S ** (-COEFFA), with COEFFA=0.31

      if(nn.eq.1) correxp(1)= aux(1)
      if(nn.gt.1) correxp(nn)= correxp(1)*dfloat(nn)**(-coeffa)

      write(50,1050) nn,aux(nn),correxp(nn)      ! OUTPUT_CORR
1050  format(i9,2e12.5)
```

```
      endif
19  continue

    write(* ,*) ' File OUT_CORR written'
    write(60,*) ' File OUT_CORR written'

    call CPU_TIME(time_corr1)

    dtime_corr = time_corr1 - time_corr0
    write(* ,*) ' dtime_corr', dtime_corr
    write(60,*) ' dtime_corr', dtime_corr

c   ---------------------------------------------------------------
    close(11)
    close(50)
    close(60)

    stop
    end

    subroutine correl(data1,data2,n,ans)
    implicit real*8(a-h,o-z)
c
    dimension data1(n), data2(n)
c    Length of FFT is the maximum anticipated FFT size.
    complex*16 FFT(n), ans(n)

c    Check that n is a power of two
    do 10 i=1,20
      ipower=2**i
      if(n.eq.ipower) go to 15
10  continue
      write(60,1000)
1000   format(' CORREL - N is not a power of two or too small, STOP')
    STOP
15  continue
c
c    FFT both data1,data2 vectors at once
    call twofft(data1,data2,fft,ans,n)

c    Compute FFT of their correlation
    ano2=dfloat(n)/2.d0
   do 11 i=1,n/2+1
   ans(i)=fft(i)*dconjg(ans(i))/ano2
11  continue
```

```
c
c   Pack first and last into one lement
    ans(1)=dcmplx(dreal(ans(1)),dreal(ans(n/2+1)))

c   Apply inverse transform to find correlation
    call realft(ans,n/2,-1)
c
    return
    end

    subroutine four1(dataa,nn,isign)
    implicit real*8(a-h,o-z)
c
c   if isign= 1 replaces dataa by its discrete fourier transform
c   if       =-1 replaces data by nn times its inverse discrete
c           fourier transform
c
c   dataa is a complex array of length nn or a real array of length
c      2*nn
c   nn must be an integer power of 2
c
    dimension dataa(*)
c
c   check that nn is a power of two
    do 5 i=1,20
    npar=2**i
    if(npar.eq.nn) go to 3
5   continue
    write(6,1003)
1003 format(' FOUR1 - , stop because nn is not a power of two')
3   continue
c
    twopi=8.*atan(1.d0)
    n=2*nn
    j=1
    do 11 i=1,n,2
     if(j.gt.i)then
      tempr=dataa(j)
      tempi=dataa(j+1)
      dataa(j)=dataa(i)
      dataa(j+1)=dataa(i+1)
      dataa(i)=tempr
      dataa(i+1)=tempi
     endif
     m=n/2
```

```
1     if ((m.ge.2).and.(j.gt.m)) then
       j=j-m
       m=m/2
     go to 1
     endif
     j=j+m
11   continue
   mmax=2
2   if (n.gt.mmax) then
     istep=2*mmax
     theta=twopi/(isign*mmax)
     wpr=-2.d0*dsin(0.5 *theta)**2
     wpi=dsin(theta)
     wr=1.d0
     wi=0.d0
     do 13 m=1,mmax,2
       do 12 i=m,n,istep
         j=i+mmax
         tempr=wr *dataa(j)  -wi *dataa(j+1)
         tempi=wr *dataa(j+1)+wi *dataa(j)
         dataa(j)=dataa(i)-tempr
         dataa(j+1)=dataa(i+1)-tempi
         dataa(i)=dataa(i)+tempr
         dataa(i+1)=dataa(i+1)+tempi
12       continue
       wtemp=wr
       wr=wr*wpr-wi*wpi+wr
       wi=wi*wpr+wtemp*wpi+wi
13     continue
     mmax=istep
   go to 2
   endif
   return
   end

   subroutine minmax(len, field, amin,amax)
C     Given a 'field' of lenght 'len', this subroutine returns
C     the minimum AMIN and maximum AMAX values in the field.
   implicit real*8 (a-h, o-z)
   dimension field(*)
C
   amin= 1.e30
   amax=-1.e30
   do 10 i=1,len
   if(field(i).le.amin) amin=field(i)
```

```
   if(field(i).ge.amax) amax=field(i)
10  continue
   return
   end

   subroutine realft(dataa,n,isign)
   implicit real*8(a-h,o-z)
c
   dimension dataa(*)
   pi=4.d0*atan(1.d0)
   theta=pi/dfloat(n)
  c1=0.5d0
  if(isign.eq.1) then
   c2=-0.5d0
   call four1(dataa,n,+1)
  else
   c2=0.5
   theta=-theta
  endif
   wpr=-2.d0*dsin(0.5d0*theta)**2
  wpi=dsin(theta)
  wr=1.d0+wpr
  wi=wpi
  n2p3=2*n+3
  do 11 i=2,n/2+1
  i1=2*i-1
  i2=i1+1
  i3=n2p3-i2
  i4=i3+1
  wrs=wr
  wis=wi
  h1r=c1*(dataa(i1)+dataa(i3))
  h1i= c1*(dataa(i2)-dataa(i4))
  h2r=-c2*(dataa(i2)+dataa(i4))
  h2i= c2*(dataa(i1)-dataa(i3))
  dataa(i1)=h1r+wrs*h2r-wis*h2i
  dataa(i2)=h1i+wrs*h2i+wis*h2r
  dataa(i3)=h1r-wrs*h2r+wis*h2i
  dataa(i4)=-h1i+wrs*h2i+wis*h2r
  wtemp=wr
  wr=wr*wpr-wi*wpi+wr
  wi=wi*wpr+wtemp*wpi+wi
11  continue
c
   if(isign.eq.1) then
```

```
     h1r=dataa(1)
     dataa(1)=h1r+dataa(2)
     dataa(2)=h1r-dataa(2)
   else
      h1r=dataa(1)
     dataa(1)=c1*(h1r+dataa(2))
     dataa(2)=c1*(h1r-dataa(2))
     call four1(dataa,n,-1)
   endif
c
   return
   end

    subroutine twofft(data1,data2,fft1,fft2,n)
    implicit real*8(a-h,o-z)
c
    dimension data1(n), data2(n)
   complex*16 fft1(n), fft2(n),h1,h2,c1,c2
   c1= dcmplx(0.5d0,  0.d0)
   c2= dcmplx(0.0d0, -0.5d0)
c
    do 11 j=1,n
   fft1(j)=dcmplx(data1(j),data2(j))
11  continue
c
    call four1(fft1,n,1)
   fft2(1)=dcmplx(dimag(fft1(1)),0.d0)
c   fft1(1)=dcmplx(dreal(fft1(1)),0.d0)
    fft1(1)=dcmplx(dreal(fft1(1)),0.d0)

   n2=n+2
   do 12 j=2,n/2+1
   h1=c1*(fft1(j)+ dconjg(fft1(n2-j)))
   h2=c2*(fft1(j)- dconjg(fft1(n2-j)))
   fft1(j)=h1
   fft1(n2-j)= dconjg(h1)
   fft2(j)=h2
   fft2(n2-j)= dconjg(h2)
12  continue
c
    return
    end
```

Information on OUTPUT files:

The code generates the files OUTPUT_CORRL and OUTPUT_PSPRO . The later can be used to plot the computed autocorrelation.

Appendix A.6

```
   PROGRAM Langevin              ! Last change: 05.06.2024

c   Program for computation of trajectories of a particle subjected to
c   random motion governed by the Langevin Equation.
c   Solution of Langevin Equation: dv/dt = - gamma * v + G(t)

c   List of subroutines and functions

c    SUBROUTINE MOMENT    Computes mean value, standard deviation, etc.
c    FUNCTION  RAND_WN    Computes WN = DSQRT(24/M) * SUM(i=1,...M)
[Ri -1/2]
c   FUNCTION  RF          Generates uniform deviates in the open interval
c    SUBROUTINE MOMENT    Computes mean value, standard deviation, etc.
c    SUBROUTINE TEST_WN  Checks that the random variables WN have a
Gaussian
c                 distribution with mean value 0 and variance 2.

c   MAIN FORTRAN VARIABLES

c   ALPHA   = Friction coefficient (kg/s)
c   AMASS   = Mass of particle (kg)
c   BOLTZ   = Boltzmann constant (J/k)
c   DELTAT  = Time step (s)
c   DIFFC   = Diffusion coefficient (m**2/s)
c   GAMMA   = ALPHA/AMASS (1/s)
c   ISBNMAX= Maximum number of realizations
c   TEMP    = Absolute temperature (K)
c   TIME    = Time (s)
c   TIMEX   = Maximum problem time (s)
c   U,V     = Velocity components (m/s)
c   X,Y     = Components of position vector of particle

   implicit real *8 (a-h, o-z)

   Parameter (imax=1 000 000, iensbmx=1000)
```

```
      open(10,file='OUT_GAUSS', status='unknown') ! For Gaussian distr.
     open(11,file='TRAJECTS' , status='unknown') ! Plotting trajectories
      open(60, file='OUTPUT'  , status='unknown')

      write(60, *) ' OUTPUT for LANGEVIN_EQ_1'
      write( *, *) ' OUTPUT for LANGEVIN_EQ_1'
      write(60, *)
      write( *, *)

      rf0=rf(-1)  ! Initialize Function RF

      call TEST_WN
      call CPU_TIME(CPUT0)

      pi=4.d0*atan(1.d0)
      alpha = 0.001d0

c     Data for aerosol particle
      sprad = 1.d-07                     ! Particle radius
      spcp  = 2140.d0                    ! Specific Heat (J/kg K)
      splamb= 0.15 d0                    ! Thermal conductivity (W/m K)
      sprho = 890.d0                     ! Density (kg/m**3)
      spvol = (4.d0*pi*sprad**3)/3.d0 ! Particle volume
      amass = spvol * sprho              ! Particle mass

      gamma = alpha/amass
      gamma = min(1.50d-04 , gamma)
      boltz = 1.380622e-23         ! Boltzmann constant (J/k)!
      temp  = 300.d0               ! Temperature (K)

      x0   = 0.d0
      y0   = 0.d0
      u0   = 1.d-03
      v0   = u0
      deltat=1.d-03
      del05 =dsqrt(deltat)

      do 800 iensb=1, 1 !iensbmx A subsequent <Realization> can
c                 be obtained repeating the loop 800
      time=0.d0
      un=u0
      vn=v0
      xn=x0
      yn=y0
```

```
      do 500 i=1,imax
        time=time + deltat

        unp1 = un * (1.d0 - gamma * deltat) + del05 * rand_wn(dummy)
        vnp1 = vn * (1.d0 - gamma * deltat) + del05 * rand_wn(dummy)

        xnp1 = xn + unp1 * deltat
        ynp1 = yn + vnp1 * deltat

        if(mod(i,1000).eq.0) then
          write(11,1011) i,time, xnp1,ynp1, unp1, vnp1   ! TRAJECTS
1011      format(i9,5e14.5)
        endif

        un=unp1
        vn=vnp1

        xn=xnp1
        yn=ynp1

500   continue

      write( *, *)
      write(60, *)
      call CPU_TIME(CPUTN)
      CPUTIME = CPUTN -CPUT0
      write( *, *) 'LANGV1 - iensb, CPUTIME', iensb, CPUTIME
      write(60, *) 'LANGV1 - iensb, CPUTIME', iensb, CPUTIME

800   continue

      close(10)
      close(11)
      close(60)

      stop
      end

      subroutine moment(data,n,ave,adev,sdev,var,skew,curt)
c     computes mean value (ave),average deviation (adev), standard
c     deviation (sdev),variance (var), skewness (skew), kurtosis (curt)
c     of values given in the array data of length n.
C
      implicit real*8 (a-h, o-z)
      dimension data(n)
```

```
c    if(n.le.1) pause 'n must be at least 2'
     s=0.d0
     do 11 j=1,n
     s=s+data(j)
11   continue
     ave=s/n
     adev=0.d0
     var=0.d0
     skew=0.d0
     curt=0.d0
     do 12 j=1,n
     s=data(j)-ave
     adev=adev+dabs(s)
     p=s*s
     var=var+p
      p=p*s
     skew=skew+p
     p=p*s
     curt=curt+p
12   continue
     adev=adev/n
     var=var/(n-1)
     sdev=dsqrt(var + 1.d-14)

    if(var.ne.0.d0) then
       skew=skew/(n*sdev**3)
       curt=curt/(n*var**2) - 3.d0
     else
      write(60, *) 'no skew or kurtosis when zero variance '
      write( *, *) 'no skew or kurtosis when zero variance '
c       pause 'no skew or kurtosis when zero variance'
     endif

     return
     en

     FUNCTION RAND_WN(dummy)
     implicit real *8 (a-h, o-z)

c    Compute WN = DSQRT(24/M) * SUM(i=1,...M) [Ri -1/2]
c    M = Large integer (100)
c    Ri = Random deviate
c    WN = Random variable with Gaussian distribution and with
c        mean value =0, variance =2.
```

```
      M= 100
      sum =0.d0
      do 100 i=1,M
      sum = sum + (rf(0) - 0.5d0)
100   continue
      wn= dsqrt(24.d0/float(m)) * sum
      RAND_WN = wn

      return
      end

      function rf(IDUM)
*--generates a uniformly distributed random fraction between 0 and 1
*----IDUM will generally be 0, but negative values may be used to
*------re-initialize the seed
c     Cfr. function RUN3 of Numerical Recipes, p. 199.
      implicit real*8 (a-h,o-z)
      SAVE MA, INEXT, INEXTP
      PARAMETER (MBIG=1000000000,MSEED=161803398,MZ=0,FAC=1.E-9)
      DIMENSION MA(55)
      DATA IFF/0/
      IF (IDUM.LT.0.OR.IFF.EQ.0) THEN
       IFF=1
       MJ=MSEED-IABS(IDUM)
       MJ=MOD(MJ,MBIG)
       MA(55)=MJ
       MK=1
       DO 50 I=1,54
        II=MOD(21*I,55)
        MA(II)=MK
        MK=MJ-MK
        IF (MK.LT.MZ) MK=MK+MBIG
        MJ=MA(II)
50      CONTINUE
       DO 100 K=1,4
        DO 60 I=1,55
         MA(I)=MA(I)-MA(1+MOD(I+30,55))
         IF (MA(I).LT.MZ) MA(I)=MA(I)+MBIG
60       CONTINUE
100     CONTINUE
       INEXT=0
       INEXTP=31
      END IF
200   INEXT=INEXT+1
      IF (INEXT.EQ.56) INEXT=1
```

```
      INEXTP=INEXTP+1
      IF (INEXTP.EQ.56) INEXTP=1
      MJ=MA(INEXT)-MA(INEXTP)
      IF (MJ.LT.MZ) MJ=MJ+MBIG
      MA(INEXT)=MJ
      RF=MJ*FAC
      IF (RF.GT.1.E-8.AND.RF.LT.0.99999999) RETURN
      GO TO 200
      END

      subroutine TEST_WN
      implicit real *8 (a-h, o-z)

c     TASK 1
c     Test of the formula WN = DSQRT(24/M) * SUM(i=1,...M) [Ri -1/2]
c     M  = Large integer (100)
c     Ri = Random deviate
c     WN = Random variable with Gaussian distribution and with
c        mean value =0, variance =2.

c     Results, using M = 100, are:

c        ISET      <WN>      VARIANCE

c          1 000     3.51 e-02    2.069
c         10 000     2.50 e-02    2.031
c        100 000     6.72 e-03    2.0068
c      1 000 000     2.41 e-03    2.0032
c     10 000 000     6.38 e-04    2.00036

c     TASK 2
c     Verify that random wariables wn have a Gaussian distribution

      Parameter (iset=10 000 000, meshes=100)
      dimension SET_WN(iset), MESHES_N(meshes)

c     TASK 1 ! Compute mean value and variance of Gaussian distribution
      write( *, *) 'Task 1'
      write(60, *) 'Task 1'

      do 200 iwn=1,iset
      SET_WN(iwn)=RAND_WN(Dummy)
200   continue

      call moment(SET_WN,iset,ave,adev,sdev,var,skew,curt)
```

```
      write( *, *) 'TEST_WN - ave, adev,var', ave,adev,var
      write( *, *) 'TEST_WN - skew, curt  ', skew,curt
      write(60, *) 'TEST_WN - ave, adev,var', ave,adev,var
      write(60, *) 'TEST_WN - skew, curt  ', skew,curt
      write( *, *)
      write(60, *)

c     TASK 2   ! Plot data for Gaussuian distribution
      write( *, *) 'Task 2'
      write(60, *) 'Task 2'

      wnmax=10.d0
      wnmin= - wnmax
      deltawn=(wnmax - wnmin) / float(meshes)

      do 300 m=1,meshes
      MESHES_N(m)= 0
300   continue

      icount=0
      do 350 iwn=1,iset
      wn=SET_WN(iwn)

      do 310 m=1,meshes
      wn1=wnmin + float((m-1))*deltawn
      wn2=wn1 + deltawn
        if(wn.gt.wn1 .and. wn.le.wn2) then
        MESHES_N(m) =MESHES_N(m) + 1
        icount = icount +1
        endif
310   continue
350   continue

      do 400 m=1,meshes
        x= wnmin + float((m-1))*deltawn + 0.5d0*deltawn
        write(10,1010) m,x, float(MESHES_N(m))          ! OUT_GAUSS
1010     format(i5,2e14.5)
400   continue
      write( *, *) 'Subroutine OUT_GAUSS written'
      write(60, *) 'Subroutine OUT_GAUSS written'

      return
      end
```

Information on OUTPUT files:

OUTPUT for LANGEVIN_EQ_1

```
 Task 1
 TEST_WN - ave, adev,var 6.37499867520348231E-004  1.1277054514612708
2.0001688019270105
 TEST_WN - skew, curt   -7.83082240871004072E-004
1.44868668438072491E-002

 Task 2
 Subroutine OUT_GAUSS written

 LANGV1 - iensb, CPUTIME       1  1.7472110000000001
```

References

Aberle, J., Brook, A., Peppler, W., Rohrbacher, H., & Schleisiek, K. (1976). *Sodium boiling experiments in a 7-pin bundle under flow run down conditions*. Kernforschungszentrum Karlsruhe. KfK 2378.

Bénard, H. (1900). Les tourbillons cellulaires dans une nappe liquide. *Revue générale des Sciences pures et appliqués, 11*, 1261–1271. et 1309–28.

Bénard, H. (1901). Les tourbillons cellulaires dans une nappe liquide transportant de la chaleur par convection en régime permanent. *Annales de Chimie et de Physiques, 23*, 62–144.

Berne, B. J. (1985). Molecular dynamics and hybrid Monte Carlo in systems with multiple time scales and long range forces: Reference system propagator algorithm. In J. U. Brackbill & B. I. Cohen (Eds.), *Multiple time scales* (pp. 419–425). Academic Press.

Bethe, H. A., & Tait, J. H. (1956). An estimate of the order of magnitude of the vigorous interaction expected should the core of a fast reactor collapse. *UKAEA-RHM, 56*, 113. UKAEA Reactor Group.

Bhatnagar, P. L., Gross, E. P., & Krook, M. (1954). A model for collision processes in gases. I. Amplitude processes in charged and neutral one-component systems. *Physical Review, 94*(3), 511–525.

Binder, K., & Heermann, D. W. (1988). *Monte Carlo simulation in statistical physics*. Springer Verlag.

Bird, G. A. (1994a). *Molecular gas dynamics and the direct simulation of gas flow*. Clarendon Press.

Bird, G. A. (1994b). *Molecular gas dynamics and the direct simulation of gas flows*. Clarendon Press.

Bottoni, M. (1998). Monte Carlo calculation of sodium vapour condensation and rewetting simulation in equivalent LMFBR bundle under hypothetical accident conditions (In Japanese). In *1998 Spring Meeting of the Japanese Atomic Power Society (JAPS)*, Osaka.

Bottoni, M. (1998a). Monte Carlo calculation of sodium vapour condensation and rewetting simulation in equivalent LMFBR bundle under hypothetical accident conditions. In *Spring Meeting of the Japanese Atomic Power Society, Osaka*. (In Japanese).

Bottoni, M. (1998b). *Monte Carlo calculation of sodium vapour condensation and rewetting simulation in equivalent LMFBR bundle under hypothetical accident conditions*. Annual Meeting of the Energy Soc. Japan. (in Japanese).

Bottoni, M. (1999). Monte Carlo calculation of sodium vapour condensation and simulation of rewetting in LMFBR bundle under hypothetical accident conditions (In Japanese). In *1999 Spring Meeting of the Japanese Atomic Power Society (JAPS)*, Hiroshima.

M. Bottoni et al., *Molecular Dynamics - Theory and Applications*, Mechanical Engineering Series, https://doi.org/10.1007/978-3-031-84567-3

Bottoni, M. (2000). Molecular approach to sodium vapour condensation, rewetting and vaporisation in LMFBR bundle under hypothetical accident conditions. In *Proceedings of the Eighth International Colloquium on Numerical Analysis and Computer Science with Applications*, Plovdiv.

Bottoni, M. (2004). Molecular approach to sodium vapour condensation, rewetting and vaporization in LMFBR bundle under hypothetical accident conditions. *Journal of Nuclear Science and Technology, 41*(5), 579–593.

Bottoni, M. (2006). Thirty years of endevour on numerical simulation of molecular dynamics. In *Invited lecture delivered at the 18th Meeting of the "International Institute for Advanced Studies (IIAS) in Systems Research and Cybernetics"*, Baden-Baden, Germany.

Bottoni, M. (2007a). Molecular dynamic simulation of thermal creep in 7-pin LMFBR bundle under hypothetical accident conditions. In *15th International Conference on Nuclear Engineering, (ICONE15), Nagoya, Japan, April 22–26, 2007.*

Bottoni, M. (2007b). Three dimensional Monte Carlo simulation method for LMFBR fuel bundle heat transfer under LOCA conditions (in Chinese). *Atomic Energy Science and Technology Journal, 41*(2).

Bottoni, M. (2007c). Molecular dynamic simulation of thermal creep in 7-pin LMFBR bundle under hypothetical accident conditions. In *15th International Conference On Nuclear Engineering, (ICONE15)*, Nagoya, Japan.

Bottoni, M. (2022). *Physical modeling and computational techniques for thermal and fluid-dynamics*. Springer.

Bottoni M., & Mantovani, S. (2008). *Molecular dynamic simulations of thermal confinement of cosmic dust particles*. Accepted at the ASME 2008 Conference, Boston, Massachusetts, USA, October 31–November 6, 2008.

Bottoni, M., & Struwe, D. (1982). *BLOW-3A: A theoretical model to describe transient two-phase conditions in LMFBR coolant channels*. Kernforschungszentrum Karlsruhe GmbH (Germany). KfK-3317.

Bottoni, M., Dorr, B., Homann, C. (1992). The tree-dimensional transient two-phase flow computer programme BACCHUS-3D/TP. KfK 4760.

Bottoni, M., Porcu, F., Prodi, F., Santachiara, G., & Travaini, S. (2005). Bi-dimensional simulation of thermophoretic forces on aerosol particle with the direct simulation Monte Carlo (DSMC) method. *Computational Fluid Dynamics JOURNAL, 14*, 255.

Bottoni, M., Bottoni, C., & Scanu, J. (2006). Molecular dynamic simulation of sodium in 7-pin LMFBR bundle under hypothetical accident conditions, In *14th International Conference on Nuclear Engineering, (ICONE14)*, Miami, USA.

Bottoni, M., Mantovani, S., & Cavicchi, M. (2018). Thermophoretic calculations with pristine and scaled variables. *Academia Journal of Scientific Research, 6*(12).

Cahalan, J. E., Tentner, A. M., & Morris, E. E. (1994). Advanced LMR safety analysis capabilities in the SASSYS-1 and SAS4A computer codes. In *Proceedings of the International Meeting on Advanced Reactor Safety, Pittsburg, Pennsilvania, April 17–21, 1994.*

Cercignani, C. (1988). *The Boltzmann equation and its applications*. Springer-Verlag.

Cercignani, C., & Lampis, M. (1974). *Rarefied gas dynamics*. In Karamcheti (Ed.), New York: Academic Press.

Chandrasekhar, S. (1943). Stochastic problems in physics and astronomy. *Reviews of Modern Physics, 15*, 1–83.

Chandrasekhar, S. (1961). *Hydrodynamic and hydromagnetic stability*. Dover Publications.

Chen, L., Yao, Z., & Bottoni, M. (2002). Three-dimensional simulation of steam condensation in water reactor containment with the Monte Carlo computer program TH3SL. In *Spring Meeting of the Japanese Atomic Power Society (JAPS)*.

Choi, S. J., Rader, D. J., & Geller, A. S. (1966). Massively parallel simulation of Brownian dynamics particle transport in low pressure parallel-plate reactors. *Journal of Vacuum Science and Technology A, 14*(2), 660.

Choi, S. J., Rader, D. J., & Geller, A. S. (1996). Massively parallel simulations of Brownian dynamics transport in low pressure parallel plate reactors. *Journal of Vacuum Science and Technology A, 14*(2), 660–665.

Chowns, T. M., & Elkins, J. E. (1974). The origin of quartz geodes and cauliflower cherts through the silicification of anhydrite nodules. *Journal of Sedimentary Petrology, 44*(3), 885–903.

Crowdy, D., & Tanveer, S. (2004). The effect of finiteness in the Saffman-Taylor viscous fingering problem. *Journal of Statistical Physics, 114*(5/6), 1501.

De Laplace, P. S. (1951). *A phylosophical essay on probabilities*. Dover.

Dunn, F. E., & Wei, T. Y. C. (1998). The role of SASSYS-1 in LMR safety analysis. In *Proceedings of the International Topical Meeting on safety of next generation power reactors, Seattle, Washington, May 1–5, 1988*.

Einstein, A. (1905). Ueber die von der molekularkinetischen Theorie der Waerme geforderte Bewegung von in ruhenden Fluessigkeiten suspendierten Teilchen. *Annalen der Physik, 17*, 549.

ESA. (2006). *Invitation to Tender: AO/1-5045/06/NL/VK, feasibility study phase B0 of the Interaction in Cosmic and Atmospheric Particle System (ICAPS) Experiment, 15th August, 2006*. Addendum of 21st August 2006.

Euken, A. (1922). Grundriss der physikalischen chemie. Leipzig.

Evans, D. J. (1983). Computer experiment for nonlinear thermodynamics of Couette flow. *The Journal of Chemical Physics, 78*, 3297–3302.

Feynman, R. (1960). There's plenty of room at the bottom. *Engineering and Science, XXIII(5)*.

Fulford, G. D. (1964). *Advances in chemical engineering* (The flow of liquids in thin films) (Vol. 5, pp. 151–236).

G.A.B. Consulting Pty Ltd. (1999). *The DS2G program user's guide, Version 3.2*.

Gallis, M. A., Rader, D. J., & Torczynski, J. R. (2002). Thermophoresis in rarefied gas flows. *Aerosol Science and Technology, 36*, 1099–1117.

Gan, X., Bottoni, M., & Zhou, Z. (2002). TH3SL: Simulation of gas distribution inside a water reactor containment (in Chinese). *Atomic Energy Science and Technology, 36*(1), 70–75.

Goyal, A., Zanghirati, G., & Bottoni, M. (2006). *Parallel version of the computer program THEPHOR*. (Internal documentation of the Physics Department of the Ferrara University).

Haenggi, P., & Marchesoni, F. (2005). Introduction: 100 years of Brownian motion. *Chaos, 15*, 026101.

Haenggi, P., Talkner, P., & Borkovec, M. (1990). Reaction-rate theory: fifty years after Kramers. *Reviews of Modern Physics, 62*(2), 251–341.

Hartmann, L. A., Medeiros, J. T. N., & Petruzzellis, L. T. (2012). Numerical simulations of amethyst geode cavity formation by ballooning of altered Paraná volcanic rocks, South America. *Geofluids, 12*(2), 133–141.

Hartmann, L. A., Medeirosm, J. T. N., & Baggio, S. B. (2015). Controls on prolate and oblate geode geometries in the Veia Alta basalt flow, largest world producer of amethyst, Paraná volcanic province, Brazil. *Ore Geology Reviews, 66*, 243–251.

Hayes, J. B. (1964). Geodes and concretions from the Mississippian Warsaw formation, Keokuk region, Iowa, Illinois, Missouri. *Journal of Sedimntary Petrology, 34*(1), 123–133.

Hele-Shaw, H. S. (1898). The flow of water. *Nature, 58*(1489), 34–36.

Hinshelwood, C. N. (1940). *The kinetic of chemical change*. Clarendon Press.

Kaburaki, H., & Yokokawa, M. (1993). Numerical simulation of near continuum gas flow by the direct simulation Monte Carlo method. In *Proceedings of the Joint International Conferenceon Mathematical Methods and Supercomputing in Nuclear Applications, M&C + SNA '93*, April 19–23, 1993, Karlsruhe, Germany, Edited by Forschungszentrum Karlsruhe (FZK).

Kaburaki, H., Yamamoto, H., & Yokokawa, M., Arisawa, T. (1990). A new collision cell system for the direct simulation Monte Carlo method. In *Proceedings of First International Conference on Supercomputing in Nuclear Applications (SNA '90)*, Mito, March 12–16, 1990, Edited by Japan Atomic Energy Research Institute (JAERI).

Kapitsa, P. (1941). The study of heat transport in Helium II. *Journal of Physics USSR, 4*.

Kondo, S., Tobita, Y., Morita, K., Brear, D. J., Kamiyama, K., Yamano, H., Fujita, S., Mashek, W., Fischer, E. A., Kiefhaber, E., Buckel, G., Coste, P., & Pigny, S. (1999). Current status and validation of the SIMMER-III LMFR safety analysis code. In *7th International Conference on Nuclear Engineering, Tokyo, Japan, April 19–23, 1999; ICONE-7249*.

Koschmieder, E. L. (1993). *Bénard cells and Taylor vortices*. Cambridge University Press.

Kramers, H. A. (1940). Brownian motion in a field of force and the diffusion model of chemical reactions. *Physica (Utrecht), 7*, 284.

Langevin, P. (1908). Comptes Rendus de Academie des. *Sciences (Paris), 146*, 530–533.

Larsen, P. S., & Borgnakke, C. (1974). Rarefied gas dynamics. In M. Becker & M. Fiebig (Eds.), *Paper A7*. DFVLR Press.

Lochart, R. W., & Martinelli, R. C. (1949). Proposed correlation of data for isothermal two-phase, two-componwent flow in pipes. *Chemical Engineering Progress, 45*(1), 39–48.

Lord, R. G. (1991). Rarefied gas dynamics. In A. E. Beylich (Ed.), *Proceedings of the seventeenth international symposium on rarefied gas dynamics*. VCH Verlagsgesellschaft mbH.

Marsaglia, G. (1972). Choosing a point from the surface of a sphere. *Annals of Mathematical Statistics, 43*, 645–646.

Matsushita, M. (1989). Experimental observation of aggregation. In A. Avnir (Ed.), *The fractal approach to heterogeneous chemistry: surfaces, colloids, polymers*. Wiley.

Matsushita, M., & Fujikawa, H. (1990). Diffusion-limited growth in bacterial colony formation. *Physica A: Statistical Mechanics and its Applications, 168*, 498–506.

Maxwell, J. C. (1871). *Theory of heat*. Longmans Green.

Maxwell, J. C. (1965). 1879, Phil. Trans. Royal Soc., I (Appendix). Reprinted in *The scientific papers of J. C. Maxwell*. New York: Dover Publications.

McDougall, J. C. (1978). *SABRE-2: A computer program for the calculation of transient three-dimensional flows in rod clusters*. AEEW-R-1104, UKAEA.

Muramatsu, T. (1995). A Direct Simulation Monte Carlo code THEMIS for the evaluation of nonstationary heat transfer phenomena in LMFR thermohydraulics. In *8-th IAHR Working Group Meeting on Advanced Nuclear Reactors Thermal Hydraulics, Nuclear Research Institute Rezplc*, Czech Republic, June 6–8.

Nambu, K. (1980a). Direct simulation scheme derived from the Boltzmann equation, (I): Monocomponent gases. *Journal of the Physical Society of Japan, 49*(5), 2042–2049.

Nambu, K. (1980b). Direct simulation scheme derived from the Boltzmann equation. II. Multicomponent gas mixtures. *Journal of the Physical Society of Japan, 49*(5), 2050–2054.

Nambu, K. (1980c). Direct simulation scheme derived from the Boltzmann equation. III. Rough sphere gases. *Journal of the Physical Society of Japan, 49*(5), 2055–2058.

Ninokata, H., & Okano, T. (1990). SABENA: Subassembly boiling evolution numerical analysis. *Nuclear Engineering and Design, 120*(2–3), 349–367.

Nose, S. (1984). A unified formulation of the constant temperature molecular-dynamics methods. *The Journal of Chemical Physics, 81*, 511–519.

Nuclear Energy Data Center (1990); (SNA '90), Tokai Mura, Ibaraki ken (Japan); 1. In *International conference on supercomputing in nuclear applications*, 1990.

Ohwada, T., & Sone, Y. (1992). Analysis of thermal stress slip flow and negative thermophoresis using the Boltzmann equation for hard-sphere molecules. *Mechanics of Fluids Journal B, 11*, 389–414.

Papadopoulos, D. H., & Rosner, D. E. (1995). Enclosure gas flow driven by non-isothermal walls. *Physics of Fluids, 7*(11), 2535–2537.

Peitgen, H.-O., Jürgens, H., & Saupe, D. (2004). *Chaos and fractals* (2nd ed.). Springer Verlag.

Press, W. H., Flannery, B. P., Teukolsky, S. A., & Vetterling, W. T. (1986). *Numerical recipes*. Cambridge University Press.

Prodi, F., Santachiara, G., Travaini, S., Vedernikov, A., Dubois, F., & Legros, J. C. (2004). Measurements of phoretic velocities of aerosol particles in microgravity conditions. In *Proceedings of the 14th International Conference on Clouds and Precipitations* (Vol. I).

Proust, D., & Fontaine, C. (2007). Amethyst-bearing lava flows in the Paraná basin (Rio Grande do Sul, Brazil): Cooling, vesiculation and formation of the geodic cavities. *Geological Magazine, 144*, 53.

Raczynski, S. (2002). The PASION simulation system. *Simulation News Europe, 35*(36), 24–28.

Rapaport, D. C. (2004). *The art of molecular dynamics simulation* (2nd ed.). Cambridge University Press.

Rayleigh, L. (1916). On convective currents in a horizontal layer of fluid when the higher temperature is on the under side. *Philosophical Magazine, 32*, 529–546.

Reynen, J. et al. (1989). EAC-European Accident Code. In *Proceedings of the 10th LMBWG meeting, Karlsruhe, 1989.*

Risken, H. (1989). *The Fokker plank equation* (2nd ed.). Springer Verlag.

Sadhal, S. S., Ayyaswamy, P. S., & Chung, J. N. (1997). *Transport phenomena with drops and bubbles*. Springer-Verlag.

Saffman P. G., & Taylor G. I. (1958). The penetration of a fluid in a porous medium of Hele-Shaw cell containing a more viscous fluid. *Proceedings of the Royal Society A: Mathematical, Physical, Engineering Sciences, 245.*

Sander, L. M., & Somfai, E. (2005). Random walks, diffusion limited aggregation in a wedge, and average conformal maps. *Chaos, 15*, 026109.

Schlechtendahl, E. G. (1970). Theoretical investigations on sodium boiling in fast reactors. *Nuclear Science and Engineering, 41*, 99–114.

Schmuck, P., Jacobs G., & Arnecke G., (1977). KADIS: A program to analyze the disassembly phase of hypothetical accidents in LMFBRs. KfK 2497.

Shaw, S. H. (1898). *Investigation of the nature of surface resistance of water and of stream-line motion under certain experimental conditions*. Inst. N.A.

Shimizu, F., Kimizuka, H., Kaburaki, H., Li, J., & Yip, S. (2000). Parallel molecular dynamic simulation on elastic properties of solid argon. In *Proceedings of the 4th International Conference SNA2000.*

Shimizu F, Kimizuka H, Kaburaki H, Arakawa C. (2002). Development of parallel molecular dynamics stencil (In Japanese). *Transactions of the Japan Society for Computational Engineering and Science*, #20020015.

Smith L. L., & Sheheen N. N (1980). *SIMMER-II: A computer program for LMFBR disassembly core analysis*. NUREG/CR-0453; LA-7515-M.

Sone, Y. (1972). Flow induced by thermal stress in rarefied gas. *Physics of Fluids, 15*, 1418.

Sone, Y., Waniguchi, Y., & Aoki, K. (1966). One-way flow of a rarefied gas induced in a channel with a periodic temperature distribution. *Physics of Fluids, 8*, 8.

Tentner, A. M., et al. (1985). The SAS-4A LMFBR whole core accident analysis code. In *Proceedings of the International Topical Meeting on Fast Reactor Safety, Vol. 2, Knoxville, Tennessee, April 21–25.*

Travis J. R., Spore J. W., & Royl P. (1998). *GASFLOW-II: A three-dimensional finite-volume fluid-dynamics code for calculating the transport, mixing and combustion of flammable gases and aerosols in geometrically complex domains, Vol. 1: Theory and computational model*. Vol. 2: User's manual. NUREG/CR-6570, LA-13357-MS, FZKA-5994.

Uhlenbeck, E. G., & Ornstein, S. L. (1930). On the theory of the brownian motion. *Physical Review, 36*, 823.

Van den Broeck, C., Meurs, P., & Kawai, R. (2005). From Maxwell demon to Brownian motor. *New Journal of Physics, 7*(1), 10. (pp. 1–24).

Vincenti, W. G., & Kruger, C. H. (1965). *Introduction to physical gas dynamics*. Krieger Publishing Company.

Waldmann, L. (1959). Ueber die Kraft eines inhomogenen Gases auf kleine suspendierte Kugeln. *Zeitschrift fuer Naturforschung, 14*(7), 589–599.

Wallis, G. B. (1969). *One-dimensional two-phase flow*. McGraw Hill.

Watanabe, T., Kaburaki, H., & Yokokawa, M. (1994). Simulation of two-dimensional Rayleigh-Bénard system using the direct simulation Monte-Carlo method. *Physical Review E, 49*(5), 4060–4064.

Weber, D. P., & Ganapol, B. D. (1976). Multi-field hydrodynamics disassembly calculations with space-time kinetics: FX2/VENUS-III. In *Proceedings of the international meeting on fast reactor safety and related physics, Conf-761001, Chicago, IL* (Vol. III).

Witten, T. A., & Sander, L. M. (1981). Diffusion-Limited Aggregation, a Kinetic Critical Phenomenon. *Physical Review Letters, 47*, 1400.

Witten, T. A., & Sander, L. M. (1983). Diffusion limited aggregation. *Physical Review B, 27*, 5686.

Yamamoto, K., & Ishihara, Y. (1988). Thermophoresis of a spherical particle in a rarefied gas of a transition regime. *Physic of Fluids, 31*, 3618–3624.

Index

M. Bottoni et al., *Molecular Dynamics - Theory and Applications*, Mechanical Engineering Series, https://doi.org/10.1007/978-3-031-84567-3

The manufacturer's authorised representative in the EU is Springer Nature Customer Service Centre GmbH, Europaplatz 3, 69115 Heidelberg, Germany. If you have any concerns regarding our products, please contact ProductSafety@springernature.com

Printed and bound by CPI Group (UK) Ltd, Croydon, CR0 4YY
21/09/2026
02220009-0016